Zhang · Chen · Glowinski · Tong (Eds.)
Current Trends in High Performance Computing and Its Applications

T0122583

Wu Zhang · Zhangxin Chen
Roland Glowinski · Weiqin Tong

Editors

Current Trends in High Performance Computing and Its Applications

Proceedings of the International Conference on
High Performance Computing and Applications,
August 8–10, 2004, Shanghai, P. R. China

With 263 Figures and 65 Tables

 Springer

Editors

Wu Zhang
Weiqin Tong
School of Computer Engineering
and Science
Shanghai University
200072 Shanghai
People's Republic of China
e-mail: wzhang@mail.shu.edu.cn
e-mail: wqtong@mail.shu.edu.cn

Roland Glowinski
Department of Mathematics
University of Houston
77204-3476 Houston, TX, USA
e-mail: roland@math.uh.edu

Zhangxin Chen
Department of Mathematics
Southern Methodist University
75275-0156 Dallas, TX, USA
e-mail: zchen@mail.smu.edu

Library of Congress Control Number: 2005924803

Mathematics Subject Classification (2000): 68W40, 68W10, 68W15, 65K05, 68U01, 68U20, 68Q25, 68N30

ISBN-10 3 -540-25785-3 Springer Berlin Heidelberg New York
ISBN-13 978-3-540-25785-1 Springer Berlin Heidelberg New York

Springer is a part of Springer Science+Business Media

springeronline.com

© Springer-Verlag Berlin Heidelberg 2005
Printed in Germany

Typesetting by the authors using a Springer TeX macro package
Cover design: Erich Kircher, Heidelberg, Germany

Printed on acid-free paper   46/3142/sz - 5 4 3 2 1 0

Preface

A large international conference on High Performance Computing and its Applications was held in Shanghai, China, August 8–10, 2004. It served as a forum to present current work by researchers and software developers from around the world as well as to highlight activities in the high performance computing area. It aimed to bring together research scientists, application pioneers, and software developers to discuss problems and solutions and to identify new issues in this area. The conference focused on the design and analysis of high performance computing algorithms, tools, and platforms and their scientific, engineering, medical, and industrial applications. It drew about 150 participants from Canada, China, Germany, India, Iran, Japan, Mexico, Singapore, South Korea, the United Kingdom, and the United States of America. More than 170 papers were received on a variety of subjects in modern high performance computing and its applications, such as numerical and software algorithm design and analysis, grid computing advance, adaptive and parallel algorithm development, distributing debugging tools, computational grid and network environment design, computer simulation and visualization, and computational language study and their applications to science, engineering, and medicine. This book contains ninety papers that are representative in these subjects. It serves as an excellent research reference for graduate students, scientists, and engineers who work with high performance computing for problems arising in science, engineering, and medicine.

This conference would not have been possible without the support of a number of organizations and agencies and the assistance of many people. It received tremendous support from the China Computer Federation, Institute of Applied Physics and Computational Mathematics in Beijing, National Natural Science Foundation of China, Shanghai Computer Society, Shanghai Jiaotong University, Shanghai Supercomputer Center, Shanghai University, Shanghai University Dingtech Co. Ltd., Sybase China, and Xi'an Jiaotong University. The warmth, enthusiasm, and hard work of the local organizers were critical to the successful completion of the conference. Special thanks go to Ms. Peijing Dai for her tremendous assistance in preparation and completion of this conference and of the conference proceedings.

Shanghai, China and Texas, USA   *Wu Zhang, Zhangxin Chen*
January 2005   *Roland Glowinski, and Weiqin Tong*

Contents

Part I Regular Papers

LES Investigation of Near Field Wakes Behind Juncture of Wing and Plate
Jiangang Cai, Shutian Deng, Hua Shan, Li Jiang, Chaoqun Liu 3

On Computation with Higher-order Markov Chains
Waiki Ching, Michael K. Ng, Shuqin Zhang 15

Dynamic Data-Driven Contaminant Simulation
Craig C. Douglas, Yalchin Efendiev, Richard Ewing, Victor Ginting,
Raytcho Lazarov, Martin J. Cole, Greg Jones, Chris R. Johnson 25

Computing the Least Squares Inverse of Sparse Matrices on a Network of Clusters
Baolai Ge 37

Numerical Simulation for Fractured Porous Media
Guanren Huan, Richard E. Ewing, Guan Qin, Zhangxin Chen 47

Multilayer Hydrodynamic Models and Their Application to Sediment Transport in Estuaries
H. Ramírez León, C. Rodríguez Cuevas, E. Herrera Díaz 59

An Application of Genetic Programming to Economic Forecasting
Kangshun Li, Zhangxin Chen, Yuanxiang Li, Aimin Zhou 71

A New Dynamical Evolutionary Algorithm Based on Particle Transportation Theory
Kangshun Li, Yuanxiang Li, Zhangxin Chen, Zhijian Wu 81

An Operator Splitting Method for Nonlinear Reactive
Transport Equations and Its Implementation Based on DLL
and COM
Jiangguo Liu, Richard E. Ewing 93

Design & Implementation of the Parallel-distributed Neural
Network Ensemble
Yue Liu, Yuan Li, Bofeng Zhang, Gengfeng Wu 103

Floating Point Matrix Multiplication on a Reconfigurable
Computing System
C. Sajish, Yogindra Abhyankar, Shailesh Ghotgalkar, K.A. Venkates 113

On Construction of Difference Schemes and Finite Elements
over Hexagon Partitions
Jiachang Sun, Chao Yang 123

Efficient MPI-I/O Support in Data-Intensive Remote I/O
Operations Using a Parallel Virtual File System
Yuichi Tsujita 135

A General Approach to Creating Fortran Interface for C++
Application Libraries
*Yang Wang, Raghurama Reddy, Roberto Gomez, Junwoo Lim, Sergiu
Sanielevici, Jaideep Ray, James Sutherland, Jackie Chen* 145

A Platform for Parallel Data Mining on Cluster System
Shaochun Wu, Gengfeng Wu, Zhaochun Yu, Hua Ban 155

Architecture Design of a Single-chip Multiprocessor
Wenbin Yao, Dongsheng Wang, Weimin Zheng, Songliu Guo 165

Metadata Management in Global Distributed Storage System
Chuanjiang Yi, Hai Jin, Yongjie Jia 175

Adaptive Parallel Wavelet Method for the Neutron Transport
Equations
Heng Zhang, Wu Zhang 185

Ternary Interpolatory Subdivision Schemes for the Triangular
Mesh
Hongchan Zheng, Zhenglin Ye 197

Part II Contributed Papers

Modeling Gene Expression Network with PCA-NN on
Continuous Inputs and Outputs Basis
Sio-Iong Ao, Michael K. Ng, Waiki Ching..........................209

A Reuse Approach of Description for Component-based
Distributed Software Architecture
Min Cao, Gengfeng Wu, Yanyan Wang.............................215

An Operation Semantics Exchange Model for Distributed
Design
Chun Chen, Shensheng Zhang, Zheru Chi, Jingyi Zhang, Lei Li........221

OGSA-based Product Lifecycle Management Model Research
Dekun Chen, Donghua Luo, Fenghui Lv, Fengwei Shi.................227

Study of Manufacturing Resource Interface Based on OGSA
Dekun Chen, Ting Su, Qian Luo, Zhiyun Xu.......................233

Hybrid Architecture for Smart Sharing Document Searching
Haitao Chen, Zunguo Huang, Xufeng Li, Zhenghu Gong..............239

ID-based Secure Group Communication in Grid Computing
Lin Chen, Xiaoqin Huang, Jinyuan You...........................245

Performance Prediction in Grid Network Environments
Based on NetSolve
Ningyu Chen, Wu Zhang, Yuanbao Li.............................251

A Hybrid Numerical Algorithm for Computing Page Rank
Waiki Ching, Michael K. Ng, Waion Yuen.........................257

A Parallel Approach Based on Hierarchical Decomposition
and VQ for Medical Image Coding
Guangtai Ding, Anping Song, Wei Huang..........................265

A High-resolution Scheme Based on the Normalized Flux
Formulation in Pressure-based Algorithm
M. H. Djavareshkian, S. Baheri Islami............................271

Aerodynamic Calculation of Multi-element High-lift Airfoil
with a Pressure-based Algorithm and NVD Technique
M. H. Djavareshkian, S. Memarpour..............................277

Active Learning with Ensembles for DOE
Tao Du, Shensheng Zhang.......................................283

Parallel Computing for Linear Systems of Equations on
Workstation Clusters
Chaojiang Fu, Wu Zhang and Linfeng Yang........................289

A Service Model Based on User Performances in Unstructured
P2P
Jianming Fu, Weinan Li, Yi Xian, Huangguo Zhang 295

DMR: A Novel Routing Algorithm in 3D Torus Switching
Fabric
Jianbo Guan, Xicheng Lu, Zhigang Sun 299

Multiagent-based Web Services Environment for E-commerce
Tianqi Huang, Xiaoqin Huang, Lin Chen, Linpeng Huang 305

Linux Cluster Based Parallel Simulation System of Power
System
Ying Huang, Kai Jiang 311

Pipeline Optimization in an HPF Compiler
Kai Jiang, Yanhua Wen, Hongmei Wei, Yadong Gui 317

Clustering Approach on Core-based and Energy-based
Vibrating
Shardrom Johnson, Daniel Hsu, Gengfeng Wu, Shenjie Jin, Wu Zhang . 325

Design and Implementation of Detection Engine Against IDS
Evasion with Unicode
Dongho Kang, Jintae Oh, Kiyoung Kim, Jongsoo Jang 333

Derivative Based vs. Derivative Free Optimization Methods
for Nonlinear Optimum Experimental Design
Stefan Körkel, Huiqin Qu, Gerd Rücker, Sebastian Sager 339

Scalable SPMD Algorithm of Evolutionary Computation
Yongmei Lei, Jun Luo 345

Checkpointing RSIP Applications at Application-level in
ChinaGrid
Chunjiang Li, Xuejun Yang, Nong Xiao 351

Fast Fourier Transform on Hexagons
Huiyuan Li, Jiachang Sun 357

LBGK Simulations of Spiral Waves in CIMA Model
Qing Li, Anping Song 363

Parallel Iterative CT Image Reconstruction on a Linux
Cluster of Legacy Computers
Xiang Li, Jun Ni, Tao He, Ge Wang, Shaowen Wang, Body Knosp 369

Grid Communication Environment Based on Multi-RP
Multicast Technology
Xiangqun Li, Xiongfei Li, Tao Sun, Xin Zhou 375

Building CFD Grid Application Platform on CGSP
*Xinhua Lin, Yang Qi, Jing Zhao, Xinda Lu, Hong Liu, Qianni Deng,
Minglu Li* 381

A New Grid Computing Platform Based on Web Services and
NetSolve
Guoyong Mao, Wu Zhang, Bing He 387

Uncertainty Analysis for Parallel Car-crash Simulation
Results
Liquan Mei, C.A. Thole 393

Distributed Partial Evaluation on Byte Code Specialization
Zhemin Ni, Hongyan Mao, Linpeng Huang, Yongqiang Sun 399

Scientific Visualization of Multidimensional Data: Genetic
Likelihood Visualization
*Juw Won Park, Mark Logue, Jun Ni, James Cremer, Alberto Segre,
Veronica Vieland* 403

Building Intrusion Tolerant Software System for High
Performance Grid Computing
Wenling Peng, Huanguo Zhang, Lina Wang, Wei Chen 409

Model Driven Information Resource Integration
Pengfei Qian, Shensheng Zhang 415

DNA-based Parallel Computation of Addition
Huiqin Qu, Hong Zhu 419

Measurements and Understanding of the KaZaA P2P
Network
Jun Shi, Jian Liang, Jinyuan You 425

An Application of Genetic Algorithm in Engineering
Optimization
Lianshuan Shi, Lin Da, Heng Fu 431

A High Performance Algorithm on Uncertainty Computing
Suixiang Shi, Qing Li, Lingyu Xu, Dengwei Xia, Xiufeng Xia, Ge Yu ... 437

Parallel Unsteady 3D MG Incompressible Flow
C. H. Tai, K. M. Liew, Y. Zhao 443

Research of Traffic Flow Forecasting Based on Grids
Guozhen Tan, Hao Liu, Wenjiang Yuan, Chengxu Li 451

A Mobile Access Control Architecture for Multiple Security
Domains Environment
Ye Tang, Shensheng Zhang, Lei Li 457

Inversion of a Cross Symmetric Positive Definite Matrix and
its Parallelization
K.A. Venkatesh 463

Intelligent File Transfer Protocol for Grid Environment
Jiazeng Wang and Linpeng Huang 469

A Parallel Algorithm for QoS Multicast Routing in
IP/DWDM Optical Internet
Xingwei Wang, Jia Li, Hui Cheng, and Min Huang 477

A Parallel and Fair QoS Multicast Routing Mechanism in
IP/DWDM Optical Internet
Xingwei Wang, Cong Liu, Jianye Cui, Min Huang 483

A Study of Multi-agent System for Network Management
Zhenglu Wang and Huaglory Tianfield 489

A Practical Partition-based Approach for Ontology Version
Zongjiang Wang, Shensheng Zhang, Yinglin Wang, Tao Du 495

Grid-based Robot Control
S.D. Wu, Y.F. Li 501

Group-based Peer-to-Peer Network Routing and Searching
Rules
Weiguo Wu, Wenhao Hu, Yongxiang Huang, Depei Qian 509

Research on the Parallel Finite Element Analysis Software
for Large Structures
Weiwei Wu, Xianglong Jin, Lijun Li 515

An Amended Partitioning Algorithm for Multi-server DVE
Yanhua Wu, Xiaoming Xu 521

High Performance Analytical Data Reconstructing Strategy
in Data Warehouse
Xiufeng Xia, Qing Li, Lingyu Xu, Weidong Sun, Suixiang Shi, Ge Yu .. 527

High Performance Historical Available Data Backup Strategy
in Data Warehouse
Xiufeng Xia, Lingyu Xu, Qing Li, Weidong Sun, Suixiang Shi, Ge Yu .. 533

A Modeling Method of Cloud Seeding for Rain Enhancement
Hui Xiao, Weijie Zhai, Zhengqi Chen, Yuxiang He, Dezhen Jin 539

**A High-level Design of Mobile Ad Hoc System for Service
Computing Compositions**
Jinkui Xie, Linpeng Huang 545

Metric Based Software Quality Assurance System
Dong Xu, Zongtian Liu, Bin Zhu, Dahong Xing 551

A Strategy for Data Replication in Data Grids
Liutong Xu, Bai Wang, Bo Ai 557

**Analysis of an IA-64 Based High Performance VIA
Implementation**
Feng Yang, Zhihui Du, Ziyu Zhu, Ruichun Tang 563

An Application and Performance Analysis Based on NetSolve
*Linfeng Yang, Wu Zhang, Chaojiang Fu, Anping Song, Jie Li,
Xiaobing Zhang* 569

**FE Numerical Simulation of Temperature Distribution in
CWR Process**
Fuqiang Ying, Jin Liu 575

**Contents Service Platforms: Stage of Symmetrical and Active
Services**
*Jie Yuan, Maogou Chen, Gengfeng Wu, James Zhang, Yu Liu, Zhisong
Chen* 581

An Efficient Planning System for Service Composition
Jianhong Zhang, Shensheng Zhang, Jian Cao 587

**A Sort-last Parallel Volume Rendering Based on Commodity
PC Clusters**
*Jiawan Zhang, Zhou Jin, Jizhou Sun, Jiening Wang, Yi Zhang,
Qianqian Han* 593

**Parallel Computation of Meshfree Methods for Extremely
Large Deformation Analysis**
Jifa Zhang, Yao Zheng 599

Solving PDEs using PETSc in NetSolve
Xiaobin Zhang, Wu Zhang, Guoyong Mao, Linfeng Yang 605

**Research on a Visual Collaborative Design System: A High
Performance Solution Based on Web Service**
*Zhengde Zhao, Shardrom Johnson, Xiaobo Chen, Hongcan Ren, Daniel
Hsu* 611

Architectural Characteristics of Active Packets Workloads
Tieying Zhu, Jiubin Ju 617

Design of Elliptic Curve Cryptography in GSI
Yanqin Zhu, Xia Lin, Gang Wang 623

Policy-based Resource Provisioning in Optical Grid Service Network
Yonghua Zhu, Rujian Lin 629

An Algorithm for Solving Computational Instability Problem in Simulation of Atmospheric Aerosol Growth and Evaporation
Yousuo J. Zou, Chiren Wu 635

List of Contributors

Yogindra Abhyankar
Hardware Technology Development Group,
Centre for Development of Advanced Computing, Pune University Campus,
Pune 411 007, India
yogindra@cdac.ernet.in

Bo Ai
Information Systems Division of China Unicom, Beijing 100032, China
aibo@chinaunicom.com.cn

Sio-Iong Ao
Department of Mathematics, The University of Hong Kong,
Hong Kong, China
siao@hkusua.hku.hk

Hua Ban
School of Computer Engineering & Science, Shanghai
University, Yanchang Road 149, Shanghai, 200072, China

Jiangang Cai
Department of Mathematics, University of
Texas at Arlington, Arlington, TX 76019, USA
jxc0622@omega.uta.edu

Jian Cao
CIT Lab, Computer Science Department,
Shanghai Jiaotong University, Shanghai 200030, China

Min Cao
School of Computer Science & Engineering, Shanghai
University, Shanghai 200072, China
mcao@mmail.shu.edu.cn

Chun Chen
Department of Computer Science and Engineering,
Shanghai Jiao Tong University, Shanghai 200030, China
chenchun@cs.sjtu.edu.cn

Dekun Chen
Shanghai University and CIMS Center, Shanghai University,
Shanghai, China
dkchen@mail.shu.edu.cn

Haitao Chen
School of Computer, National University of Defense Technology,
Changsha 410073, China
nchrist@163.com

Jackie Chen
Sandia National Laboratories,
Livermore, CA 94550-0969, USA

Lin Chen
Department of Computer Science
and Engineering,
Shanghai Jiao Tong University,
Shanghai 200030, China
chenlin@sjtu.edu.cn

Maogou Chen
School of Computer Engineering and
Science, Shanghai
University, Shanghai 200072, China

Ningyu Chen
School of Computer Engineering and
Science, Shanghai
University, Shanghai 200072, China
cny314@sohu.com

Wei Chen
School of Computer, Wuhan
University, Wuhan
430079, Hubei, China

Xiaobo Chen
School of Computer Engineering and
Science, Shanghai
University, Shanghai 200072, China

Zhangxin Chen
Center for Scientific Computation
and Department of
Mathematics, Southern Methodist
University,
Dallas, TX 75275-0156, USA
zchen@mail.smu.edu

Zhengqi Chen
Shaanxi Province Weather Mod-
ification Center, Xi'an 710015,
China

Zhisong Chen
Denso Corporation, Japan

Hui Cheng
Computing Center, Northeastern
University, Shenyang
110004, China

Zheru Chi
CMSP, Dept. of EIE, PolyU.,
HongKong, China
enzheru@inet.polyu.edu.hk

Waiki Ching
Department of Mathematics, The
University of Hong Kong,
Pokfulam Road, Hong Kong, China
wkc@maths.hku.hk

Martin J. Cole
Scientific Computing and Imaging
Institute, University of Utah,
Salt Lake City, UT, USA
mjc@sci.utah.edu

James Cremer
Center for Statistical Genetics
Research,
University of Iowa, Iowa City, IA
52242, USA
james-cremer@uiowa.edu

C. Rodríguez Cuevas
Université Aix-Marseille III.
CNRS, France
cuevas@L3m.univ-mrs.fr

Jianye Cui
Computing Center, Northeastern
University, Shenyang
110004, China

Lin Da
Beijin Jiaotong University, Beijing
010021, China

Qianni Deng
Department of Computer Science &
Engineering, Shanghai
JiaoTong University, Shanghai
20030, China
deng-qn@cs.sjtu.edu.cn

Shutian Deng
Department of Mathematics,
University of
Texas at Arlington, Arlington, TX
76019, USA

E. Herrera Díaz
Instituto
Politécnico Nacional, Mexico
enrique_herrera@att.net.mx

Guangtai Ding
School of Computer Engineering and
Science, Shanghai University,
Yanchang Rd. 149, Shanghai 200072,
China
dgt@mail.shu.edu.cn

M. H. Djavareshkian
Faculty of Mechanical Engineering,
University of
Tabriz, Tabriz, Iran
Djavaresh@tabrizu.ac.ir

Craig C. Douglas
University of Kentucky, Department
of Computer Science,
773 Anderson Hall, Lexington, KY
40506-0046, USA and
Yale University, Department of
Computer Science,
P.O. Box 208285
New Haven, CT 06520-8285, USA
douglas-craig@cs.yale.edu

Tao Du
Department of Computer Science
and Engineering,
Shanghai Jiao Tong University,
Shanghai 200030, China
dutao@sjtu.edu.cn

Zhihui Du
Department of Computer Science
and Technology, Tsinghua University,
100084, Beijing, China

Yalchin Efendiev
Institute for Scientific Computation,
Texas A&M University, College
Station, TX 77843-3404, USA
efendiev@math.tamu.edu

Richard E. Ewing
Institute for Scientific Computation,
Texas A&M University, College
Station, TX 77843-3404, USA
richard-ewing@tamu.edu

Chaojiang Fu
School of Computer Engineering and
Science, Shanghai
University, Shanghai 200072, China
fcj@mail.shu.edu.cn

Heng Fu
Information College, Tianjin
University, Tianjin 300222, China

Jianming Fu
School of Computer, Wuhan
University, Wuhan
430072, China
fujms@public.wh.hb.cn

Baolai Ge
SHARCNET,
The University of Western Ontario,
London, Canada N6A 5B7
bge@sharcnet.ca

Shailesh Ghotgalkar
Hardware Technology Development
Group,
Centre for Development of Ad-
vanced Computing, Pune University
Campus,
Pune 411 007, India

Victor Ginting
Institute for Scientific Computation,
Texas A&M University, College
Station, TX 77843-3404, USA

Roberto Gomez
Pittsburgh Supercomputing Center,
Carnegie Mellon
University, Pittsburgh, PA 15213,
USA

Zhenghu Gong
School of Computer, National
University of Defense Technology,
Changsha 410073, China

Jianbo Guan
School of Computer Science,
National University of
Defense Technology, Changsha
410073, China
guanjb@nudt.edu.cn

Yadong Gui
Shanghai Supercomputer Center,
Shanghai, China

Songliu Guo
Department of Computer Science
and Technology, Tsinghua
University, China

Qianqian Han
High Performance Computing
Center, Shanghai Jiaotong
University, Shanghai 200030, China

Bing He
School of Computer Engineering and
Science, Shanghai
University, Shanghai 200072, China
he.bing@huawei.com

Tao He
Center for Statistical Genetics
Research,
University of Iowa, Iowa City, IA
52242, USA
tao-he@uiowa.edu

Yuxiang He
Laboratory of Cloud-Precipitation
Physics and Severe
Storms (LACS), Institute of
Atmospheric Physics, Chinese
Academy
of Sciences, Beijing 100029, China

Daniel Hsu
School of Computer Engineering and
Science, Shanghai
University, Shanghai 200072, China

Wenhao Hu
Department of Computer Science
and Technology, Xi'an
Jiaotong University, Xi'an 710049,
China

Guanren Huan
Institute for Scientific Computation,
Texas A&M University, College
Station, TX 77843-3404, USA
ghuan@isc.tamu.edu

Linpeng Huang
Department of Computer Science
and Engineering,
Shanghai Jiao Tong University,
Shanghai 200030, China
huang-lp@cs.sjtu.edu.cn

Min Huang
School of
Information Science and Engineering,
Northeastern University,
Shenyang 110004, China

Tianqi Huang
Department of Computer Science
and Engineering,
Shanghai Jiao Tong University,
Shanghai 200030, China

Wei Huang
School of Computer Engineering and
Science, Shanghai University,
Yanchang Rd. 149, Shanghai 200072,
China
planewalker@graduate.shu.edu.cn

Xiaoqin Huang
Department of Computer Science
and Engineering,
Shanghai Jiao Tong University,
Shanghai 200030, China
huangxq@sjtu.edu.cn

Ying Huang
Zhejiang University, Hangzou,
Zhejiang 310027, China

Yongxiang Huang
Department of Computer Science
and Technology, Xi'an
Jiaotong University, Xi'an 710049,
China

Zunguo Huang
School of Computer, National
University of Defense Technology,
Changsha 410073, China

S. Baheri Islami
Name and Faculty of Mechanical
Engineering, University of Tabriz,
Tabriz, Iran
baheri@tabrizu.ac.ir

Jongsoo Jang
Security Gateway System Team,
Electronics and Telecommunications
Research Institute,
161 Gajeong-Dong, Yuseoung-Gu,
Daejeon, 305-350, KOREA
jsjang@tri.re.kr

Yongjie Jia
Cluster and Grid Computing Lab
Huazhong University of Science and
Technology Wuhan, 430074, China

Kai Jiang
Shanghai Supercomputer Center,
Shanghai, China
kjiang@ssc.net.cn

Li Jiang
Department of Mathematics,
University of
Texas at Arlington, Arlington, TX
76019, USA

Dezhen Jin
Jilin Province Weather Modification
Center, Changchun 130062,
China

Hai Jin
Cluster and Grid Computing Lab
Huazhong University of Science and
Technology Wuhan, 430074, China
hjin@hust.edu.cn

Shenjie Jin
Global Delivery China
Center, Hewlett-Packard De-
velopment Company, Shanghai
201206,
China

Xianglong Jin
High Performance Computing
Center, Shanghai Jiaotong
University, Shanghai 200030, China
jxlong@sjtu.edu.cn

Zhou Jin
IBM Computer Technology Center,
Department of Computer
Science, School of Electronic and
Information Engineering, Tianjin
300072, China

Chris R. Johnson
Scientific Computing and Imaging
Institute, University of Utah,
Salt Lake City, UT, USA
crj@cs.utah.edu

Shardrom Johnson
School of Computer Engineering and
Science, Shanghai
University, Shanghai 200072, China
jshardrom@staff.shu.edu.cn

Greg Jones
Scientific Computing and Imaging
Institute, University of Utah,
Salt Lake City, UT, USA
gjones@sci.utah.edu

Jiubin Ju
School of Computer Science, North-
east Normal University, Shanghai,
China

Dongho Kang
Security Gateway System Team,
Electronics and Telecommunications
Research Institute,
161 Gajeong-Dong, Yuseoung-Gu,
Daejeon, 305-350, KOREA
dhkang@etri.re.kr

Kiyoung Kim
Security Gateway System Team,
Electronics and Telecommunications
Research Institute,
161 Gajeong-Dong, Yuseoung-Gu,
Daejeon, 305-350, KOREA
kykim@etri.re.kr

Body Knosp
Center for Statistical Genetics
Research,
University of Iowa, Iowa City, IA
52242, USA
boyd-knosp@uiowa.edu

Stefan Körkel
Interdisciplinary Center for Scientific
Computing, University of
Heidelberg, Im Neuenheimer Feld
368, D-69120 Heidelberg, Germany

Raytcho Lazarov
Institute for Scientific Computation,
Texas A&M University, College
Station, TX 77843-3404, USA
lazarov@math.tamu.edu

Yongmei Lei
School of Computer Engineering and
Science, Shanghai
University, Shanghai 200072, China
ymlei@mail.shu.edu.cn

H. Ramírez León
Instituto Mexicano del Petróleo.
Mexico
hrleon@imp.mx

Chengxu Li
Department of Computer Science
and Engineering,
Dalian University of Technology,
Dalian 116024, China

Chunjiang Li
School of Computer Science,
National University of Defense
Technology, Changsha 410073, China
lcj@hnxinmao.com

Huiyuan Li
Laboratory of Parallel Computing,
Institute of
Software, Chinese Academy of
Sciences, Beijing 100080,
China
hynli@mail.rdcps.ac.cn

Jia Li
Computing Center, Northeastern
University, Shenyang
110004, China

Jie Li
Department of Computer Science,
Guangxi
Polytechnic, Guangxi, China
janelee@gxzjy.com

Kangshun Li
State Key Laboratory of Software
Engineering, Wuhan
University, Wuhan 430072, and
School of Information Engineering,
Jiangxi University of Science &
Technology, Jiangxi 341000, China
lks@public1.gzptt.jx.cn

Lei Li
CIT Lab, Computer Science
Department, Shanghai Jiaotong
University, Shanghai 200030, China,
lilei@cs.sjtu.edu.cn

Lijun Li
High Performance Computing
Center, Shanghai Jiaotong
University, Shanghai 200030, China

Minglu Li
Department of Computer Science &
Engineering, Shanghai
JiaoTong University, Shanghai
20030, China
li-ml@cs.sjtu.edu.cn

Qing Li
School of Computer Science &
Engineering, Shanghai University,
Shanghai 200072, China

Weinan Li
School of Computer, Wuhan
University, Wuhan
430072, China

Xiang Li
Center for Statistical Genetics
Research,
University of Iowa, Iowa City, IA
52242, USA
xiang-li@uiowa.edu

Xiangqun Li
College of Computer Science and
Technology, Jilin University,
Changchun 130025, China

Xiongfei Li
College of Computer Science and
Technology, Jilin University,
Changchun 130025, China
lxf@jlu.edu.cn

Xufeng Li
School of Computer, National
University of Defense Technology,
Changsha 410073, China

Y.F. Li
Dept. of Manufacturing Eng. and
Eng. Management,
City University of Hong Kong,
Kowloon, Hong Kong
meyfli@cityu.edu.hk

Yuan Li
School of Computer Engineering &
Science, Shanghai
University, Yanchang Road 149,
Shanghai, 200072, China
yuanli415@tom.com

Yuanbao Li
School of Computer Engineering and
Science, Shanghai
University, Shanghai 200072, China
leon_lyb@163.com

Yuanxiang Li
State Key Laboratory of Software
Engineering, Wuhan
University, Wuhan 430072, China

Jian Liang
Department of Computer and
Information Science, Polytechnic
Univ., USA
jliang@photon.poly.edu

K. M. Liew
School of Mechanical and
Production Engineering, Nanyang
Technological University, Nanyang
Avenue Singapore 639798, Republic
of Singapore
mkmliew@ntu.edu.sg

Junwoo Lim
Pittsburgh Supercomputing Center,
Carnegie Mellon
University, Pittsburgh, PA 15213,
USA

Rujian Lin
Communication and Inf. Eng. Tech.
Lab, Shanghai University,
Yanchang Rd. 149, Shanghai 200072,
China
rujianlin@sina.com

Xia Lin
Computer Science & Technology
School, Soochow
University, Soochow, China
linx166@163.com

Xinhua Lin
Department of Computer Science &
Engineering, Shanghai
JiaoTong University, Shanghai
20030, China
lin-xh@sjtu.edu.cn

Chaoqun Liu
Department of Mathematics,
University of
Texas at Arlington, Arlington, TX
76019, USA
cliu@uta.edu

Cong Liu
Computing Center, Northeastern
University, Shenyang
110004, China

Hao Liu
Department of Computer Science
and Engineering,
Dalian University of Technology,
Dalian 116024, China

Hong Liu
Department of Engineering Mechanics,
Shanghai JiaoTong University,
Shanghai
20030, China
hongliu@sjtu.edu.cn

Jiangguo Liu
Institute for Scientific Computation,
Texas A&M University, College
Station, TX 77843-3404, USA
jliu@isc.tamu.edu

Jin Liu
School of Mechatronical Engineering
and Automation,
Shanghai University, Shanghai
200072, China
liujin@public1.sta.net.cn

Yu Liu
Shanghai University-DingTech
Software Co., Ltd, Shanghai
University, Shanghai 200072, China

Yue Liu
School of Computer Engineering &
Science, Shanghai
University, Yanchang Road 149,
Shanghai, 200072, China
yliu@staff.shu.edu.cn

Zongtian Liu
School of Computer Engineering and
Science, Shanghai
University, Shanghai 200072, China

Mark Logue
Center for Statistical Genetics Research,
University of Iowa, Iowa City, IA 52242, USA
mark-logue@uiowa.edu

Xicheng Lu
School of Computer Science,
National University of Defense
Technology, Changsha 410073, China

Xinda Lu
Department of Computer Science & Engineering, Shanghai
JiaoTong University, Shanghai 20030, China
lu-xd@cs.sjtu.edu.cn

Donghua Luo
Shanghai University and CIMS
Center, Shanghai University,
Shanghai, China

Jun Luo
School of Computer Engineering and Science, Shanghai
University, Shanghai 200072, China
luojun9803@163.com

Fenghui Lv
Shanghai University and CIMS
Center, Shanghai University,
Shanghai, China

Guoyong Mao
School of Computer Engineering and Science, Shanghai
University, Shanghai 200072, China
gymao@mail.shu.edu.cn

Hongyan Mao
Department of Computer Science
and Engineering,
Shanghai Jiaotong University,
Shanghai 200030, China
mhy@sjtu.edu.cn

Liquan Mei
School of Science, Xi'an
Jiaotong University, Xi'an 710049,
China
lqmei@mail.xjtu.edu.cn

S. Memarpour
Faculty of Mechanical Engineering,
University of
Tabriz, Tabriz, Iran
shmemarpour@yahoo.com

Michael K. Ng
Department of Mathematics, The
University of Hong Kong,
Pokfulam Road, Hong Kong, China
mng@maths.hku.hk

Jun Ni
Center for Statistical Genetics Research,
University of Iowa, Iowa City, IA 52242, USA
un-ni@uiowa.edu

Zhemin Ni
Department of Computer Science
and Engineering,
Shanghai Jiaotong University,
Shanghai 200030, China
zhmni@sjtu.edu.cn

Jintae Oh
Security Gateway System Team,
Electronics and Telecommunications
Research Institute,
161 Gajeong-Dong, Yuseoung-Gu,
Daejeon, 305-350, Korea

Juw Won Park
Center for Statistical Genetics Research,
University of Iowa, Iowa City, IA 52242, USA
juw-park@uiowa.edu

Wenling Peng
School of Computer, Wuhan
University, Wuhan
430079, Hubei, China
peng_wenling@yahoo.com.cn

Yang Qi
Department of Engineering Mechanics,
Shanghai JiaoTong University,
Shanghai
20030, China
luvanaki@sjtu.edu.cn

Depei Qian
Department of Computer Science
and Technology, Xi'an
Jiaotong University, Xi'an 710049,
China

Pengfei Qian
CIT Lab, Computer Science & Tech
Department, Shanghai
Jiaotong University, Shanghai
200030, China

Guan Qin
Institute for Scientific Computation,
Texas A&M University, College
Station, TX 77843-3404, USA
guan.qin@neo.tamu.edu

Huiqin Qu
Intelligent Information Processing
Laboratory, Fudan University, China
huiqin.qu@fudan.edu.cn

Jaideep Ray
Sandia National Laboratories,
Livermore, CA 94550-0969, USA

Raghurama Reddy
Pittsburgh Supercomputing Center,
Carnegie Mellon
University, Pittsburgh, PA 15213,
USA

Hongcan Ren
School of Computer Engineering and
Science, Shanghai
University, Shanghai 200072, China

Gerd Rücker
Deutsche Börse AG, 60485 Frankfurt
am Main, Germany

Sebastian Sager
Interdisciplinary Center for Scientific
Computing, University of
Heidelberg, Im Neuenheimer Feld
368, D-69120 Heidelberg, Germany

C. Sajish
Hardware Technology Development
Group,
Centre for Development of Advanced Computing, Pune University
Campus,
Pune 411 007, India

Sergiu Sanielevici
Pittsburgh Supercomputing Center,
Carnegie Mellon
University, Pittsburgh, PA 15213,
USA

Alberto Segre
Center for Statistical Genetics
Research,
University of Iowa, Iowa City, IA
52242, USA
alberto-segre@uiowa.edu

Hua Shan
Department of Mathematics,
University of
Texas at Arlington, Arlington, TX
76019, USA

Fengwei Shi
Shanghai University and CIMS
Center, Shanghai University,
Shanghai, China

Jun Shi
Department of Computer Science
and Engineering,
Shanghai Jiao Tong University,
Shanghai 200030, China
shijun@cs.sjtu.edu.cn

Lianshuan Shi
Computer Department, Tianjin
University of Technology and
Education,
Tianjin 300222, China
shilianshuan@263.net

Suixiang Shi
School of Information Science &
Engineering,
Northeastern University, Shenyang
110004, China
ssx@mail.nmdis.gov.cn

Anping Song
School of Computer Engineering and
Science, Shanghai
University, Shanghai 200072, China
apsong@mail.shu.edu.cn

Ting Su
Shanghai University and CIMS
Center, Shanghai University,
Shanghai, China

Jiachang Sun
Laboratory of Parallel Computing,
Institute of
Software, Chinese Academy of
Sciences, Beijing 100080,
China
sun@mail.rdcps.ac.cn

Jizhou Sun
IBM Computer Technology Center,
Department of Computer
Science, School of Electronic and
Information Engineering, Tianjin
300072, China

Tao Sun
College of Computer Science and
Technology, Jilin University,
Changchun 130025, China

Weidong Sun
School of Computer Science
& Engineering, Shenyang Institute
of Aeronautical Engineering,
Shenyang 110034, China

Yongqiang Sun
Department of Computer Science
and Engineering,
Shanghai Jiaotong University,
Shanghai 200030, China
sun-yg@sjtu.edu.cn

Zhigang Sun
School of Computer Science,
National University of Defense
Technology, Changsha 410073, China

James Sutherland
Sandia National Laboratories,
Livermore, CA 94550-0969, USA

C. H. Tai
Nanyang Centre for Supercomputing
and Visualization,
Nanyang Technological University,
Nanyang Avenue
Singapore 639798, Republic of
Singapore
jontai73@pmail.ntu.edu.sg

Guozhen Tan
Department of Computer Science
and Engineering,
Dalian University of Technology,
Dalian 116024, China
gztan@dlut.edu.cn

Ruichun Tang
Department of Computer Science
and Technology, Tsinghua University,
100084, Beijing, China

Ye Tang
CIT Lab, Computer Science
Department, Shanghai Jiaotong
University, Shanghai 200030, China,
tangye@cs.sjtu.edu.cn

C.A. Thole
Fraunhofer Institute
for Algorithms and Scientific
Computing,
Schloss Birlinghoven, 53754, St.
Augustin, Germany

Huaglory Tianfield
School of Computing and Mathe-
matical Sciences Glasgow
Caledonian University, 70 Cow-
caddens Road Glasgow, G4 0BA,
United
Kingdom
h.tianfield@gcal.ac.uk

Yuichi Tsujita
Department of Electronic Engineer-
ing and Computer
Science,
Faculty of Engineering, Kinki
University
1 Umenobe, Takaya, Higashi-
Hiroshima, Hiroshima
739-2116, Japan
tsujita@hiro.kindai.ac.jp

K.A.Venkatesh
Post Graduate Department of
Computer Applications,
Alliance Business Academy, Banga-
lore 560 076, India
venki@rediffmail.com

Veronica Vieland
Center for Statistical Genetics
Research,
University of Iowa, Iowa City, IA
52242, USA
veronica-vieland@uiowa.edu

Bai Wang
School of Computer Science and
Technology,
Beijing University of Posts and
Telecommunications,
Beijing 100876, China
wangbai@bupt.edu.cn

Dongsheng Wang
Research Institute of
Information Technology, Tsinghua
University, China
wds@mail.tsinghua.edu.cn

Gang Wang
Computer Science & Technology
School, Soochow
University, Soochow, China
trulywang@163.com

Ge Wang
Center for Statistical Genetics
Research,
University of Iowa, Iowa City, IA
52242, USA
ge-wang@uiowa.edu

Jiazeng Wang
Department of Computer Science,
Shanghai Jiao Tong University,
Shanghai 200030, China
wangjz@sjtu.edu.cn

Jiening Wang
Civil Aviation
University of China, Dongli District,
Tianjin, China
jieningwang@cauc.edu.cn

Lina Wang
School of Computer, Wuhan
University, Wuhan
430079, Hubei, China

Shaowen Wang
Center for Statistical Genetics
Research,
University of Iowa, Iowa City, IA
52242, USA
shaowen-wang@uiowa.edu

Xingwei Wang
Computing Center, Northeastern
University, Shenyang
110004, China
wangxw@mail.neu.edu.cn

Yang Wang
Pittsburgh Supercomputing Center,
Carnegie Mellon
University, Pittsburgh, PA 15213,
USA
ywg@psc.edu

Yanyan Wang
Information & Scientific
Computation Groups, Zhongyuan
Technical College, Zhengzhou
475004,
China

Yinglin Wang
Department of Computer Science
and Engineering,
Shanghai Jiao Tong University,
Shanghai 200030, China

Zhenglu Wang
School of Computing and Mathe-
matical Sciences Glasgow
Caledonian University, 70 Cow-
caddens Road Glasgow, G4 0BA,
United
Kingdom
zhengluw@hotmail.com

Zongjiang Wang
Department of Computer Science
and Engineering,
Shanghai Jiao Tong University,
Shanghai 200030, China
microw@sjtu.edu.cn

Hongmei Wei
Jiangnan Computing
Technology Institute, Jiangnan,
China

Yanhua Wen
Jiangnan Computing
Technology Institute, Jiangnan,
China

Gengfeng Wu
School of Computer Engineering &
Science, Shanghai
University, Yanchang Road 149,
Shanghai 200072, China
gfwu@staff.shu.edu.cn

Chiren Wu
Guangxi University of
Technology, Liuzhou, Guangxi
545006, China
chcldw@163.com

S.D. Wu
Dept. of Manufacturing Eng. and
Eng. Management,
City University of Hong Kong,
Kowloon, Hong Kong
s.d.wu@student.cityu.edu.hk

Shaochun Wu
School of Computer Engineering &
Science, Shanghai
University, Yanchang Road 149,
Shanghai, 200072, China
scwu@mail.shu.edu.cn

Weiguo Wu
Department of Computer Science
and Technology, Xi'an
Jiaotong University, Xi'an 710049,
China

Weiwei Wu
Institute of Automobile Engineering,
Shanghai University
of Engineering Science, Shanghai
200336, China
wuzilin@263.net

Yanhua Wu
Department of Computer Science,
Shanghai Jiaotong University,
Shanghai 200030, China
wyhross@sjtu.edu.cn

Zhijian Wu
State Key Laboratory of Software
Engineering, Wuhan
University, Wuhan 430072, China
zjwu@public.wh.hb.cn

Dengwei Xia
School of Information Science &
Engineering,
Northeastern University, Shenyang
110004, China

Xiufeng Xia
School of Information Science &
Engineering,
Northeastern University, Shenyang
110004, China

Yi Xian
School of Computer, Wuhan
University, Wuhan
430072, China

Hui Xiao
Laboratory of Cloud-Precipitation
Physics and Severe
Storms (LACS), Institute of
Atmospheric Physics, Chinese
Academy
of Sciences, Beijing 100029, China
hxiao@mail.iap.ac.cn

Nong Xiao
School of Computer Science,
National University of Defense
Technology, Changsha, China

Jinkui Xie
Department of Computer Science
and Engineering,
Shanghai Jiao Tong University,
Shanghai 200030, China
jkxie@cs.sjtu.edu.cn

Dahong Xing
School of Computer Engineering and
Science, Shanghai
University, Shanghai 200072, China

Dong Xu
School of Computer Engineering and
Science, Shanghai
University, Shanghai 200072, China
d.xuu@163.com

Lingyu Xu
School of Computer Science &
Engineering, Shanghai University,
Shanghai 200072, China

Liutong Xu
School of Computer Science and
Technology,
Beijing University of Posts and
Telecommunications,
Beijing 100876, China
xliutong@bupt.edu.cn

Xiaoming Xu
Department of Computer Science,
Shanghai Jiaotong University,
Shanghai 200030, China
xmx@sjtu.edu.cn

Zhiyun Xu
Shanghai University and CIMS
Center, Shanghai University,
Shanghai, China

Chao Yang
Laboratory of Parallel Computing,
Institute of
Software, Chinese Academy of
Sciences, Beijing 100080,
China
yc@mail.rdcps.ac.cn

Feng Yang
Department of Computer Science
and Technology, Tsinghua University,
100084, Beijing, China
yfeng99@mails.tsinghua.edu.cn

Linfeng Yang
School of Computer Engineering and
Science, Shanghai
University, Shanghai 200072, China
LFYang98@163.com

Xuejun Yang
School of Computer Science,
National University of Defense
Technology, Changsha, China

Wenbin Yao
Department of Computer Science
and Technology, Tsinghua
University, China
yao-wb@mail.tsinghua.edu.cn

Zhenglin Ye
Department of Applied Mathematics,
Northwestern
Polytechnical University, Xi'an,
Shaanxi 710072, China

Chuanjiang Yi
Cluster and Grid Computing Lab
Huazhong University of Science and
Technology Wuhan, 430074, China

Fuqiang Ying
School of Mechatronical Engineering
and Automation,
Shanghai University, Shanghai
200072, China
motor@zjut.edu.cn

Jinyuan You
Department of Computer Science
and Engineering,
Shanghai Jiao Tong University,
Shanghai 200030, China
you-jy@cs.sjtu.edu.cn

Ge Yu
School of Information Science &
Engineering,
Northeastern University, Shenyang
110004, China

Zhaochun Yu
School of Computer Engineering &
Science, Shanghai
University, Yanchang Road 149,
Shanghai, 200072, China

Jie Yuan
School of Computer Engineering and
Science, Shanghai
University, Shanghai 200072, China
jyuan@staff.shu.edu.cn

Wenjiang Yuan
Department of Computer Science
and Engineering,
Dalian University of Technology,
Dalian 116024, China

Waion Yuen
Department of Mathematics, The
University of Hong Kong,
Pokfulam Road, Hong Kong, China
h9923304@hkusua.hku.hk

Weijie Zhai
Louhe Occupation Technical College,
Luohe 450003, China

Bofeng Zhang
School of Computer Engineering &
Science, Shanghai
University, Yanchang Road 149,
Shanghai, 200072, China
bfzhang@staff.shu.edu.cn

Heng Zhang
Department of Mathematics, Shihezi
University,
Shihezi 832000, China
zhheng01@163.com

Huangguo Zhang
School of Computer, Wuhan
University, Wuhan
430072, Hubei, China

James Zhang
Shanghai University-DingTech
Software Co., Ltd, Shanghai
University, Shanghai 200072, China

Jianhong Zhang
CIT Lab, Computer Science
Department,
Shanghai Jiaotong University,
Shanghai 200030, China
zhang-jh@cs.sjtu.edu.cn

Jiawan Zhang
IBM Computer Technology Center,
Department of Computer
Science, School of Electronic and
Information Engineering, Tianjin
300072, China
jwzhang@tju.edu.cn

Jifa Zhang
Center for Engineering and Scientific
Computation, and
College of Computer Science,
Zhejiang University, Hangzhou,
Zhejiang 310027, China
jifa_zhang@yahoo.com.cn

Jingyi Zhang
Department of Computer Science
and Engineering,
Shanghai Jiao Tong University,
Shanghai 200030, China

Shensheng Zhang
CIT Lab, Computer Science
Department, Shanghai Jiaotong
University, Shanghai 200030, China
sszhang@cs.sjtu.edu.cn

Shuqin Zhang
Department of Mathematics, The
University of Hong Kong,
Pokfulam Road, Hong Kong, China
sqzhang@hkusua.hku.hk

Wu Zhang
School of Computer Engineering and
Science, Shanghai
University, Shanghai 200072, China
wzhang@staff.shu.edu.cn

Xiaobin Zhang
School of Computer Engineering and
Science, Shanghai
University, Shanghai 200072, China
zxb0412@hotmail.com

Yi Zhang
High Performance Computing
Center, Shanghai Jiaotong
University, Shanghai 200030, China
jxlong@sjtu.edu.cn

Jing Zhao
Department of Computer Science &
Engineering, Shanghai
JiaoTong University, Shanghai
20030, China
zhaojing@sjtu.edu.cn

Y. Zhao
School of Mechanical and
Production Engineering, Nanyang
Technological University, Nanyang
Avenue Singapore 639798, Republic
of Singapore

Zhengde Zhao
School of Computer Engineering and
Science, Shanghai
University, Shanghai 200072, China
zhdzhao@163.com

Hongchan Zheng
Department of Applied Mathematics,
Northwestern
Polytechnical University, Xi'an,
Shaanxi 710072, China

Weimin Zheng
Department of Computer Science
and Technology, Tsinghua
University, China
zwm-cs@mail.tsinghua.edu.cn

Yao Zheng
Center for Engineering and Scientific
Computation, and
College of Computer Science,
Zhejiang University, Hangzhou,
Zhejiang 310027, China
yao.zheng@yahoo.com.cn

Aimin Zhou
State Key Laboratory of Software
Engineering, Wuhan
University, Wuhan 430072, China

Xin Zhou
College of Computer Science and
Technology, Jilin University,
Changchun 130025, China

Bin Zhu
School of Computer Engineering and
Science, Shanghai
University, Shanghai 200072, China

Hong Zhu
Intelligent Information Processing
Laboratory, Fudan University, China
hzhu@fudan.edu.cn

Tieying Zhu
School of Computer Science, Jilin
University, China
zhuty@nenu.edu.cn

Yanqin Zhu
Computer Science & Technology
School, Soochow
University, Soochow, China
yqzhu@suda.edu.cn

Yonghua Zhu
School of Computer Science and
Engineering,
Shanghai University, Yanchang Rd.
149, Shanghai 200072, China
yhzhu@mail.shu.edu.cn

Ziyu Zhu
Department of Computer Science
and Technology, Tsinghua University,
100084, Beijing, China

Yousuo J. Zou
University of Guam, Mangilao,
Guam 96923, USA.
yjzou@guam.uog.edu

Part I

Regular Papers

LES Investigation of Near Field Wakes Behind Juncture of Wing and Plate

Jiangang Cai[1], Shutian Deng[1], Hua Shan[1], Li Jiang[1], and Chaoqun Liu[1]

Department of Mathematics, University of Texas at Arlington, Arlington, TX 76019, USA jxc0622@omega.uta.edu, cliu@uta.edu

Abstract. This paper is a preliminary work on the numerical study of the near field wake and wingtip vortex behind the juncture of wing and plate. The object is to develop a LES-based code to simulate the near field wake and wingtip vortex, which will help understanding near field wakes and wake control.

The filtered structure function subgrid model was applied on the simulation. The computational approach involves a fully implicit time-marching solver for the Navier-Stokes equation in the generalized curvelinear coordinates with a sixth order compact scheme and an eighth order filter. A one-block high-quality mesh is generated in the complex computation domain. The coordinate transformation metrics are calculated with the high-order compact scheme that is consistent with the flow solver. The parallel computation is accomplished by using domain decomposition and message passing interface (MPI).

The code is validated for simulation of flow transition on a flat plate. The code is also used for simulation of 2-D airfoil vortex shedding. Then the flow field around NACA0014 airfoil is simulated with 8 million grid points. The preliminary results were obtained. To study the near field wake and wingtip vortex behind the juncture of wing and plate, a LES simulation with 25 million grid points is undergoing, in which the angle of attach of the wing is $10°$.

1 Introduction

The wake behind a juncture of the control surface and sub off body is considered as one source of the noises, vibration, and detectable signals. It also results in a noisy and unsteady inflow for the propeller. The interaction between the wake and the vortices generated by the propeller increases the complexity of the flow structure. The simulation, understanding, and control of wakes behind control surface become an important issue in order to improve the submarine performance and to achieve the optimal design of the propeller.

However, most numerical studies to date are still focused on low Reynolds number or laminar flows around a circular cylinder or sphere, which is important to oil industry but still pretty far away from the real applications in Naval Hydrodynamics. In real Naval Hydrodynamics, the Reynolds number is high and inflow is most likely turbulent. Therefore, the numerical simulation of wakes behind juncture of the control surface and sub off body with high Reynolds number turbulent flow is required to gain more understanding of the complex flow.

There are some numerical simulations for flow passing control surface and propeller based on the Reynolds averaged Navier-Stokes equations (RANS, see Pankajakshan et al. 2001 [1]). However, since the flow pattern and vortex structure are strongly time dependent and extremely complicated, the simulation with RANS is not able to predict the flow accurately. The alternative approaches include direct numerical simulation (DNS) and large eddy simulation (LES). There are some reports regarding DNS and LES for near field turbulent wake (Mittal et al. 1996 [2];Xia et al. 1997 [3]), but only for 2-D circular cylinder at a Reynolds number of 3900. For high Reynolds number flow, there are some reports from CTR (Wang 1997, 1999, 2001 [4]) for trailing edge or circular cylinder. The LES for turbulent flow past a juncture of a wing and a flat plate is rarely found in literature. The present work will focus on LES of wakes and wingtip vortices behind a juncture of a wing and flat plate. Though the configuration is different from the real case in Naval Hydrodynamics, yet it is able to catch the most of physics.

Large Eddy Simulation (LES) is a technique that has been successfully applied to the study of turbulent flows, and has became a very important method of numerical simulation, especially for the simulations of flows with high Reynolds numbers. In LES only the large energy-carrying scales are resolved, while the influence of small or subgrid scales must be modeled appropriately. A filtering process is usually used to separate the large- and small-scale motions, large-scale structures are resolved by the filtered equations, but a model is employed to formulate the contributions from the subgrid scale fluctuations, such as the subgrid scale stress and heat flux terms. LES is able to solve problems within complex geometry (Ducros et al., 1996 [5]). Furthermore, compared with DNS, LES is more favorable in simulations of high Reynolds number flows, since it needs much fewer grid points.

The objective of this work is to develop a LES-based code to simulate the turbulent flow around the control surface as well as the near field wakes behind the control surface of a submarine. The filtered structure function subgrid model (Ducros et al., 1996 [5]) is employed in the simulation. A fully implicit time-marching solver for the Navier-Stokes equation in the generalized curvelinear coordinates is used with a sixth order compact scheme and an eighth order filter. A one-block high-quality mesh is generated using an elliptic grid generation method, first proposed by Spekreijse (1995) [6]. The Jacobian of the coordinate transformation is calculated with the high-order compact scheme that is consistent with the flow solver. The code also integrates the message passing interface (MPI) parallel computation whose performance scales almost linearly over a large number of processors.

The code is first validated for flow transition on a flat plate as a result of the unstable modes imposed at inflow boundary. The fully developed turbulent flow velocity profile and logarithm law is checked and accurately achieved. Then the code is used to simulate vortex shedding and wakes around a 2-D airfoil.

The turbulent flow passing a juncture of a wing and flat plate is simulated by LES with filtered structure function subgrid models. The near-field wakes behind the juncture at $Re = 4.6 \times 10^6$ was simulated. In this case, 8 million grid points are used. The preliminary results are reported in this paper while the simulation is not completed. Another similar case with $Re = 4.6 \times 10^6$, $Ma_\infty = 0.15$, attack angle $= 10°$ and 25 million grid points currently is still being simulated although the computation is not completed and data are not enough for any subtle conclusion.

This paper is organized as follows: Section 2 introduces the mathematical model and the numerical methods used in the simulation. The numerical grid generation is presented in Section 3. Section 4 shows some of the preliminary simulation results. Section 5 includes the concluding remarks.

2 Mathematical Model and Numerical Methods

2.1 Governing Equations

In large eddy simulation of fluid flow, the large-scale structures are simulated and resolved by numerical method while the effect of the small-scale structures is modeled. The large scales are extracted from the dependent variables by applying a filtering operation to the governing equations.

Although our object is to study the incompressible flow field, the current LES code still use compressible equations at the low Mach number flows, where flow could be considered as incompressible. High-order incompressible flow solver will be developed in the near future.

For the Navier-Stokes equations, the Favre-filtering operation is used. The resolved velocity and temperature fields, written in terms of the Favre-filtered quantities, can be defined as

$$\tilde{F} = \overline{\rho F}/\bar{\rho} \tag{1}$$

where the "˜" denotes the spatial filtering. The non-dimensional Favre-filtered governing equations of continuity, momentum, and temperature are described as follows:

$$\frac{\partial \bar{\rho}}{\partial t} + \frac{\partial}{\partial x_l}(\bar{\rho}\tilde{u}_k) = 0 \tag{2}$$

$$\frac{\partial \bar{\rho}\tilde{u}_k}{\partial t} + \frac{\partial}{\partial x_l}(\bar{\rho}\tilde{u}_k\tilde{u}_l) = -\frac{\partial \bar{p}}{\partial x_k} + \frac{1}{Re}\frac{\partial \tilde{\sigma}_{kl}}{\partial x_l} + \frac{\partial \tau_{kl}}{\partial x_l} \tag{3}$$

$$\frac{\partial \bar{\rho}\tilde{T}}{\partial t} + \frac{\partial}{\partial x_l}(\bar{\rho}\tilde{u}_k\tilde{T}) = -\gamma(\gamma-1)M_\infty^2 \bar{p}\frac{\partial \tilde{u}_k}{\partial x_k} + \frac{\gamma(\gamma-1)M_\infty^2}{Re}\tilde{\sigma}_{kl}\frac{\partial \tilde{u}_k}{\partial x_l}$$
$$+ \frac{\partial}{\partial x_k}(\frac{\gamma\tilde{\mu}}{PrRe}\frac{\partial \tilde{T}}{\partial x_k}) + \frac{\partial q_k}{\partial x_k} \tag{4}$$

where ρ is the density, u_k the velocity component in the kth direction, p the pressure, and T the temperature. The viscous stress is

$$\tilde{\sigma}_{kl} = \tilde{\mu}[(\frac{\partial \tilde{u}_k}{\partial x_l} + \frac{\partial \tilde{u}_l}{\partial x_k}) - \frac{2}{3}\frac{\partial \tilde{u}_m}{\partial x_m}\delta_{kl}] \tag{5}$$

In the nondimensionalization, the reference values for length, density, velocity, and temperature are δ_{in}, ρ_∞, U_∞, and T_∞, respectively. δ_{in} is the displacement thickness of inflow. The Mach number, the Reynolds number, the Prandtl number, and the ratio of specific heats, are defined respectively as follows:

$$M_\infty = \frac{U_\infty}{\sqrt{\gamma RT}}, \quad Re = \frac{\rho_\infty U_\infty \delta_{in}}{\mu_\infty}, \quad Pr = \frac{C_p\mu_\infty}{\kappa_\infty}, \quad \gamma = \frac{C_p}{C_v}$$

where R is the ideal gas constant, C_p and C_v are the specific heats at constant pressure and constant volume. Through this paper, $Pr = 0.7$, and $\gamma = 1.4$. The viscosity is determined according to Surtherland's law, in dimensionless form

$$\mu = \frac{T^{3/2}(1+S)}{T+S}, \quad S = \frac{110.3K}{T_\infty}$$

In Eq. (3) and (4), the sub grid model scale stress and heat flux are denoted by

$$\tau_{kl} = -\bar{\rho}(\widetilde{u_k u_l} - \tilde{u}_k\tilde{u}_l) \tag{6}$$

$$q_k = -\bar{\rho}(\widetilde{u_k T} - \tilde{u}_k\tilde{T}) \tag{7}$$

which are needed to be modeled.

2.2 Filtered Structure Function Model

The filtered structure-function model is developed by Ducros et al.(1996) [5]. The subgrid scale shear stress and heat flux can be modeled as

$$\tau_{kl} = \bar{\rho}\mu_t[(\frac{\partial \tilde{u}_k}{\partial x_l} + \frac{\partial \tilde{u}_l}{\partial x_k})] - \frac{2}{3}\frac{\partial \tilde{u}_m}{\partial x_m}\delta_{kl})] \tag{8}$$

$$q_k = \frac{\gamma\bar{\rho}\mu_t}{Pr_t}\frac{\partial \tilde{T}}{\partial x_k} \tag{9}$$

Here, Pr_t is the turbulent Prandtl number taken equal to 0.6 as in isotropic turbulence, μ_t is the turbulent kinetic viscousity defined as

$$\nu_t(x,t) = 0.0014 C_k^{-3/2} \Delta [\tilde{F}_2^{(3)}(x,t)]^{1/2} \tag{10}$$

where $F_2^{(3)}$ is the the filtered structure function. In this case, takes the fourth-neighbor formulation proposed by Normand & Lesieur (1992).

$$\tilde{F}_2^{(3)} = \tfrac{1}{4}[\|\tilde{u}_{i+1,j,k}^3 - \tilde{u}_{i,j,k}^{(3)}\|^2 + \|\tilde{u}_{i-1,j,k}^3 - \tilde{u}_{i,j,k}^{(3)}\|^2 + \|\tilde{u}_{i,j+1,k}^3 - \tilde{u}_{i,j,k}^{(3)}\|^2$$
$$+ \|\tilde{u}_{i,j-1,k}^3 - \tilde{u}_{i,j,k}^{(3)}\|^2 + \|\tilde{u}_{i,j,k+1}^3 - \tilde{u}_{i,j,k}^{(3)}\|^2 + \|\tilde{u}_{i,j,k-1}^3 - \tilde{u}_{i,j,k}^{(3)}\|^2] \tag{11}$$

where $\tilde{u}_{i,j,k}^{(3)} = HP^{(3)}(\tilde{u}_{i,j,k})$.

$\Delta = \sqrt{\Delta_x \Delta_y \Delta_z}$ is used to characterize the grid size. C_k is the Kolmogorov constant taking the value of 1.4. $HP^{(3)}$ is a discrete Laplacian filter iterated 3 times, which is served as a high-pass filter before computing the filtered structure function. The first iteration of the Laplacian filter $HP^{(1)}$ is defined by

$$\tilde{u}_{i,j,k}^{(1)} = HP^{(1)}(\tilde{u}_{i,j,k}) = \tilde{u}_{i+1,j,k} - 2\tilde{u}_{i,j,k} + \tilde{u}_{i-1,j,k}$$
$$+ \tilde{u}_{i,j+1,k} - 2\tilde{u}_{i,j,k} + \tilde{u}_{i,j-1,k} + \tilde{u}_{i,j,k+1} - 2\tilde{u}_{i,j,k} + \tilde{u}_{i,j,k-1} \tag{12}$$

2.3 Numerical Methods

The compact scheme (Lele 1992) has been widely used in simulation of transition and turbulent flows (Visbal et al. 1998 [7]; Jiang et al. 1999a [8]; Shan et al. 1999 [10]). In the present work, a sixth order compact scheme is used for spatial discretization in streamwise direction and wall normal direction. The LU-SGS implicit scheme (Yoon & Kwak 1992 [11]) is used to solve equations. To avoid possible non-physical wave reflection from the boundaries, the non-reflecting boundary conditions (Jiang et al. 1999b [9]) are specified at the far field and the outflow boundaries.

The parallel computation is accomplished through the Message Passing Interface (MPI) together with a domain decomposition approach. The computational domain is partitioned into n equal-sized sub-domains along the streamwise direction. n is the number of processors used in the parallel computation and up to 64 processors are used for parallel computation in the present work.

3 Numerical Grid Generation

In this paper, we concentrate on the wake and tip vortex behind a juncture of wing and flat plate. The physical configuration shown in Figure 1 will be replaced by a simplified computational domain, which includes a rectangular wing with a NACA0014 airfoil section and a flat tip, as shown in the Figure 2. The span-chord ratio of the wing is 0.27. In Figure 2, the bottom plane is corresponding to the flat plate. The spanwise direction of the wing is perpendicular to the flat plate.

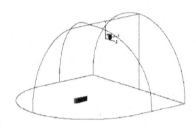

Fig. 1. SUBOFF with Propulsors   **Fig. 2.** Physical domain

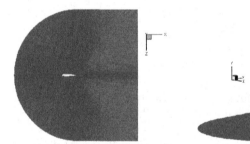

Fig. 3. C-type grid on the flat plate   **Fig. 4.** H-type grid on the exit

In this paper, we use a one-block mesh for the computation. Generally a multi-block mesh is more flexible, but it may involve some additional special treatments at the boundary of each block that may introduce more numerical errors and reduce the accuracy of the computation. Therefore, the one-block mesh approach is used in the present work. For a configuration of the juncture of the wing and flat plate, a single C-H topology is adopted for the grid generation. First the 3-D grid is generated algebraically, then an elliptic grid generation method followed that of Spekreijse(1995) [6] is used to redistribution and smooth the grid inside the domain. Figure 3 shows the C-type grid surrounding the wing on the flat plate. The grid lines in this figure are corresponding to the lines and the lines in the transformed computational domain (the ξ-η -ζ space).

In the directions that are corresponding to the η and ξ directions in the transformed computational domain, the grid is of H-type, such as the vertical plane shown in Figure 4 , where the grid at the downstream boundary of the domain is displayed.

The sharp edge of the wingtip is replaced by a rounded corner with a radius of no more than 10^{-5} of the chord length in order to avoid any singularity in the computation of the Jacobian of the coordinate transformation. The detail of the rounded corner is shown in Figure 5

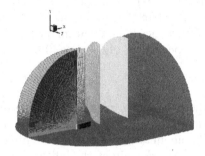

Fig. 5. Rounded corner of wingtip   **Fig. 6.** Grid topology

The overall grid geometry is shown in Figure 6 . The grid we generated in this case has size of 1280×80×80, for the streamwise, spanwise and wing surface normal directions respectively.

4 Results and Discussions

4.1 Code Validation

In order to validate the code, DNS of the flow transition on a flat plate is shown as the first case here. In this case, the LES model is switched off. The computational domain is displayed in Figure 7, where the boundaries of the computation domain are shown.

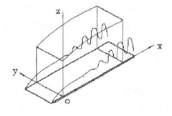

Fig. 7. Computational domain of flow over a flat plate

In the test case, the Mach number $M_\infty = 0.5$, the Reynolds number based on the freestream velocity and the displacement thickness at the inflow boundary $Re_{\delta_{in}} = 1000$. Here δ_{in} is the inlet displacement thickness. The Prandtl number $Pr = 0.7$. The x-coordinate of the inflow boundary measured from the leading edge of the flat plate is $x_{in} = 300.8\delta_{in}$. The linear spatial evolution of the small disturbance imposed at the inlet is simulated in the test case. At the inlet boundary, the most amplified eigenmode of the 2-D Tollmien-Schlichting(T-S) waves is enforced. The eigenfunction has a frequency $\omega = 0.0957$, and a space wavenumber $\alpha = 0.25792 - i6.72054 \times 10^{-3}$. The disturbance imposed at inflow boundary is $A_{2d}q'_{2d}$. The amplitude of T-S wave is specified as $A_{2d} = 5 \times 10^{-4}$, and

$$q'_{2d} = \phi_r \cos(\omega t) + \phi_i \sin(\omega t)$$

where ϕ_r and ϕ_i are the respective real and imaginary parts of the eigenfunction obtained from linear stability theory (LST). The length of computational domain along the streamwise direction includes 8 T-S wavelengths, and height of the domain at the inflow boundary is $40\delta_{in}$. For a 2-D simulation, the grid size is of 128×61 representing the number of grids in the streamwise (x), and wall normal (z) directions.

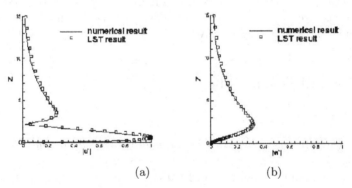

(a)   (b)

Fig. 8. Comparison of the numerical and LST velocity profiles at $Re_x = 394300$

After eight time periods of calculation, the T-S wave has been fully established over the entire domain, the Fourier transform in time is conducted on physical variables. The amplitude of disturbance is defined as Fourier amplitude. At a certain streamwise location, the Fourier amplitude is a function of z, which is the wall direction coordinate. The profiles of the disturbance amplitude of the streamwise velocity u, and the wall normal velocity w, are shown in Figure 8 by solid lines, while the LST results are plotted by square symbols in the same figure for comparison.

For a 3-D DNS of flow transition over the flat plate, the flow parameters are similar to the previous case. The length of computational domain along

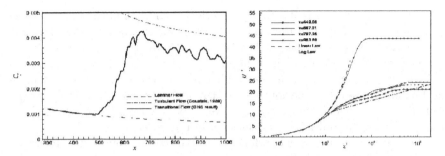

Fig. 9. Streamwise evolution of the time- and spanwise -averaged skin-friction coefficient

Fig. 10. Log plots of the averaged velocity profiles

the streamwise direction is $800\delta_{in}$, the width along the spanwise direction is $22\delta_{in}$, and height at the inflow boundary is $40\delta_{in}$. The grid size is $640\times64\times60$ representing the number of grids in the streamwise (x), spanwise (y), and wall normal (z) directions. The inflow consists of a 2-D T-S wave eigenmode and a 3-D random disturbance. The frequency of the T-S eigenmode is $\omega = 0.114027$. The space wavenumber of the mode is given by $\alpha = 0.29919 - i5.00586\times10^{-3}$. The amplitude of the 2-D mode is specified as $A_{2d} = 0.02$, which is much larger than that used in the 2-D simulation. The 3-D random disturbance at inflow is given by white noise with amplitude of 0.01.

The skin friction coefficient calculated from the time- and spanwise-averaged velocity profile is displayed in Figure 9. The spatial evolution of skin friction coefficient of laminar flow and that of turbulent flow from Couteix(1989) are also plotted in the same figure for comparison.

In Figure 9, the streamwise evolution of the time- and spanwise -averaged skin-friction coefficientthe time- and spanwise-averaged velocity profiles at different streamwise locations are plotted in logarithmic wall unit. In Figure 10, the Log plots of the averaged velocity profiles are shown. The curves of the linear law near the wall and the log law are also plotted for comparison.

The test case shows our DNS code can discribe the flow transitions and fully developed turbulent flow.

4.2 Flow around 2-D Airfoil

(a)t=0.107   (b) t=10.56

Fig. 11. Instantaneous vortex contour

The second test case is the direct numerical simulation of the flow passing an asymmetric airfoil. The Reynolds number is 5×10^5, and the free stream Mach number is 0.15, attack angle is zero, grids size 1401×181. The airfoil geometry and the instant vortex contour are shown in Figure 11.

Instability is observed near the trailing edge of the airfoil, where at first the boundary layer became much more thicken, as shown in Figure 11(a). Soon after that, separation occurred. Vortex shedding from the wake near the trailing edge moves downstream, as shown in Figure 11(b).

These test cases show good performance of the code through the numerical simulation of 2-D and 3-D flow field.

4.3 Flow over Wing-Flat Plate Juncture

For the case of flow over the juncture of a wing and a flat plate, a preliminary simulation is carried out for a Reynolds number of 4.6×10^6 based on the chord length of the wing. The angle of attack is zero. The inflow mach number is 0.15.

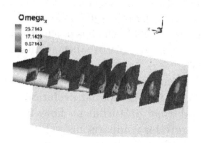
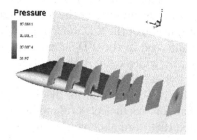

Fig. 12. Instantaneous ω_x contour   **Fig. 13.** Instantaneous pressure coutour

Although computation is not completed yet, some preliminary data will be shown in this section because they are also helpful for understanding the flow.

Figure 12 shows the instantaneous ω_x contour of the wingtip vortex. Figure 13 shows the instantaneous pressure contour of the wingtip vortex. The wingtip vortex is clearly visible in this figure.

The large flow structure of wake is also observed in our simulation. However, in this 3-D case, the wake is much more complicated than the 2-D flow wake especially near the wingtip. The structures similar to Karman vortex street as appeared in 2-D flow can be observed at the low attitude surface parallel to the plate. At the attitude above about 0.25 chord length (the span-chord ratio of the wing is 0.27), this kind of 2-D vortex shedding disappears. The flow structure also varies significantly at different attitude along the wingspan direction. Figure 14 shows the surface instant streamline at the two different attitudes at the same time. It can be attribute to the 3-D nature of the flow over the wing-flat plate juncture.

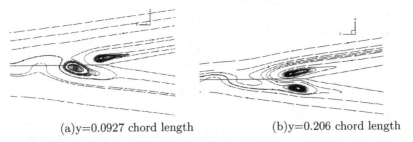

(a)y=0.0927 chord length   (b)y=0.206 chord length

Fig. 14. Instantaneous surface streamline at different attitude

Our LES results have the similar flow structure with the flow in the experiment by Bradshaw, et al, 1995 [12, 13].

5 Conclusions

In this paper the LES based code was developed to study the wingtip vortex and near field wake in a configuration of a wing-flat plate juncture.

The computer code was validated for the linear instability and the flow transition simulation over a flat plate. Vortex shedding and wakes around a 2-D airfoil has also been simulated. The results show the capability of this code in the numerical simulation of 2-D and 3-D flow fields.

The current study focuses on near field wake and vortex shedding behind the juncture of a wing and a flat plate. A one-block high quality grid is generated. The preliminary results show some interesting features of the flow field. Our experience in the simulation indicates that more grid points are needed for future simulation. The results with 8 millions of grids is still not satisfactory due to insufficient resolution, especially near the edge of the wingtip, where the tip vortices develop. The simulation with 25 million grid points may give much better results although it costs much more in computation time.

Acknowledgments

This work was sponsored by the Office of Naval Research (ONR) and monitored by Dr. Ron Joslin under grant number N00014-03-1-0492. The authors also thank the High Performance Computing Center of US Department of Defense for providing supercomputer hours.

References

[1]   Pankajakshan, R., Taylor, L. K., Sheng, C., Jiang, M. J., Briley, W. R., and D. L. Whitfield, 2001. Parallel Efficiency in Implicit Multi-block, Multigrid Simulations, with Application to Submarine Maneuvering, AIAA Paper 2001-1093

[2]   Mittal, R. and Moin, P. 1996. Large-eddy simulation of flow past a circular cylinder, APS Bulletin, Vol 41 **(9)**, 49th DFD meeting.

[3]   Xia, M. and Karniadakis, G., The spectrum of the turbulent near-wake: a comparison of DNS and LES, Second AFOSR International Conference on DNS/LES, Ruston, LA, 1997.

[4]   Wang, M., Progress in large-eddy simulation of trailing-edge turbulence and aeroacoustics, Center for Turbulence Research, Annual Research Briefs, 1997.

[5]   Ducros, F., Comte, P. and Lesieur, M. 1996. Large-eddy simulation of transition to turbulence in a boundary layer developing spatially over a flat plate. J. Fluid Mech. **326**, pp1-36.

[6]   Spekreijse, S.P. 1995. Elliptic grid generation based on Laplace equations and algebraic transformation. J. Comp. Phys. **118**, **38**.

[7]   Visbal, M.R., Gaitonde, D.V. 1998. High-order accurate methods for unsteady vortical flows on curvilinear meshes. AIAA paper 98-0131

[8]   Jiang, L., Shan, H., Liu, C. 1999a. Direct numerical simulation of boundary-layer receptivity for subsonic flow around airfoil. Recent Advances in DNS and LES, Proceedings of the Second AFOSR International Conference on DNS/LES, Rutgers-The State University of New Jersey, New Brunswick, U.S.A., June 7-9, 1999.

[9]   Jiang, L., Shan, H., Liu, C. 1999b. Non-reflecting boundary conditions for DNS in curvilinear coordinates. Recent Advances in DNS and LES, Proceedings of the Second AFOSR International Conference on DNS/LES, Rutgers-The State University of New Jersey, New Brunswick, U.S.A., June 7-9, 1999.

[10]   Shan, H., Jiang, L., Liu, C. 2000. Numerical simulation of complex flow around a 85 delta wing. DNS/LES Progress and Challenges, Proceedings of the Third AFOSR International Conference on DNS/LES, University of Texas at Arlington, Arlington, TX, U.S.A., Aug. 5-9, 2000.

[11]   Yoon, S., Kwak, D. 1992. Implicit Navier-Stokes solver for three-dimensional compressible flows. AIAA Journal, **30**, 2653.

[12]   Dacles-Mariani. J., Zilliac G.G., Chow J.S., Bradshaw P., 1995. Numerical experimental study of a wingtip vortex in the near field. AIAA J. 33, 1561

[13]   Bradshaw, et al, 1995 Chow, J.S., Zilliac, G.G., Bradshaw, P. 1997. Mean and Turbulence Mearements in the Near Field of a wingtip vortex. AIAA J.35, 1561

On Computation with Higher-order Markov Chains

Waiki Ching[1], Michael K. Ng[2], and Shuqin Zhang[3]

Department of Mathematics, The University of Hong Kong, Pokfulam Road, Hong Kong, China.
{wkc,mng}@maths.hku.hk, sqzhang@hkusua.hku.hk

Summary. Categorical data sequences occur in many real world applications. The major problem in using higher-order Markov chain model is that the number of parameters increases exponentially with respect to the order of the model. In this paper, we propose a higher-order Markov chain model for modeling categorical data sequences where the number of model parameters increases linearly with respect to the order of the model. We present efficient estimation methods based on linear programming for the model parameters. The model is then compared with other existing models with simulated sequences and DNA data sequences of mouse.

1 Introduction

Categorical data sequences occur in many real world applications. For categorical data sequences, there are many situations that one would like to employ higher-order Markov chain models, see for instance [4, 5, 6]. A number of applications can be found in literatures, examples include sales demand prediction [1], webpage prediction [2], Alignment of sequences in DNA sequence analysis [8]. It has been shown that higher-order Markov chain models can be a promising approach for these applications.

For simplicity of discussion, in the following we let X_t be a data point of a categorical data sequence at time t and X_t takes values in the set $\mathcal{M} \equiv \{1, 2, \cdots, m\}$ where m is finite, i.e., it has m possible categories or states. The conventional model for a n-th order Markov chain has $O(m^{n+1})$ model parameters. The major problem in using such kind of model is that the number of parameters (the transition probabilities) increases exponentially with respect to the order of the model. This large number of parameters discourages people from using a higher-order Markov chain directly.

In this paper, we present a higher-order Markov chain model based on models of Raftery [6, 7] and Ching et al. [1, 2] for modeling categorical data sequences. The rest of the paper is organized as follows. In Section 2, we present the existing higher-order Markov chain models and discuss some properties of these models. We then present our proposed model. In Section 3, we propose estimation methods for the model parameters required in our higher-order Markov chain model. In Section 4, numerical examples on simulated sequences and DNA sequences are given to demonstrate effectiveness of the proposed model. Finally, concluding remarks are given in Section 5.

2 Higher-order Markov Chain Models

Raftery [6] proposed a higher-order Markov chain model which involves only one additional parameter for each extra lag. The model can be written as follows:

$$P(X_t = k_0 \mid X_{t-1} = k_1, \ldots, X_{t-n} = k_n) = \sum_{i=1}^{n} \lambda_i q_{k_0 k_i} \tag{1}$$

where $\sum_{i=1}^{n} \lambda_i = 1$ and $Q = [q_{ij}]$ is a transition matrix with column sums equal to one, such that

$$0 \leq \sum_{i=1}^{n} \lambda_i q_{k_0 k_i} \leq 1, \quad k_0, k_i \in \mathcal{M}. \tag{2}$$

The constraint in (4) is to guarantee that the right-hand-side of (4) is a probability distribution. The total number of independent parameters in his model is of $O(n + m^2)$. Raftery showed that (4) is analogous to the standard AR(n) model in the sense that each additional lag, after the first is specified by a single parameter and the autocorrelations satisfy a system of linear equations similar to the Yule-Walker equations. Moreover, the parameters $q_{k_0 k_i}, \lambda_i$ can be estimated numerically by maximizing the log-likelihood of (4) subject to the constraints in (4). But this approach involves solving a highly non-linear optimization problem. The proposed numerical method neither guarantees convergence nor a global maximum.

Later Ching et al. extended the Raftery model [6] to a more general higher-order Markov model by allowing Q to vary with different lags. Here they assume that the weighting λ_i is non-negative such that

$$\sum_{i=1}^{n} \lambda_i = 1. \tag{3}$$

We first notice that (4) can be re-written as

$$\mathbf{X}_{t+n+1} = \sum_{i=1}^{n} \lambda_i Q \mathbf{X}_{t+n+1-i} \tag{4}$$

where $\mathbf{X}_{t+n+1-i}$ is the probability distribution of the states at time $(t + n + 1 - i)$. Using (3) and the fact that Q is a transition probability matrix, we note that each entry of $\mathbf{X}_{t+n+1}$ is in between 0 and 1, and the sum of all entries is equal to one. In their model, they generalized Raftery's model in (4) as follows:

$$\mathbf{X}_{t+n+1} = \sum_{i=1}^{n} \lambda_i Q_i \mathbf{X}_{t+n+1-i}. \tag{5}$$

The total number of independent parameters in the new model is $O(n+nm^2)$. We note that if $Q_1 = Q_2 = \ldots = Q_n$ then (5) is just the Raftery model in (4). In their model, they assume that $\mathbf{X}_{t+n+1}$ depends on $\mathbf{X}_{t+i}$ $(i = 1, 2, \ldots, n)$ via the matrix Q_i and weight λ_i. One may relate Q_i to the ith step transition matrix of the sequence and this idea is then used to estimate Q_i. They also proved the following proposition.

Proposition 1. *(Ching et al. [2]) If Q_n is irreducible and $\lambda_n > 0$ such that*

$$0 \leq \lambda_i \leq 1 \quad \text{and} \quad \sum_{i=1}^{n} \lambda_i = 1$$

then the model in (5) has a stationary distribution $\bar{\mathbf{X}}$ when $t \to \infty$ independent of the initial state vectors $\mathbf{X}_0, \mathbf{X}_1, \ldots, \mathbf{X}_{n-1}$. The stationary distribution $\bar{\mathbf{X}}$ is also the unique solution of the following linear system of equations:

$$(I - \sum_{i=1}^{n} \lambda_i Q_i)\bar{\mathbf{X}} = \mathbf{0} \quad \text{and} \quad \mathbf{1}^T \bar{\mathbf{X}} = 1.$$

Here I is the $m \times m$ identity matrix (m is the number of possible states taken by each data point) and $\mathbf{1}$ is an m-vector of ones.

In their proof, they let $\mathbf{Y}_{n+1} = (\mathbf{X}_{t+n+1}, \mathbf{X}_{t+n}, \ldots, \mathbf{X}_{t+2})^T$ be an $nm \times 1$ vector. Then one may write $\mathbf{Y}_{n+1} = R\mathbf{Y}_n$ where

$$R = \begin{pmatrix} \lambda_1 Q_1 & \lambda_2 Q_2 & \cdots & \lambda_{n-1} Q_{n-1} & \lambda_n Q_n \\ I & 0 & \cdots & 0 & 0 \\ 0 & I & 0 & & \vdots \\ \vdots & \ddots & \ddots & \ddots & 0 \\ 0 & \cdots & 0 & I & 0 \end{pmatrix} \tag{6}$$

is an $nm \times nm$ square matrix. By using the Perron-Frobenius Theorem, they showed that under the assumptions in the proposition, all the eigenvalues of $\tilde{R}$ lie in the interval $[0, 1]$ and there is exactly one eigenvalue equals to one. This implies that $\lim_{s\to\infty} \overbrace{R \ldots R}^{s} = \lim_{s\to\infty} (R)^s = \mathbf{V}\mathbf{U}^T$. Therefore we have

$$\lim_{s\to\infty} \mathbf{Y}_{s+n+1} = \lim_{s\to\infty} (R)^s \mathbf{Y}_{n+1} = \mathbf{V}(\mathbf{U}^T \mathbf{Y}_{n+1}) = \alpha \mathbf{V}.$$

Here α is a positive number because $\mathbf{Y}_{n+1} \neq \mathbf{0}$ and is non-negative. This implies that X_s also tends to a stationary distribution as s goes to infinity. Hence we have $\lim_{s\to\infty} \mathbf{X}_{s+n+1} = \lim_{s\to\infty} \sum_{i=1}^{n} \lambda_i Q_i \mathbf{X}_{s+n+1-i}$ and therefore we have $\bar{\mathbf{X}} = \sum_{i=1}^{n} \lambda_i Q_i \bar{\mathbf{X}}$. The stationary distribution vector $\bar{\mathbf{X}}$ satisfies

$$(I - \sum_{i=1}^{n} \lambda_i Q_i)\bar{\mathbf{X}} = \mathbf{0} \quad \text{with} \quad \mathbf{1}^T \bar{\mathbf{X}} = 1. \tag{7}$$

The normalization constraint is necessary as the matrix $(I - \sum_{i=1}^{n} \lambda_i Q_i)$ has an one-dimensional null space. We remark that the model (5) is a stationary model as it has a stationary distribution.

However, the non-negativity assumption on λ_i is restrictive, see the numerical examples in Section 4. Here we propose to replace the constraints

$$0 \leq \lambda_i \leq 1, \quad i = 1, 2, \dots, n \quad \text{and} \quad \sum_{i=1}^{n} \lambda_i = 1$$

by

$$0 \leq \sum_{i=1}^{n} \lambda_i q_{k_0 k_i}^{(i)} \leq 1, \quad k_0, k_i \in \mathcal{M} \quad \text{and} \quad \sum_{i=1}^{n} \lambda_i = 1$$

Our new higher-order Markov chain model is then obtained by revising the model in Ching et al. [2] as above. We expect this new model will have better prediction accuracy when appropriate order of model is used, see the numerical results in Sections 4 for instance.

3 Parameters Estimation

In this section, we present efficient methods to estimate the parameters Q_i and λ_i for $i = 1, 2, \dots, n$. We will demonstrate these methods by a simple example. To estimate Q_i, we regard Q_i as the ith step transition matrix of the categorical data sequence $\{X_t\}$. Given the categorical data sequence $\{X_t\}$, one can count the transition frequency $f_{jk}^{(i)}$ in the sequence from state k to state j in the ith step. Hence one can construct the ith step transition matrix for the sequence $\{X_t\}$ as follows:

$$F^{(i)} = \begin{pmatrix} f_{11}^{(i)} & \cdots & \cdots & f_{m1}^{(i)} \\ f_{12}^{(i)} & \cdots & \cdots & f_{m2}^{(i)} \\ \vdots & \vdots & \vdots & \vdots \\ f_{1m}^{(i)} & \cdots & \cdots & f_{mm}^{(i)} \end{pmatrix}. \tag{8}$$

From $F^{(i)}$, we get the estimates for $Q_i = [q_{kj}^{(i)}]$ as follows:

$$
\hat{Q}_i = \begin{pmatrix} \hat{q}_{11}^{(i)} & \cdots & \cdots & \hat{q}_{m1}^{(i)} \\ \hat{q}_{12}^{(i)} & \cdots & \cdots & \hat{q}_{m2}^{(i)} \\ \vdots & \vdots & \vdots & \vdots \\ \hat{q}_{1m}^{(i)} & \cdots & \cdots & \hat{q}_{mm}^{(i)} \end{pmatrix} \quad \text{where} \quad \hat{q}_{kj}^{(i)} = \begin{cases} \dfrac{f_{kj}^{(i)}}{\sum\limits_{k=1}^{m} f_{kj}^{(i)}} & \text{if } \sum\limits_{k=1}^{m} f_{kj}^{(i)} \neq 0 \\ \\ 0 & \text{otherwise.} \end{cases} \tag{9}
$$

The computational cost for the construction of $F^{(i)}$ is of $O(L)$ operations and the construction cost of $Q^{(i)}$ from $F^{(i)}$ is $O(m^2)$, where L is the length of the given data sequence. Hence the total computational cost for the construction of $\{Q^{(i)}\}_{i=1}^{n}$ is of $O(n(L + m^2))$ operations.

3.1 Linear Programming Formulation for Estimation of λ_i

Suppose that Proposition 1 holds then it gives a sufficient condition for the sequence $\mathbf{X}_s$ to be convergent to a stationary distribution $\bar{\mathbf{X}}$. Then $\bar{\mathbf{X}}$ can be estimated from the observed sequence $\{X_s\}$ by computing the proportion of the occurrence of each state in the sequence. Let us denote it by $\hat{\mathbf{X}}$. From (7) one would expect that

$$
\sum_{i=1}^{n} \lambda_i \hat{Q}_i \hat{\mathbf{X}} \approx \hat{\mathbf{X}}. \tag{10}
$$

In [2], it suggests one possible way to estimate the parameters $\lambda = (\lambda_1, \ldots, \lambda_n)$ as follows. In view of (10) one can consider the following optimization problem:

$$
\min_{\lambda} \left\| \sum_{i=1}^{n} \lambda_i \hat{Q}_i \hat{\mathbf{X}} - \hat{\mathbf{X}} \right\|_{\infty} = \min_{\lambda} \max_{k} \left[\left| \sum_{i=1}^{n} \lambda_i \hat{Q}_i \hat{\mathbf{X}} - \hat{\mathbf{X}} \right| \right]_k
$$

subject to

$$
\sum_{i=1}^{n} \lambda_i = 1, \quad \text{and} \quad 0 \leq \sum_{i=1}^{n} \lambda_i q_{k_0 k_i}^{(i)} \leq 1, k_0, k_i \in \mathcal{M}.
$$

Here $[\cdot]_k$ denotes the kth entry of the vector. We see that the above optimization problem can be re-formulated as a linear programming problem:

$$
\min_{\lambda} w
$$

subject to

$$
\begin{pmatrix} w \\ w \\ \vdots \\ w \end{pmatrix} \geq \hat{\mathbf{X}} - \left[\hat{Q}_1 \hat{\mathbf{X}} \mid \hat{Q}_2 \hat{\mathbf{X}} \mid \cdots \mid \hat{Q}_n \hat{\mathbf{X}} \right] \begin{pmatrix} \lambda_1 \\ \lambda_2 \\ \vdots \\ \lambda_n \end{pmatrix},
$$

$$\begin{pmatrix} w \\ w \\ \vdots \\ w \end{pmatrix} \geq -\hat{\mathbf{X}} + \left[\hat{Q}_1 \hat{\mathbf{X}} \mid \hat{Q}_2 \hat{\mathbf{X}} \mid \cdots \mid \hat{Q}_n \hat{\mathbf{X}} \right] \begin{pmatrix} \lambda_1 \\ \lambda_2 \\ \vdots \\ \lambda_n \end{pmatrix},$$

$$w \geq 0, \quad \sum_{i=1}^{n} \lambda_i = 1, \quad \text{and} \quad 0 \leq \sum_{i=1}^{n} \lambda_i q_{k_0 k_i}^{(i)} \leq 1, k_0, k_i \in \mathcal{M}.$$

Instead of solving an min-max problem, one can also formulate the following optimization problem:

$$\min_{\lambda} \left\| \sum_{i=1}^{n} \lambda_i \hat{Q}_i \hat{\mathbf{X}} - \hat{\mathbf{X}} \right\|_1 = \min_{\lambda} \sum_{k=1}^{n} \left[\left| \sum_{i=1}^{n} \lambda_i \hat{Q}_i \hat{\mathbf{X}} - \hat{\mathbf{X}} \right| \right]_k$$

subject to

$$\sum_{i=1}^{n} \lambda_i = 1, \quad \text{and} \quad 0 \leq \sum_{i=1}^{n} \lambda_i q_{k_0 k_i}^{(i)} \leq 1, k_0, k_i \in \mathcal{M}.$$

The corresponding linear programming problem is given as follows:

$$\min_{\lambda} \sum_{k=1}^{m} w_k$$

subject to

$$\begin{pmatrix} w_1 \\ w_2 \\ \vdots \\ w_m \end{pmatrix} \geq \hat{\mathbf{X}} - \left[\hat{Q}_1 \hat{\mathbf{X}} \mid \hat{Q}_2 \hat{\mathbf{X}} \mid \cdots \mid \hat{Q}_n \hat{\mathbf{X}} \right] \begin{pmatrix} \lambda_1 \\ \lambda_2 \\ \vdots \\ \lambda_n \end{pmatrix},$$

$$\begin{pmatrix} w_1 \\ w_2 \\ \vdots \\ w_m \end{pmatrix} \geq -\hat{\mathbf{X}} + \left[\hat{Q}_1 \hat{\mathbf{X}} \mid \hat{Q}_2 \hat{\mathbf{X}} \mid \cdots \mid \hat{Q}_n \hat{\mathbf{X}} \right] \begin{pmatrix} \lambda_1 \\ \lambda_2 \\ \vdots \\ \lambda_n \end{pmatrix},$$

$$w_i \geq 0, \quad \forall i, \quad \sum_{i=1}^{n} \lambda_i = 1, \quad \text{and} \quad 0 \leq \sum_{i=1}^{n} \lambda_i q_{k_0 k_i}^{(i)} \leq 1, k_0, k_i \in \mathcal{M}.$$

Remark 1. However, Proposition 1 is not necessarily true for our new model when the parameters λ_i are free. If all the modulus of the eigenvalues of the matrix R are strictly less than one except that 1 is the unique eigenvalue of R such that it has modulus of one. Then one can follow the argument in Section 2 or see [1], our new model is still a stationary model and Proposition 1 still holds. The only extra work here is to solve an $mn \times mn$ eigenvalue problem.

In the following, we give a simple example to demonstrate our estimation methods.

3.2 An Example

We consider a sequence $\{X_t\}$ of two states ($m = 2$) given by

$$\{1, 1, 2, 2, 1, 2, 2, 1, 2, 2, 1, 2, 2, 1, 2, 2, 1, 2, 2, 2\}. \tag{11}$$

The sequence $\{X_t\}$ can be written in vector form

$$\mathbf{X}_1 = (1, 0)^T, \quad \mathbf{X}_2 = (1, 0)^T, \quad \mathbf{X}_3 = (0, 1)^T, \quad \cdots \quad , \mathbf{X}_{20} = (0, 1)^T.$$

We consider $n = 2, 3, 4$, then from (11) we have the transition frequency matrices

$$F^{(1)} = \begin{pmatrix} 1 & 5 \\ 6 & 7 \end{pmatrix}, \quad F^{(2)} = \begin{pmatrix} 0 & 5 \\ 7 & 6 \end{pmatrix}, \quad F^{(3)} = \begin{pmatrix} 5 & 0 \\ 2 & 10 \end{pmatrix} \text{ and } F^{(4)} = \begin{pmatrix} 1 & 4 \\ 5 & 6 \end{pmatrix} \tag{12}$$

Therefore from (12) we have the i-step transition matrices ($i = 1, 2, 3, 4$) as follows:

$$\hat{Q}_1 = \begin{pmatrix} \frac{1}{7} & \frac{5}{12} \\ \frac{6}{7} & \frac{7}{12} \end{pmatrix}, \quad \hat{Q}_2 = \begin{pmatrix} 0 & \frac{5}{11} \\ 1 & \frac{6}{11} \end{pmatrix}, \quad \hat{Q}_3 = \begin{pmatrix} \frac{5}{7} & 0 \\ \frac{2}{7} & 1 \end{pmatrix} \text{ and } \hat{Q}_4 = \begin{pmatrix} \frac{1}{6} & \frac{4}{10} \\ \frac{5}{6} & \frac{6}{10} \end{pmatrix}$$

and $\hat{\mathbf{X}} = (0.35, 0.65)^T$. For this example, the model parameters can be obtained by solving a linear programming problem. It turns out that the parameters obtained are identical the same for both $\|\cdot\|_1$ and $\|\cdot\|_\infty$. We report the model parameters for the cases of $n = 2, 3, 4$. For $n = 2$, we have $(\lambda_1^*, \lambda_2^*) = (1.4583, -0.4583)$. For $n = 3$, we have $(\lambda_1^*, \lambda_2^*, \lambda_3^*) = (1.25, 0, -0.25)$. For $n = 4$, we have $(\lambda_1^*, \lambda_2^*, \lambda_3^*, \lambda_4^*) = (0, 0, -0.3043, 1.3043)$.

4 Some Numerical Examples

In this section, we present two methods for comparing the performance of the new method and the model by Ching et al. based on two sets of data: the data set in Section 3 and the DNA data sequences in [6].

We first propose the prediction method as follows. Given the state vectors $\mathbf{X}_i$, $i = t - n, t - n + 1, \cdots, t - 1$, the state probability distribution at time t can be estimated as follows:

$$\hat{\mathbf{X}}_t = \sum_{i=1}^{n} \lambda_i \hat{Q}_i \mathbf{X}_{t-i}.$$

In many applications, one would like to make use of the higher-order Markov models for the purpose of prediction. According to this state probability distribution, the prediction of the next state $\hat{X}_t$ at time t can be taken as the state with the maximum probability, i.e., $\hat{X}_t = j$, if $[\hat{\mathbf{X}}_t]_i \leq [\hat{\mathbf{X}}_t]_j, \forall 1 \leq i \leq m$. To evaluate the performance and effectiveness of our higher-order Markov chain model, a prediction result is measured by the prediction accuracy r defined as

$$r = \frac{\sum_{t=n+1}^{T} \delta_t}{T}, \quad \text{where} \quad \delta_t = \begin{cases} 1, & \text{if } \hat{X}_t = X_t \\ 0, & \text{otherwise} \end{cases}$$

and T is the length of the data sequence.

We then present the second method, the χ^2 statistics for our new model when it is stationary. From the observed data sequence, one can obtain the distribution of states $(O_1, O_2, \ldots, O_m)$. From the model parameters Q_i and λ_i, by solving:

$$\mathbf{X} = \sum_{i=1}^{n} \lambda_i \hat{Q}_i \mathbf{X} \quad \text{with} \quad \mathbf{1}^T \mathbf{X} = 1$$

one can obtain the theoretical distribution of the states $(E_1, E_2, \ldots, E_m)$. Then the χ^2 statistics is defined as

$$\chi^2 = L \sum_{i=1}^{m} \frac{(E_i - O_i)^2}{E_i}.$$

The smaller this value is the better the model will be.

We now present the numerical comparisons with the data set in the previous section, (let us denote it by "Sample") and also the DNA data set of 3-state sequence from the mouse αA-crystallin gene see for instance [7], (let us denote it by "DNA"). The length of the sequence of "Sample" is 20 and the length of the sequence of "DNA" is 1307. The results are reported in Table 1 below. All the computations here are done by MATLAB with a PC.

Table 1. Prediction accuracy and χ^2 value.

$n = 2$	Sample (2-state)	DNA (3-state)
New Model ($\|\cdot\|_\infty$)	0.3889 ($\chi^2 = 1.2672$)	0.4858 ($\chi^2 = 7.09E - 4$)
New Model ($\|\cdot\|_1$)	0.3889 ($\chi^2 = 1.2672$)	0.4858 ($\chi^2 = 7.09E - 4$)
Ching's Model ($\|\cdot\|_\infty$)	0.6842 ($\chi^2 = 3.1368$)	0.4858 ($\chi^2 = 7.09E - 4$)
Ching's Model ($\|\cdot\|_1$)	0.6842 ($\chi^2 = 3.1368$)	0.4858 ($\chi^2 = 7.09E - 4$)
Randomly Chosen	0.5000	0.3333
$n = 3$	Sample (2-state)	DNA (3-state)
New Model ($\|\cdot\|_\infty$)	0.3529 ($\chi^2 = 0.3265$)	0.4946 ($\chi^2 = 4.24E - 4$)
New Model ($\|\cdot\|_1$)	0.3529 ($\chi^2 = 0.3265$)	0.4893 ($\chi^2 = 8.44E - 5$)
Ching's Model ($\|\cdot\|_\infty$)	0.6842 ($\chi^2 = 3.1368$)	0.4858 ($\chi^2 = 7.09E - 4$)
Ching's Model ($\|\cdot\|_1$)	0.6842 ($\chi^2 = 3.1368$)	0.4858 ($\chi^2 = 7.09E - 4$)
Randomly Chosen	0.5000	0.3333
$n = 4$	Sample (2-state)	DNA (3-state)
New Model ($\|\cdot\|_\infty$)	0.9375 ($\chi^2 = 0.2924$)	0.4666 ($\chi^2 = 1.30E - 4$)
New Model ($\|\cdot\|_1$)	0.9375 ($\chi^2 = 0.2924$)	0.4812 ($\chi^2 = 4.55E - 5$)
Ching's Model ($\|\cdot\|_\infty$)	0.6842 ($\chi^2 = 3.1368$)	0.4858 ($\chi^2 = 7.09E - 4$)
Ching's Model ($\|\cdot\|_1$)	0.6842 ($\chi^2 = 3.1368$)	0.4858 ($\chi^2 = 7.09E - 4$)
Randomly Chosen	0.5000	0.3333

We note that the new model can improve the prediction accuracy by raising the order of the model from $n = 2$ to $n = 3$ in the "DNA" data set. Significant improvement in prediction accuracy is also observed in the "Sample" data set when the order is increased from $n = 3$ to $n = 4$. These are not possible for Ching's model. From the tested results, the best model for the "Sample" sequence is our new model with order 4 and the best model for the "DNA" sequence is our new model with order 3. Moreover, the model obtained through $\|.\|_1$ minimization is a bit better than the model obtained through $\|.\|_\infty$ minimization.

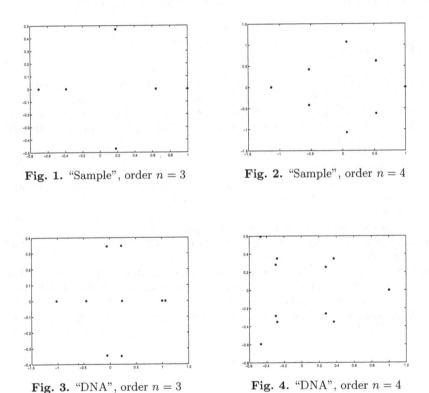

Fig. 1. "Sample", order $n = 3$ **Fig. 2.** "Sample", order $n = 4$

Fig. 3. "DNA", order $n = 3$ **Fig. 4.** "DNA", order $n = 4$

5 Concluding Remarks

In this paper, we proposed and developed a higher-order Markov chain model for categorical data sequences. The number of model parameters increases linearly with respect to the number of lags. Efficient estimation methods for the model parameters are also proposed by making use of the observed transition frequencies and the steady state distribution. The condition on the matrix R

in Remark 1 is a sufficient condition for the model to be stationary. We give the eigenvalue distributions of R for our new models for both the "Sample" and "DNA" sequences when $n = 3$ and $n = 4$ (when $\| \cdot \|_\infty$ is used). It is observed that the sufficient condition in Remark 1 is satisfied in Fig. 1, but not in Fig. 2, Fig. 3 and Fig. 4. From the numerical results, we conjecture that our higher-order Markov chain model presented here is stationary. Further research and numerical tests can be done in this direction.

Acknowledgment: Research supported in part by RGC Grant No. HKU 7126/02P and HKU CRCG Grant Nos. 10204436 and 10205105.

References

1. Ching, W., Fung, E., Ng, M.: A Higher-order Markov Model for the Newsboy's Problem Journal of Operational Research Society **54** (2003) 291-298
2. Ching, W., Fung, E., Ng, M.: Higher-order Markov Chain Models for Categorical Data Sequences, International Journal of Naval Research Logistics **51** (2004) 557-574
3. Fang, S., Puthenpura, S.: Linear Optimization and Extensions, Prentice-Hall, New Jersey (1993)
4. Logan, J.: A Structural Model of the Higher-order Markov Process Incorporating Reversion Effects, J. Math. Sociol. **8** (1981) 75-89
5. MacDonald, I., Zucchini, W.: Hidden Markov and Other Models for Discrete-valued Time Series, Chapman & Hall, London (1997)
6. Raftery, A.: A Model for High-order Markov Chains, J. R. Statist. Soc. B **47** (1985) 528-539
7. Raftery, A., Tavare, S.: Estimation and Modelling Repeated Patterns in High Order Markov Chains with the Mixture Transition Distribution Model, Appl. Statist. **43** (1994) 179-199
8. Waterman, M.:Introduction to Computational Biology, Chapman & Hall, Cambridge (1995)

Dynamic Data-Driven Contaminant Simulation

Craig C. Douglas[1,2], Yalchin Efendiev[3], Richard Ewing[3], Victor Ginting[3], Raytcho Lazarov[3], Martin J. Cole[4], Greg Jones[4], and Chris R. Johnson[4]

[1] University of Kentucky, Department of Computer Science, 773 Anderson Hall, Lexington, KY 40506-0046, USA
[2] Yale University, Department of Computer Science, P.O. Box 208285
New Haven, CT 06520-8285, USA. `douglas-craig@cs.yale.edu`
[3] Texas A&M University, College Station, TX, USA
`{efendiev,lazarov}@math.tamu.edu` and `richard-ewing@tamu.edu`
[4] Scientific Computing and Imaging Institute, University of Utah, Salt Lake City, UT, USA. `{mjc,gjones}@sci.utah.edu` and `crj@cs.utah.edu`

Summary. In this paper we discuss a numerical procedure for performing dynamic data driven simulations (DDDAS). In dynamic data driven simulations our goal is to update the solution as well as input parameters involved in the simulation based on local measurements. The updates are performed in time. In the paper we discuss (1) updating the solution using multiscale interpolation technique (2) recovering as well as updating initial conditions based on least squares approach (3) updating the permeability field using Markov Chain Monte Carlo techniques. We test our method on various synthetic examples.

1 Introduction

Dynamic data driven simulations are important for many practical applications. Consider an extreme example of a disaster scenario in which a major waste spill occurs in a subsurface near clean water aquifer. Sensors can now be used to measure where the contamination is, where the contaminant is going to go, and to monitor the environmental impact of the spill. One of the objectives of dynamic data driven simulations is to incorporate the sensor data into the real time simulations. A number of important issues are involved in DDDAS and they are described in [6]. In this paper our goal is to discuss a numerical procedure and its components that incorporates the sensor data at sparse locations into the simulator.

As the data injected into the simulator, we propose updating the solution as well as the input parameters of the problems, such as permeability field and initial conditions. The update is performed based on the sensor measurements, which streamed from few spatial locations. The schematic description of the update is presented in Figure 1. As data is injected, we update (1) the solution (2) the initial condition and (2) the permeability data. Because of the heterogeneities of the porous media, we employ multiscale interpolation technique for updating the solution. This is done in the context of general non-linear parabolic operators that include subsurface processes. The main idea of this interpolation is that we do not alter the heterogeneities of the random field that drives the contaminant. Rather based on the sensor data we rescale the solution in a manner that it preserves the heterogeneities. This rescaling uses the solution of the local problems. We compare the numerical results for simulations that employ both updating the data at sensor location and the simulations that do not update the locations. The simulation results that do not use the update or use it less frequently produces large errors. This was observed in our numerical studies.

The errors in the simulations will persist if one does not change the input parameters. In the paper we consider modifying both the local permeability field as well as the initial data. To update the permeability field we use Markov Chain Monte Carlo (MCMC) technique. The coarse-scale permeability field can be computed using gradient based approached. Based on the value of the coarse-scale permeability field we attempt to find the underlying fine-scale permeability. This is a downscaling problem. Using MCMC one can write the posterior distribution for the fine-scale permeability field. Furthermore, using classical upscaling techniques and Metropolis-Hasting criteria we update the local fine-scale permeability permeability field. One of the other error sources in dynamic data driven simulation is associated with initial conditions. Moreover, correcting initial data allows us to better identify the spill locations. The initial data update is performed using least squares approach. Each time the data is injected we use the previous information about the initial condition as a prior information in order to update it. The objective function is penalized using the prior information about the initial data. The penalization constant can be modified throughout simulations. Numerical examples are presented in the paper showing the improvement of the predictions as new data is taken into account.

2 Solution update using multiscale interpolation

Our goal in this section is to discuss the mapping of the sensor data to the finite dimensional space where the solution is calculated. This procedure is nontrivial in general, because the solution space usually has high dimension, while the sensors are located only at few locations. Our simplified approach presented in this paper consists of passing the sensor data to the simulations and its use for the next time step simulations. Since the sensor data represents the solution only at few coarse locations one has to modify the solution conditioned to this data. This step we call multiscale interpolation which consists of mapping the sensor data to the solution space. At each time step the sensor data is received by the simulator. There are two options to handle the

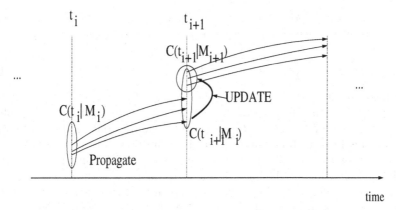

Fig. 1. Schematic description of data driven simulations. $C(t_i|M_i)$ designates the quantity of interest, such as the concentration, at time t_i, conditioned to some data M_i. In our basic methodology we consider M_i to be the sensor data. With each data injection the solution as well as input parameters are updated.

data. We can treat it as hard data or as soft data. The latter means that the data contains some noise and not needed to be imposed exactly. In this paper the first approach, "hard constraint", will be considered. At the beginning of each time step we need to map the sensor data to the solution space. This is performed using DDDAS mapping operator, the main feature of which is not to alter the heterogeneous field.

The proposed mapping for the sensor data is general and applicable to various classes of equations. To demonstrate this we consider general nonlinear parabolic equations

$$\frac{\partial}{\partial t}u_\epsilon = \nabla \cdot (a_\epsilon(x,t,u_\epsilon,\nabla u_\epsilon)) + a_{0,\epsilon}(x,t,u_\epsilon,\nabla u_\epsilon), \ \in \ \Omega \times [0,T], \quad (1)$$

where ϵ indicates the presence of the small scales heterogeneities. This equation includes various physical process that occur in subsurfaces. In the next section numerical examples for particular cases of (1) will be discussed. Assume the domain is divided into the coarse grid such that the sensor points are the nodal points of the coarse grid. Note that we do not require all nodal points to be sensor locations. Further denote by S^h the space of piecewise linear functions on this partition,

$$S_h = \{v_h \in C^0(\overline{\Omega}) : \text{the restriction } v_h \text{ is linear for each triangle } K \in \Pi_h\}.$$

Our objective now is to map the function defined on S^h to the fine grid that represents the heterogeneities. This grid is obtained from *apriori* information about the field using geostatistical packages. Denote by the operator E the mapping from the coarse dimensional space into the fine grid,

$$E : S^h \to V_\epsilon^h,$$

which is constructed as follows. For each element in $u_h \in S^h$ at a given time t_n we construct space time function $u_{\epsilon,h}(x,t)$ in $K \times [t_n, t_{n+1}]$ such that it satisfies

$$\frac{\partial}{\partial t} u_{\epsilon,h}(x,t) = \nabla \cdot (a_\epsilon(x,t,\eta,\nabla u_{\epsilon,h})) \tag{2}$$

in each coarse element K, where η is the average of u_h. $u_{\epsilon,h}(x,t)$ is calculated by solving (2) on the fine grid, and thus it is a fine scale function. To complete the construction of E we need to set boundary and initial conditions for (2). One can set different boundary and initial conditions and this will give rise to different maps. In our numerical simulations we will take the boundary and initial condition for the local problems to be linear with prescribed nodal values. These values are obtained from the sensor data, if available. If the sensor data is not available at some location we use the values obtained from the simulations. Different local boundary conditions can be also imposed and will be discussed later. Mathematical aspects of this interpolation operator, such as convergence and etc, are described in [4].

Once the solution at time $t = t_n$ is computed its values with sensor data at the sensor locations are compared. After changing the values of the solution we interpolate it to the fine grid and use it for the next time step. At the last step we use multiscale approach which is computationally efficient. In particular, the solution at the next time step is calculated based on

$$\int_\Omega (u_h(x,t_{n+1}) - u_h(x,t_n))v_h dx + \sum_K \int_{t_n}^{t_{n+1}} \int_K ((a_\epsilon(x,t,\eta,\nabla u_{\epsilon,h}),\nabla v_h)$$
$$+ a_{0,\epsilon}(x,t,\eta,\nabla u_{\epsilon,h})v_h)dxdt = \int_{t_n}^{t_{n+1}} \int_\Omega f dxdt.$$
$$\tag{3}$$

Here Ω refers to the spatial domain and K are the coarse elements. We would like to note that our approach has limitations and it is not applicable when there are large deviations between sensor data and the solution.

We now present some numerical results that demonstrate the accuracy and limitations of our proposed method. The systems we consider are intended to represent cross sections (in $x - z$) of the subsurface; for that reason we take the system length in x (L_x) to be five times the length in z (L_z). All of the fine grid fields used in this study are 121×121 realizations of prescribed overall variance (σ^2) and correlation structure. The fields were generated using GSLIB algorithms [1] with an exponential covariance model. In the examples below, we specify the dimensionless correlation lengths l_x and l_z, where each correlation length is nondimensionalized by the system length in the corresponding direction. For example, $l_x = 0.3$ means that the actual correlation length of the permeability field is $0.3L_x$. In the calculation below, we consider (1) with a fixed concentration and at the inlet edge of the model ($x = 0$) and a fixed concentration at the outlet edge ($x = L_x$) of the model. The top and bottom boundaries are closed to flow. Initially zero contaminant concentration is assumed in the domain. For comparison purposes results are presented in terms of cross sectional averages and l_2 norm errors. The computations are carried out till the final time $t = 0.1$ with different frequency of updating.

For the numerical test we consider linear heterogeneous diffusion $\frac{\partial}{\partial t} u = \nabla \cdot (a_\epsilon(x) \nabla u)$, where the original "true" diffusion coefficient $a_\epsilon(x) = \exp(\alpha_\epsilon(x))$, where $\alpha_\epsilon(x)$ is a realization of the random field with $l_x = 0.3$, $l_z = 0.02$ and $\sigma = 0.75$. For the simulation purposes we consider the diffusion coefficients to be the same realization of the random field but with $\sigma = 1.5$. Thus we assume that the heterogeneities have the same nature and the only difference between the true field and the one used in the computations is associated with the scaling. Objective of this numerical results is to demonstrate how frequency of updating sensor data in the simulations improves the accuracy of the method. In this example the sensor data is used four times during simulations, i.e., the frequency of updating is 4. In Figure 2 we plot the average of the solutions. Solid line designates the fine scale (true) simulation results, while dotted line represents the results obtained using our methodology with 4 updating. The dashed line represents the simulation results with no updating. As we see from this figure the simulation with data updating performs better compare to that with no updating. The l_2 error between the true solution and the one corresponding with 4 updating is about 6 percent, while l_2 error corresponding with no updating is almost 9 percent. Simulations with more frequent update indicate that the frequent updating improve the accuracy of the predictions and thus important for DDDAS. This is also observed in various numerical examples for both linear and nonlinear equations which we have tested (see [8]).

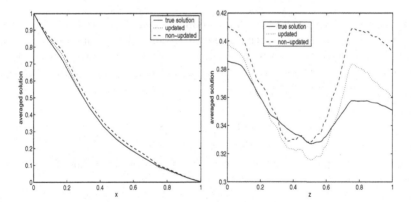

Fig. 2. Comparisons of the average solutions across x and z directions for linear case. The true field is a realization of a random field with $l_x = 0.2$, $l_z = 0.01$, $\sigma = 1$, while random field used for the simulations has $\sigma = 2$. Solid line designates the true solution, dotted line designates the solution obtained using our simulations with 4 updates, and the dashed line designates the solution that has not been updated. The l_2 error between true and updated solution is 5 percent, while the error between the true solution and non-updated solution is 9 percent.

3 Updating initial data

In this section we discuss updating initial conditions. The incorrect initial condition can cause the errors in the simulations. Our goal is to update initial condition each time new data is obtained from the measurements. This update can be performed locally in time, i.e., we do not need to solve the equations from initial time, but rather can use previous time information for the initial condition updates. We consider a model problem

$$\frac{\partial C}{\partial t} + v \cdot \nabla C - \nabla \cdot (D\nabla C) = 0 \quad \text{in } \Omega \tag{4}$$

where v is the Darcy velocity for a heterogeneous permeability field, which is generated using certain statistical variogram and correlation structure, and D is the diffusion coefficient. Next we will discuss updating the initial condition $C^0(\mathbf{x})$ given a set of spatially sparse concentration measurements at certain times.

Before presenting the procedure, we shall introduce several pertaining notations. Let N_s be the number of sensors placed in various points in the porous medium and $\{\mathbf{x}_j\}_{j=1}^{N_s}$ denote such points. Let N_t be the number of how many times the concentration is measured in time and $\{t_k\}_{k=1}^{N_t}$ denote such time levels. Furthermore let $\gamma_j(t_k)$ denotes the measured concentration at sensor located in $\mathbf{x}_j$ and at time t_k. We set

$$M(\gamma) = \{\gamma_j(t_k), j = 1, \cdots, N_s, k = 1, \cdots, N_t\}. \tag{5}$$

Suppose we have a set of N_c possible initial conditions $\{\tilde{C}_i^0(\mathbf{x})\}_{i=1}^{N_c}$, where $\tilde{C}_i^0(\mathbf{x})$ are basis functions with support determined a priori, or $\tilde{C}_i^0(\mathbf{x})$ can be functions of certain form determined a priori. Furthermore, we designate $\tilde{C}_i(\mathbf{x},t)$ the solution of (4) using an initial condition $\tilde{C}_i^0(\mathbf{x})$. Let $\alpha = (\alpha_1, \alpha_2, \cdots, \alpha_{N_c})$ be a vector of real numbers, and write

$$\tilde{C}^0(\mathbf{x}) = \sum_{i=1}^{N_c} \alpha_i \tilde{C}_i^0(\mathbf{x}). \tag{6}$$

Then obviously the solution of (4) with initial condition (6) has the following form:

$$\tilde{C}(\mathbf{x},t) = \sum_{i=1}^{N_c} \alpha_i \tilde{C}_i(\mathbf{x},t). \tag{7}$$

Having written the representation of the initial condition as in (6), we may transform the task of recovering the initial condition of (4) into a problem of finding the "best" α such that $\tilde{C}^0(\mathbf{x}) \approx C^0(\mathbf{x})$. This is achieved through the minimization of the following target function:

$$F(\alpha) = \sum_{j=1}^{N_s} \left(\sum_{i=1}^{N_c} \alpha_i \tilde{C}_i(\mathbf{x}_j, t) - \gamma_j(t) \right)^2 + \sum_{i=1}^{N_c} \kappa_i (\alpha_i - \beta_i)^2, \tag{8}$$

where $\kappa = (\kappa_1, \kappa_2, \cdots, \kappa_{N_c})$ is the penalty coefficients for an a priori vector $\beta = (\beta_1, \beta_2, \cdots, \beta_{N_c})$. Minimization of the target function (8) is done by setting

$$\frac{\partial F(\alpha)}{\partial \alpha_m} = 0, \qquad m = 1, \cdots, N_c, \tag{9}$$

which gives the linear system

$$A\alpha = R, \tag{10}$$

where $A_{mn} = \sum_{j=1}^{N_s} \tilde{C}_m(\mathbf{x}_j, t)\tilde{C}_n(\mathbf{x}_j, t) + \delta_{mn}\kappa_m$, $\quad m, n = 1, \cdots, N_c$, with $\delta_{mn} = 1$ if $m = n$ and zero otherwise, $R_m = \sum_{j=1}^{N_s} \tilde{C}_m(\mathbf{x}_j, t)\gamma_j(t) + \kappa_m\beta_m$, $\quad m = 1, \cdots, N_c$. We note that the penalty term $\sum_{i=1}^{N_c} \kappa_i (\alpha_i - \beta_i)^2$ is needed to regularize the problem. With $\kappa = 0$, the problem can be ill-posed if the number of sensors N_s is less than the number of possible initial conditions N_c.

The least squares approach described above heavily relies on some prior knowledge inherent in the predicted initial condition. This set of knowledge is referred to as *priors*. In the formulation above, the priors are quantified by the variables N_c, β and κ. In the first place, one needs to determine possible location of the initial condition and its support. This can be controlled by the number N_c. If there is a reasonable level of certainty on its location and support, then a smaller number of N_c may be used. On the other hand, a larger N_c has to be used if one has a low certainty level of the location, which results in predicted initial condition whose support occupies a larger portion of the porous medium. Closely connected to the initial condition location is initial condition values at the specified points. These values are represented by the variable β. Finally, the level of certainty of the closeness of the presumed β to the exact values of the initial condition (represented by variable α) is controlled by the variable κ.

The procedure that we propose in this paper is related to the sequence of updating sensor data described in [8]. We assume that multiple sets of measurement data are available. To explain the idea in a rigorous way, we concentrate on updating β in $n > 1$ steps real time simulations, namely at time $t_1, t_2, t_3, \cdots, t_n$. We note that in this case there are n sets of measurement data at our disposal. At the very beginning of the simulation (at time level t_1), we come up with prior values of $\beta = \beta^0$. Running the simulation at time level t_1 gives us the predicted α^1. Now the aim is to update β to be used in the simulation at time level t_2, and for this we use the already obtained values of α^1. The updating sequence is then similarly proceeded up to time level t_n. Obviously, in conjunction with updating β, we may also update the value of κ with the assumption that the updated β are converging to the predicted α.

In practical applications, the data acquired from measurement contain inherent errors/noise which add up the uncertainty in the whole simulation process. This noise affects the accuracy of the initial condition recovery. One way to quantify this uncertainty is to sample the measurement data into several realizations and use them in the simulation to obtain the corresponding set of realizations of initial conditions. Another way is to transform the initial data recovery into a Bayesian approach. This method will be presented elsewhere [10].

Below we present a synthetic examples to test the performance of the method described in the previous section. More examples and discussion can be found in [9]. We use $\Omega = [0,1] \times [0,1]$. Typical boundary conditions in the subsurface flow for the pressure equation are given pressure at the inlet and outlet edges (i.e., $x = 0$ and $x = 1$, respectively), and no flow at the bottom and top edges (i.e., $z = 0$ and $z = 1$, respectively). The permeability k is generated with given correlation length $l_x = 0.2$ and $l_z = 0.02$, with a spherical variogram.

For the convection-diffusion equation (4), we set the diffusion coefficient $D = 0.1$ over all domain. We assume zero concentration at the inlet, bottom, and top edges, and a zero diffusion, i.e., $(D\nabla C) \cdot \mathbf{n} = 0$, at the outlet edge, with $\mathbf{n}$ being the unit normal vector pointing outward on the outlet edge. The initial condition $C^0(x, z)$ is set to be nonzero in the region $(0.2, 0.4) \times (0.2, 0.4)$ and zero elsewhere. The convection-diffusion equations are solved by the finite volume method using rectangular grids. For all cases we discretize the domain into 100×100 elements, i.e., 100 elements in each direction. The sensors information are obtained from the concentration solved with the given "true" initial condition. We use time step $\Delta t = 0.01$. As mentioned earlier, the initial condition is nonzero in the region $(0.2, 0.4) \times (0.2, 0.4)$ and zero elsewhere in the domain. For all examples presented in this paper we use initial condition shown on the left side of Figure 3. It has the form $C^0(x, z) = \sum_{i=1}^{9} c_i \phi_i(x, z)$, where $\phi_i(x, z)$ are the standard bilinear basis functions for nine nodes in the rectangular region $(0.2, 0.4) \times (0.2, 0.4)$.

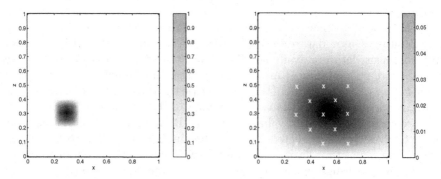

Fig. 3. Left: The initial condition profile. Right: Concentration at time $t = 0.2$. The (x) indicates the sensor location.

The data from measurement $M(\gamma)$ are taken from multiple set of numerical simulations with the initial condition mentioned earlier. The measurement is assumed to be conducted at time level $t_1 = 0.1$, $t_2 = 0.2$, $t_3 = 0.3$, and

$t_4 = 0.4$. A number of sensors are installed at various locations in the porous medium. Figure 3 (right Figure) shows the concentration profile at $t = 0.2$ along with the sensor locations which are denoted by (x) indicator.

We assume that we do not know exactly the support of the exact initial condition. We use $N_c = 27$, where the nodes now consist not only on the whole portion of the exact initial condition support, but also on the outside of it, namely, on the left and right boundaries of the exact initial condition support. In addition to updating β we might as well update κ. Here we propose the following procedure. With the assumption that we know exactly the concentration value will never be negative or greater than one, then as in the previous examples every-time negative value occurs at certain node i we truncate it to zero (or if it is greater than one we truncate it to one). Correspondingly we update κ_i by increasing it to be used in the next time level. Figure 4 shows the updated initial condition with both β and κ updated for the case of larger support. The prior for κ is $\kappa_i^0 = 2 \times 10^{-12}$ for all i, and when updated it is multiplied by ten. The figure shows very good agreement on the predicted initial condition. The value of κ can be related to the relative error associated with the sensor data. In the next Figure, we choose the true initial condition to have two-peaks and two distinct supports. The prior initial condition is chosen such that its support covers the supports of both peaks. The numerical results presented in Figure 5 shows that updating both κ and β gives significant improvement and we obtain good agreement. These issues and more examples are presented in [9].

4 Remark on updating permeability field

In this section we discuss updating the permeability field. One of the error sources in dynamic data driven simulations is the incorrect heterogeneous permeability field. Based on the local errors we first determine the region where the permeability will be updated. Then by considering the neighborhood of the sensor location as a coarse block we determine the effective permeability, k_{sim}^*. This quantity is determined such that the local problem with a single permeability coefficient, k_{sim}^*, will give the same response as the underlying fine-scale field. Next our goal is to find, k_{true}^*, that will give us the same respond as the one from the sensor data. This is done using, for example, gradient based approaches. Since we only have to fit one parameter this problem can be easily handled without extensive computations. Now we can impose k_{true}^* both as soft as well as hard constraint. For the hard constraint we will rescale the local heterogeneous field based on the ratio k_{sim}^*/k_{true}^*. Note that the hard constraint does not change the structure of the local heterogeneities and can be easily implemented since it only requires local rescaling. Next we address how to impose the coarse scale information as a soft constraint.

For the soft constraint we use Bayesian framework following to [5]. Assume the *apriori* distribution for the local coarse heterogeneous field is given.

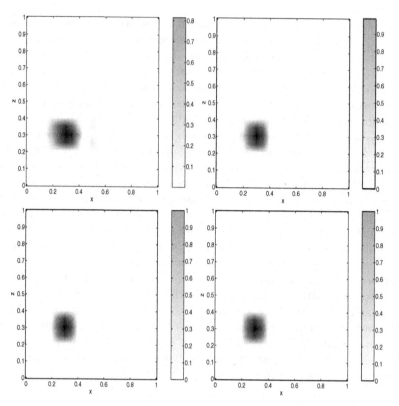

Fig. 4. Updated initial condition: $t = 0.1$ (top left), $t = 0.2$ (top right), $t = 0.3$ (bottom left), $t = 0.4$ (bottom right). The prior for β assumes a larger support. Both β and κ are updated.

Our objective to generate fine-scale random permeability fields conditioned on the coarse-scale data. Denote k^c and k^f to be coarse and fine scale fields respectively. Then using Bayes theorem we have $P(k^f, k^c) = P(k^f)P(k^c|k^f)$, where $P(k^f)$ is the probability distribution for the fine scale field, $P(k^f, k^c)$ is the joint probability distributions. From this equation we have

$$P(k^f|k^c) \propto P(k^f)P(k^c|k^f). \tag{11}$$

In this equation $P(k^f)$ is a prior distribution of the fine-scale field which is prescribed. To impose the coarse-scale data as a soft constraint we take $k^c = J(k^f) + \epsilon$, where J is a local upscaling operator and ϵ is a random noise, $\epsilon \sim N(0, \sigma^2)$. The local upscaling operator involves the solution of the local partial differential equations similar to the one described above (see also [2, 3]). Soft constraint assumes that the coarse scale information is not accurate. Note that letting $\sigma \to 0$ we can get the hard constraint.

We can generate multiple realizations from the posterior distribution using Markov Chain Monte Carlo (MCMC) (see [7]). This approach is known

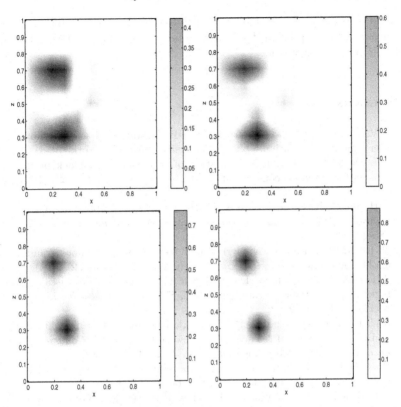

Fig. 5. Updated initial condition for two-peak initial condition: $t = 0.1$ (top left), $t = 0.2$ (top right), $t = 0.3$ (bottom left), $t = 0.4$ (bottom right). The prior for β assumes a larger support that covers the support of both peaks. Both β and κ are updated.

to be very general and can handle complex posterior distributions as the ones considered here. The main idea of MCMC is the use of Markov Chain with a specified stationary distribution. The random drawing from the target distribution can be accomplished by a sequence of draws from full conditional distribution. Since the full conditionals have non-explicit form involving local problems Metropolis-Hastings algorithm is used. We have implemented this procedure and the obtained numerical results show that the fine-scale permeability can be recovered based on the coarse-scale permeability and the prior information.

5 Conclusions

In this paper we presented numerical methods that incorporate the sensor data from sparse locations into the simulations and updating the solution.

We have considered three problems (1) updating the solution based on local measurements (2) updating the initial condition and (3) updating the local fine-scale permeability field.

References

1. C.V. Deutsch and A. G. Journel, GSLIB: Geostatistical software library and user's guide, 2nd edition, Oxford University Press, New York, 1998.
2. L.J. Durlofsky, Numerical calculation of equivalent grid block permeability tensors for heterogeneous porous media, *Water Resour. Res.* **27** (1991), 699-708.
3. X.H. Wu, Y.R. Efendiev and T. Y. Hou, Analysis of upscaling absolute permeability, *Discrete and Continuous Dynamical Systems, Series B* **2** (2002), 185-204.
4. Y. Efendiev A. Pankov, Numerical homogenization of nonlinear random parabolic operators, *SIAM Multiscale Modeling and Simulation*, **2** (2004), 237-268.
5. S.H. Lee, A. Malallah, A. Datta-Gupta and D. Higdon, Multiscale Data Integration Using Markov Random Fields, *SPE Reservoir Evaluation and Engineering*, February 2002.
6. C.C. Douglas, Y. Efendiev, R. Ewing, R. Lazarov, M.R. Cole, C.R. Johnson, and G. Jones, Virtual telemetry middleware for DDDAS, *Computational Sciences* ICCS 2003, P.M. Sllot *et al.*, eds., **4** (2003), 279–288.
7. W.R. Gilks, S. Richardson and D.J. Spegelhalter, Markov Cain Monte Carlo in Practice, Chapman and Hall/CRC, London, 1996.
8. C.C. Douglas, C.E. Shannon, Y. Efendiev, R.E. Ewing, V. Ginting, R. Lazarov, M.J. Cole, G. Jones, C.R. Johnson, and J. Simpson, A Note on Data-Driven Contaminant Simulation, Lecture Notes in Computer Science, vol. 3038, Springer-Verlag, 2004, 701–708.
9. C.C. Douglas, Y. Efendiev, R.E. Ewing, V. Ginting, and R. Lazarov, Least-Square Approach for Initial Data Recovery in Dynamic Data-Driven Simulations, in preparation.
10. C.C. Douglas, Y. Efendiev, R.E. Ewing, V. Ginting, and R. Lazarov, Bayesian Approaches for Initial Data Recovery in Dynamic Data-Driven Simulations, in preparation.

Computing the Least Squares Inverse of Sparse Matrices on a Network of Clusters

Baolai Ge

SHARCNET, The University of Western Ontario, London, Canada N6A 5B7
bge@sharcnet.ca

Summary. This paper describes the calculation of the least square inverse of sparse matrices and the use of load balancing schemes for parallel processing in a heterogeneous environment. Due to the variation of number of non zero entries to be calculated row wise, as well as the difference of processor speeds, load imbalance may occur and have an impact on the performance. To improve the performance by keeping processors as busy as possible the redistribution of tasks and data is needed. We present an architecture and implementation outlines of a few load balancing schemes featured with one-sided communications in a framework of multithreading. We show through our tests that the use of load balancing schemes can improve the performance in some cases.

1 Introduction

Consider the solution of nonsingular linear system using stationary iterative methods of the form

$$x^{(k+1)} = (I - ZA)x^{(k)} + Zb \tag{1}$$

with $k = 0, 1, \ldots$, where Z is some approximate inverse of A. Intuitively if Z is "close" enough to A^{-1}, then one can expect that the iteration (1) converges fast to the solution of the linear system. In practice, however, the approximate inverse Z is barely used directly. It is often either used in some multigrid schemes [3] [1] or more recently in the Krylov subspace methods as a preconditioner [13, 7, 17, 8].

The quality of the approximate inverse Z can be measured in certain matrix norm. The use of Frobenius norm leads to the least squares approximation of A^{-1}. Following the definitions in [1], the least squares approximate inverse is defined as follows

Definition 1. *Let A, M be $n \times n$ real matrices, and let P denote the set of all matrices that have the same sparsity pattern. The matrix Z is said to be the least squares approximate inverse of A if*

$$\|I - ZA\|_F = \min_{M \in P} \|I - MA\|_F. \tag{2}$$

Extensions to singular cases are discussed by Benson and Frederickson [4].

Equation (2) leads to n least squares problems

$$\min \|e_i - z_i G_i^T\|_2^2, \quad i = 1, \ldots, n \tag{3}$$

where G_i is the transpose of a submatrix consisting of chunks of rows from A determined by the non zero entries in the ith row b_i in Z, and e_i is a subvector of length equal to the row size of G_i, extracted from the ith row in the identity matrix I. It is easy to show that for nonsingular matrix A, G_i is full rank and overdetermined. Equation 3 is best solved using QR factorization method, and can be solved independently, thus suitable for parallel processing.

The idea of the least squares approximate inverse was first introduced by Benson [2] in the early 1970s. Earlier work by Benson and several other people has demonstrated the success of employing such approximate inverses in the development of multigrid algorithms (see [11], [5], [4]). The least squares approximate inverse based methods did not receive much attention in the early days possibly due to its computational cost and implementation difficulties. It has recently regained the interests of computational scientists. New usages as preconditioners in the framework of Krylov subspace methods have been studied extensively, for example in [13], [17], and [8].

In the next section, we discuss an algebraic approach to choosing sparsity patterns for the approximate inverse. In Section 3 we describe the design and implementation details of a few load balancing schemes for calculating the least squares inverse in a heterogeneous environment. We present some test results in Section 4, followed by the conclusion in Section 5.

2 The Sparsity of Approximate Inverses

One key step in computing the least squares inverses of a matrix is to determine a sparsity pattern. The selection of the sparsity pattern has been discussed by many people (e.g. [17], [8]). Various schemes fall into two categories. One is based on the (geometric) property of the underlying numerical discretization (see for example [1], [5]) and the other is based the algebraic property of the matrix itself (see for example [12], [8] and [7]). A common choice is the pattern of A^q, $q \geq 2$, as it tends to have more fill-ins. In [12] the relation between the powers of A and the directed graph associated with A was discussed. The equivalence of finding the powers of A and expanding the index set of a row of A was mentioned independently in [13] and [8]. In the following we elaborate in details on the selection of sparsity in a formal way.

Consider the ith equation of a linear system $Au = b$

$$a_{ii} u_i + \sum_{j \neq i} a_{ij} u_j = b_i. \quad i = 1, \ldots, n. \tag{4}$$

Assuming that $a_{ii} \neq 0$ (we will always assume this hereafter), then u_i can be represented by the linear combination of others u_j, $j \neq i$. This reveals some relation between the ith component and rest reflected in the entries of A. Following notations in graph theory, we define the neighborhood of a point (or *vertex* in the context of a directed graph associated with A) in the following way

Definition 2. *For an $n \times n$ real matrix A define the (direct) neighborhood of a point i as a set of indices (See [6] and [15]) subject to a constant θ*

$$N_i(\theta) = \{j : |a_{ij}| = \theta|a_{ii}|, \quad j \neq i, \quad a_{ii} \neq 0\}, \tag{5}$$

and the distance $d(i,j) = 1$. In particular, denote $N_i^{(1)} = N_i$ and $N_i^{(0)} = \{i\}$, and $d(i,i) = 0$.

In the following we omit θ for short notation. As a natural extension, we define the l-local neighborhood of i as follows

Definition 3. *For $i \in N_s$ and $j \in N_t$, i and j are connected or there exists a path from i to j, if there exists a set of points $k_1, \ldots, k_\ell$, such that $k_1 \in N_s \cap N_1$, $k_2 \in N_1 \cap N_2$, $\ldots$, $k_\ell \in N_{\ell-1} \cap N_t$. Define $d(i,j) = d(i,k_1) + \cdots + d(k_{\ell-1}, j)$.*

Definition 4. *Given an $n \times n$ real matrix A, define the ℓ-local neighborhood of i as a set of indices $N_i^{(\ell)}$ such that for each $j \in N_i^{(\ell)}$ there exists a path from i to j of distance $d(i,j) \leq \ell$.*

Definition 4 essentially says "neighbor's neighbors are also neighbors" within certain range. More precisely we make the following statements

Proposition 1. *For an $n \times n$ real matrix A, the ℓ-local neighborhood of i can be determined recursively as*

$$N_i^{(\ell)} = \{k : k \in N_i^{(\ell-1)} \cup N_j, \; j \in N_i^{(\ell-1)}\}. \tag{6}$$

Proof: This can be proved by induction. It is trivial for case $\ell = 0$. Assume $\ell = q > 1$ (6) holds. We need to show that 1) for all $k \in N_i^{(q+1)}$, $d(i,k) \leq q+1$ and 2) for any k for which $d(i,k) \leq q+1$, $k \in N_i^{(q+1)}$. We omit the formal proof.

The following pseudo code implements a way of collecting all the neighbors in an ℓ-local neighborhood of i.

Algorithm 1 Determining q-level neighborhood

```
procedure collect(i, N_i, ℓ, q)
    if q > ℓ return;
    for all j ∈ N_i do
        collect(j, N_j, ℓ, q + 1);
        if j ∉ N_i^(ℓ), N_i^(ℓ) := N_i^(ℓ) ∪ j;
    end do
    return;
end procedure
```

We now relate the concept of neighborhood to the pattern of powers of A by introducing the following notations. Similar to the concept of adjacent matrix, we introduce the following definition of an ℓ-local connection matrix $C^{(\ell)}$ as follows

Definition 5. *For the digraph associated with a given $n \times n$ real matrix A, define an ℓ-local connection matrix $C^{(\ell)}$ by*

$$c_{ij} = \begin{cases} 1 & \text{if } j \in N_i^{(\ell)}, \\ 0 & \text{otherwise.} \end{cases} \tag{7}$$

With the help of the definitions above, we have the following conclusion

Proposition 2. *For the digraph associated with the given $n \times n$ real matrix A, the ℓ-local connection matrix $C^{(\ell)}$ satisfies*

$$C^{(\ell)} = P(P(A)^{\ell}) \tag{8}$$

where P is a boolean operator such that for a matrix M, $P(M)$ maps the entries to

$$p_{ij} = \begin{cases} 1 & \text{if } m_{ij} \neq 0, \\ 0 & \text{otherwise.} \end{cases} \tag{9}$$

Proof: The proposition can be proved by induction and is omitted.

Definition 5 and Proposition 2 give us a choice to determine the sparsity for approximate inverses. There is a simple geometric interpretation for the pattern of a q-local approximate inverse Z. Consider the $2D$ five point finite difference stencil of Laplacian. By including neighboring points of each point of the stencil we obtain a 2-local neighborhood, which is identical to a 13-point stencil for the biharmonic operator [12]. The sparsity pattern of 2-local "expansion" can also be obtained through the multiplication of the Laplacian by itself.

In general the inverse of a sparse matrix is full. Those that are derived from PDEs often have a decay in the magnitude of off-diagonal entries [17].

This suggests that an approximate inverse with the same sparsity (level 1) or a denser level 2 sparsity might be a good choice.

The pattern $P(A)^q$, however, is not always satisfactory. It is easy to see that the 2-local pattern of a lower-right-pointing "arrow-shaped" matrix, with non zeros on the diagonal, the last row and the last column, is full and therefore not a suitable choice at all.

If we take a look at the inverse of such a matrix, for instance, a 256×256 matrix with tridiagonal entries repeated by $(-1, 2, -1)$ and all -1's in its last column and last row, we will see its inverse has a very special pattern as shown in Figure 1(a), which may not suggest a sparsity pattern at all.

The following example shows that even for a perfectly tridiagonal matrix, its inverse may not suggest any sparsity at all (see also [17])

(a)   (b)

Fig. 1. The sparsity patterns: (a) The bitmap of the inverse of a 256×256 tridiagonal matrix with tridiagonal entries $(-1,\ 2,\ -1)$ and all -1's in its last column and last row. (b) The bitmap of the inverse of $T(0.98)$ of order 256.

$$T(\alpha) = \begin{bmatrix} \alpha & -1 & & & \\ -1 & 2 & -1 & & \\ & & \ddots & & \\ & & & \ddots & \\ & & -1 & 2 & -1 \\ & & & -1 & 2 \end{bmatrix}, \; T^{-1}(1) = \begin{bmatrix} n & n-1 & n-2 & \cdots & 1 \\ n-1 & n-1 & n-2 & \cdots & 1 \\ & & \ddots & & \\ & & & \ddots & \\ 2 & \cdots & & 2 & 1 \\ 1 & \cdots & & 1 & 1 \end{bmatrix}.$$

When α is close to 1, $T^{-1}(\alpha)$ has a pattern shown in Figure 1(b).

3 Load Balancing

Load imbalance in computing the least squares inverses may occur from two sources. One is the non uniform problem size during the process of solving a sequence of least squares problems. The other is the different speeds of processors on which the least squares problems are solved. A few authors have mentioned their work on load balancing in computing the approximate inverse for preconditioners. The author of SPAI [13] used a simple round robin type of scheme. The implementation of ParaSails [8] uses a repartitioning scheme based on the estimate cost obtained during the computation of the sparsity pattern for the inverse. With this scheme, an improvement in the performance is seen in some test cases.

3.1 Dynamic Load Balancing Schemes

Various load balancing schemes for distributed processing have been well studied (e.g. [18], [9], [10], [16], [14]). In this paper we consider only two dynamic balancing schemes. The first approach employs (asynchronous) load monitoring (ALM). The load information of a compute process – the number of rows pending calculation – is sent back periodically to a coordinating process or coordinator. When a compute process becomes idle it asks the coordinator for more work. The coordinator keeps a load table updated every time when a job-done response from a compute process arrives . Upon receiving a request for more work, the coordinator sends the rank of the most loaded peer to the requestor. The drawback of this scheme is it requires extra, and perhaps excessive communications during the computation.

The second approach uses a global round robin (GRR) style scheme to (randomly) determine the donor. When a compute process asks for more work, the coordinator chooses a peer and sends the rank of that peer to the starving process. In this approach, there is no communication until a compute process becomes idle.

3.2 Architecture and Implementation

The architecture we used involves a coordinating processes or coordinator and a pool of compute processes as shown in Figure 2. The coordinator initiates the computation process by sending divided work – chunk of rows of B to compute – to compute precesses. Each compute process is composed of two threads, a thread that does the computation (compute thread) and a thread that handles incoming messages (receive thread). The use of a dedicate thread for messages is to facilitate the one sided communication required. Such an arrangement allows us to use a simple set of MPI functions to handle all the communications and computation in parallel in an asynchronous fashion.

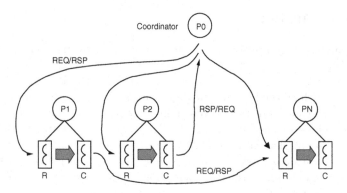

Fig. 2. The organization of processes: a coordinator process of compute processes. Each compute process has a receive thread (R) a compute thread (C).

The transaction flows of messages (Figure 3) and actions are summarized as follows:

- The coordinator P_0 sends a work order along with data (cached rows in A and rows of indices of B) to each compute processes P_i, $i = 1, \ldots, N$.
- The receive thread of process P_i receives the work order, creates a work event and pushes it on an event queue shared by the compute thread. The compute thread is in suspension of its execution until there's an event on the queue.
- During the computation, the compute process optionally sends load information to the coordinator at an adjustable load feedback rate.
- When a compute process is done with the computation, it sends a message back to the coordinator P_0, asking for more work.
- The coordinator P_0 sends the rank of a compute process P_j chosen according to certain rules.
- The compute process P_i receives the rank j of its peer P_j, then sends a job request to P_j.
- The receive thread of process P_j receives the job request without the awareness of the compute thread. If there are still remaining rows to process, it sends a portion of the remaining data[1] to the requesting process P_i and update the total number of rows to process used by the compute process atomically. If there is no outstanding work, it responds to P_i with a job request failed message.
- If P_i's job request is satisfied, it continues the computation and then repeats the above steps. If the job request fails, it sends a peer request to the coordinator.
- If P_o receives a peer request, it will pick a peer and responds back to the requestor.

This process continues until a termination condition is met.

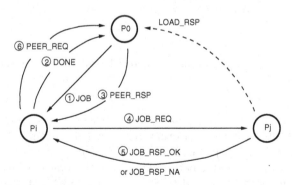

Fig. 3. Transaction flow of load balancing process among the coordinator and compute processors. The circled numbers indicate the order of occurrence.

---

[1]This is determined by a parameter, set as an environment variable. Also a threshold is set such that if the number of remaining rows lower than the "watermark", the request shall be rejected.

4 Experiment

We present in Figure 5 and 6 the timings of computing the least squares inverses of three matrices (Figure 4) selected from the Matrix Market published at Netlib[2]. Three cases are tested for each matrix: (1) without load balancing (WOL); (2) with asynchronous load monitoring (ALM) and (3) with global round robin (GRR). For all the tests, the same sparsity of A is used.

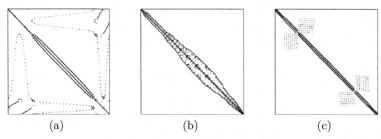

Fig. 4. Sparsity patterns of test matrices selected from the Matrix Market: (a) BCSSTK24 (3562×3562), (b) BCSSTK18 (11948×11948) and (c) FIDAP011 (16614×16614).

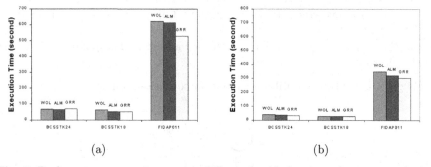

Fig. 5. Performance comparisons using different load balancing schemes on a single cluster A with 833MHz Alpha processors: using (a) four and (b) eight processors.

It is evident that using the load balancing algorithms mentioned above in principle results in better performances. Our experiments also show that the performance varies depending upon the matrix patterns. In the case of FIDAP011, GRR gives better performance in both homogeneous and heterogeneous environments, while for BCSSTK18 and BCSSTK24, GRR and ALM perform almost the same. In the case of BCSSTK18, slightly better performance of ALM is observed with the load feedback rate set to 50 rows and more.

---

[2]http://www.netlib.org/

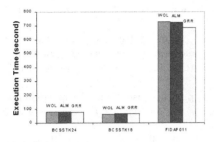
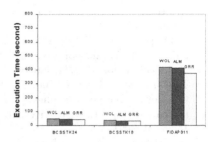

(a)   (b)

Fig. 6. Performance comparisons using different load balancing schemes across cluster A with 833MHz Alpha processors and cluster B with 667MHz Alpha processors (with same number of processors on each): using (a) four and (b) eight processors.

It is suggested from the tests that load balancing will only become effective when the work load – size of least squares problem determined by the bandwidths (span of non zero indices in a row) – varies significantly across rows, typically as shown in Figure 4(c).

5 Conclusion

In this paper we have introduced a practical approach to choosing the pattern of sparse approximate inverses. We have also presented an architecture and the implementation outlines of load balancing schemes for parallel processing on heterogeneous systems. Applying these load balancing schemes can improve the performance when the work load varies significantly across rows and/or the speeds vary among processors. Future studies on more load balancing schemes are being planned.

References

1. R. N. Banerjee and M. W. Benson. "An approximate inverse based multigrid approach to biharmonic problem". *Inter. J. Computer Math.*, 40:201–210, 1991.
2. M. W. Benson. "Approximate inverses for solving large sparse linear systems". Master's thesis, Lakehead University, 1973.
3. M. W. Benson and P. O. Frederickson. "Iterative solution of large sparse linear systems arising in certain multidimensional approximation problems". *Utilitas Mathematica*, 22:127–140, 1982.
4. M. W. Benson and P. O. Frederickson. "Fast parallel algorithms for the Moore-Penrose pseudo inverse". In M. T. Heath, editor, *Hypercube Multiprocessors*, pages 597–604. SIAM, 1987.
5. M. W. Benson, J. Krettmann, and M. Wright. "Parallel algorithms for the solution of certain large sparse linear systems". *Int. J. Comput. Math.*, 16:245–260, 1984.

6. A. Brandt. "Algebraic multigrid theory: Symmetric cases". *App. Math. Comp.*, 19:23–56, 1986.

7. B. Carpentieri, I. S. Duff, and L. Giraud. "Sparse pattern selection strategies for robust Frobenius norm minimization preconditioners in electromagnetism". Technical Report TR/PA/00/05, CERFACS, 42 Avenue G. Coriolis, 31057 Toulouse Celex, France, 2000.

8. E. Chow. "parallel implementation and practical use of sparse approximate inverses with a priori sparsity patterns". *Int. J. High Perf. Comput. Appl.*, 15:56–74, 2001.

9. G. Cybenko. "Dynamic load balancing for distributed memory multiprocessors". *J. Parallel and Distributed Computing*, 7:279–301, 1989.

10. R. Elsasser, B. Monien, and R. Preis. "Diffusive Load Balancing Schemes on Heterogeneous Networks". In *Proc. SPAA 2000*, Maine, 2000.

11. P. O. Frederickson. "Fast approximate inverse of large sparse linear systems". Technical Report Mathematics Report #7-75, Department of Mathematical Sciences, Lakehead University, Canada, 1975.

12. B. Ge. "On the patterns of the least squares inverses of sparse matrices", 1992. Essay, Lakehead University, Canada.

13. M. Grote and T. Huckle. "Parallel preconditioning with sparse approximate inverses". *SIAM J. Sci. Comput.*, 18(3):838–853, 1997.

14. Z. Lan, V. E. Taylor, and G. Bryan. "Dynamic load balancing of SAMR applications on distributed systems". In *Proc. 2001 ACM/IEEE conference on Supercomputing*, Denver, CO, 2001.

15. J. W. Ruge and K. Stüben. "Algebraic Multigrid". In S. F. McCormic, editor, *Multigrid Methods*, volume 66. SIAM, Philadelphia, 1987.

16. K. Schloegel, G. Karypis, and V. Kumar. "Multilevel diffusion schemes for repartitioning of adaptive meshes". *J. Parallel and Distributed Computing*, 47:109–124, 1997.

17. W-P. Tang. "Toward an effective sparse approximate inverse preconditioner". *SIAM J. Matrix Anal. Appl.*, 20(4):970–986, 1999.

18. R. V. van Nieuwpoort, T. Kielmann, and H. E. Bal. "Efficient load balancing for wide-area divide-and-conquer applications". In *Proceedings of the eighth ACM SIGPLAN symposium on Principles and practices of parallel programming*, pages 34–43, Snowbird, Utah, 2001. ACM.

Numerical Simulation for Fractured Porous Media

Guanren Huan[1], Richard E. Ewing[1], Guan Qin[1], and Zhangxin Chen[2]

[1] Institute for Scientific Computation, Texas A&M University, College Station, TX 77843-3404, USA. ghuan@isc.tamu.edu, richard-ewing@tamu.edu, guan.qin@neo.tamu.edu
[2] Center for Scientific Computation and Department of Mathematics, Southern Methodist University, Dallas, TX 75275-0156 USA. zchen@mail.smu.edu

Abstract

Dual porosity and dual porosity/permeability models are considered for multidimensional, multiphase flow in fractured porous media. Different approaches for the treatment of matrix-fracture flow transfer terms are discussed; special attention is paid to the inclusion of capillary pressure, gravity, and viscous forces and pressure gradients across matrix blocks in these transfer terms. Numerical experiments are reported for the benchmark problems of the sixth comparative solution project organized by the society of petroleum engineers.

1 Introduction

There has been an increased interest in the numerical simulation of multiphase flow in fractured porous media in recent years. A fractured porous medium has throughout its extent a system of interconnected fractures dividing the medium into a series of essentially disjoint blocks of porous rock, called the matrix blocks. It has two main length scales of interest: the microscopic scale of the fracture thickness (about 10^{-4} m) and the macroscopic scale of the average distance between fracture planes, i.e., the size of the matrix blocks (about 0.1-1 m). Since the entire porous medium is about 10^3-10^4 m across, flow can be mathematically simulated only in some average sense. The concept of dual-porosity [1, 2, 4, 6, 7, 10] has been utilized to model the flow of fluids on its various scales. In this concept, the fracture system is treated as a porous structure distinct from the usual porous structure of the matrix itself. The fracture system is highly permeable, but can store very little fluid, while the matrix has the opposite characteristics. When developing a dual-porosity model, it is crucial to treat the flow transfer terms between the fracture and matrix systems.

There have been two approaches to treat a matrix-fracture flow transfer term. In the conventional approach, this term for a particular fluid phase is directly related to a shape factor, the fluid mobility, and the potential difference between these two systems, and the capillary pressure, gravity, and viscous forces are properly incorporated into this term. In this paper, this approach will be reviewed. Moreover, the inclusion of a pressure gradient across a matrix block in this term in a general fashion is also studied. A more recent approach is to treat the flow transfer term explicitly through boundary conditions on the matrix blocks. This approach avoids the introduction of *ad hoc* parameters (e.g., the shape factor) in the conventional approach, and is more general. However, this latter approach appears applied only to a dual porosity model, not to a dual porosity/permeability model.

The formulation of the mass balance equation for each fluid phase in a fractured porous medium follows the classical one with an additional matrix-fracture transfer term. The two overlapping continua, fractures and matrix blocks, are allowed to co-exist and interact with each other. Furthermore, there are matrix-matrix connections. In this case, a dual porosity/permeability model is required for the fractured porous medium. If the matrix blocks act only as a source term to the fracture system and there is no matrix-matrix connection, a dual porosity (and single permeability) model is applied. These two models will be described in this paper.

Multidimensional numerical simulators are developed for dual porosity and dual porosity/permeability models of multiphase flow in fractured porous media. The spatial discretization scheme of these simulators is based on a block-centered finite difference method (equivalently, a mixed finite element method on rectangular parallelepipeds [8]). The solution scheme is fully implicit. Numerical results are reported for the benchmark problems of the sixth comparative solution project (CSP) organized by the society of petroleum engineers (SPE) [5], and show that these numerical simulators perform very well for depletion, gas injection, and water injection applications in fractured petroleum reservoirs.

2 Flow Equations

The fluid flow equations are based on a three-component, three-phase black oil model. In this model, it is assumed that the hydrocarbon components are divided into a gas component and an oil component in a stock tank at the standard pressure and temperature, and that no mass transfer occurs between the water phase and the other two phases (oil and gas). The gas component mainly consists of methane and ethane.

To reduce confusion, we carefully distinguish between phases and components. We use lower-case and upper-case letter subscripts to denote the phases and components, respectively. Furthermore, a superscript f is used to denote fracture variables.

2.1 Dual porosity/permeability model

Let ϕ and $\mathbf{k}$ denote the porosity and permeability of a matrix system, and S_α, μ_α, p_α, $\mathbf{u}_\alpha$, ρ_α, and $k_{r\alpha}$ be the saturation, viscosity, pressure, volumetric velocity, density, and relative permeability of the α phase, $\alpha = w, o, g$, respectively. Because of mass interchange between the oil and gas phases, mass is not conserved within each phase, but rather the total mass of each component must be conserved. Thus, for the matrix system, the mass balance equations are

$$\frac{\partial(\phi\rho_w S_w)}{\partial t} = -\nabla \cdot (\rho_w \mathbf{u}_w) - q_{Wm}, \tag{1}$$

for the water component,

$$\frac{\partial(\phi\rho_{Oo} S_o)}{\partial t} = -\nabla \cdot (\rho_{Oo} \mathbf{u}_o) - q_{Oom}, \tag{2}$$

for the oil component, and

$$\frac{\partial}{\partial t}\left(\phi(\rho_{Go} S_o + \rho_g S_g)\right) = -\nabla \cdot (\rho_{Go} \mathbf{u}_o + \rho_g \mathbf{u}_g) - (q_{Gom} + q_{Gm}), \tag{3}$$

for the gas component, where ρ_{Oo} and ρ_{Go} indicate the partial densities of the oil and and gas components in the oil phase, respectively, and q_{Wm}, q_{Oom}, q_{Gom}, and q_{Gm} represent the matrix-fracture transfer terms. Equation (3) implies that the gas component may exist in both the oil and gas phases.

Darcy's law for each phase is written in the usual form

$$\mathbf{u}_\alpha = -\frac{k_{r\alpha}}{\mu_\alpha}\mathbf{k}\left(\nabla p_\alpha - \rho_\alpha \wp \nabla z\right), \qquad \alpha = w,\, o,\, g, \tag{4}$$

where $\wp$ is the magnitude of the gravitational acceleration and z is the depth. The saturation constraint reads

$$S_w + S_o + S_g = 1. \tag{5}$$

Finally, the phase pressures are related by capillary pressures

$$p_{cow} = p_o - p_w, \quad p_{cgo} = p_g - p_o. \tag{6}$$

For the fracture system, the mass balance equations are

$$\frac{\partial(\phi\rho_w S_w)^f}{\partial t} = -\nabla \cdot (\rho_w \mathbf{u}_w)^f + q_{Wm} + q_W,$$

$$\frac{\partial(\phi\rho_{Oo} S_o)^f}{\partial t} = -\nabla \cdot (\rho_{Oo} \mathbf{u}_o)^f + q_{Oom},$$

$$\frac{\partial}{\partial t}\left(\phi(\rho_{Go} S_o + \rho_g S_g)\right)^f = -\nabla \cdot (\rho_{Go} \mathbf{u}_o + \rho_g \mathbf{u}_g)^f$$

$$\qquad\qquad + (q_{Gom} + q_{Gm}) + (q_{Go} + q_G), \tag{7}$$

where q_W, q_{Oo}, q_{Go}, and q_G denote the external sources and sinks. We have assumed that these external terms interact only with the fracture system. This is reasonable since flow is much faster in this system than in the matrix blocks. Equations (4)–(6) remain valid for the fracture quantities.

The matrix-fracture transfer terms for the dual porosity/permeability model are defined following Warren-Root [10] and Kazemi [6]. The transfer term for a particular component is directly related to the shape factor σ, the fluid mobility, and the potential difference between the fracture and matrix systems. The capillary pressure, gravity, and viscous forces must be properly incorporated into this term. Furthermore, the contributions from a pressure gradient across each matrix block and the molecular diffusion rate for each component must be included as well. For the brevity of presentation, we neglect the diffusion rate, and discuss the contribution from the pressure gradient.

The treatment of a pressure gradient across a block is based on the following observation: For an oil matrix block surrounded with water in the fractures, we see that

$$\Delta p_w = 0, \quad \Delta p_o = \wp(\rho_w - \rho_o).$$

Analogously, for an oil block surrounded with gas fractures and a gas block surrounded with water fractures, we see, respectively, that

$$\Delta p_g = 0, \quad \Delta p_o = \wp(\rho_o - \rho_g),$$

and

$$\Delta p_w = 0, \quad \Delta p_g = \wp(\rho_w - \rho_g).$$

In general, we introduce the global fluid density in the fractures

$$\rho^f = S_w^f \rho_w + S_o^f \rho_o + S_g^f \rho_g,$$

and define the pressure gradients

$$\Delta p_\alpha = \wp \left| \rho^f - \rho_\alpha \right|, \quad \alpha = w, o, g.$$

Now, the transfer terms that include the contributions from the capillary pressure, gravity, and viscous forces and the pressure gradients across matrix blocks are defined by

$$q_{Wm} = T_m \frac{k_{rw}\rho_w}{\mu_w} \left(\Phi_w - \Phi_w^f + L_c \Delta p_w \right),$$

$$q_{Oom} = T_m \frac{k_{ro}\rho_{Oo}}{\mu_o} \left(\Phi_o - \Phi_o^f + L_c \Delta p_o \right),$$

$$q_{Gom} = T_m \frac{k_{ro}\rho_{Go}}{\mu_o} \left(\Phi_o - \Phi_o^f + L_c \Delta p_o \right), \tag{8}$$

$$q_{Gm} = T_m \frac{k_{rg}\rho_g}{\mu_g} \left(\Phi_g - \Phi_g^f + L_c \Delta p_g \right),$$

where Φ_α is the phase potential

$$\Phi_\alpha = p_\alpha - \rho_\alpha \wp z, \qquad \alpha = w, o, g,$$

L_c is the characteristic length for the matrix-fracture flow, and T_m is the matrix-fracture transmissibility

$$T_m = k\sigma \left(\frac{1}{l_x^2} + \frac{1}{l_y^2} + \frac{1}{l_z^2} \right),$$

with σ being the shape factor and l_x, l_y, and l_z being the matrix block dimensions [3, 6]. When the matrix permeability k in the three coordinate directions is different, the matrix-fracture transmissibility is modified by

$$T_m = \sigma \left(\frac{k_x}{l_x^2} + \frac{k_y}{l_y^2} + \frac{k_z}{l_z^2} \right).$$

2.2 Dual porosity model

For the derivation of a dual porosity model, we assume that fluid does not flow directly from one matrix block to another block. Rather, it first flows into the fractures, and then it flows into another block or remain in the fractures. This is reasonable since fluid flows more rapidly in the fractures than in the matrix. Therefore, the matrix blocks act as a source term to the fracture system, and there is no matrix-matrix connection for the dual porosity model. In this case, there are two approaches to derive this model: the conventional one as in Section 2.1 and a more recent one to be defined later.

Conventional approach

In this approach, the mass balance equations in the matrix are given by

$$\frac{\partial(\phi \rho_w S_w)}{\partial t} = -q_{Wm},$$

$$\frac{\partial(\phi \rho_{Oo} S_o)}{\partial t} = -q_{Oom}, \tag{9}$$

$$\frac{\partial}{\partial t}(\phi(\rho_{Go} S_o + \rho_g S_g)) = -(q_{Gom} + q_{gm}),$$

where q_{Wm}, q_{Oom}, q_{Gom}, and q_{gm} are given by (8).

A recent approach

For a dual porosity model, the matrix-fracture transfer terms can be modeled explicitly through boundary conditions on the matrix blocks, following [1, 7].

Let the matrix system be composed of disjoint blocks $\{\Omega_i\}$. On each block $\{\Omega_i\}$, the mass balance equations hold

$$\frac{\partial(\phi\rho_w S_w)}{\partial t} = -\nabla \cdot (\rho_w \mathbf{u}_w),$$

$$\frac{\partial(\phi\rho_{Oo} S_o)}{\partial t} = -\nabla \cdot (\rho_{Oo}\mathbf{u}_o),$$

$$\frac{\partial}{\partial t}(\phi(\rho_{Go} S_o + \rho_g S_g)) = -\nabla \cdot (\rho_{Go}\mathbf{u}_o + \rho_g\mathbf{u}_g). \tag{10}$$

The total mass of water leaving the ith matrix block Ω_i per unit time is

$$\int_{\partial\Omega_i} \rho_w \mathbf{u}_w \cdot \boldsymbol{\nu} da(\mathbf{x}),$$

where $\boldsymbol{\nu}$ is the outward unit normal to the surface $\partial\Omega_i$ of Ω_i. The divergence theorem and the first equation of (10) imply that

$$\int_{\partial\Omega_i} \rho_w \mathbf{u}_w \cdot \boldsymbol{\nu} da(\mathbf{x}) = \int_{\Omega_i} \nabla \cdot (\rho_w \mathbf{u}_w) d\mathbf{x} = -\int_{\Omega_i} \frac{\partial(\phi\rho_w S_w)}{\partial t} d\mathbf{x}. \tag{11}$$

Now, we define q_{Wm} by

$$q_{Wm} = -\sum_i \chi_i(\mathbf{x}) \frac{1}{|\Omega_i|} \int_{\Omega_i} \frac{\partial(\phi\rho_w S_w)}{\partial t} d\mathbf{x}, \tag{12}$$

where $|\Omega_i|$ denotes the volume of Ω_i and $\chi_i(\mathbf{x})$ is its characteristic function, i.e.,

$$\chi_i(\mathbf{x}) = \begin{cases} 1 & \text{if } \mathbf{x} \in \Omega_i, \\ 0 & \text{otherwise.} \end{cases}$$

Similarly, q_{Oom} and $q_{Gom} + q_{Gm}$ are given by

$$q_{Oom} = -\sum_i \chi_i(\mathbf{x}) \frac{1}{|\Omega_i|} \int_{\Omega_i} \frac{\partial(\phi\rho_{Oo} S_o)}{\partial t} d\mathbf{x}, \tag{13}$$

and

$$q_{Gom} + q_{Gm} = -\sum_i \chi_i(\mathbf{x}) \frac{1}{|\Omega_i|} \int_{\Omega_i} \frac{\partial(\phi(\rho_{Go} S_o + \rho_g S_g))}{\partial t} d\mathbf{x}. \tag{14}$$

With the definition of q_{Wm}, q_{Oom}, and $q_{Gom} + q_{Gm}$, boundary conditions on the surface of each matrix block can be imposed in a general fashion, where gravitational forces and pressure gradient effects across the block are incorporated into these conditions [2]. The present approach of defining the transfer terms avoids the introduction of *ad hoc* parameters (e.g., the shape factor).

Table 1. Basic physical and fluid data

$k = 1$ (md), $\phi = 0.29$, $\phi^f = 0.01$
$N_x = 10$, $N_y = 1$, $N_z = 5$
$\Delta x = 200$, $\Delta y = 1000$, $\Delta z = 50$ (ft)
z-direction transmissibility: multiply computed values by 0.1
Initial pressure: 6014.7 (psia), saturation pressure: 5559.7 (psia)
Water viscosity: 0.35 (cp), water compressibility: 3.5×10^{-6} (psi^{-1})
Water formation volume factor: 1.07
Rock and oil compressibility: 3.5×10^{-6}, 1.2×10^{-5} (psi^{-1})
Temperature: 200°F, datum: 13400 (ft), depth to the top: 13400 (ft)
Densities of stock tank oil and water: 0.81918 and 1.0412 (gm/cc)
Gas specific gravity: 0.7595
Rate=$\dfrac{k_r PI}{B\mu}$, Δp in psi, μ in cp, B in RB/STB, and rate in STB/D

Table 2. Reservoir layer description

Layer	k^f (md)	Matrix block height (ft)	$PI \left(\dfrac{RB\ cp}{D\ psi} \right)$
1	10	25	1
2	10	25	1
3	90	5	9
4	20	10	2
5	20	10	2

Table 3. Matrix block shape factors

Block size (ft)	Water-oil (ft^{-2})	Gas-oil (ft^{-2})
5	1.00	0.08
10	0.25	0.02
25	0.04	0.0032

Table 4. Fracture rock data

S_w	k_{rw}	k_{row}	p_{cow}
0.0	0.0	1.0	0.0
1.0	1.0	0.0	0.0

S_g	k_{rg}	k_{rog}	p_{cgo}
0.0	0.0	1.0	0.0375
0.1	0.1	0.9	0.0425
0.2	0.2	0.8	0.0475
0.3	0.3	0.7	0.0575
0.4	0.4	0.6	0.0725
0.5	0.5	0.5	0.0880
0.7	0.7	0.3	0.1260
1.0	1.0	0.0	0.1930

Table 5. Matrix rock data

S_w	k_{rw}	k_{row}	p_{cow}
0.2	0.0	1.0	1.0
0.25	0.005	0.860	0.5
0.30	0.010	0.723	0.3
0.35	0.020	0.600	0.15
0.40	0.030	0.492	0.0
0.45	0.045	0.392	-0.2
0.50	0.060	0.304	-1.2
0.60	0.110	0.154	-4.0
0.70	0.180	0.042	-10.0
0.75	0.230	0.000	-40.0
1.0	1.0	0.0	-100.0
S_g	k_{rg}	k_{rog}	p_{cgo}
0.0	0.0	1.0	0.075
0.1	0.015	0.70	0.085
0.2	0.050	0.45	0.095
0.3	0.103	0.25	0.115
0.4	0.190	0.11	0.145
0.5	0.310	0.028	0.255
0.55	0.420	0.0	0.386
0.6	0.553	0.0	1.0
0.8	1.0	0.0	100.0

3 Numerical Experiments

The experimental problems are chosen from the benchmark problems of the sixth CSP [5, 9]. Ten organizations participated in that comparative project. In these problems, various aspects of the physics of multiphase flow in fractured petroleum reservoirs are examined. The question of a fracture capillary pressure and its influence on reservoir performance is addressed by including zero and nonzero gas-oil capillary pressures in the fractures. The nonzero capillary pressure is not based on any actual measurements, but is intended as a parameter for sensitivity studies. The variation of gas-oil interfacial tension with pressure is also incorporated in these problems. The gas-oil capillary pressure is directly related to interfacial tension, and thus this pressure should be adjusted according to the ratio of interfacial tension at pressure and that at the pressure at which capillary pressures are specified.

The example under consideration is a cross-sectional example, and is designed to simulate depletion, gas injection, and water injection in fractured petroleum reservoirs. Table 1 states the basic physical and fluid property data, Table 2 shows the reservoir layer description, Table 3 gives the matrix block shape factors, Tables 4 and 5 indicate the fracture and rock data (relative permeabilities and capillary pressures), and Tables 6 and 7 represent the oil and gas PVT data, where B_o and B_g are the oil and gas formation volume factors,

Table 6. Oil PVT data

p_b (psia)	R_{so} (SCF/STB)	μ_o (cp)	C_{vo} (psi^{-1})	B_o (RB/STB)
1688.7	367	0.529	0.0000325	1.3001
2045.7	447	0.487	0.0000353	1.3359
2544.7	564	0.436	0.0000394	1.3891
3005.7	679	0.397	0.0000433	1.4425
3567.7	832	0.351	0.0000490	1.5141
4124.7	1000	0.310	0.0000555	1.5938
4558.7	1143	0.278	0.0000619	1.6630
4949.7	1285	0.248	0.0000694	1.7315
5269.7	1413	0.229	0.0000751	1.7953
5559.7	1530	0.210	0.0000819	1.8540
7014.7	2259	0.109	0.0001578	2.1978

Table 7. Gas PVT data

p_g (psia)	μ_g (cp)	B_g (RB/STB)	IFT (dyne/cm)
1688.7	0.0162	1.98	6.0
2045.7	0.0171	1.62	4.7
2544.7	0.0184	1.30	3.3
3005.7	0.0197	1.11	2.2
3567.7	0.0213	0.959	1.28
4124.7	0.0230	0.855	0.72
4558.7	0.0244	0.795	0.444
4949.7	0.0255	0.751	0.255
5269.7	0.0265	0.720	0.155
5559.7	0.0274	0.696	0.090
7014.7	0.0330	0.600	0.050

$$\sigma_1 = IFT(p)/IFT(p_{ref}), \ p_{cgo}(S_g) = p_{cgo,ref}(S_g)\sigma_1.$$

R_{so} is the gas solubility factor, and C_{vo} is the oil viscosity compressibility. In all the experiments, the injector is located at $I = 1$, and the producer is located at $I = 10$. The input data for each experiment are given below.

Depletion. Depletion runs are performed to a maximum of ten years or whenever production is less than 1 STB/D. The producer has a maximum rate of 500 STB/D, and it is constrained by a maximum drawdown of 100 psi. This well is perforated only in the bottom layer (layer 5). Two cases are studied: zero and nonzero fracture capillary pressures. The nonzero capillary data are reported in Table 4. These data are given at the bubble point pressure p_b of 5,545 psig and are adjusted for the effect of pressure on interfacial tension, as noted.

Gas injection. In this experiment 90% of the gas produced from the previous time step is reinjected. The injector is perforated in layers 1–3. The producer is perforated in layers 4 and 5, and is constrained by a maximum drawdown of 100 psi. A maximum rate of 1,000 STB/D is applied, and the minimum cut-off rate is 100 STB/D. Again, the zero and nonzero fracture capillary pressures are studied, with the latter data given in Table 4.

Water injection. In this study water is injected initially at a maximum rate of 1,750 STB/D and constrained by a maximum injection pressure of 6,100 psig. The production rate is set at 1,000 STB/D fo the total fluid (water and oil). The injector is perforated in layers 1–5, and the producer is perforated in layers 1–3. The final time of runs is 20 years.

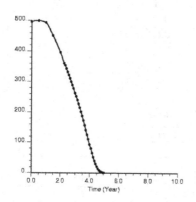

Figure 1. Q_o (depletion, $p_{cgo} = 0$).   **Figure 2.** GOR (depletion, $p_{cgo} = 0$).

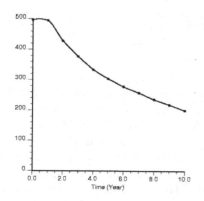
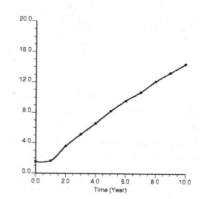

Figure 3. Q_o (depletion, $p_{cgo} \neq 0$).   **Figure 4.** GOR (depletion, $p_{cgo} \neq 0$).

Numerical results are reported for the oil production rate (Q_o in STB/D) and gas-oil ratio (GOR in SCF/STB) versus time (year) in the first two studies (depletion and gas injection), and for the oil production rate and water cut (percent) in the last study (water injection). These results are shown in Figs. 1–10, where the zero and nonzero fracture capillary pressure cases are illustrated. A comparison of these two cases indicates that the capillary continuity has a major influence on the numerical results. The reason for this influence is that in the depletion study, for example, when the capillary pressure force is stronger than the gravity drainage force, the oil flow from the matrix blocks decreases since interfacial tension increases with a decrease in pressure. Also, notice that there is a stable water-cut curve after the 10th year. This phenomenon occurs because the entire fracture system contains water after the 10th year; the major flow exchange mechanism between the matrix and fractures depends on imbibition (minus the value of p_{cow}) with a small flow rate for a long time.

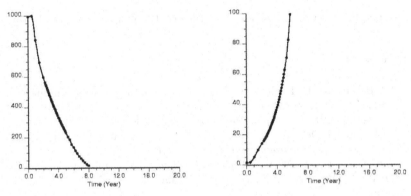

Figure 5. Q_o (gas recycling, $p_{cgo} = 0$).   **Figure 6.** GOR (gas recycling, $p_{cgo} = 0$).

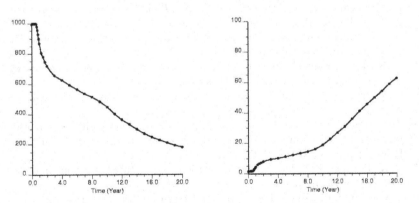

Figure 7. Q_o (gas recycling, $p_{cgo} \neq 0$).   **Figure 8.** GOR (gas recycling, $p_{cgo} \neq 0$).

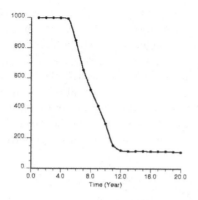

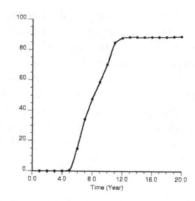

Figure 9. Q_o (water flooding).   **Figure 10.** Water cut (water flooding).

References

1. G. I. Barenblatt, Iu. P. Zheltov, and I. N. Kochina, Basic concepts in the theory of seepage of homogeneous liquids in fissured rocks [strata], *Prikl. Mat. Mekh.* **24** (1960), 852–864; *J. Appl. Math. Mech.* **24** (1960), 1286–1303.

2. Z. Chen and J. Douglas, Jr., Modelling of compositional flow in naturally fractured reservoirs, *Environmental Studies: Mathematical, Computational, and Statistical Analysis* (M. F. Wheeler, ed.), The IMA Volumes in Mathematics and its Applications, Springer-Verlag, Berlin and New York, Vol. 79, 1995, 65–96.

3. K. H. Coats, Implicit compositional simulation of single-porosity and dual-porosity reservoirs, paper SPE 18427, *SPE Symposium on Reservoir Simulation*, Houston, Texas, Feb. 6–8, 1989.

4. J. Douglas, Jr. and T. Arbogast, Dual-porosity models for flow in naturally fractured reservoirs, *Dynamics of Fluids in Hierarchical Porous Media*, J. H. Cushman, ed. Academic Press, London, 1990, 177–221.

5. A. Firoozabadi and L. K. Thomas, Sixth comparative solution project: Dual porosity simulators, *JPT* **42** (1990), 710–715.

6. H. Kazemi, Pressure transient analysis of naturally fractured reservoirs with uniform fracture distribution, *SPEJ* **9** (1969), 451–462.

7. S. J. Pirson, Performance of fractured oil reservoirs, *Bull. Amer. Assoc. Petroleum Geologists* **37** (1953), 232–244.

8. T. F. Russell and M. F. Wheeler, Finite element and finite difference methods for continuous flows in porous media, the Mathematics of Reservoir Simulation, R. E. Ewing, ed., SIAM, Philadelphia, pp. 35–106, 1983.

9. L. K. Thomas, T. N. Dixon, and R. G. Pierson, Fractured reservoir simulation, *SPEJ* (Feb. 1983), 42–54.

10. J. Warren and P. Root, The behavior of naturally fractured reservoirs, *SPEJ* **3** (1963), 245–255.

Multilayer Hydrodynamic Models and Their Application to Sediment Transport in Estuaries

H. Ramírez León[1], C. Rodríguez Cuevas[2], and E. Herrera Díaz[3]

[1] Instituto Mexicano del Petróleo. Mexico. hrleon@imp.mx
[2] Université Aix-Marseille III. CNRS. cuevas@L3m.univ-mrs.fr
[3] Instituto Politécnico Nacional. Mexico. enrique_herrera@att.net.mx

Abstract. Sedimentation in estuaries and coastal seas is an important concern due to the rapid morphological changes induced in the coastal zone. When the sediments coming from the continent are important, this problem is increased because they introduce morphological consequences, but also, they can transport several kinds of organic and inorganic pollutants due to anthropogenic activities in the hydrological basin. In this fact sediments play an important role with the consequently drop of the water quality in the environment. In this paper, the hydrodynamics is solved with a multilayer approximation and afterwards coupled with the resolution of total load of sediments (composed by suspended, bed and transition layer). Hydrodynamics calculation take into account tidal flow, wind driven influences, mass balance between fresh and salt water, and the hydrological influence on the river. Horizontal plane is considered with either a homogeneous or non-homogeneous grid; in the vertical, the model considers a numerical integration by layers.

Key words: Multilayer circulation model, shallow water equations, semi-implicit difference method, load and bottom sediments transport

1 Introduction

In this paper a multilayer hydrodynamic model is coupled with the resolution of total load of sediments (composed by suspended, bed and transition) and applied to interaction ocean-continent. The hydrodynamic model can provide us with the hydrodynamics taking into account tidal flow, wind driven circulation, mass balance between fresh and salt water, and the corresponding hydrological influence into the river. The application of the 3D dynamical model situations is not yet feasible because of excessive computer demand. Therefore, sometimes it has been a common practice to divide the mathematical flow models into different classes according to the dimension of the phenomena involved. We are considering the shallow water approximation in order to solve the 3D equations. A finite difference formulation is used for the discretization of the horizontal plane (x, y) with either a homogeneous or non-homogeneous grid in order to have a finer resolution over specific areas. The vertical plane is discretized by several layers in such a way that vertical velocity gradients can be simulated. The turbulence terms are evaluated using a mixing length turbulence model adjusted according to the horizontal and vertical variation of the numerical grid. The solution of sediments transport equation is coupled to the hydrodynamics model each time step considering the suspended and bottom transport equations into the algorithm. At each time step we solve a particular approximation in several control volumes in order to guarantee the conservation of total load of sediments in a water column in time and in space. Specific initial and boundary conditions are introduced in both river an ocean systems. Particular application is given on the Grijalva River in Mexico; basin of Grijalva River, is the most important for explotation and production of petroleum and derives in Mexico. The model reproduces appropriately the hydrodynamic currents and transport of sediments from the river up to coastal interaction. Global results are compared with settelite images.

2 PHYSICAL AND MATHEMATICAL FORMULATION

2.1 Shallow water equations

The governing equations for shallow water flows with density constant and free surface can be derived from the Navier-Stokes-Reynolds equations; after turbulent averaging and under simplifying assumption that the pressure is hydrostatic we can write:

Continuity equation

$$\frac{\partial U}{\partial x} + \frac{\partial V}{\partial y} + \frac{\partial W}{\partial z} = 0 \tag{1}$$

Velocity equations

$$\frac{\partial U}{\partial t} + U\frac{\partial U}{\partial x} + V\frac{\partial U}{\partial y} + W\frac{\partial U}{\partial z} = -g\frac{\partial \eta}{\partial x} + \nu_{Txy}\Delta_{(x,y)}U + \frac{\partial}{\partial z}\left(\nu_z\frac{\partial U}{\partial z}\right) + fV \tag{2}$$

$$\frac{\partial V}{\partial t} + U\frac{\partial V}{\partial x} + V\frac{\partial V}{\partial y} + W\frac{\partial V}{\partial z} = -g\frac{\partial \eta}{\partial y} + \nu_{Txy}\Delta_{(x,y)}V + \frac{\partial}{\partial z}\left(\nu_z\frac{\partial V}{\partial z}\right) - fU \tag{3}$$

$$\frac{1}{\rho}\frac{\partial p}{\partial z} = -g \tag{4}$$

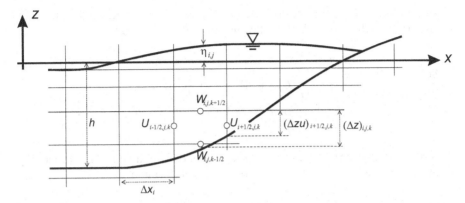

Fig. 1. Schematical diagram of the vertical grid for a multilayer approximation

Water surface elevation η is measured from undisturbed water surface, Fig. 1, [CCh92], [Vre94]:

Integrating the continuity equation over the depth and using a kinematic condition at the free surface leads to the free surface equation:

$$\frac{\partial \eta}{\partial t} + \frac{\partial}{\partial x}\left(\int_{-h}^{\eta} U\, dz\right) + \frac{\partial}{\partial y}\left(\int_{-h}^{\eta} V\, dz\right) = 0 \qquad (5)$$

Total water depth is defined as $H = h + \eta$ (Fig. 3)

The boundary conditions at the free surface are specified by the prescribed wind stresses τ_{wx} and τ_{wy}[ChPD76]:

$$\tau_{wx} = \nu_{Tz}\left.\frac{\partial U}{\partial z}\right|_{surface}, \qquad \tau_{wy} = \nu_{Tz}\left.\frac{\partial V}{\partial z}\right|_{surface} \qquad (6)$$

$$\tau_{bx} = \nu_{Tz}\left.\frac{\partial U}{\partial z}\right|_{bottom} = \frac{g\sqrt{U^2+V^2}}{Cz^2}U, \quad \tau_{by} = \nu_{Tz}\left.\frac{\partial V}{\partial z}\right|_{bottom} = \frac{g\sqrt{U^2+V^2}}{Cz^2}V \qquad (7)$$

Where Cz is the Chézy friction coefficient [Gar77].

So the system of equations to solve is composed by vertically averaged momentum eqs. (1)-(4) integered from the bed $z = -h$ to the free surface $z = \eta$, plus free surface eq. (5) and the boundary conditions, eqs. (6) and (7).

2.2 Sediments transport

The total sediment transport is evaluated considering separately coastal zone and a river system; in both cases we consider total sediments as an addition of suspended and bottom transport. On the river, the suspended sediments transport is evaluated with the following formulation [Rij93]:

$$\frac{\partial \bar{C}_s}{\partial t} + U \frac{\partial}{\partial x}\left(\bar{C}_s\right) + V \frac{\partial}{\partial y}\left(\bar{C}_s\right) + (W - w_s)\frac{\partial}{\partial z}\left(\bar{C}_s\right) =$$
$$- \frac{\partial}{\partial x}\left(\varepsilon_{s,x}\frac{\partial \bar{C}_s}{\partial x}\right) - \frac{\partial}{\partial y}\left(\varepsilon_{s,y}\frac{\partial \bar{C}_s}{\partial y}\right) - \frac{\partial}{\partial z}\left(\varepsilon_{s,z}\frac{\partial \bar{C}_s}{\partial y}\right) + E_r - D_p \quad (8)$$

Where w_s is the particle fall velocity of suspended sediment. The horizontal sediment mixing coefficient is evaluated with the formulation:

$$\varepsilon_s = \beta \Phi \varepsilon_j; \qquad \varepsilon_j = k u_* z \tag{9}$$

The vertical mixing coefficient is evaluated by the following parabolic function:

$$\varepsilon_{s,c} = \varepsilon_{s,\max} - \varepsilon_{s,\max}\left(1 - \frac{2z}{h}\right)^2; \qquad \varepsilon_{s,\max} = 0.25\beta k u_* h \tag{10}$$

Erosion and deposition parameter are evaluated, respectively, with:

$$E_r = \alpha\left(\frac{\tau_o}{\tau_c} - 1\right) \qquad D_p = w_s c_b\left(1 - \frac{\tau_o}{\tau_c}\right) \tag{11}$$

Where τ_o is the shear stress given by the sediments and τ_c is critical shear stress given by the flow. Bottom sediments transport is obtained from a derivation of eq. (8) on the (x,y) plane considering height of the boundary layer (δ_b) is approximately $2D_{50}$, with this considerations the equation results to:

$$\frac{\partial \bar{C}_b}{\partial t} + U \frac{\partial}{\partial x}\left(\bar{C}_b\right) + V \frac{\partial}{\partial y}\left(\bar{C}_b\right) = 0 \tag{12}$$

On the other hand, the coastal transport sediments are evaluated with the following formulations [Ell03]:

$$Q_S = (0.0188)\, E_{X-\text{día}}; \quad E_{X-\text{día}} = (86400) E_X; \quad E_X = \left(\frac{\gamma\left(H_r\right)^2 C_g}{16}\right)\sin(2\alpha_r) \tag{13}$$

Where Q_s is the total solid material flow transported by the current. In the simulations we are considering together coastal and river sediments motion.

3 Numerical model description

Horizontal velocities are estimated following recommendations given by Casulli and Cheng [Cch92], that's means a semi-implicit numeric method for the solution of the 3D free surface flow equations in which the gradient of surface elevation in the momentum equations and the velocity in the free surface equations will be discretized implicitly. The convective, Coriolis and horizontal viscosity terms in the momentum equations will be discretized explicitly, but in order to eliminate a stability condition due to the vertical eddy viscosity, the vertical mixing terms will be also discretized implicitly. So, the U-momentum equation, expressed in finite differences result:

$$
U_{i-1/2,j,k}^{n+1} = (FU)_{i-1/2,j,k}^{*} - g \frac{\Delta t}{(\Delta x_i + \Delta x_{i-1})0.5} \left(\eta_{i,j}^{n+1} - \eta_{i-1,j}^{n+1} \right)
$$
$$
\frac{\Delta t}{(\Delta z)_{i,j,k}} \left\{ \begin{array}{c} \frac{(\nu_{Tz}u)_{i-1/2,j,k+1/2}}{(\Delta zu)_{i-1/2,j,k+1/2}} \left(U_{i-1/2,j,k+1}^{n+1} - U_{i-1/2,j,k}^{n+1} \right) - \\ \frac{(\nu_{Tz}u)_{i-1/2,j,k-1/2}}{(\Delta zu)_{i-1/2,j,k-1/2}} \left(U_{i-1/2,j,k}^{n+1} - U_{i-1/2,j,k-1}^{n+1} \right) \end{array} \right\} \tag{14}
$$

Where the function (FU) is evaluated as

$$
(FU)_{i-1/2,j,k}^{*} = U_{i-1/2,j,k}^{*} + \Delta t \,(difux + coriou) \tag{15}
$$

Diffusion and Coriolis terms are calculated in the following way, respectively

$$
difux = (\nu_{Txy}) \left\{ \frac{1}{\Delta x_{i-1/2}} \left(\frac{U_{i+1/2,j,k}^{*} - U_{i-1/2,j,k}^{*}}{\Delta x_i} - \frac{U_{i-1/2,j,k}^{*} - U_{i-3/2,j,k}^{*}}{\Delta x_{i-1}} \right) \right.
$$
$$
\left. + \frac{1}{\Delta y_j} \left(\frac{U_{i-1/2,j+1,k}^{*} - U_{i-1/2,j,k}^{*}}{\Delta y_{j+1/2}} - \frac{U_{i-1/2,j,k}^{*} - U_{i-1/2,j-1,k}^{*}}{\Delta y_{j-1/2}} \right) \right\} \tag{16}
$$

$$
coriou = f \overline{\overline{V}}_{i-1/2,j,k}^{*} \tag{17}
$$

Velocities are evaluated in a cell using the next interpolation formulation following a Lagrangian approximation [Abb85], [RREM04]:

$$
U_{i-\frac{1}{2},j,k}^{*} = U_{i-a,j-b,k-d}^{n} = (1-r) \left\{ (1-p) \left[(1-q) U_{i-l,j-m,k-n}^{n} + \right. \right.
$$
$$
\left. qU_{i-l,j-m-1,k-n}^{n} \right] + p \left[(1-q) U_{i-l-1,j-m,k-n}^{n} + qU_{i-l-1,j-m-1,k-n}^{n} \right] \right\}
$$
$$
r \left\{ (1-p) \left[(1-q) U_{i-l,j-m,k-n-1}^{n} + qU_{i-l,j-m-1,k-n-1}^{n} \right] + \right.
$$
$$
\left. p \left[(1-q) U_{i-l-1,j-m,k-n-1}^{n} + qU_{i-l-1,j-m-1,k-n-1}^{n} \right] \right\} \tag{18}
$$

Where coefficients are defined by the following definitions

$$a = U^n_{i-\frac{1}{2},j,k} \frac{\Delta t}{\Delta x} \quad b = \overline{\overline{V}}^n_{i-\frac{1}{2},j,k} \frac{\Delta t}{\Delta y} \quad d = \overline{\overline{W}}^n_{i-\frac{1}{2},j,k} \frac{\Delta t}{\Delta z}$$
$$a = l + p; \qquad b = m + q; \qquad d = n + r. \tag{19}$$

And velocities are considered as the average:

$$\overline{\overline{V}}^n_{i-\frac{1}{2},j,k} = (V^n_{i,j,k} + V^n_{i,j+1,k} + V^n_{i-1,j+1,k} + V^n_{i-1,j,k})/4 \tag{20}$$

$$\overline{\overline{W}}^n_{i-\frac{1}{2},j,k} = (W^n_{i,j,k} + W^n_{i,j,k+1} + W^n_{i-1,j,k} + W^n_{i-1,j,k+1})/4 \tag{21}$$

A similar discretization is made for the equation in the y-direction. The following step is to estimate the free surface, which is determined by the velocity field, for each cell of the (i,j) plane. Eq. (5) is solved as

$$\eta^{n+1}_{i,j} = \eta^n_{i,j} - \frac{\Delta t}{\Delta x_i} \left\{ \sum_{k=nnku}^{nk} \left[(\Delta z)_{i+1,j,k} U^{n+1}_{i+1/2,j,k} - (\Delta z)_{i,j,k} U^{n+1}_{i-1/2,j,k} \right] \right\}$$
$$- \frac{\Delta t}{\Delta y_j} \left\{ \sum_{k=nnku}^{nk} \left[(\Delta z)_{i,j+1,k} V^{n+1}_{i,j+1/2,k} - (\Delta z)_{i,j,k} V^{n+1}_{i,j-1/2,k} \right] \right\}$$

$$\tag{22}$$

This system is recursive, so with this new free surface value, the velocity field (U and V) is calculated again along with the free surface until the difference between $\eta^{n+1}_{i,j}$ and $\eta^n_{i,j}$ are below an established tolerance ($\xi \to 0$) .

Finally, it is necessary to calculate the third component of the velocity by means of representing the continuity eq. (1) as

$$W^{n+1}_{i,j,k} = W^{n+1}_{i,j,k-1} - \frac{\left[(\Delta zu)_{i+1/2,j,k} U^{n+1}_{i+1/2,j,k} - (\Delta zu)_{i-1/2,j,k} U^{n+1}_{i-1/2,j,k} \right]}{\Delta x_i}$$
$$- \frac{\left[(\Delta zv)_{i,j+1/2,k} V^{n+1}_{i,j+1,k} - (\Delta zv)_{i,j-1/2,k} V^{n+1}_{i,j-1/2,k} \right]}{\Delta y_j} \tag{23}$$

Horizontal turbulent viscosity is evaluated with the formulation [KV73]:

$$\nu_{Txy} = \frac{5.9 H^n_{i,j,k} \sqrt{g} \left[\left(U^n_{i,j,k} \right)^2 + \left(V^n_{i,j,k} \right)^2 \right]^{1/2}}{Cz} \tag{24}$$

4 Application to Grijalva River in Mexico

Erosion, transport and sedimentation introduce a very big problem in estuaries systems, Grijalva River, the biggest in Mexico, isn't an exception. In rain season this river carries about 2,500 m³/s and consequently an estimated load of sediments of 141Kgf/s.m. In Fig. 2 we show the total simulated area, including a San Pedro River on the right of Grijalva. The region of study contains a grid with 151 cells on the x-direction and 141 cells on y-direction; layers on the vertical dimension vary depending on the topography and bathymetry of specifically zone. Total area involved is 2,976 km².

Fig. 2. Ecosystem below study, plan view

Simulations where carried out with the conditions showed in Table 1:

Table 1. Hydrodynamics and sediments simulation conditions

Simulation period	1st to 15th March 2003
Rate flow of Grijalva River:	2,510.55 m³/s
Rate flow of San Pedro River:	83.69 m³/s
Wind velocity and direction:	16 m/s, SE
Tidal month:	March
Velocity and direction of surface littoral current:	0.36 m/s, NE
Surface temperature:	29° C
Mean salinity:	35 ups
Sediment concentration:	100 gr/m³

In Fig. 3, we show the result of the hydrodynamics calculation; stability of the flow is reached after 24 hr of calculation [Her04]. Maximal velocities are reached when the tidal flow influences is lower than the hydrological flow coming from the river.

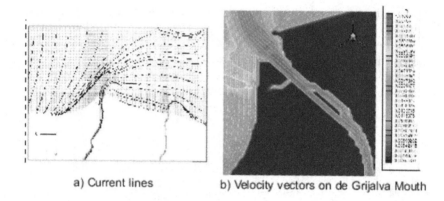

a) Current lines   b) Velocity vectors on de Grijalva Mouth

Fig. 3. Velocity field on the surface

Concentrations of bottom sediments are a behavior following the simulation time and have a stronger dependency on the fall velocity (Fig. 4); in this figure we can show that the behavior depends on the bathymetry too.

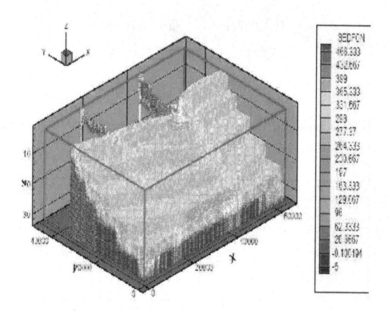

Fig. 4. Bottom sediments distribution (gr/m^3)

Suspended sediments come down from water column when they reach the ocean (Fig. 5) due to effects of bottom concentration, shear stress balance and fall velocity, as was pointed out by mechanism described by the eq. (8)

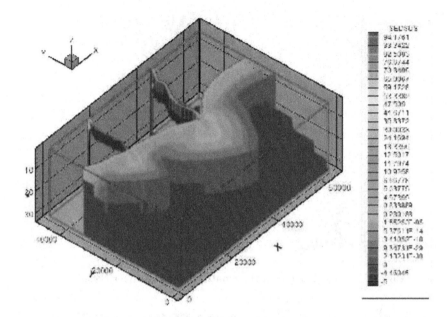

Fig. 5. Suspended sediments distribution $(\mathrm{gr/m^3})$

The resultant, of both suspended and bottom simulation results, mainly by the Grijalva River, constitute the continental contribution over a marine field; in this zone sediments plume have higher environmental impact compared with coastal sediments transport because sediments carry anthropogenic pollution associated with them, sediments impact water quality in several senses. On Fig. 6 we show a close-up of the impact of Grilava River mouth.

Model developed provides us evolution in time and space of the hydrodynamics and total sediments, in Fig. 7 we can see two stages of plume dispersion of both Grijalva and San Pedro Rivers.

In Fig 8 we show the total sediments transport profiles on the water column of a river for both March and September.

On the mouth (0 Km) in Fig. 8 we can see the tidal influences on the transport sediments. In dry season, on March, the tidal influences is bigger and penetrates into the river; in raining season, September, the river carries most flow and consequently most sediment load and have a remarkably influence on the sea. Finally, Fig. 9 shows the comparison of simulation results with a satellite image for 18th March 2003. As we can see results obtained with the model is very similar to patterns showed by satellite image.

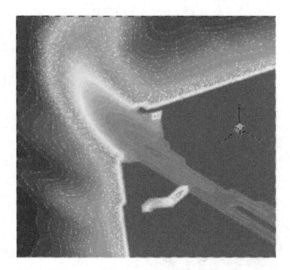

Fig. 6. Total sediments on the mouth of Grijalva River (gr/m^3)

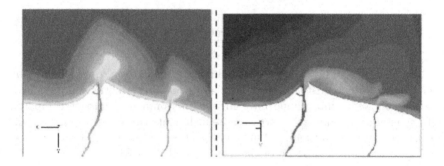

Fig. 7. Two steps of the transition calculation

5 Conclusions

A numerical model was developed to simulate the hydrodynamic of coastal systems; the model solved the shallow water equations with a multilayer approach. Several numerical and experimental tests were carried out in order to validate the code [RREM04]. In regards of Grijalva estuary their dynamics effects are well represented by the model; in the same way, sediments modules, simulating suspended, bottom and transition interface, reproduce the total load transportation into the river and their interaction with the coastal sediment transport. The model takes into account the interaction between suspended and bottom sediments by the way of particle fall velocity and their influences on the erosion and sedimentation parameters analyzed by means of shear stress balance between particles dynamics and the flow on the bottom

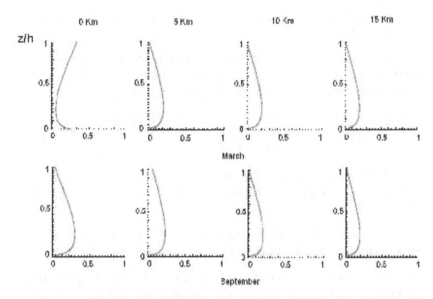

Fig. 8. Dimensionless profiles for total transport in last 15 Km of the Grijalva River

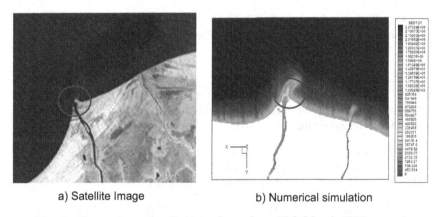

a) Satellite Image   b) Numerical simulation

Fig. 9. Comparison of sediments plume for a 18th March 2003 scenario

boundary layer. Final results are compared with satellite images for specific periods, the global behavior of sediments plume dispersion are satisfactory and encouraging.

6 References

[Abb85] Abbott, M. B.: Computational Hydraulics elements of the theory of free surface flows, International Institute for Hydraulics and Environmental Engineering, Delf and Danish Hydraulic Institute, Hørsholm, 1985.

[CCh92] Casulli, V., Cheng, R.T.: Semi-implicit finite difference methods for three-dimensional shallow water flow, International Journal for Numerical Methods in Fluids, vol. 15 629-648, 1992.

[ChPD76] Cheng, R. T., Powell, T. M., Dillon T. M.: Numerical models of wind-driven circulation in lakes, Applied Mathematical Modeling, 1, 141, December. 1976.

[Ell03] Elliott, A.: Suspended Sediment Concentrations in the Tamar Estuary United Kingdown, Estuarine, Coastal and Shelf Science, 57: 679-688, 2003.

[Gar77] Garrat, J. R.: Review of drag coefficients over oceans and continents, Monthly Weather Review, Vol 105, pp 915-929, 1977.

[Her04] Herrera, E. Dynamics modeling of transport sediments on the mouth of Grijalva river. Master These in Sciences. Polytechnic Institute of Mexico, 2004.

[KV73] Kuipers, J., Vreugdenhil, C. B.: "Calculations of two-dimensional horizontal flow", Res. Rep S163, Part 1, Delft Hydraulics Laboratory, 1973.

[RREM04] Rodríguez, C. Ramírez-León, H., Escalante, M., Morales, R. A multilayer circulation model to simulate currents patterns in lagoons. International Association of Hydraulic Research. Submitted in 2004, reference number P2670.

[Rij93] Van Rijjn, L. C.: Principles of Sediment Transport in River, Estuaries and Coastal Seas. Holanda, University of Utrecht, DELFT Hydraulics, Ed. Aqua Publications, 634pp, 1993.

[Vre94] Vreugdenhil, C. B.: Numerical Methods for Shallow-Water Flow. Kluwer Academic Publishers. The Netherlands, 1994.

An Application of Genetic Programming to Economic Forecasting

Kangshun Li[1,2*], Zhangxin Chen[3], Yuanxiang Li[1], and Aimin Zhou[1]

[1] State Key Laboratory of Software Engineering, Wuhan University, Wuhan 430072, China. lks@public1.gzptt.jx.cn
[2] School of Information Engineering, Jiangxi University of Science & Technology, Jiangxi 341000, China.
[3] Center for Scientific Computation and Department of Mathematics, Southern Methodist University, Dallas, TX 75275-0156 USA. zchen@mail.smu.edu

Abstract

In this paper, we propose an application of genetic programming to economic forecasting that can obviously improve traditional economic forecasting methods; the latter can only obtain rough fitting curves with unsatisfactory results. Forecasted and estimated standard errors are also computed and analyzed. Using practical historical data from Statistical Yearbooks of the People's Republic of China in recent years, an automatically generated mathematical model of economic forecasting by genetic programming is established. Forecasting results indicate that the accuracy obtained by genetic programming is obviously higher than traditional methods such as linear, exponential, and parabolic regression methods.

1 Introduction

To facilitate a healthy economic development of a country or region, it is necessary to perform global adjustments and controls of economy according to economic and historical data. The basis of performing such global adjustments and controls is to make an accurate prediction and analysis of economic situations in the future. However, the most commonly used method has so far been to construct linear, exponential, or parabolic regression equations to fit an economic curve of a development trend (called traditional forecasting models). Because of various types of economic trends, it is difficult to fit accurately the curves of all the economic trends using fixed regression models. We have

*This work is partly supported by the National Natural Science Key Foundation of China with the Grant No.60133010 and the National Research Foundation for the Doctoral Program of Higher Education of China with the Grant No.20030486049.

done a lot of experiments using these traditional models to fit economic trends such as population, agricultural, and industrial curves. All these experiments have indicated that such curves are very rough and the standard errors of correspondent data are quite large. To overcome these shortcomings, we present a method of evolutionary modeling using genetic programming (GP). This method can generate a dynamic regression model according to the characteristic of dynamic time-series at random. It is different from the traditional prediction methods, and both its accuracy of predicting results and estimating the standard error are obviously superior to the traditional methods.

2 Genetic Programming

GP [3] is a branch of an evolutionary computational algorithm [5], a method for finding the most fitted computer programs by means of artificial evolution. Genetic programs are usually represented by trees that consist of functions and terminal symbols. The population of computer programs is generated at random, and they are evolved to better programs using genetic operators. The ability of a program to solve a problem is measured by its fitness value. GP extends genetic algorithms to computer programming, and breeds computer programs hereditarily to solve problems like system identification, optimum controlling, gaming, optimizing, and planning. We replace a character string by a computer program of multi-layers to express such problems in a more natural way and to explain the problem complexity according to the structure and size of the computer program. GP has expanded the application of artificial intelligence, machine learning, and symbol disposing. Its main idea is: n-ary trees represent the solution space of a problem, terminal nodes denote independent variables of the problem, intermediate nodes represent a function combined with the independent variables, and root notes denote the final output function. This paper describes a new approach to produce a function of economic forecasting of a new generation through crossover and mutation of notes (subtrees) based on GP. This approach produces a forecasting function that satisfies the terminal given condition.

3 Description of Genetic Programming

Given a symbol set $m = \{+, -, *, /, \sin, \cos, \exp, \ln\}$ and a datum set $P = \{a, -10 < a < 10\}$, we define a space of program structures, PS, which is a combination of an operator symbol (function) set (O) and a datum set (D) [2]; i.e., $PS = O \cdot D$ (called a PS space). GP is to search for the most suitable self-adaptive function at random. The procedure of automatically generated functions of GP is given as follows:

3.1 Confirming the representation of programs

There exist many representations of a program, such as an original code expression of a program language and a binary string [3], among which the most common method is a syntax tree [1]. The individuals who make up a swarm adopt a dynamic tree structure. The notes consist of terminal notes, primitive functions, and operators. Root notes and intermediate notes are called internal notes, which are functions assembling the original variables; for example, the function $\sin x * \ln y + 6$ can be represented by a syntax tree (Fig. 1).

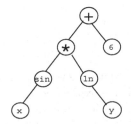

Figure 1. A syntax tree.

3.2 Designing genetic operators

The designing of genetic operators is related to the representation of computer programs. As far as a syntax tree is concerned with, we often exchange two subtrees to crossover two parents as in Fig. 2 and reproduce a subtree to mutate a parent as in Fig. 3 [4].

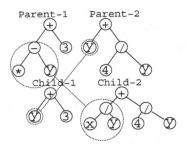

Figure 2. A crossover operation.

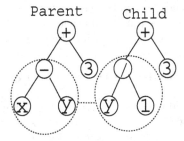

Figure 3. A mutating operation.

3.3 Constructing an evaluation function

An evaluation function is used to measure an approximate degree of a problem, which generally combines all environments of parameters and the convergence analysis of an algorithm, and is denoted by a norm. For example, the evaluation function between a program P and its original program P_0 can be defined by

$$\text{fitness}(P) = \|P - P_0\|,$$

where $\|\cdot\|$ denotes the "distance" between P and P_0; the shorter the distance, the closer between P and P_0. If P can realize all functions in P_0, then the distance will become zero. In economic forecasting, we often use the ℓ_2-norm to calculate the distance; i.e., $\text{fitness}(Y) = \|\widehat{Y} - Y\|_2$, where $\widehat{Y}$ is a vector of prediction values of Y and Y is a vector of practical values.

3.4 Realization of the algorithm

The steps for the realization of our algorithm are as follows:

1. Confirm the operator and datum sets of a model.
2. Initialize a population of tree expressions.
3. Evaluate each expression in the population.
4. Create new expressions (children) by mating current expressions, and apply mutation and crossover to the parent tree expressions.
5. Replace the members of the population with the child trees.
6. A local hill climbing mechanism (called "relabelling") is executed periodically to relabel the nodes of the population trees.
7. If a termination criterion is satisfied, stop; otherwise, go back to Step 2.
8. Check with the observed data.

Compared with the traditional modeling methods, this automatically generated evolutionary modeling method has many advantages. First, before the evolutionary modeling, it is not necessary to confirm a fixed structure of a model as in the traditional economic forecasting methods (e.g., linear, exponential, and parabolic regressions). Second, the evolutionary modeling does not utilize a mathematical method to solve a problem, but rather uses the GP algorithm to search for the optimum solution of the problem.

3.5 Analyzing the standard error

The purpose of automatic modeling by GP is to make the standard error between the prediction values of a model and the practical values of the model as small as possible. In economic forecasting, we usually estimate the standard error and forecast it in economic statistics to check the accuracy of the model. Estimating the standard error refers to the standard error between the estimated values obtained using a forecasting model and its practical values, and forecasting it refers to the standard error between the forecasted values to future data obtained using a forecasting model and its practical values in the future. The formula of the standard error is given by

$$\sigma = \sqrt{\sum_{i=1}^{n}(y_i - \widehat{y}_i)^2 f_i / \sum_{i=1}^{n} f_i},$$ where $\widehat{y}_i$ is the estimated value using the prediction model, y_i is the practical value, and f_i is the weighted value of y_i, $i = 1, 2, \ldots, n$.

4 Traditional Models of Economic Forecasting

In the traditional economic forecasting, one usually uses the least mean square approach to construct a fixed forecast model [6] by the next methods.

4.1 Linear regression

The forecasting model of linear regression in an economic statistics book is written as $Y = X\beta + \varepsilon$, where $Y = \mathrm{col}(y_1, y_2, \cdots, y_n)$ stands for n estimated values, $X = \begin{pmatrix} x_1 \\ x_2 \\ \vdots \\ x_n \end{pmatrix} = \begin{pmatrix} 1 & x_{11} & \cdots & x_{1p} \\ 1 & x_{21} & \cdots & x_{2p} \\ \vdots & \vdots & \vdots & \vdots \\ 1 & x_{n1} & \cdots & x_{np} \end{pmatrix}$ is the matrix of the values of predictor variables, $x_i = (1, x_{i1}, \cdots, x_{ip}), i = 1, 2, \ldots, n$, is the ith vector of the values of the p predictor variables, $\beta = \mathrm{col}(\beta_0, \beta_1, \cdots, \beta_p)$ is the vector of unknown parameters, and $\varepsilon = \mathrm{col}(\varepsilon_1, \varepsilon_2, \cdots, \varepsilon_n)$ is the error vector.

4.2 Exponential regression

The forecasting model of exponential regression is given by $y_i = \beta_0 \beta_1^{x_{i1}} \beta_2^{x_{i2}} \cdots \beta_p^{x_{ip}} \varepsilon_i, i = 1, 2, \ldots, n$. We take logarithm on both sides of this equation and transfer it to linear regression as follows:

$$\widehat{Y} = \widehat{X}\beta + \varepsilon,$$

where $\widehat{Y} = \mathrm{col}(\log(y_1), \log(y_2), \cdots, \log(y_n))$ are the logarithmic values of n estimated values, $\widehat{X} = \begin{pmatrix} \widehat{x}_1 \\ \widehat{x}_2 \\ \vdots \\ \widehat{x}_n \end{pmatrix} = \begin{pmatrix} 1 & x_{11} & \cdots & x_{1p} \\ 1 & x_{21} & \cdots & x_{2p} \\ \vdots & \vdots & \vdots & \vdots \\ 1 & x_{n1} & \cdots & x_{np} \end{pmatrix}$ is the matrix of the values of predictor variables, $\widehat{x}_i = (1, x_{i1}, \cdots, x_{ip}), i = 1, 2, \ldots, n$, is the ith vector of the values of the p predictor variables, $\widehat{\beta} = \mathrm{col}(\log(\beta_0), \log(\beta_1), \cdots, \log(\beta_p))$ is the logarithmic vector of the unknown parameters, and $\widehat{\varepsilon} = \mathrm{col}(\log(\varepsilon_1), \log(\varepsilon_2), \cdots, \log(\varepsilon_n))$ is the logarithmic vector of the error vector.

4.3 Parabolic regression

The parabolic regression equation is defined by

$$y_i = \beta_0 + \beta_1 x_i + \beta_2 x_i^2 + \varepsilon_i, \qquad i = 1, 2, \ldots, n.$$

Let $A = \mathrm{col}(y_1, y_2, \cdots, y_n)$, $B = \mathrm{col}(\beta_0, \beta_1, \beta_2)$,

$E = \mathrm{col}(\varepsilon_1, \varepsilon_2, \cdots, \varepsilon_n)$, and $P = \begin{pmatrix} 1 & x_1 & x_1^2 \\ 1 & x_2 & x_2^2 \\ \vdots & \vdots & \vdots \\ 1 & x_n & x_n^2 \end{pmatrix}$.

Then the parabolic regression equation becomes $A = PB + E$.

According to the least mean square method, all the above parameters $\beta_0, \beta_1, \cdots, \beta_p$ of the regression equations can be obtained by minimizing the quantity $Q = \sum_{i=1}^{n} \varepsilon_i^2$, and they can be found by solving the partial differential equations

$$\frac{\partial Q}{\partial \beta_0} = 0, \quad \frac{\partial Q}{\partial \beta_1} = 0, \quad \cdots, \quad \frac{\partial Q}{\partial \beta_p} = 0.$$

5 Numerical Experiments

In this section, we will employ practical data to perform two experiments to demonstrate advantages of the evolutionary modeling based on GP. Using the GP method and the trend of an original economic problem, we can construct an economic forecasting model. Assume the data of the original problem: (x_i, y_i), $i = 1, 2, \ldots, m$, where $x_i \in \Re$, $y = f(x)$, and the solution $f \in M$, where M is the model space determined by an operator set O and the depth of a tree, so that the modeling problem can be transformed to an optimization problem:

$$\min_{f \in M} d(f(x_i), y_i) = \min_{f \in M} \sqrt{\sum_{i=1}^{m} (f(x_i) - y_i)^2}.$$

Experiment 1: We construct an automatic model and fit a curve using GP according to the gross domestic product (GDP) of [7] from 1978 to 1995; we then forecast the data from 1996 to 2002 and compare them with the results of three traditional forecasting models to study their standard errors.

We set the function type: complex function, tree depth= 8, population size= 60, breeding times= 40,000, and fitness$(Y) = \|\widehat{Y} - Y\|_2 < 2,000$. The evolved forecasting model by running the GP program is as follows:

$$Y = \exp(\ln(\ln((t^2 + t^3) + (t^3 + t^2))) + \sin((\sin(246.68111)(t^2/224.58571)))$$
$$+ \ln(\ln(t^7) + t^2 + 679.37254))/\cos(\cos(\cos(512.86355 * 258.82748$$
$$/(t^6 - 373.94330) + \ln(|t^4 - 34.76058|))) + \exp(\exp(\sin(t^7)/(t^2 + t))$$
$$+ \ln|(t^3 + 646.41255)/(-1270)| + \sin(\cos(704.73342)) + \ln|\sin(t^7)$$
$$+ t^2 + 679.37254)|/\cos(\cos(\sin(1439.92431 - \log(t^5))(\ln(985.77837)$$
$$+ \sin(497.78741))/(\cos(257.48466) - \ln(t^6))))).$$

Although this generated function is very long, its structure is very simple and can be easily calculated by a computer. The results, the estimated standard errors, and the forecasted standard errors obtained by the evolutionary model are compared with the corresponding results of the linear, exponential, and parabolic regression models in Table 1.

Table 1. The models of gross domestic product (GDP) from 1981 to 1995

Year(t)	GDP(100 million yuan)	Value of GP regression	Value of linear regression	Value of exponential regression	Value of parabolic regression
1981(1)	4901.40	4838.22	-3069.36	4375.44	8072.45
1982(2)	5489.20	5621.51	137.47	5206.77	6504.21
1983(3)	6076.30	5883.88	3344.30	6196.06	5670.59
1984(4)	7164.40	7277.72	6551.13	7373.31	5571.59
1985(5)	8792.10	8762.39	9757.96	8774.24	6207.21
1986(6)	10132.80	9685.83	12964.79	10441.35	7577.45
1987(7)	11784.70	11626.37	16171.62	12425.20	9682.31
1988(8)	14704.00	14779.08	19378.45	14785.99	12521.79
1989(9)	16466.00	16362.37	22585.28	17595.33	16095.89
1990(10)	18319.50	18037.35	25792.11	20938.44	20404.61
1991(11)	21280.40	21147.69	28998.94	24916.74	25447.95
1992(12)	25863.70	26609.18	32205.77	29650.93	31225.91
1993(13)	34500.70	34748.73	35412.60	35284.60	37738.49
1994(14)	46690.70	46194.46	38619.43	41988.68	44985.69
1995(15)	58510.50	58420.24	41826.26	49966.53	52967.51
Estimating standard error σ		292.39	6769.97	2970.41	2954.12
Forecasting standard error $\hat{\sigma}$ from 1996 to 2002		468.30	31772.80	27843.16	12870.26

Table 1 shows that the estimated standard error obtained by the prediction model of GP is less than 0.1 of that of the other three models, and the forecasted standard error by the forecasting values from 1996 to 2002 by GP is less than 0.05 that of the other three models. In other words, the fitted precision of the curve by GP is 10 times more than that of the other three models, and the precision of forecasted data by GP is 20 times more than that of the other three models. Up to now, this is a regression prediction model whose fitted precision of the curve is the best and the predicted precision is the most accurate. We can easily analyze from Figs. 4–7 that the curve of Fig. 7 fitted by GP is obviously closer to the practical data than the other three models because the points plotted by the GP forecasting model and the points plotted using the practical data locate almost in the same positions in Fig. 7.

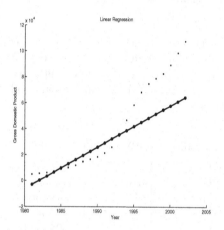

Figure 4. Linear regression.

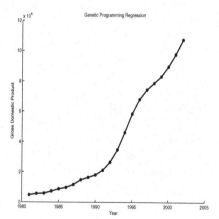

Figure 5. Exponential regression.

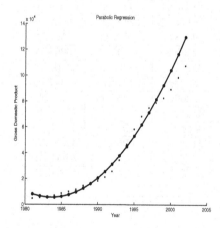

Figure 6. Parabolic regression.

Figure 7. GP regression.

Experiment 2: We construct an automatic evolutionary model and fit a curve using GP according to the fixed assets investment (FAI) of [7] from 1985 to 1999, and then forecast the data from 2000 to 2002 and compare the results with the three traditional forecasting models to study their standard errors.

We set the function type: complex function, tree depth= 5, population size= 60, breeding times= 30,000, and fitness$(Y) = \|\widehat{Y} - Y\|_2 < 1,800$. The evolved forecasting model by executing the GP program is given by

$$Y = \exp(\ln|\ln(t^2)|/(\sin t \ln|\cos(t^3)|) - (t(-910.73641)$$
$$+(-675.77746)/t + (-932.35572) + t^2 - t^5/t^2)))) - \ln|(t^5 + 109.64690$$
$$+88.92178t^5 - \exp(t^2 + t^2))|/\exp(12.35389/(-13.46782 - t^2))$$
$$*\ln(\ln(96.68874))\sin(\sin(-316.00391))\ln|((-258.44295 + t^2)$$
$$*\cos(t^4)(-311.43834t^5))|\cos(\exp(t^4/(-608.03247) + \ln(t^4))).$$

The results, the estimated standard errors, and the forecasted standard errors obtained by the evolutionary model are compared with the corresponding results of the linear, exponential, and parabolic regression models in Table 2.

Table 2. The models of fixed assets investment (FAI) from 1985 to 1999

Year(t)	FAI(100 million yuan)	Value of GP regression	Value of linear regression	Value of exponential regression	Value of parabolic regression
1985(1)	2543.20	2607.97	-2145.20	2366.18	2299.16
1986(2)	3120.60	3027.65	-0.10	2872.07	2539.56
1987(3)	3791.70	3760.40	2145.00	3486.12	3073.00
1988(4)	4753.80	4755.90	4290.10	4231.45	3899.48
1989(5)	4410.40	4320.74	6435.20	5136.14	5019.00
1990(6)	4517.00	4707.22	8580.30	6234.24	6431.56
1991(7)	5594.50	5627.05	10725.40	7567.12	8137.16
1992(8)	8080.10	8013.68	12870.50	9184.98	10135.80
1993(9)	13072.30	13562.20	15015.60	11148.72	12427.48
1994(10)	17042.10	17217.23	17160.70	13532.32	15012.20
1995(11)	20019.30	19407.76	19305.80	16425.53	17889.96
1996(12)	22913.50	22632.25	21450.90	19937.31	21060.76
1997(13)	24941.10	25082.53	23596.00	24199.91	24524.60
1998(14)	28406.20	28141.17	25741.10	29373.85	28281.48
1999(15)	29854.70	30275.40	27886.20	35653.98	32331.40
Estimating standard error σ	265.14	2872.40	2337.46	1531.8	
Forecasting standard error $\widehat{\sigma}$ from 2000 to 2002	170.01	6270.21	15836.3	3577.78	

Table 2 shows that both the estimated and forecasted standard errors by the prediction model of GP are less than 0.2 of those of the other three models, and that the result by GP is much better than that of the other three models for the fitted curve and forecasted future data. Furthermore, we have done some experiments on other practical data of [7, 8], such as population, agricultural output, and the total amount of merchandise retails, and the fitted curves can be analyzed similarly. The results indicate that the precision of the curve fitted automatically by GP is much higher than that of the other curves fitted by the traditional models, and the forecasting result in future is much closer to the practical data.

6 Conclusions

In the first experiment we have used the data in the time period 1981-1995 to construct the prediction model by GP we proposed here; we have done many experiments by GP using the data collected some time ago to evolve the prediction function. All the experimental results indicated that we can use the data collected some time ago to fit curves by GP, and the estimated values obtained by this method are much closer to practical values than the traditional methods. So are the forecasted values obtained by this method, which are difficult to obtain for the traditional models of economic forecasting. A lot of experiments have proved that if we evolve prediction models using the latest historical data in GP to predict future data, the results will be better. Meanwhile, we can use this model to estimate the missing historical data precisely. Therefore, GP not only produces more precise curves of an economic trend, but also can be used to construct a model of economic forecasting to predict future data and fill the historically missing data more accurately than other traditional economic prediction models.

References

1. K. D. Jong, L. Fogel, H. P. Schwefel, *et al.*, The Hand Book of Evolutionary Computations, IOP Publishing Ltd, Oxford University Press, 1997.
2. L. Kang, Y. Li, and Y. Chen, A tentative research on complexity of automatic programming, Wuhan University Journal of Natural Science, **6** (2001), 59–62.
3. J. R. Koza, Genetic Programming I: On the Programming of Computers by Means of Natural Selection, MIT Press, Cambridge, MA, 1992.
4. J. R. Koza. Genetic Programming III: Darwinian Invention and Problem Solving, San Francisco, Morgan Kaufmann Publishers, 1999.
5. Z. Pan, L. Kang, and Y. Chen, Evolutionary Computations, Tsinghua University Press, Beijing, China, 1998.
6. State Statistical Bureau of PRC, Theory of Statistics, Statistical Press of China, 2002.
7. State Statistical Bureau of PRC, Statistical Yearbook of China, Statistical Press of China, 2002.
8. Statistical Bureau of Jiangxi Province, Statistical Yearbook of Jiangxi Province, Statistical Press of China, 2002.

A New Dynamical Evolutionary Algorithm Based on Particle Transportation Theory

Kangshun Li[1,2] *, Yuanxiang Li[1], Zhangxin Chen[3], and Zhijian Wu[1]

[1] State Key Laboratory of Software Engineering, Wuhan University, Wuhan 430072, China. lks@public1.gzptt.jx.cn

[2] School of Information Engineering, Jiangxi University of Science & Technology, Jiangxi 341000, China.

[3] Center for Scientific Computation and Department of Mathematics, Southern Methodist University, Dallas, TX 75275-0156 USA. zchen@mail.smu.edu

Abstract

In this paper, a new dynamical evolutionary algorithm is presented based on a particle transportation theory according to the principle of energy minimization and the law of entropy increasing in the phase space of particles. In numerical experiments we use this algorithm to solve optimization problems, which are difficult to solve using traditional evolutionary algorithms (e.g., the minimization problem of six-hump camel back functions). Compared with the traditional evolutionary algorithms, this new algorithm not only solves linear and nonlinear optimization problems more quickly, but also more easily finds all the points that reach the global solutions of these problems because it drives almost all the individuals to have chances to participate in crossing and mutating. The results of numerical experiments show that this dynamical evolutionary algorithm obviously improves the computing performance of the traditional evolutionary algorithms in that its convergent speed is faster and it is more reliable.

1 Introduction

An evolutionary algorithm [7] (EA), which imitates biological evolution and natural genetic selection mechanism at random, has been used to solve practical optimization problems. Compared with traditional algorithms for these problems, EA has an intellective property and parallel character. After setting up an encoding plan, fitness function, and genetic operator, this algorithm

---

*This work is partly supported by the National Natural Science Key Foundation of China with the Grant No.60133010 and the National Research Foundation for the Doctoral Program of Higher Education of China with the Grant No.20030486049.

will use the information obtained through evolving population to self-organize and search. Because the strategy based on the natural selection is the survival of the fittest and discarding of the non-fittest, the individuals whose fitness values are better have higher survival probability. In general, the individuals with better fitness values have more adaptable gene structures to participate in crossing and mutating, and may generate the next generation even more adaptable to the environment. Its self-organization and self-adaptability character endow itself the ability to discover the property of the environment automatically according to the environment change. Furthermore, the natural selection of EA eliminates the most serious obstacle in the traditional algorithm designing for which one must describe all the properties of an algorithm and take a different measure for a different algorithm (before its designing). Nowadays, one uses EA to solve problems of complex structures, such as the problem of finding the minimum value of a multi-modal function and the problem of recovering the legal-code of generated cyclic codes [3]. Since the early 1990's, a new research topic in which evolutionary computations (EC) incorporate biology, computer science, mathematics, physics, and other disciplines has attracted many researchers; it has expanded the application of EC to a greater extent, and at the same time has improved the computing performance of EC.

However, the earlier EA (called the traditional EA) has a fatal shortcoming; it cannot resolve the premature phenomenon of a problem and cannot find all the points reaching the global optimal solution of the problem. To overcome this shortcoming, we construct a new particle dynamical evolutionary algorithm (PDEA). PDEA is based on a transportation theory of particles in the phase space according to the transportation equation built by the location and momentum of every particle (these particles can be described as matters like vehicles in the street and residents in cities) moving in the phase space and built by processes in which the change of eigenvalues is caused by the dynamic change of every relative physical quantity with the dynamic change of time and space according to the principle of energy minimization and the law of entropy increasing of a particle swarm. This algorithm guarantees all the individuals to have chances to participate in the crossing and mutating of population all the time, and changes the traditional algorithm which only replaces the worse-function-fitness-value individuals with the better-function-fitness-value individuals to cross and mutate (that is why it is difficult for this algorithm to find all the points reaching the global optimal solution of a problem). This dynamical EA can not only speed up convergence of an algorithm, but also can avoid the phenomenon of premature and more quickly find all the points reaching the global optimal solution of a problem.

2 Foundation of Transportation Theory

2.1 The canonical Hamiltonian equation and transportation equation of particles

A transportation theory [1] is to study and analyze a transportation process of particles based on statistical mechanics. Assume that a system consists of N particles and these N particles have $6N$ generalized coordinate variables in the three-dimensional space. These variables include $3N$ coordinates $q_1, q_2, \ldots, q_{3N}$ and $3N$ generalized conjugate momentums $p_1, p_2, \ldots, p_{3N}$. We regard these $3N$ coordinates and $3N$ generalized conjugate momentums as the coordinates of a point in the $6N$-dimensional space. The $6N$ dimensional phase space includes the microstate of the entire system of N particles at a certain time. The N particles in the $6N$ dimensional phase space make the microstate change as time goes, which then causes the representation points to move. Denote the energy function of the system by $H(\mathbf{q}, \mathbf{p})$, the representation point of the N particles in the phase space by

$$\Gamma_N = \{q_r, p_r : r = 1, 2, \ldots, 3N\}.$$

When particles move continuously, the representation point of the phase space generates a moving orbit in the phase space. This moving orbit can be described by the canonical Hamiltonian equation of motion

$$\frac{\partial H}{\partial t} = -\frac{\partial H}{\partial p_i} = \{q_i, H\}, \quad \frac{dp_i}{dt} = -\frac{\partial H}{\partial q_i} = \{p_i, H\}, \quad i = 1, 2, \ldots, n, \quad (1)$$

where $\{q_i, H\}$ and $\{p_i, H\}$ are the *Poisson brackets*. The *Hamiltonian quantity* of the system is

$$H = H(\Gamma_N), \quad (2)$$

and the transportation of N particles is given by

$$\frac{\partial n}{\partial t} + \mathbf{v} \cdot \frac{\partial n}{\partial \mathbf{r}} + \frac{\mathbf{F}}{m} \cdot \frac{\partial n}{\partial \mathbf{v}} = \left(\frac{\partial n}{\partial t}\right)_c + \left(\frac{\partial n}{\partial t}\right)_s, \quad (3)$$

where $n(\mathbf{r}, \mathbf{p}, t) = N \int d\mathbf{q}_j d\mathbf{p}_j \delta(\mathbf{r} - \mathbf{q}_j(t)) \delta(\mathbf{p} - \mathbf{p}_j(t)) f(\mathbf{q}_j, \mathbf{p}_j, t)$ is the ensemble average of the density operator of particles in the phase space (i.e., it is a density distribution function of the particles in the phase space), $\left(\frac{\partial n}{\partial t}\right)_c$ is the change rate generated by colliding between the particles, and $\left(\frac{\partial n}{\partial t}\right)_s$ is the change rate generated by a source or sink (i.e, a negative source).

2.2 Ensemble and probability

An ensemble is a collection of many configurations of systems that all describe the same thermodynamic state; i.e., there are the same particles, chemical property, and macrostate in every system of the phase space. A microcanonical ensemble represents a collection of configurations of isolated systems that have reached a thermal equilibrium. A system will be isolated from its environment if it does not exchange either particles or energy with its surroundings. The volume, internal energy, and number of particles in such a system are constant, and so are its configurations, which are part of the same microcanonical ensemble. Namely, a microcanonical ensemble is an isolated system with a fixed energy E and number of particles N.

A microcanonical ensemble describes the statistical property of an isolated system. Therefore, the energy of this system is between E and $E + \Delta E$ and lies in a thin shell of volume ΔW in the phase space, and the total number of configurations is $\Delta \Omega$. For a system belonging to a microcanonical ensemble, the probability for it in a configuration (q_i, p_i) with the energy $E'(q_i, p_i)$ is

$$p(q_i, p_i) = \begin{cases} 1/\Delta\Omega(E) & \text{if } E \leq E'(q_i, p_i) \leq E + \Delta E, \\ 0 & \text{otherwise,} \end{cases} \tag{4}$$

where $\Delta\Omega(E)$ is the number of the configurations that have energy in the interval $E \leq E'(q_i, p_i) \leq E + \Delta E$. Thus all configurations (q_i, p_i) falling into this energy interval are occupied with the equal probability $p(q_i, p_i)$.

2.3 The law of entropy increasing and principle of energy minimizing

We consider an isolated system composed of two subsystems that are contacted with each other so that they can exchange energy and particles. As a result of the contact between the two subsystems, the entropy is an additive quantity, so

$$S = S_1 + S_2,$$

where S_1 and S_2 are the entropies of the two subsystems. We assume that entropy is a function of the number of the accessible configurations in a microcanonical ensemble:

$$S = f(\Delta\Omega).$$

Since the two subsystems are independent of each other, the total number of the configurations is

$$\Delta\Omega = \Delta\Omega_1 \Delta\Omega_2.$$

Thus

$$S_1 + S_2 = f(\Delta\Omega_1 \Delta\Omega_2),$$

which leads to the conclusion that S and $\Delta\Omega$ are related by

$$S = k_B \ln(\Delta\Omega). \tag{5}$$

This relation was discovered by Boltzmann and is thus called the Boltzmann equation. The proportionality constant k_B is the Boltzmann constant. According to the entropy equilibrium equation and Boltzman H-theorem, we know that the entropy function is an increasing function of time in an isolated system; i.e., $dS/dt \geq 0$. Therefore, the entropy is irreversible in the thermo-insulated system, which is the law of the entropy increasing. Now, we introduce the Helmholtz free energy equation

$$F(T, V, N) = -k_B T \ln(Z(T, V, N)) = \langle E(S, V, N) \rangle - TS, \qquad (6)$$

where $\langle E \rangle$, Z, T, and S are the average internal energy, partition function, temperature, and entropy of the system, respectively. From (5) and (6) we see that the entropy of an isolated system will never decrease; if the Helmholtz free energy is $F = E - TS$, then $\Delta F = \Delta E - T \Delta S$.

If a system begins with a non-equilibrium state or some constraints within it are removed, its Helmholtz free energy will decrease and F will be a minimum. Therefore, the decreasing of free energy and increasing of entropy are useful for the particle system to change its state from the non-equilibrium to the equilibrium state. In other words, the equilibrium state of the system is the result of the competition between the decreasing of energy and the increasing of entropy. This is the principle of energy decreasing, and the relationship of unity and opposites between the entropy increasing and the free energy decreasing.

3 Evolving Process of PDEA

As noted, PDEA is based on a particle transportation theory. We can associate the N particles $n_1, n_2, \ldots, n_N$ of a system from the phase space $\Gamma_N = \{q_r, p_r : r = 1, 2, \ldots, 3N\}$ with the N individuals $p_1, p_2, \ldots, p_N$ of a population in EA, and the closed equilibrium system of the particles in the phase space with the evolving pool of the population. From (1) we see that the Hamiltonian equation of the particles represents the kinetic direction of the particles according to time and momentum, respectively, and we can create a crossing method of the population in PDEA that simulates the Hamiltonian particle equation based on the Hamiltonian direction to search for multiple points reaching the optimal solution of an optimization problem and to change the past premature phenomenon of the traditional EA, so that it is easier to find the solution of the problem. The selection of the traditional EA is based on Darwin's evolving theory of the survival of the fittest [2]; namely, the individuals with better fitness values may have more chances to cross and mutate to generate new individuals, and the individuals with worse fitness values do not have a chance at all since they are removed at the first time. In this situation it may be difficult to find all the points reaching the optimal solution of the problem (the traditional EA cannot even find any point or can only find local solutions that lead to the premature phenomenon because all the points reaching the optimal solution may sometimes happen to fall into the area where the fitness values are worse at the first selection). However, the

PDEA is based on the transportation theory and can resolve all the above problems because this algorithm can always drive all the individuals (particles) to simulate particles' activities in the phase space to cross and mutate just like the transportation equation (3) that is affected by $(\frac{\partial n}{\partial t})_c$ and $(\frac{\partial n}{\partial t})_s$ and can reach the state of equilibrium of a system in the phase space.

An iterative stopping criterion and convergent speed are very important to every algorithm. PDEA combines the traditional EA with the law of entropy increasing and the principle of energy minimization to determine a more adaptable iterative stopping criterion, so one can generate an iterative stopping criterion of PDEA which simulates the Boltzmann equation (5) of entropy and the Helmholtz free energy equation (6) to speed up convergence of this algorithm and improve its computing performance. All these give the brief idea of PDEA, and its concrete stopping criterion through simulating the law of entropy increasing and the principle of energy minimization. PDEA will be defined in detail in the next section.

4 Description of PDEA

4.1 Basic problem for EA

EA is a branch of EC, and it searches for the optimal solution of a problem using biological natural selection and natural genetic mechanism. It is a new method to solve the problem through combining natural genetics with computer science.

We consider an optimization problem of the following type [4]:

$$\min\{f(\mathbf{x}) : \mathbf{x} \in \mathbf{X}\},$$

subject to the constraint

$$g_i(\mathbf{x}) = 0, \quad i = 1, 2, \ldots, n,$$

where $f(\mathbf{x})$ is an objective function in the set $\mathbf{X}$ that is a feasible set of solutions. $\mathbf{X}$ can be a finite set (e.g., a combination optimization problem), or a subset of a real space $\Re^n$ (e.g., a continuous optimization problem).

4.2 Fitness function and iterative stopping criterion of PDEA

We can define the stopping criterion of PDEA according to its idea as follows: Given the population size N, we call the N individuals $\mathbf{x}_1, \mathbf{x}_2, \ldots, \mathbf{x}_N$ as N particles, and the number of iterations of continuous evolving is defined as the dynamic time t of particles in the phase space. We convert the optimization problem $\min\limits_{\mathbf{x}\in\mathbf{X}} f(\mathbf{x})$ into the dynamic evolving optimization problem $\min\limits_{\mathbf{x}\in\mathbf{X}, t>0} f(t, \mathbf{x})$. Then we have the following two definitions:

Definition 4.2.1. Assume that $f(\mathbf{x})$ is defined on a set $\mathbf{X}$ of real numbers. Then

$$\nabla p(t, \mathbf{x}_i) = \nabla f(t, \mathbf{x}_i) - \nabla f(t - 1, \mathbf{x}_i) \qquad (7)$$

is called the transportation equation of N particles $\mathbf{x}_1, \mathbf{x}_2, \ldots, \mathbf{x}_N$ of PDEA, where $\mathbf{x}_i \in \mathbf{X}, i = 1, 2, \ldots, n$ and ∇ denotes the gradient operator in $\mathbf{x}$.

Definition 4.2.2. Function $\alpha(t, \mathbf{x}_i)$ is termed the activity of particles $\mathbf{x}_1, \mathbf{x}_2, \ldots, \mathbf{x}_N$ at time t provided that if $\mathbf{x}_i$ is selected to evolve at time t, then $\alpha(t, \mathbf{x}_i) = \alpha(t - 1, \mathbf{x}_i) + 1$; otherwise, $\alpha(t, \mathbf{x}_i) = \alpha(t - 1, \mathbf{x}_i)$.

According to the Boltzmann equation (5) of entropy, the Helmholtz free energy equation (6), and definitions 4.2.1 and 4.2.2, we can define the fitness function and the iterative stopping criterion of PDEA as follows:

$$\begin{cases} \text{Fitness function}: \ \text{select}(t, \mathbf{x}_i) = (1 - \lambda) \sum_{k=0}^{t} \|\nabla p(t, \mathbf{x}_i)\| \\ \qquad\qquad\qquad\qquad +\lambda \ln(\alpha(t, \mathbf{x}_i) + 1), \qquad (8) \\ \text{Stopping criterion}: \ \sum_{i=1}^{N} \|\nabla p(t, \mathbf{x}_i)\| < \varepsilon, \end{cases}$$

where $\lambda \in [0, 1]$ is a weighted coefficient (called a simulating Boltzmann constant) depending on the significance of $\sum_{k=0}^{t} \|\nabla p(t, \mathbf{x}_i)\|$ and $\ln(\alpha(t, \mathbf{x}_i) + 1)$ in the right-hand side of (8); i.e., if $\sum_{k=0}^{t} \|\nabla p(t, \mathbf{x}_i)\|$ in the first term is more significant than $\ln(\alpha(t, \mathbf{x}_i) + 1)$ in the second, then $1 - \lambda \geq \lambda$; otherwise, $1 - \lambda \leq \lambda$.

Because the first term in the right-hand side of (8) is constructed based on the energy minimization principle and the second term is constructed based on the entropy increasing law, these two terms are like the Helmholtz free energy and the entropy of particles in the phase space, and always compete with each other when the system state changes from non-equilibrium to equilibrium spontaneously at the same temperature. The more the iteration numbers of EA are, the more intensely they compete with each other. We use this strategy to select the individuals to cross and mutate according to the fitness value of function $\text{select}(t, \mathbf{x}_i)$ in the order from small to large.

Therefore, from equation (8) we easily see that the individuals that are not selected in the previous generation have more probability to be selected to take part in the evolving operation in the next generation because the fitness values that are calculated by non-selected individuals in the previous generation are less than other individuals' fitness values that are calculated by the selected individuals in the previous generation. In this way, we guarantee that all the individuals in the population have a chance to take part in crossing and mutating all the time; this is the main factor that contributes to PDEA's finding all the points that have reached the optimal solutions of a problem. Thus the procedure of PDEA can be defined as in the next subsection.

4.3 The PDEA procedure

1. Initialize a dynamical system of particles in the phase space to generate N individuals $\Gamma_N = \{x_1, x_2, \ldots, x_N\}$ of an initial population at random; t:=0.
2. Calculate all the function values of the particles in Γ_N and set $\nabla p(t, x_i) = 0, \alpha(t, x_i) = 0, x_i \in \Gamma_N$; then calculate the fitness values of functions $select(t, x_i)$, which are in the order from small to large.
3. Save the best particle and its function value in the system Γ_N.
4. Begin to iterate: t:=t+1.
5. Select n particles $x'_i, i = 1, 2, \ldots, n$ on the forefront of $select(t - 1, x_i)$; if all the values of $select(t-1, x_i)$ are the same, select n particles at random.
6. Implement evolutionary operations on the n particles of the dynamical system, and generate n random numbers $\alpha_i \in [-1, 1], i = 1, 2, \ldots, n$, that satisfy $-1.5 \leq \sum_{i=1}^{n} \alpha_i \leq 1.5$ and $\widehat{x} = \sum_{i=1}^{n} \alpha_i x'_i \in X$; if the function value at the point $\widehat{x}$ is better than the worst function value at the point $\widetilde{x}$, then we replace the individual $\widetilde{x}$ by $\widehat{x}$; otherwise, repeat this evolving operation.
7. Save the best particle and its function value in the system Γ_N.
8. Renew all the values of $select(t, x_i)$ and re-sort in an ascending order.
9. Apply the stopping criterion; if $\sum_{i=1}^{N} \|\nabla p(t, x_i)\| < \varepsilon$, stop; otherwise, go to Step 4.

Compared with the traditional EA, the advantages of PDEA are that it uses the particle transportation theory to describe a dynamical algorithm, and the principle of free energy minimization and the law of entropy increasing to construct the stopping criterion to drive all the particles (individuals) of a dynamical system to participate in the evolving operation, and it avoids some particles' not having any chance to cross and mutate. In other words, this algorithm can avoid the premature phenomenon.

5 Numerical Experiments

In this section, three typical optimization problems that are difficult to solve using the traditional EA [4] will be experimented to test the performance of PDEA. In the first experiment, we use PDEA to solve the minimization problem of the function

$$\min_{x \in S} f(x_1, x_2) = \left(4 - 2.1x_1^2 + \frac{1}{3}x_1^4\right) x_1^2 + x_1 x_2 + (-4 + 4x_2^2)x_2^2, \quad (9)$$

where $-3 \leq x_1 \leq 3$ and $-2 \leq x_2 \leq 2$. This function is called a six-hump camel back function [5]. It is known that the global minimum value of this problem is -1.031628 at two different points $(-0.089842, 0.712656)$ and

(0.089842, −0.712656). Because there exist two points that reach the minimum value of the function, only the local optimal solutions can be solved using the traditional EA in general.

We set the population size $N = 80$, the weighted coefficient $\lambda = 0.7$, $\varepsilon = 10^{-11}$, and the maximal number of the iterative steps $T = 10000$, and we select four particles that are located in front of the fitness values of the function select$(t, \mathbf{x}_i)$ in the order from small to large to cross and mutate. Then we run 10 times consecutively; each time, we can get the optimal point that is given in Table 1, and the iteration steps are less than 10,000. We can see from Table 1 that the probability of the appearing of the two points reaching the global minimum value of the function is similar. This experiment shows that PDEA can find the global optimal solutions more easily and quickly, the convergent speed is much faster, and the results are more accurate than the traditional EA.

Table 1. Six-hump camel back function

Min value f	x_1	x_2	$step$
-1.031628	-0.0898500	0.712657	1324
-1.031628	-0.0898269	0.712667	7738
-1.031628	0.0898540	-0.712663	9172
-1.031628	0.0898422	-0.712666	2231
-1.031628	-0.0898359	0.712643	6846
-1.031628	0.0898150	-0.712631	2463
-1.031628	-0.0898390	0.712652	5191
-1.031628	-0.0898231	0.712628	5824
-1.031628	0.0897980	-0.712531	4585
-1.031628	0.0898509	-0.712671	3055

In the second experiment, we use PDEA to test the axis-parallel hyperellipsoid function [6]:

$$\min_{x \in S} f(x_1, x_2, x_3) = \sum_{i=1}^{3} i x_i^2, \tag{10}$$

where $-5.12 \leq x_i \leq 5.12, i = 1, 2, 3$. This function is similar to de Jong's function and is convex, continuous, and unimodal. From [6] we see that this function has the global minimum value 0 at the point $(0, 0, 0)$. Generally, this type of simple convex function is difficult to check the performance of EA, but [6] indicates that if one wants to test functions to exhibit the strength of EA, he/she does not use convex functions. However, many experiments have shown that we can use PDEA to test almost all the convex functions to exhibit the strength of an algorithm. Here we just set $\varepsilon = 10^{-3}$, and other parameters are unchanged. We execute the program consecutively for 10 times to get the results listed in Table 2. When the program reaches 200 iteration steps, we get the approximate global minimum value $|f(\tilde{\mathbf{x}})| \leq 10^{-6}$, and when it reaches the stopping criterion (about 10,000 iteration steps), we get the global minimum value of the function almost at the point $(0, 0, 0)$.

Table 2. Axis parallel hyper-ellipsoid function

Min value f	x_1	x_2	x_3	$step$
1.12756E-234	-6.68584E-118	-5.81156E-118	4.10999E-119	9306
6.55069E-235	-5.99085E-118	1.92814E-118	2.71914E-118	8947
2.67126E-234	6.31116E-118	1.04782E-117	-1.60322E-118	8316
8.27800E-236	-1.32622E-118	1.57727E-119	1.46849E-118	9968
6.80792E-236	-1.00731E-118	-1.54099E-118	-5.89904E-119	9997
1.61351E-236	7.58502E-119	-1.15062E-119	-5.80719E-119	9949
1.66717E-234	4.33493E-118	-1.53254E-118	-6.90960E-118	9974
1.63894E-235	-1.70187E-118	-1.83837E-118	1.49821E-118	9989
1.43078E-235	-9.47395E-119	1.53325E-118	-1.70378E-118	9999
1.77483E-236	-1.26561E-118	1.75461E-119	-1.92772E-119	9983

In the third experiment, we find the minimum value of Griewangk's function [8]:

$$\min_{x \in S} f(x_1, x_2) = \sum_{i=1}^{2} \frac{x_i^2}{4000} - \prod_{i=1}^{2} \cos\left(\frac{x_i}{\sqrt{i}}\right) + 1, \tag{11}$$

where $-600 \leq x_i \leq 600, i = 1, 2$. From [6], the global minimum value $f(\mathbf{x}) = 0$ occurs at $(0,0)$. This function, which has many widespread local minima, is similar to Rastrigin's function. If we use the traditional EA, we can only find the local minimum values, but not the global minimum value of the function. Here we use PDEA to find the global minimum value of this function. Set $\varepsilon = 10^{-11}$ and other parameters are the same as above. PDEA is executed 10 times consecutively to obtain the results in Table 3. Every time, we can reach the stopping criterion in iterations less than 880, and get the global minimum value zero. Therefore, it indicates that the convergent speed is very fast and the accuracy is very high.

In addition to the above experiments, we have also used PDEA to test many other functions [10, 9], which are difficult to find optimal values using the traditional EA, and we have obtained very satisfactory results. The above experiments and all the other experiments have all proved that PDEA can easily obtain the optimal solution of problems and get all the points that reach the optimal solution (e.g., the six-hump camel back function), it can find the optimal solution of convex functions (e.g., the axis parallel hyper-ellipsoid function and de Jong's function), and it can solve the optimal problems that include many widespread local minima (e.g., Griewangk's function). Moreover,

Table 3. Griewangk's function

Min value f	x_1	x_2	step
0	-3.23219E-09	-4.83134E-09	816
0	1.09074E-09	-3.99652E-09	741
0	1.06804E-09	-1.07859E-09	861
0	4.30993E-09	-2.35365E-09	823
0	6.90104E-10	-6.78510E-10	864
0	9.00902E-10	4.46025E-10	872
0	-1.92239E-09	7.04812E-09	826
0	-3.84783E-09	-4.83865E-09	848
0	-2.13062E-09	5.57251E-09	832
0	2.03903E-09	8.46366E-09	871

these experiments have confirmed the reliability and safety of the stopping criterion and the fitness function of PDEA.

6 Conclusions

Through a theoretical and experimental analysis of PDEA, we conclude that PDEA has obviously improved the performance of the traditional EA. Because PDEA is based on the particle transportation theory, the principle of energy minimization, and the law of entropy increasing in the phase space of particles, the crossing and mutating of individuals in population has simulated the kinetic particles in this phase space. PDEA allows worse particles to have chances to evolve, drives all the particles to cross and mutate, and reproduces new individuals of the next generation from the beginning to the end. In other words, PDEA can search for the global optimal solution in all the particles of a system and avoid the premature phenomenon of the algorithm; this is also the reason why PDEA can easily find the global minimum value of the function at the two points in the first experiment, of a convex function in the second experiment, and of a function with many widespread local minima in the third experiment. Therefore, PDEA has resolved the long-existing problems in the traditional EA.

References

1. Z. Huang, Transportation Theory, Chinese Science Press, 1986.
2. D. L. Lack, Darwin's Finches, Cambridge University Press, Cambridge, England, 1947.
3. K. Li, Y. Li, and Z. Wu, Recovering legal-codes of generated cyclic codes by evolutionary computations, *Computer Engineering and Applications* **40** (2004).

4. Y. Li, X. Zou, L. Kang, and Z. Michalewicz, A new dynamical evolution algorithm based on statistical mechanics, *Computer Science & Technology* **18** (2003).
5. Z. Michaelwicz, Genetic Algorithms+Data Structures=Evolution Programs, Springer-Verlag, Berlin, New York, 1996.
6. M. Mitchell, S. Forrest, and J. H. Holland, The royal road for genetic algorithms: Fitness landscapes and GA performance, in Proc. the first European Conference on Artificial Life, F. J. Varela and P. Bourgine (eds.), MIT Press, Cambridge, Massachusetts, 1992, pp.245–254.
7. Z. Pan, L. Kang, and Y. Chen, Evolutionary Computations, Tsinghua University Press, Beijing, China, 1998.
8. Z. Wu, L. Kang, and X. Zou, An elite subspace evolutionary algorithm for solving function optimization problems, *Computer Applications* **23** (2003), 13–15.
9. X. Zou, L. Kang and Y. Li, A aynamical evolutionary algorithm for constrained optimization problems, Proceedings of the IEEE Congress on Evolutionary Computations, the TEEE Press, vol. 1, 2002, pp. 890–895.
10. X. Zou, Y. Li, L. Kang, and Z. Wu, An efficient dynamical evolutionary algorithm for global optimization, *International Journal of Computer and Mathematics* **10** (2003).

An Operator Splitting Method for Nonlinear Reactive Transport Equations and Its Implementation Based on DLL and COM

Jiangguo Liu[1] and Richard E. Ewing[1]

Institute for Scientific Computation, Texas A&M University, College Station, TX 77843-3404 jliu@isc.tamu.edu,richard-ewing@tamu.edu

In this paper, we propose an operator splitting method for convection-dominated transport equations with nonlinear reactions, which model groundwater contaminant biodegradation and many other interesting applications. The proposed method can be efficiently implemented by applying software integration techniques such as dynamical link library (DLL) or component object model (COM). Numerical results are also included to demonstrate the performance of the method.

1 Introduction

The growing concern of groundwater contamination demands accurate description and thorough understanding of contaminant transport in porous media. The cleanup of contaminated groundwater can be effective only after complete knowledge is gained about the mechanisms in transport and chemical and biological reactions. In particular, understanding of the microbial degradation of organic contaminants in groundwater provides valuable guidelines for *in-situ* remediation strategies. Injection of oxygen or various nutrients into the subsurface might be encouraged or enhanced to stimulate the microbial activities.

The mathematical models describing contaminant transport with biodegradation are time-dependent convection-diffusion-reaction equations. The nonlinear reaction terms may describe the reactions among all the species and may be coupled to the growth equations for the bacterial populations. These models are often complicated due to the heterogeneous geological structure in porous media, through which the contaminants are transported. The nonlinearity in the reaction terms also needs to be treated carefully. Therefore, solutions to these equations present serious challenges to numerical methods.

In [WEC1995], Eulerian-Lagrangian localized adjoint methods (ELLAM) based on different operator splittings of the adjoint equation are developed to solve linear transport equations. Then specific linearization techniques are discussed and are combined with the ELLAM schemes to solve the nonlinear, multi-species transport equations. Other operator splitting methods for convection-diffusion-reaction equations with other applications, e.g, air pollution modeling, can be found in [KR2000, KR1997, LV1999] and the references therein.

In this paper, we propose an operator splitting method for the governing equation to handle the nonlinear reactions. Roughly speaking, we split the transport equation into a convection-reaction part and a diffusion part. The diffusion problem is more or less a standard parabolic problem, whereas the convection-reaction problem becomes a nonlinear ordinary differential equation (ODE) along characteristics. We shall focus on the implementation issues of the splitting method. We shall treat a finite difference/element/volume method-based solver for conventional (linear) problems as a stand-alone object, which can be integrated with our own codes in a flexible manner. To be specific, this integration is done through dynamical link library (DLL) and component object model (COM). The CPU intensive numerics will be implemented with low-level programming languages such as C/C++/FORTRAN and code design will adhere to the object-oriented programming paradigm.

Recently, object-oriented programming (OOP) is becoming a popular practice in designing numerical (computing) software. Here we give [LM2001, BM2002] as just two interesting examples among many other excellent papers in this aspect.

2 Transport Equations with Nonlinear Reactions

A general single-species reactive transport equation can be written as

$$
\begin{cases}
u_t + \nabla \cdot (\mathbf{v}u - \mathbf{D}\nabla u) = R(u) + f(\mathbf{x},t), & \mathbf{x} \in \Omega,\ t \in (0,T) \\
u|_{\partial\Omega} = 0 \\
u(\mathbf{x},0) = u_0(\mathbf{x})
\end{cases}
\tag{1}
$$

where $[0,T]$ is a time period, Ω a domain in $\mathbf{R}^d (d = 1, 2, \text{or}, 3)$, $u(\mathbf{x},t)$ is the unknown (population) density or concentration of the species, $\mathbf{v}$ is a divergence-free velocity field, $\mathbf{D}$ is a diffusion tensor (it is assumed that $|\mathbf{D}| << |\mathbf{v}|$), $R(u)$ is a nonlinear reaction term, and $u_0(\mathbf{x})$ is an initial condition.

Some very interesting cases for the reactions are

- Radioactive decay with $R(u) = -au$;
- Logistic model with $R(u) = au - bu^2$;
- Biodegradation model with $R(u) = au/(u + b)$.

It should be pointed out that the third case is just a simplified model. More realistic models and detailed discussions on biodegradation can be found in [CKH1989].

3 An Operator Splitting Method for Nonlinear Reactive Transport Equations

Splitting methods come in two basic flavors: dimensionality reduction and operator decomposition. Both have been adopted in solving partial differential equations. A brief discussion on applying operator splitting techniques to solve transport equations in porous media can be found in [E2002]. In [LV1999], there is a very interesting theoretical analysis of operator splitting for convection-diffusion-reaction equations from the viewpoint of Lie algebra.

The transport equation (1) can be rewritten as

$$u_t = A(u) + B(u)$$

with

$$A(u) = -\mathbf{v} \cdot \nabla u + R(u),$$

$$B(u) = \nabla \cdot (\mathbf{D}\nabla u) + f(\mathbf{x}, t).$$

Let N be a positive integer, $\Delta t := T/N$, and $t_n = n\Delta t$ $(n = 0, 1, \cdots, N)$ be a uniform partition of the time period $[0, T]$. We split equation (1) into two in each small time period $[t_n, t_{n+1}]$:

$$u_t + \mathbf{v} \cdot \nabla u = R(u), \tag{2}$$

$$u_t = \nabla \cdot (\mathbf{D}\nabla u) + f(\mathbf{x}, t). \tag{3}$$

For any $\mathbf{x} \in \Omega$, the characteristic $\mathbf{y}(s; \mathbf{x}, t_{n+1})$ passing through $(\mathbf{x}, t_{n+1})$ is determined by

$$\begin{cases} \dfrac{d\mathbf{y}}{ds} = \mathbf{v}(\mathbf{y}, s), & s \in (t_n, t_{n+1}) \\ \mathbf{y}(t_{n+1}; \mathbf{x}, t_{n+1}) = \mathbf{x}. \end{cases} \tag{4}$$

Let $(\mathbf{x}^*, t_n)$ be the image of exact backtracking of $(\mathbf{x}, t_{n+1})$. If the characteristic hits the boundary, then we use $(\mathbf{x}^*, t_n^*)$, where $\mathbf{x}^* \in \partial\Omega, t_n \leq t_n^* \leq t_{n+1}$. Notice that exact tracking of characteristics is usually unavailable in practice and we have to resort to numerical means. All commonly used numerical methods for solving ODEs, e.g., Euler and Runge-Kutta methods, can be applied to problem (4).

Along characteristics, the convection-reaction equation becomes a nonlinear ODE

$$\begin{cases} \dfrac{du^{(1)}}{dt} = R(u^{(1)}), & t \in (t_n^*, t_{n+1}), \\ u^{(1)}(\mathbf{x}^*, t_n^*) = u^{(2)}(\mathbf{x}^*, t_n^*), \end{cases} \quad (5)$$

where $u^{(2)}$ is the solution for the parabolic problem and will be explained later. When $n = 0$ or $\mathbf{x}^* \in \partial\Omega$, we should replace $u^{(2)}(\mathbf{x}^*, t_n^*)$ by $u_0(\mathbf{x}^*)$ or the boundary condition.

Problem (5) can be solved numerically by, e.g., Euler and Runge-Kutta methods. This way the nonlinearity in the reaction term can be well resolved if $R(u)$ is Lipschitz with respect to u, which is true for the logistic and biodegradation models.

The other part is an initial boundary value problem for a typical parabolic equation

$$\begin{cases} u_t^{(2)} = \nabla \cdot (\mathbf{D}\nabla u^{(2)}) + f(\mathbf{x}, t), & \mathbf{x} \in \Omega, \ t \in (t_n, t_{n+1}), \\ u^{(2)}|_{\partial\Omega} = 0, \\ u^{(2)}(\mathbf{x}, t_n) = u^{(1)}(\mathbf{x}, t_{n+1}). \end{cases} \quad (6)$$

It is conventional and can be solved by finite difference, element, or volume methods.

Let h be the spatial mesh size in the numerical scheme (finite difference or element or volume) for solving the parabolic part (6), δt the temporal step size. Since $|\mathbf{D}| << 1$, the stability condition $|\mathbf{D}|\delta t/h^2 \leq 1/2$ are readily satisfied.

4 Techniques of Software Integration

4.1 DLL: Dynamic Link Library

A dynamic link library(DLL) is a collection of functions and data that are available for use by one or more applications running on a computer system. At run-time, the DLL is loaded into memory and made accessible to all applications. In other words, executable code modules in a DLL are loaded on demand and linked at run time, and then unloaded when they are no longer needed.

The modules calling a DLL could be an executable (EXE) or even another DLL. When a DLL is loaded, it is mapped into the address space of the calling process. So DLLs also help reduce memory overhead when several applications use the same functionality at the same time; because although each application gets its own copy of the data, they can share the code.

DLLs can define two kinds of functions: exported and internal. Exported functions can be called by other modules. Internal functions can only be called from within the DLL where they are defined. Although a DLL can export data, its data is usually only used by its own functions.

Moreover, DLLs provide a way to modularize applications so that functionality can be updated and reused more easily. This enables the library code to be updated automatically and be transparent to applications.

The DLL standard is now supported by many high level programming languages such as C++, Fortran, and Matlab.

4.2 COM: Component Object Model

Component Object Model (COM) is the Microsoft's binary standard for object interoperability and has been widely accepted for integration of external functionality into Microsoft applications. COM is a set of object-oriented technologies and tools, which allows software developers to integrate application-specific components from different vendors into their own applications. These assembled reusable components communicate via COM. By applying COM to build applications from preexisting components, developers enjoy benefits of good maintainability and adaptability.

COM allows clients to invoke services provided by COM-compliant components (COM objects). COM objects are implemented as either Dynamic Link Libraries (DLLs) or executables (EXEs). COM objects implemented as DLLs are called in-process servers, while those implemented as EXEs are called local servers. An in-process COM server exposes its functions to outside applications, whereas a local server does not.

There are three ways in which a client can access COM objects:

- In-process server: The client can link directly to a library containing the server. The client and server execute in the same process. Communication is accomplished through function calls;
- Local Object Proxy: The client can access a server running in a different process (but on the same machine) through an inter-process communication mechanism;
- Remote Object Proxy: The client can access a remote server running on another machine. The mechanism supporting access to remote servers is called DCOM (D means Distributed).

If the client and server are in the same process, sharing data between the two is simple. However, when the server process is separated from the client process, as in a local server or remote server, COM must format and bundle the data in order to share it. This process of preparing data is called marshalling. For more details about marshalling, readers are refereed to [MsCOM].

Again, the COM standard is now supported by many high level programming languages such as C++, Fortran, and even Matlab.

5 Implementation of the Operator Splitting Method

In this section, we discuss some issues related to implementation of the operator splitting method.

Mesh Generation

Both the convection-reaction part and the parabolic part in the operator splitting method need a spatial mesh, which can be generated via applying most of those good quality mesh generators. The mesh information will be saved as data files and then read into memory later. A mesh generator can be integrated with our own program via COM and mesh data input can be done through flat file transfer.

Splitting Alternately

In our implementation, we actually adopt an alternating approach. That is, we solve the convection-reaction problem first, then the parabolic problem. For the following time step, we switch the order: solving the parabolic part first and then the convection-reaction part. Our numerical experiments indicate that this alternating approach has a better control over the splitting error.

Tracking Characteristics

Characteristic tracking is an important part of this operator splitting method. Usually, approximate characteristics instead of exact characteristics are used in numerical schemes, although the latter will be utilized whenever they are available. An approximate characteristic is understood as a chain of line segments. These line segments are adaptively formulated according to the magnitude of the velocity field. Small time steps are used where velocity magnitude is large, whereas relatively large time steps can be used when velocity is small. The number of points on each (approximate) characteristic varies. In addition, both forward and backward tracking of characteristics will be performed in practice. Therefore, doubly linked list seems to be an efficient data structure for approximate characteristics. It is this part (adaptive tracking of characteristics) where we have to develop our own code.

Solving Nonlinear ODEs along Characteristics

Now the nonlinear ODE (5) along characteristics are actually solved along approximate characteristics. Euler and Runge-Kutta methods can be applied. Note that there might be different time steps on different characteristics.

Solving Parabolic Problem via FEM

Numerically solving an initial boundary value problem to a linear parabolic equation via the finite difference/element/volume method is now more or less a conventional task. There are plenty of commercial or free softwares for doing just that. As we know, libMesh developed at The University of Texas at Austin [libMesh] and OFELI (Object Finite Element Library) developed in France [OFELI] provide free C++ source code and are easy to use. The PDE toolbox in Matlab is a basic finite element package and does not require users to have C++ programming experience.

DLL Inside and COM Outside

The operator-splitting method is actually a time-stepping procedure. Before the time-stepping loop, a spatial mesh is generated via a mesh generator. Within each time step, a convection-reaction problem and a parabolic problem are solved. After the loop, numerical solutions can be visualized by calling graphics utilities like the Visualization ToolKit (VTK). Then another spatial mesh might be needed to re-run the whole process. Taking the overhead in calling COM objects into consideration, we implement the two functionality modules, namely, solving a convection-reaction problem and solving a parabolic problem, through DLL. The mesh generator and visualization modules outside the time-stepping loop are implemented via COM.

6 Numerical Experiments

6.1 Example 1: Linear Reaction

To examine our method, we first consider a 2-dimensional problem with a linear reaction, to which we can find the exact solution so that we can compare the numerical and exact solutions. In particular, we have a rotating velocity $\mathbf{v} = (-4y, 4x)$, a constant scalar diffusion $D > 0$, a linear reaction $R(u) = Ku$ with K being a constant, and a null source/sink, i.e., $f \equiv 0$. Assume the substance is initially normally distributed, i.e., the initial condition is specified as a Gaussian hill

$$u_0(x, y) = \exp\left(-\frac{(x - x_c)^2 + (y - y_c)^2}{2\sigma^2}\right),$$

where (x_c, y_c) and $\sigma > 0$ are the center and standard deviation, respectively. Then the exact solution is given by

$$u(x, y, t) = \frac{2\sigma^2}{2\sigma^2 + 4Dt} \exp\left(Kt - \frac{(x^* - x_c)^2 + (y^* - y_c)^2}{2\sigma^2 + 4Dt}\right),$$

where $(x^*, y^*, 0)$ is the backtracking foot of the characteristic from (x, y, t), that is,

$$\begin{cases} x^* = (\cos 4t)x + (\sin 4t)y, \\ y^* = -(\sin 4t)x + (\cos 4t)y. \end{cases}$$

For simplicity, we use a uniform triangular mesh. The 2nd order Runge-Kutta (or Heun) method is used for characteristic tracking even though exact tracking is available. The finite element solver for the parabolic part (as a DLL) is derived from the source code in OFELI.

For numerical runs, we choose $T = \pi/2$, $\Omega = [-1, 1] \times [-1, 1]$, $D = 10^{-4}$, $K = 0.1$, $(x_c, y_c) = (-0.5, -0.5)$, and $\sigma^2 = 0.01$. For the parabolic solver, we use 20 micro steps within each global time step $[t_n, t_{n+1}]$. Accordingly, we set the maximal number of time steps in characteristic tracking to 20 also. Table 1 lists some results for the numerical solution at the final time. We still obtain very good numerical solutions, even though large global time steps are used.

Table 1. Numerical results of example 1 with $\Delta t = \pi/8$

Mesh size h	L^∞-error	L^1-error	L^2-error
1/20	1.266×10^{-2}	1.247×10^{-4}	3.138×10^{-4}
1/40	1.031×10^{-2}	5.061×10^{-5}	2.085×10^{-4}
1/50	9.984×10^{-3}	4.153×10^{-5}	1.923×10^{-4}
1/60	9.796×10^{-3}	3.613×10^{-5}	1.825×10^{-4}

6.2 Example 2: Nonlinear Reaction

The second example is a simplified model for single-species biodegradation: $R(u) = au/(u + b)$. We consider a 2-dimensional problem with a constant velocity filed (V_1, V_2), a scalar diffusion $D > 0$, and no source/sink. The initial condition is a normal distribution (Gaussian hill). Again we use the parabolic solver (DLL) compiled from OFELI.

Table 2. Numerical results of example 2

Δt	h	U_{min}	U_{max}	Δt	h	U_{min}	U_{max}
0.25	1/20	0.0	1.5159	0.125	1/40	0.0	1.5251
0.25	1/40	0.0	1.5176	0.10	1/20	0.0	1.5248
0.25	1/60	0.0	1.5179	0.10	1/50	0.0	1.5268

For numerical runs, we choose $T = 1$, $\Omega = [-1, 1] \times [-1, 1]$, $(V_1, V_2) = (1, 1)$, $D = 10^{-4}$, $a = b = 1$, $(x_c, y_c) = (-0.5, -0.5)$, and $\sigma^2 = 0.01$. Table 2 lists some results of the numerical solution at the final time.

No exact solution is known for this problem. But from Table 2, we can observe that the operator splitting method is stable, and keeps positivity of the solution.

7 Concluding Remarks

In this paper, we propose an operator splitting method for transport equations with nonlinear reactions. In the implementation of the proposed method, we incorporate some existing commercial and free software components into our own program. These components include mesh generators, finite element packages, and graphics utilities. By integrating functionalities of existing software components, application developers do not have to create every thing from scratches. So less code has to be written and the time period of software development is significantly shortened. Of course, this is based on the re-usability of software components, which requires all developers to abide software development standards. Moreover, the case study presented here shows that we are moving towards to reuse of binary objects (DLL or COM), instead of merely source code. We also want to point out that customization of existing software components to some extent might still be necessary.

This paper focuses on strategies for efficient implementation of the operator splitting method. Numerical results indicate that the method works very well. Error analysis on the method will be presented in our future work. In addition, extension of the proposed operator splitting method to coupled systems is also under our investigation.

References

[BM2002]   Bertolazzi, E., Manzini, G.: P2MESH: Generic objected-oriented interface between 2-d unstructured meshes and FEM/FVM-based PDE solvers. ACM Transactions on Mathematical Softwares, **28**, 101–131 (2002)

[CKH1989] Celia, M.A., Kindred, J.S., Herrera, I.: Contaminant transport and biodegradation 1. a numerical model for reactive transport in porous media. Water Resources Research, **25**, 1141–1148 (1989)

[DER1995] Dahle, H.K., Ewing, R.E., Russell, T.F.: Eulerian-Lagrangian localized adjoint methods for a nonlinear advection-diffusion equation. Comput. Meth. Appl. Mech. Eng., **122**, 223–250 (1995)

[E2002]   Ewing, R.E.: Upscaling of biological processes and multiphase flow in porous media. In: Fluid Flow and Transport in Porous Media: Mathematical and Numerical Treatment, Contemp. Math. **295**, 195–215 (2002)

[E1990]   Ewing, R.E.: Operator splitting and Eulerian-Lagrangian localized adjoint methods for multiphase flow. In: The mathematics of Finite Elements and Applications, VII (Uxbridge). Academic Press, London (1991)

[KR2000]   Karlsen, K.H., Risebro, N.H.: Corrected operator splitting for nonlinear parabolic equations. SIAM J. Numer. Anal., **37**, 980–1003 (2000)

[KR1997]   Karlsen, K.H., Risebro, N.H.: An operator splitting method for nonlinear convection-diffusion equations. Numer. Math., **77**, 365–382 (1997)

[LM2001]   Langtangen, H.P., Munthe, O.: Solving systems of partial differential equations using object-oriented programming techniques with coupled heat and fluid flow as example. ACM Transactions on Mathematical Softwares, **27**, 1–26 (2001)

[LV1999]   Lanser, D., Verwer, J.G.: Analysis of operator splitting for advection-diffusion-reaction problems from air pollution modelling. J. Comput. Appl. Math., **111**, 201–216 (1999)

[WEC1995] Wang, H., Ewing, R.E., Celia, M.A.: Eulerian-Lagrangian localized adjoint methods for reactive transport with biodegradation. Numer. Meth. Partial Diff. Eqn., **11** , 229–254 (1995)

[MsCOM]   Website: Component Object Model: http://www.microsoft.com/com/

[libMesh]   Website: libMesh: http://libmesh.sourceforge.net/

[OFELI]   Website: OFELI: http://ofeli.sourceforge.net/

[VTK]   Website: VTK: http://www.vtk.org/

Design & Implementation of the Parallel-distributed Neural Network Ensemble

Yue Liu, Yuan Li, Bofeng Zhang, and Gengfeng Wu

School of Computer Engineering & Science, Shanghai University, Yanchang Road 149, Shanghai, 200072, China yliu@staff.shu.edu.cn, yuanli415@tom.com,{bfzhang,gfwu}@staff.shu.edu.cn

Abstract Neural network ensemble is a recently developed technology, which trains a few of neural networks and then combines their prediction results. It significantly improves the generalization ability of neural network system and relieves the trial-by-error process of tuning architectures. However, it is time-consuming. In order to overcome the disadvantage, a parallel-distributed neural network ensemble named PDNNE is proposed in this paper. The design and implementation of the PDNNE are presented through discussing the main issues such as partitioning, communication, and the component neural network. The experiments show both the generalization ability and time efficiency are significantly improved.

Keywords Neural Network Ensemble, RBF neural network, Parallel Computing, Distributed Computing

1 Introduction

The artificial neural network ensemble, proposed originally by Hansen and Salamon in 1990 [HS90], is a collection of a number of neural networks trained for the same task [SK96]. It has the ability to relieve the trial-by-error process of tuning architectures and significantly improve the performance of a system where a single artificial neural network is used. Since it can be used easily, neural network ensemble has been successfully applied in many areas such as face recognition [GW96], optical character recognition [Mao98], and medical diagnosis [ZJY02]. Bagging and Boosting are two prevailing ensemble methods. They follow a two-stage design process [Sha96]: (1) obtain the component neural networks; (2) combine the component predictions. Bagging was proposed by Breiman [Bre96] based on Bootstrap sampling [ET93]. It generates several training sets from the original training set, trains a component neural network on each of those training sets, and aggregates the results. Boosting was introduced by Schapire [Sch90], which generates a series of component neural networks sequentially. The training set of new component neural network is determined by the performance of former ones. But Bagging and Boosting need a lot of time to train many component neural networks and many storage spaces to store them. Parallel and distributed computing are introduced to overcome this disadvantage.

Bagging can easily be broken into discrete pieces of work that can be solved simultaneously. Boosting is essentially sequential and its parallelization requires granularity. As far as we know, parallel Boosting and Bagging computation has been done for OC1 (Oblique Classifier 1) decision tree classifiers [YS01] by using the BSP (Bulk Synchronous Parallel) library as the parallel programming environment. Parallel computation of kernel density estimates classifiers and their ensembles [LA03] have developed by using MPI (Message Passing Interface) library. In this paper, a parallel-distributed computation of Radial Basis Function neural network (RBFNN) ensemble, named PDNNE is proposed. The PDNNE builds an ensemble of n RBFNNs that are trained independently on m processors, and then combines their prediction results in the main process. In the PDNNE, bootstrap datasets are also generated parallelly in slave processes to avoid too much data communication and enhance the diversity among component RBFNNs. The PDNNE is realized by using Master-Slave model in MPICH [DL01], a portable MPI implementation and tested in several UCI datasets [BKM98]. The results show that both the generalization ability and the time efficiency are significantly improved.

The rest of this paper is organized as follows. In Section 2, the design and implementation of the PDNNE are introduced through discussing the main issues such as partitioning, communication, and the component RBFNN. In section 3, experiments are presented. Finally, conclusions are drawn and several issues for future works are indicated.

2 The Parallel-distributed Neural Network Ensemble

Master-Slave model is employed to implement the PDNNE. The number of tasks in each node should be specified before the PDNNE program is invoked. Then the main process and slave processes cooperate to complete the prediction task as the flow shown in Fig.1.

(1) The pre-processing of the initial dataset centralizes in the main process. The data are distributed to slave nodes after being divided to a training dataset and a testing dataset;

(2) Each slave process corresponds to one task, generates one bootstrap dataset, and trains one component RBFNN in parallel manner. The trained component RBFNN then is used to predict independently;

(3) Slave processes send the prediction results and corresponding messages to the main process;

(4) The main process collects the messages, and then combines the results using major voting method or simply averaging method.

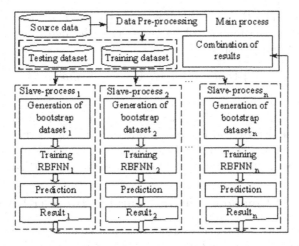

Fig. 1. Flow chart of the parallel-distributed neural network ensemble

Below we will discuss the main issues on the design and implementation of the PDNNE, including the assignment of the number of tasks in each slave nodes, the generation of the bootstrap datasets, the training of component RBFNN on each bootstrap dataset, and the collection and combination of the prediction results through the communication between the main process and slave processes.

2.1 Assignment of Tasks and Generation of Bootstrap Datasets

The diversity among the component neural networks is one of the most important factors related to the performance of the neural network ensemble. It can be guaranteed by bootstrap sampling algorithm to a great extent. Bootstrapping is based on random sampling with replacement. Therefore, taking a bootstrap dataset $BS_i = (X_{i1}, X_{i2}, \ldots, X_{in})$ (random selection with replacement) of the training set $TS = (X_1, X_2, \ldots, X_n)$, one can sometimes get less misleading training objects in the bootstrap training set. The indices $(i_1, i_2, \ldots, i_n)$ are decided by the random function. Obviously, the generation of bootstrap datasets is a parallel process, which can be distributed to different processes in different nodes.

We know the generalization of the neural network ensemble has close relation with the number of bootstrap dataset. However, the number of bootstrap datasets is not invariable for different problem domains. Generally, it should be specified before the program is invoked. We specify the total number of bootstrap datasets statically while the number of the processes in each node is adjustable according to the computational environment. The time required to train a neural network mainly depends on the number of instances in the training dataset, the number of classes, the number of features, and the training epoch. Because the size of each bootstrap dataset is same, the time cost to train component RBFNN is roughly same. Therefore, in order to guarantee the load balance, the number of processes in each node that has the same computational ability is specified by formula (1).

$$u_i = \begin{cases} \lfloor n \ div \ m \rfloor + 1, & 0 < i \le (n \ mod \ m) \\ \lfloor n \ div \ m \rfloor, & (n \ mod \ m) < i \le m \end{cases} \tag{1}$$

where n is the total number of bootstrap datasets; m is the number of available nodes. m can be determined before the program is invoked through checking the available nodes. The name and the number of processes of each available node must be written to a configuration file used in the PDNNE.

2.2 Training of Component RBF Neural Network (RBFNN)

In order to improve the generalization and time efficiency of the neural network ensemble, the component neural network is another important factor. In the PDNNE, RBFNN with three layers is employed as the component neural network. RBFNN can approximate any given nonlinear function with any accuracy. Its speed of learning is very quickly due to the avoidance of tedious redundancy computing of reverse direction propagation between input layer and hidden layer. The RBFNN employed in the PDNNE is similar to that developed by Moody and Darken [MD89]. But the centers are decided by using the nearest neighbor-clustering algorithm instead of K-Means algorithm. Each process trains a component RBFNN on one bootstrap dataset as the following steps.

Step 1 Initialize the parameters and construct the topology of the RBFNN.

- Specify the learning rate η, goal error and maximum epoch. Acquire the number of nodes in input layer m and that of the output layer q according to the samples.
- Acquire the centers $C=\{C_i, i=1,2,...,m\ \}$ and the width of each center σ_i using nearest neighbor clustering algorithm [HK00] on one bootstrap dataset $X=\{X_i|X_i \in R^N, i=1,2,...,m\ \}$. As the bootstrap dataset is different from each other, the number of the centers of each RBFNN is also different. Therefore, the diversity of component neural network is increased.

Step 2 For epoch = 1: maximum epoch, train the RBFNN neural network.
(1) Update the outputs of hidden nodes R(x)=[$R_1(x)$, $R_2(x)$, , $R_m(x)]^T$ according to formula (2).

$$R_i(x) = \exp[-\frac{\|x - c_i\|^2}{2\sigma_i^2}] \quad i = 1, 2, ..., m \tag{2}$$

(2) Calculate output y_j by formula (3).

$$y_i = \sum_{i=1}^{m} \omega_{ij} R_i(x) \quad j = 1, 2, ..., q \tag{3}$$

(3) Update weights ω_{ij} between hidden and output layer using the Delta rule shown as formula (4), where y'_i denotes the i-th dimension teaching value.

$$\omega_{ij}(l+1) = \omega_{ij}(l) + \eta[y'_i - y_i(l)]R(x)^T/R(x) \quad i = 1, 2, ..., q \qquad (4)$$

(4) Calculate error E according to formula (5), where $y_i^{k\prime}$ denotes the t-th dimension teaching value of k-th sample.

$$e = \sum_{k=1}^{n} E_k = \sum_{k=1}^{n} \sum_{t=1}^{q} (y_t^{k\prime} - y_t^k)^2/2n \qquad (5)$$

(5) If $E >$ goal error, goto (1) else goto step 3.

Step 3 If the error E never changes during the whole training process, delete it in order to avoid over-fitting; else store the trained RBFNN to establish the prediction committee.

2.3 Combination of Results and Communication

It will take lots of time to train the component neural networks. In the PDNNE, these time-consuming tasks will be executed concurrently and independently in slave processes without communication. However, the combination of the results in the main process will require data associated with the slave processes. Data must be transferred between each slave process and the main process. This kind of communication can be implemented based on message passing library, in which a collection of processes executes programs written in a standard sequential language augmented with calls to a library of functions for sending and receiving messages. At present, there are different message-passing systems, such as P4, PVM, Express, PARMACS and MPI. MPI is the most popular one. Therefore, MPICH, a portable MPI implementation is selected to realize the the PDNNE.

There are four kinds of messages between the main process and slave processes, including test result message(M_{98}), validation result message (M_{99}), time cost message (M_2), and shutdown message of slave process (M_1). The communication is described in Fig.2.

In each slave process, one component RBFNN is trained on one bootstrap dataset and then used to predict independently. After that, the message tags are specified to identify four different kinds of messages. At last, the messages are sent to the main process. The main process receives the messages, does different operations according to the message tags and then calculates the final result by taking the majority of the votes of the component RBFNNs when these are categorical; or the average of the outputs when these are numerical. We use MPI functions to implement the communication as follows.

```
MPI_Init(&argc,&argv); //Initiate MPI computation
MPI_Comm_rank(MPI_COMM_WORLD,&myrank);//Determine my
```

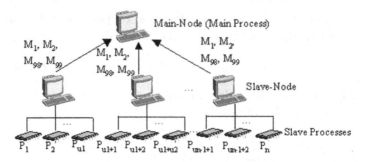

Fig. 2. The communication between the main process and slave processes

```
                                                    process id.
MPI_Comm_size(MPI_COMM_WORLD,&size);//Determine the
                                                number of processes
If (myrank==0){//Code for process 0, i.e. the main process
  ...//Generate a bootstrap dataset
  ...//Train & test a component RBF neural network
  while (numSlave>0){//Repeat until termination
    //Receive a message from slave process
    MPI_Recv(message,500,MPI_FLOAT,MPI_ANY_SOURCE,
             MPI_ANY_TAG,MPI_COMM_WORLD,&status);
    m=status.MPI_SOURCE;//Find process id of source process
    n=status.MPI_TAG;//Find message tag
    switch (n){
      case 1:...//Shutdown message
      case 2:...//Collect time cost message
      case 98:...//Collect test results
      case 99:...//Collect validation results
      }
    ...//Combine the results
} else{ //Code for slave processes
  ...//Generate a bootstrap dataset
  ...//Train & test a component RBF neural network
  ...//Bundle test result to a buffer, mess1
  //send corresponding messages to main process
  MPI_Send(mess1,numTestSample,MPI_FLOAT,
                  0,98,MPI_COMM_WORLD);
  MPI_Send(mess2, numValidateSample,MPI_FLOAT,
                  0,99,MPI_COMM_WORLD);
  MPI_Send(time,1,MPI_FLOAT,0,2,MPI_COMM_WORLD);
  MPI_Send(end,0,MPI_FLOAT,0,1,MPI_COMM_WORLD);
}
MPI_Finalize();//Terminate the computation
```

3 Experiments

Two data sets of experiments are performed to validate the performance of the PDNNE. The data sets are obtained from the UCI repository[BKM98]. The information of data sets removed missing values is tabulated in Table 1.

Table 1. Data sets summary

Data set	# Samples	# Variables	# Classes	# Test dataset
Wine	178	13	3	20
Breast cancer	683(699)	9	2	250

Each data set is divided into two sets randomly: a training set and a testing set. The random division of the data is repeated 10 times. All the reported results are the averages over the 10 iterations. The single neural network and component neural network of ensemble have the same architecture, i.e., the RBFNN just discussed in section 2.2 with the same learning rate =0.2, error e=0.001 and the maximum epoch= 1000.

Firstly, we train single neural network and get the misclassifications rate Es. For the purpose of comparison, we generate 30 bootstrap replicates to train neural network ensemble in serial and parallel-distributed (PDNNE) manners respectively. The misclassification rate for the serial neural network ensemble is denoted by Eb, and the PDNNE by Ep. As shown in Table 2, the prediction accuracy of the PDNNE Ep is similar to that of the serial neural network ensemble Eb and better than that of the single neural network Es. Fig.3 and Fig.4 give the comparisons on the prediction accuracy between the single neural network and the PDNNE for ten runs. When the single neural network cannot work well, the PDNNE will significantly improve the performance.

Table 2. Misclassification Rates List

Data set	Es (%)	Eb (%)	Ep (%)
Training dataset of Wine	36.0	0.23	0.23
Training dataset of Breast cancer	45.6	2.4	2.3
Testing dataset of Wine	37.1	0.02	0.02
Testing dataset of Breast cancer	48.4	2.6	2.6

Another major advantage of the PDNNE is that it requires much less computational time. Table 3 shows the time required for training the serial neural network ensemble Tb and that for the PDNNE Tp when simulated on a network connected 30 nodes with 100 Bt Ethernet. Each node has the same computational capability with 1.80G MHz processor and 128 Mb memory.

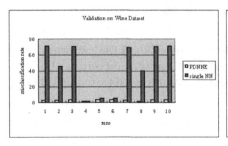
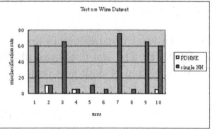

Fig. 3. Comparison of results on wine dataset

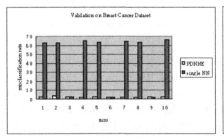
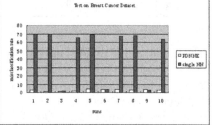

Fig. 4. Comparison of results on breast cancer dataset

The result indicates that the PDNNE is obviously better than the serial neural network ensemble in time efficiency.

Table 3. Time cost for the serial neural network ensemble and PDNNE

Data set	$Tb(s)$	$Tp(s)$
Wine	1234.2360	40.151
Breast cancer	5228.6020	201.3136

In order to give some ideas of the relationships among the misclassification rate, time cost, the number of bootstrap replicates and the number of available nodes, we ran the PDNNE on breast cancer dataset again. The number of bootstrap replicates varies from 1 to 50, and the number of available nodes ranges from 1 to 30.

The experiments of Fig.5 shows that the prediction accuracy is improved with the increase of the number of bootstrap replicates. We are getting most of the improvement using only 10 bootstrap replicates. More than 20 bootstrap replicates is labor lost. The time required to build the PDNNE depends on the number of bootstrap replicates, the number of available nodes and the communication cost. It can be clearly seen from Fig.6 that 30 nodes are perfect and 10 nodes are sufficient for training an ensemble of 30 component RBFNNs.

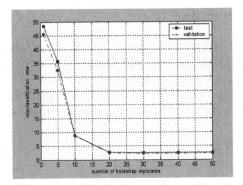

Fig. 5. The relationship between misclassification rate and number of bootstrap replicates

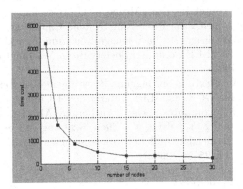

Fig. 6. The relationship between time cost and number of available nodes

It is evident that we should establish a satisfactory trade off among time cost, generalization capability, and the number of available nodes.

4 Conclusions and Future Works

Neural network ensemble minimizes the difficulty in tuning the architecture of the neural network and improves the performance of the neural network system. But it needs more computational and storage cost. In this paper, we design and implement a parallel-distributed neural network ensemble named PDNNE. The prediction accuracy and time efficiency of the neural network ensemble are increased through adopting parallel computing and distributed computing technologies. However, we only realize the parallelization of Bagging. The parallel distributed processing of component neural network is going to be investigated. The application to data mining in large dataset is also worth to be studied furthermore.

5 Acknowledgements

We are grateful to the National Natural Science Foundation of P. R. China (Grant No. 60203011), the Science and Technology Development Foundation of Shanghai Municipal Education Commission (Grant NO. 02AK13) and the foundation of fourth period of key subject of Shanghai Municipal Education Commission (Project No. B682) for their financial support to our work.

References

[HS90]   Hansen, L.K., Salamon, P.: Neural network ensembles. IEEE Trans Pattern Analysis and Machine Intelligence, **12**,993–1001 (1990)

[SK96]   Sollich, P., Krogh, A.: Learning with ensembles: when over-fitting can be useful. In: Touretzky, D.S., Mozer, M.C., Hasselmo, M.E. (ed) Advances in Neural Information Processing System. Cambridge, MA(1996)

[GW96]   Gutta, S.,Wechsler, H.: Face recognition using hybrid classifier systems. In: ICNN-96. IEEE Computer Society Press, Washington, DC,Los Alamitos, CA (1996)

[Mao98]   Mao, J.: A case study on Bagging, Boosting and basic ensembles of neural networks for OCR. In: IEEE International Conference on Neural Networks. IEEE Computer Society, Anchorage, AK (1998)

[ZJY02]   Zhou, Z., Jiang, Y., Yang, Y., Chen, S.: Lung cancer cell identification based on artificial neural network ensembles. Artificial Intelligence in Medicine, **24**, 25–36 (2002)

[Sha96]   Sharkey, A.J.C.: On combining artificial neural nets. Connection Science, **8**, 299–313 (1996)

[Bre96]   Breiman, L.: Bagging predictors. Machine Learning, **24**, 123–140 (1996)

[ET93]   Efron, B.,Tibshirani, R.: An Introduction to The Bootstrap. Chapman & Hall, New York (1993)

[Sch90]   Schapire, R.E.: The strength of weak learnability. Machine Learning, **5**, 197–227 (1990)

[YS01]   Yu, C., Skilliom, D.B.: Parallelizing Boosting and Bagging. Technical Report, Queen's University, Kingston, CA (2001)

[LA03]   Lozano, E.,Acuna, E.: Parallel computation of kernel density estimates classifiers and their ensembles. In: International Conference on Computer, Communication and Control Technologies (2003)

[DL01]   Du, Zhihui, Li, Sanli: High Performance Computing and Parallel Programming Technologies MPI Parallel Programming. Tsinghua University Press,Beijing (2001)

[BKM98]   Blake, C., Keogh, E., Merz, C.J.: UCI repository of machine learning databases. http://www.ics.uci.edu/ mlearn/MLRepository.html. Department of Information and Computer Science, University of California, Irvine, California (1998)

[MD89]   Moody, J., Darken, C.: Learning with localized receptive fields. In: the 1988 Connectionist Models Summer School. San Mateo, CA (1989)

[HK00]   Han, J., Kamber, M.: Data Mining: Concepts and Techniques. Morgan Kaufmann Publishers Inc. (2000)

Floating Point Matrix Multiplication on a Reconfigurable Computing System

C. Sajish[1], Yogindra Abhyankar[1], Shailesh Ghotgalkar[1], and K.A. Venkates[2]

[1] Hardware Technology Development Group, Centre for Development of Advanced Computing, Pune University Campus, Pune 411 007, India
[2] Alliance Business Academy, Bangalore 560 076, India yogindra@cdac.ernet.in

Summary. Matrix multiplication is one of the most fundamental and computationally intense operation that is used in a variety of scientific and engineering applications. There are many implementations of this normally $O(n^3)$ operation. These implementations differ mainly in terms of algorithms or the platforms. In this paper we present our experimentation of using a reconfigurable computing platform for calling such a routine. This routine use our own developed IEEE-754 compliant double precision hardware library elements implemented on our own developed FPGA based reconfigurable platform to provide acceleration.

Key words: Matrix Multiplication, FPGA, Reconfigurable Computing System.

1 Introduction

Matrix multiplication is a frequently used routine having a wide range of applications including seismic data processing, weather forecasting, structural mechanics, image processing etc. In the high performance computing area, this kernel appears as a part of the linear algebra package. Due to its popularity among the user community, this kernel is always in the focus of researchers for finding different implementations that will improvise on the normally $O(n^3)$ operation. These implementations differ mainly in terms of algorithms or the hardware platforms used. A few tricky algorithms like Strassen [Str69, JV98], has reduced the complexity from $O(n^3)$ to $O(n^{2.807})$ and a few modified versions including that by winograd [Win73] even reducing it further.

In recent years, there has been a considerable interest in high performance computing. For this, there are many emerging approaches including Reconfigurable Systems [CH02, AG02].

This paper presents an implementation of a double precision matrix multiplication on a Field Programmable Gate Array (FPGA) based reconfigurable computing platform. Our implementation uses a broadcast algorithm.

In our experimentation, an application that makes use of the matrix multiplication operation runs on a host machine with the add-on FPGA based reconfigurable computing card. During the code execution, whenever a matrix multiplication is required, a call to the double precision matrix multiplication kernel routine on the reconfigurable hardware is made.

The matrix multiplication kernel for this experimentation allowed us to test our own developed IEEE-754 compliant [IEE85] double precision hardware library elements such as adders and multipliers implemented on our own developed FPGA based reconfigurable platform to provide acceleration.

We begin this paper by first discussing the matrix multiplication algorithm. Section 3 outlines reconfigurable systems and the one developed by Centre for Development of Advanced Computing (C-DAC). In Section 4, we discuss the scheme used for our implementation of matrix multiplication on the FPGA based reconfigurable hardware. Finally, we summarize this paper and indicate some directions for future improvements.

2 Matrix Multiplication

The matrix multiplication C = A B of two matrices A and B is conformable, if the number of columns of A is equal to the number of rows of B. There are three basic algorithms for matrix multiplication. The first one is simple and most commonly used.

The ijth element of C is given by,

$$C_{ij} = \sum_{k=1}^{n} a_{ik} b_{kj} \tag{1}$$

For matrices of size n x n, this algorithm requires n^3 multiplications.

The second is a 'divide and conquer' algorithm, used mainly in distributed computing. Recursively, it divides the matrices into 4 sub-matrices, allowing the smaller parts to be calculated in parallel on multiple machines. This also has O(n^3) complexity.

The third algorithm is Strassen's, reducing a multiplication at each step and bringing down the asymptotic complexity to O(n$^{2.807}$).

While considering matrix multiplication on FPGAs, device architecture, resources and speed are important factors. Due to the recent advancements in the FPGAs, the actual implementation has become possible.

Multipliers and adders are the basic building blocks of the matrix multiplication kernel. Earlier experiments were focused on implementations with integer units, non-pipelined followed by pipelined designs [AB02]. Next class of devices made it possible to show the feasibility of having single precision floating point units [SGD00]. With the introduction of Virtex-II and Virtex-II Pro class of devices from Xilinx [Xil04] and other vendors, it has become possible to realize the double precision floating-point units [ZP04, UH04] and pack multiple MAC units in a single device.

The long awaited double precision units in FPGAs are instrumental in bringing back the attention of scientific and engineering applications community toward reconfigurable or adaptive computing.

For the present implementation of matrix multiplication, we have chosen the first algorithm out of the three algorithms mentioned in the beginning of this section. The application multiplies a 1024 X 1024 matrix A with another 1024 X 1024 matrix B. The inputs are 64-bit wide and are in the standard IEEE-754 double precision floating-point format. The Matrix multiplication operation is performed by broadcasting rows of A and multiplying the corresponding column elements of B. Fully pipelined double precision floating point units i.e. floating-point adder and multipliers designed by us are used for the matrix multiplication.

The work by other authors as sited above, generally excludes the use of denormal numbers that are useful in many scientific applications. Our floating-point units support these numbers.

3 Reconfigurable Computing System (RCS)

The reconfigurable or adaptive computing systems are machines that can modify their own hardware at runtime under software control to meet the needs of an application.

RCS blurs traditional boundaries between hardware and software. Traditionally hardware is fixed while the software is considered as a flexible component. RC explores hw/sw solutions where the underlying hardware is also flexible.

An enabling technology for RC is the FPGA, a VLSI chip whose hardware functionality is user-programmable. Putting FPGAs on a PC add-on card or on a motherboard allows FPGAs to serve as compute-intensive co-processors. Unlike a fixed floating-point co-processor, these FPGAs can be re-configured again, to perform a variety of operations.

The following subsection summarizes the reconfigurable hardware used in this experimentation.

3.1 C-DAC's RCS Card

We have designed this card, at the hardware technology development group of C-DAC. It is a FPGA based card that can be plugged to a host computer via the 64-bit, 66 MHz PCI bus. This card has two Xilinx FPGAs. Out of these, the larger device, XC2V3000 is mainly used as a compute engine, while the XC2V1000 device serves as a control engine. It holds the controller for PCI and other logic. When plugged into a PCI slot, the RCS card can be assumed to work as a co-processor to the host.

There is on-board 256MB of SDRAM and 4MB of ZBT RAM. The SDRAM is useful for storing the input, intermediate as well as the final results. The ZBT is suitable for applications where caching is required. The card supports DMA operations. The control and compute FPGAs can be configured through a variety of options. Input and output data to the card may be supplied from the host using the PCI interface. Figure 1 shows the RCS card's block diagram.

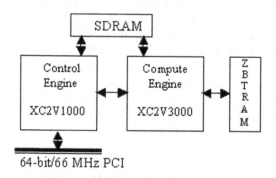

Fig. 1. RCS Card Block Diagram

3.2 System Software Interface, RCSL

The system software interface or the RCS Library (RCSL) for the card is implemented over Red Hat Linux 9.0 operating system. It provides all the basic functionalities in terms of the data transfer and card control irrespective of the intended application. RCSL consists of two basic components: RCS Kernel agent and RCS user agent. The RCS kernel agent is a device driver that performs resource and interrupts management. It typically provides services to allocate/free DMA buffers, enable/disable interrupt support etc. RCS user agent is a library that provides basic services to configure, setup/free resources, send input data, receive output data, initiate computation etc.

4 Matrix Multiplication on RCS

The complete matrix multiplication application that resides in the XC2V3000 compute engine consists of three main components: matrix multiplication block, a cover around the matrix multiplication block and the XC2V3000-XC2V1000 interface.

In the following subsections, we describe these three main components of the matrix multiplication application.

4.1 Matrix Multiplication Block

The matrix multiplication block performs the multiplication of two matrices. The operations performed are, read columns of B and store them in internal buffers, read individual row elements of A, do multiplication on column and row elements, accumulate the multiplier output and write back the results to the output buffers.

The matrix multiplication block is made of three sub blocks: MAC Units (floating point multiplier and adder), MAC Control and memory to store a group of columns at a time. A simplified diagram of the matrix multiplication block is shown in Figure 2. This block contains four MAC units that in turn

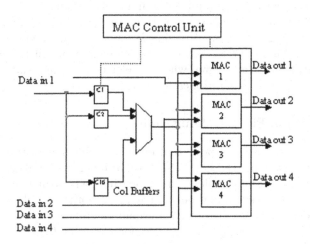

Fig. 2. Matrix Multiplication Block

consist of one floating-point adder and one floating-point multiplier. The row and column elements are supplied as the two inputs to the floating-point multiplier. The output of the floating-point multiplier is directly given to the floating-point adder as one of the inputs. The previous output of the floating-point adder is feedback as the second input to the adder. Thus the whole unit act as a multiply accumulate unit. Both the floating-point adder and multiplier are fully pipelined for achieving a high throughput. The pipeline stages of the floating-point adder and multiplier are kept same. Because of this, no buffers are required for storing the adder output.

Multiplication of the two 1K x 1K matrices is performed by dividing the matrices into smaller chunks. This is required as the XC2V3000 compute engine cannot store and process all the elements at a time. By considering the internal memory available with the compute engine and the individual column and row size of 1024 elements, we arrive at a chunk size of 16 columns and 4 rows that are suitable for processing at a time.

Initially the matrix multiplication block reads the first 16 columns of B and stores them in internal buffers. After this, the row elements of A are broadcasted one at a time. A particular row element is multiplied with the corresponding elements of all the 16-columns, i.e. the first element of a row with the first element of all the columns, then the second element of a row with the second element of all the columns and so on. After broadcasting all the 1024 elements of a row, 16 results of a row of the result matrix are available from the matrix multiplication block. This is the output from one MAC unit. A total of 64 results from 4 MAC units corresponding to 4 rows of the result matrix are available after the broadcast. Once all the rows are broadcasted, 16K results are obtained. Then the columns stored in the internal buffers are replaced by the next set of 16 columns of B. The whole operation is repeated and the process continues until all the columns are exhausted.

The MAC control unit is a state machine that controls the matrix multiplication block. It generates control signals for the internal buffers i.e. read-address, write-address, write-enable etc. It keeps a count on the number of elements multiplied, flags a signal when the elements of the result matrix are available from the matrix multiplication block.

Double Precision Floating Point Units

The double precision floating-point multiplier and adders constitute the MAC unit. Multiplier is a pipelined floating-point multiplier. Its inputs and outputs are in the standard 64-bit double precision floating point format. At the input, the numbers are checked to find out if they are denormal numbers. If any of the input is denormal, it is fed to the normalizer unit, otherwise by default, the mantissa part of input A, goes to the normalizer and the mantissa part of the input B goes through a series of registers. The multiplier takes same number of clocks for multiplication irrespective of the inputs. The normalized mantissa part of the two inputs is fed to the integer-pipelined multiplier that uses the Virtex-II's 18x18 multipliers. The integer multiplier produces full-length result. Since both of the inputs to the integer multiplier are normalized, the output can be made normalized by looking at the two MSBs of the integer multiplier output. The multiplier implements IEEE 'Round to Nearest' rounding mode.

Adder is a pipelined floating-point adder. The inputs and outputs are all in the standard IEEE-754 floating-point format. Initially the difference of the exponents of the two inputs is calculated. The mantissa of the number having smaller exponent is fed to a barrel-shifter. The shift is made by an amount equal to the modulus of the exponent difference. In order to adjust the pipeline, the mantissa part of the other number is fed through a series of registers. Resulting mantissas are fed into an integer adder.

The rounding is performed after normalizing the adder output. The Output exponent is determined according to the difference of the input exponents. The larger exponent is given to the normalizer. After normalizing the adder output, the corrected exponent is generated. Table 4, gives the area utilization of the double precision floating point units.

Table 1. Area utilization of floating point units

	Adder	Multiplier
Number of occupied slices	1,894	890
Frequency (MHz)	100	100
Number of MULT18X18s	0	16

4.2 Cover around Matrix Multiplication Block

The cover around the matrix multiplication block consists of control logic and other supporting logic that performs different data transactions with the SDRAM, generates required signals for the matrix multiplication block and keeps track on the number of rows of A and columns of B that have been multiplied. Figure 3 shows the cover around the matrix multiplication block. The elements of the input matrices are read from the SDRAM and stored into

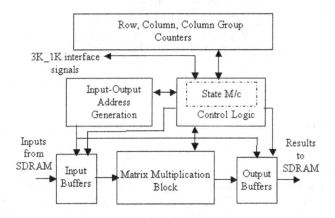

Fig. 3. Cover around Matrix Multiplication Block

the input buffer. The columns of the B matrix are read from the SDRAM, stored into the input buffer, and then transferred to the matrix multiplication block's internal buffer.

In case of matrix A, the matrix block for computation, directly uses the row elements stored in the input buffer. For the purpose of prefetching data from the SDRAM, the buffers are configured as 2K x 64-bit size. Because of the prefetching of data, the matrix computation is not stopped while the results are being written to the SDRAM. The matrix A and B are stored row and column-wise respectively in the SDRAM. The results are stored in the output buffers that are constructed using the device memory available in the FPGA.

The controller state machine generates control signals for data transaction with the SDRAM and also for the matrix multiplication block. The state machine talks to the SDRAM via the XC2V3000-XC2V1000 interface. Different counters are used to keep a track on the rows to be read, group of columns being processed and the SDRAM read and write addresses.

4.3 XC2V3000-XC2V1000 Interface

The XC2V3000-XC2V1000 interface allows communication between the compute and control engine. There are a set of control and data lines, a set of registers and a well-defined protocol that allows communication through the interface. Table 2 shows the area utilization for the complete matrix multiplication application that resides in the XC2V3000 compute engine.

Table 2. Area Utilization of Matrix Multiplication Application

Number of Slice Flip Flops	15,167
Number of occupied Slices	13,413
Total Number 4 input LUTs	18,042
Number of Block RAMs	96
Number of MULT18X18s	64

4.4 Experimentation

The experimentation with the matrix multiplication application was carried out in two phases. In the first phase, we ran the purely software version of the code using Intel Math Kernel Library (MKL) [Int04] with BLAS Level 3 support on a HP-Kayak P-III 800 MHz host under Red Hat Linux 9.0.

For the second phase, we modified the application code by replacing the Intel library call with the matrix multiplication call directed on the reconfigurable hardware attached to the host. Table 3 shows the execution time comparison between the above-mentioned phases for 1K x 1K matrix multiplication. Although this comparison is not to the mark, we got a fairly good idea on how to have a better speed-up by enhancing the FPGAs and logic from the present RCS platform.

Table 3. Execution Time Comparison for 1K x 1K matrix multiplication

Method	Execution Time (Sec)
Intel Math Kernel Library	3.763
RCS card attached to the host	2.863

5 Conclusion

In this paper, we have presented our experimentation for implementing matrix multiplication on the FPGA based reconfigurable computing system. For this purpose, we have used our own designed IEEE-754 compliant double precision hardware library units and the FPGA based reconfigurable platform.

We compared the performance of matrix multiplication code when running exclusively on the host machine with the code running on the host and calling the matrix multiplication implemented on the RCS. In the later case, we obtained 1.3 times speedup.

In terms of the accuracy of the results, it was quite encouraging to note that we obtained highly accurate results on the RCS using our own double precision library elements.

The performance of the matrix multiplication in hardware can be improved by two ways. One of them is by accommodating more number of MAC units and the other, by increasing the frequency of operation. Both of these are constrained by the device that we are using. Right now four MAC units are put in the XC2V3000 device. By reducing some of the hardware from the control logic, a maximum of five MAC units can be fitted in the device. However in this case, attaining a frequency of 100 MHz will become more difficult. This is why in our present experimentation we decided to use only four MAC units.

Matrix multiplication for a larger matrix size can be done, but the number of block rams available in the device will become a constraint. Currently some of the buffers are made using distributed memories because of the lack of block rams. Matrix multiplication using block floating point operations can be evaluated. This will allow for additional MAC units within the device.

Feasibility study of other algorithms including Strassen with $O(n2.8)$ and Winograd for matrix multiplication on FPGA is being carried out. The Strassen algorithm uses block operations on matrices, making it possible to span different matrix sizes.

The computationally intense, matrix multiplication core as described in this paper will be useful for accelerating scientific and engineering applications on the reconfigurable computing system.

Acknowledgement

We acknowledge the Ministry of Information Technology, Government of India and C-DAC for the funding. We are also thankful to all of our colleagues from

HTDG and members of the beta testing group for their direct and indirect contribution to this work.

References

[Str69] Strassen, V: Gaussian Elimination is not Optimal. Num. Math., **13**, 354-356 (1969)

[JV98] Jelfimova, L., and Vajtersik, M.: A New Fast Parallel version of Strassen Matrix Multiplication Algorithm and its Efficient Implementation on Systolic-type Arrays. In: Third International Conference on Massively Parallel Computing Systems, Colorado USA,(1998)

[Win73] Winograd, S.: Some remarks on fast multiplication of polynomials. In: J. F. Traub, (ed) Complexity of Sequential and Parallel Numerical Algorithms Academic Press, New York (1973)

[CH02] Compton Katherine, Hauck Scott: Reconfigurable computing: a survey of systems and software. ACM Computing Surveys (CSUR), Vol. **34** No. **2**, 171-210 (2002)

[AG02] Abhyankar Yogindra and Ghotgalkar Shailesh : Reconfigurable Computing System (RCS). In: Arora RK, Sinha PK, Purohit SC, Dixit SP, Mohan Ram N (Eds) Initiatives in high performance computing. 29-34, Tata McGraw-Hill, New Delhi (2002)

[IEE85] IEEE Standard Board. IEEE Standard for Binary Floating-Point Arithmetic, ANSI/IEEE Std 754-1985, August 1985

[AB02] Amira, A. and Bensaali, F.: An FPGA based Parametrisable System for Matrix Product Implementation. In: IEEE Workshop on signal Processing Systems Design and Implementation, California USA (2002)

[SGD00] Sahin, I., Gloster, C. and Doss, C.: Feasibility of Floating-Point Arithmetic in Reconfigurable Computing Systems. In: Military and Aerospace Applications of Programmable Devices and Technology Conference (MAPLD), USA (2000)

[Xil04] Xilinx Incorporation. Application Note, Virtex-II Series and Xilinx ISE 6.1i Design Environment, http://www.xilinx.com, 2004

[ZP04] Zhuo Ling and Prasanna, V.K.: Scalable and Modular Algorithm for Floating-Point Matrix Multiplication on FPGAs. In: International Parallel and Distributed Processing Symposium (IPDPS'04), New Mexico (2004)

[UH04] Underwood Keith, D., Hemmert Scott, K.: Closing the gap: CPU and FPGA Trends in sustainable floating-point BLAS performance. In: Field Programmable Custom Computing Machines (FCCM 2004), Napa California (2004)

[Int04] Intel Corporation. Intel Math Kernel Library, http://www.intel.com/software/products/mkl/, 2004

On Construction of Difference Schemes and Finite Elements over Hexagon Partitions

Jiachang Sun[1] * and Chao Yang[1]

Lab. of Parallel Computing, Institute of Software, Chinese Academy of Sciences
sun, yc@mail.rdcps.ac.cn

In this paper we proposed a so-called coupled 4-point difference scheme for Laplacian operator over hexagon partition. It is shown that the scheme has the same order accuracy to the usual 7-point scheme in 3-direction mesh and 5-point scheme in rectangle mesh, though the local truncation error only has first order accuracy. Several hexagonal finite elements, such as piecewise quadratic and cubic, rational functions, are also investigated. Some numerical tests are given.

Mathematics subject classification: 65N25, 65T50

Keywords: Numerical PDE, Non-traditional Partition, Hexagon Partition, 4-point Difference Scheme, Hexagonal Finite Element

1 Introduction

It is well known that structured box partition has been widely used in traditional numerical PDE methods for quite a long time. Recently, as the complexity of the problem is getting much higher than before, unstructured mesh is proposed to meet the requirement of mesh generation. However, in some application areas, for example, in reservoir simulation which we have involved for ten years, totally unstructured mesh may cause some troubles, such as hard to be controlled, lack of fast algorithms, exponential working complexity, and so on.

Therefore, in our point of view, it is necessary to investigate numerical PDE partitions between global unstructured grids and traditional structured box partitions. It is the purpose of our present work to labor on filling this gap partly in 2-D case. And the related work in 3-D case will be presented in our forthcoming papers.

---

*Project supported by National Natural Science Foundation of China (Major Project No: 10431050 and Project No.60173021).

In 2-D case, besides triangular and quadrilateral tessellations, which are widely used in application, there is another regular tessellation, named hexagonal tessellation. Inspired by this consideration, we have done some elementary work on this type of non-traditional partition, which will be explored here.

In the next section, we will introduce some definitions first and then give a new coupled 4-point scheme over regular hexagon partition. In Section 3, the scheme will be generalized to a class of irregular hexagon partitions. In Section 4, we will introduce some hexagonal elements. At the end of this paper, some numerical results and some conclusion remarks will also be given.

2 Finite difference schemes on a hexagon partition

2.1 Preliminaries

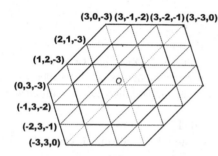

Fig. 1. 3-direction grids in 2-D   **Fig. 2.** basic hexagon domain Ω

To explore with symmetry property along three direction more deeply, we introduce a 3-direction coordinates system [PM01], instead of usual 2-D Cartesian coordinates or barycentric coordinates.

Given two independent vectors e_1 and e_2, set $e_3 = -e_1 - e_2$. Denote

$$n_1 = \frac{(e_1,e_1)e_2 - (e_1,e_2)e_1}{(e_1,e_1)(e_2,e_2) - (e_1,e_2)^2}, \quad n_2 = \frac{(e_2,e_2)e_3 - (e_2,e_3)e_2}{(e_2,e_2)(e_3,e_3) - (e_2,e_3)^2}, \quad n_3 = -(n_1 + n_2),$$

so that $(n_1,e_1) = 0$, $(n_1,e_2) = 1$, $(n_1,e_3) = -1$ and so on.

Definition 1. *For any 2-D point P, we define its 3-direction coordinates as $P = (t_1,t_2,t_3)$, where $t_l = (P,n_l)$, $l = 1,2,3$. And it is obvious that $t_1 + t_2 + t_3 = 0$.*

Definition 2. *The basic parallel hexagon domain is defined as (see Fig. 2)*

$$\Omega = \{P|P = (t_1,t_2,t_3\} \quad t_1 + t_2 + t_3 = 0, -1 \le t_1,t_2,t_3 \le 1\} \qquad (1)$$

Definition 3. *A function $f(P)$, defined in the 3-direction coordinates, is called periodic function with periodicity $Q = (n_1,n_2,n_3)$, if for all $P = (t_1,t_2,t_3)$, the following equality holds*

$$f(P + Q) = f(P)$$

Definition 4. *In terms of the 3-direction form, an elliptic PDE operator is defined by*

$$\mathcal{L} = -(\frac{\partial}{\partial t_1} - \frac{\partial}{\partial t_2})^2 - (\frac{\partial}{\partial t_2} - \frac{\partial}{\partial t_3})^2 - (\frac{\partial}{\partial t_3} - \frac{\partial}{\partial t_1})^2 \qquad (2)$$

where t_1, t_2, t_3 are taken as three independent variables.

In fact, it is elliptic. We may call the operator as Laplacian-like. Moreover, if the hexagon is regular with unit side length, say $e_1 = \{0, -1\}$, $e_2 = \{\frac{\sqrt{3}}{2}, \frac{1}{2}\}$, the operator turns to be Laplacian operator $\mathcal{L} = -\frac{2}{3}\Delta$.

At first it is natural to define a 7-point difference scheme for Laplacian operator in the 3-direction mesh

Definition 5.

$$L_7[u(P)] := 6u(P) - \sum_{i=1}^{3}(u(P + he_i) + u(P - he_i)). \qquad (3)$$

A simple Taylor expansion leads to

$$L_7[u(P)] = h^2 \sum_{i=1}^{3}(e_i, \nabla)^2 u(P) + \frac{h^4}{12}\sum_{i=1}^{3}(e_i, \nabla)^4 u(P) + O(h^6). \qquad (4)$$

In particular, for regular hexagon partition, the truncation error becomes

$$\frac{2}{3h^2}L_7[u(P)] = -\Delta u(P) - \frac{h^2}{16}\Delta^2 u(P) + O(h^4). \qquad (5)$$

Adding suitable boundary conditions on boundary $\partial\Omega$, such as periodic boundary conditions in three direction or Dirichlet boundary condition, the resulting matrix is with diagonal dominant, and the norm of the related inverse matrix is uniformly bounded. Hence we obtain

Lemma 1. *With a regular 3-direction uniform mesh, for 1-st boundary condition or periodic boundary conditions, the 7-point centered scheme (3) has second order accuracy for smooth functions.*

2.2 A new coupled 4-point scheme over regular hexagon partition

Now let us introduce how to construct difference scheme for Laplacian over hexagon partition. We note that in 3-direction mesh all integer nodes are divided into three different kinds, and all the neighbor nodes for each kind are beyond the same kind. That means all nodes can divided into three colors, say red, blue and yellow, see Fig. 3. Once we collect two colors, say red and blue, and exclude another one , say yellow, we may derive a hexagon partition from the 3-direction mesh, see Fig. 4. With this meshing, all three neighbors of a red node are blue, and all three neighbors of a blue node are red. So the three colors in 3-direction mesh now turn to two colors in hexagon partition. Therefore, we may set up a coupled 4-point scheme for Laplacian operator instead of 7-point scheme in the last section.

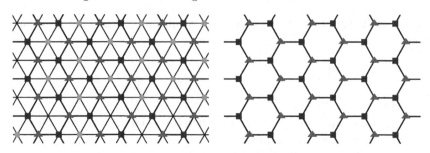

Fig. 3. 3-direction mesh   **Fig. 4.** hexagon partition

Definition 6. *Define the coupled 4-point scheme as*

$$L_4^R[u(P)] := 3u(P) - \sum_{i=1}^{3} u(P + he_i), \tag{6a}$$

$$L_4^B[u(P)] := 3u(P) - \sum_{i=1}^{3} u(P - he_i). \tag{6b}$$

For the regular hexagon partition, the truncation error becomes

$$L_4^R[u(P)] = -\frac{3}{4}h^2[\Delta u(P) + hru(P) + \frac{h^2}{16}\Delta^2 u(P) + O(h^3)], \tag{7a}$$

$$L_4^B[u(P)] = -\frac{3}{4}h^2[\Delta u(P) - hru(P) + \frac{h^2}{16}\Delta^2 u(P) + O(h^3)], \tag{7b}$$

where r is a third order partial differential operator. For the case of $e_1 = (0, -1)$, $e_2 = (\sqrt{3}/2, 1/2)$, $r = (\partial_x^3 - 3\partial_x\partial_y^2)/6$.

Hence, for each inner point, the local truncation error only has first order accuracy. However, for 1-st boundary condition or periodic boundary conditions our numerical experiment, listed in Section 5, shows that coupled 4-point scheme also has second order accuracy despite the local first order accuracy.

To explain this observation, we rewrite the difference equation for the coupled 4-point scheme in the following matrix form via a red-blue reordering

$$A \begin{pmatrix} U_R \\ U_B \end{pmatrix} = -\frac{3}{4}h^2 \begin{pmatrix} \Delta u_R \\ \Delta u_B \end{pmatrix}, \quad \text{where } A = \begin{bmatrix} 3I_R & A_{RB} \\ A_{BR} & 3I_B \end{bmatrix} \tag{8}$$

and for the true solution u, due to (7a) and (7b)

$$A \begin{pmatrix} u_R \\ u_B \end{pmatrix} = -\frac{3}{4}h^2 \left[\begin{pmatrix} \Delta u_R \\ \Delta u_B \end{pmatrix} + h \begin{pmatrix} ru_R \\ -ru_B \end{pmatrix} + \frac{h^2}{16} \begin{pmatrix} \Delta^2 u_R \\ \Delta^2 u_B \end{pmatrix} + O(h^3) \right]. \tag{9}$$

Hence, for the error term $e = U - u$ we have

$$A \begin{pmatrix} e_R \\ e_B \end{pmatrix} = -\frac{3}{4}h^3 \begin{pmatrix} ru_R \\ -ru_B \end{pmatrix} - \frac{3h^4}{64} \begin{pmatrix} \Delta^2 u_R \\ \Delta^2 u_B \end{pmatrix} + O(h^5). \tag{10}$$

Lemma 2.

$$A^{-1} = \begin{bmatrix} 3A_{7R}^{-1} & -A_{7R}^{-1}A_{RB} \\ -A_{7B}^{-1}A_{BR} & 3A_{7B}^{-1} \end{bmatrix} \tag{11}$$

where

$$A_{7R} = 9I_R - A_{RB}A_{BR}, \quad A_{7B} = 9I_B - A_{BR}A_{RB}. \tag{12}$$

It is worth to be noted that each row or column in matrix $A_{RB} = A'_{BR}$ has three non-zero elements -1 at most. Thus, from the above lemma and estimations (7a) and (7b), we have

$$A^{-1} \begin{pmatrix} ru_R \\ -ru_B \end{pmatrix} = \begin{pmatrix} A_{7R}^{-1}(3ru_R + A_{RB}ru_B) \\ -A_{7B}^{-1}(3ru_B + A_{BR}ru_R) \end{pmatrix} = O(h^{-2}) \begin{pmatrix} L_4^R[ru(P_R)] \\ -L_4^B[ru(P_B)] \end{pmatrix} = O(1).$$

Besides, as is well known $||A^{-1}|| = O(h^{-2})$ for second order elliptic problems, substituting there derivations into the error estimation (10) leads to the following result

Theorem 1. *The coupled 4-point scheme (6a) and (6b) over regular hexagon partition has the same second order accuracy to the 7-point scheme (3) for smooth functions, though each separate one color scheme only has first order local truncation error.*

3 A generalized coupled 4-point scheme

If this section, we generalize the coupled 4-point scheme for Laplacian to a class of irregular hexagon partition cases. Let $e_1 = (h\cos\theta_1, 0)$, $e_2 = (-h\cos\theta_2\cos\theta_3, h\cos\theta_2\sin\theta_3)$, $e_3 = (-h\cos\theta_3\cos\theta_2, -h\cos\theta_3\sin\theta_2)$, where $0 < \theta_i < \pi/2$, $i = 1, 2, 3$ and $\theta_1 + \theta_2 + \theta_3 = \pi$. From these three vectors, a hexagon partition can be constructed, see Fig. 9.

For each red point P_R, three blue neighbors $P_{B_i}(P_R)$, $i = 1, 2, 3$ satisfy $P_{B_i}(P_R) = P_R + e_i$, and for each blue point P_B, three red neighbors $P_{R_i}(P_B)$, $i = 1, 2, 3$ satisfy $P_{R_i}(P_B) = P_B - e_i$, see Fig. 5.

Definition 7. *Define the generalized coupled 4-point scheme as*

$$\Gamma_4^R[u(P)] := (\sum_{i=1}^{3}\tan\theta_i)u(P) - \sum_{i=1}^{3}(\tan\theta_i u(P + e_i)), \tag{13a}$$

$$\Gamma_4^B[u(P)] := (\sum_{i=1}^{3}\tan\theta_i)u(P) - \sum_{i=1}^{3}(\tan\theta_i u(P - e_i)). \tag{13b}$$

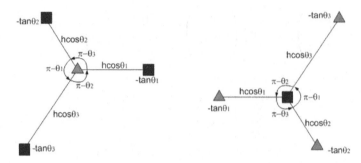

Fig. 5. the neighborhood of two color points.

If $\theta_1 = \theta_2 = \theta_3 = \pi/3$, we will get the same 4-point scheme as (6a) (6b).

The truncation error of the scheme becomes

$$\Gamma_4^R[u(P)] = -\frac{h^2}{2}(\prod_{i=1}^{3} \sin\theta_i \Delta u(P) + hru(P) + O(h^2)), \tag{14a}$$

$$\Gamma_4^B[u(P)] = -\frac{h^2}{2}(\prod_{i=1}^{3} \sin\theta_i \Delta u(P) - hru(P) + O(h^2)) \tag{14b}$$

where r is a differential operator.

We give the following convergence theorem, which is just like in the regular case, without detailed proof.

Theorem 2. *The generalized coupled 4-point scheme (13a) and (13b) has second order accuracy, despite the first order local truncation error.*

4 Hexagonal finite elements

Let Ω_0 be a basic parallel hexagon defined in Definition 2. To construct hexagonal finite elements, we need to find shape functions ϕ_k for each vertex A_k on Ω_0, satisfying

$$\phi_k(A_l) = \delta_{kl}, \quad k,l = 1,2,\cdots,6 \tag{15}$$

where $A_1 = (0,-1,1)$, $A_2 = (1,-1,0)$, $A_3 = (1,0,-1)$, $A_4 = (0,1,-1)$, $A_5 = (-1,1,0)$, $A_6 = (-1,0,1)$.

It seems natural that the six values (15) at vertices might exactly determine a quadratic surface

$$\phi_k(x,y) = c_0 + c_1 x + c_2 y + c_3 x^2 + c_4 xy + c_5 y^2 \tag{16}$$

for the six vertex A_k, $k = 1,2,\cdots,6$.

Unfortunately, the solution of (16) is not unique in general, because the resulting coefficient matrix might be singular. This singularity occurs from the symmetry of the six vertices of Ω_0, which are on a same quadratic curve. It is just like in triangular element case, a linear interpolation can be determined by values at the three vertices which are not collinear.

4.1 Incomplete polynomial interpolation

Since a quadric surface can't satisfy vertices condition (15), we may consider higher-degree polynomial interpolation. This means some extra terms of freedom will be added, so we also need some artificial restrictions to form shape functions ϕ_k on Ω_0. As an example, we consider a cubic surface

$$
\begin{aligned}
S(t_1, t_2, t_3) =& c_0 + c_1 t_1 + c_2 t_2 + c_3 t_3 + c_4 (t_1 - t_2)^2 + c_5 (t_2 - t_3)^2 + c_6 (t_3 - t_1)^2 \\
&+ c_7 (t_1 - t_2)(t_2 - t_3)(t_3 - t_1)
\end{aligned}
\tag{17}
$$

with two constrains $c_1 + c_2 + c_3 = 0$, and $c_4 + c_5 + c_6 = 0$.

It has unique solution on Ω_0, for example, the shape function

$$
\begin{aligned}
\phi_1(t_1, t_2, t_3) =& (1 - t_2 + t_3)/6 - (t_1 - t_2)^2/18 + (t_2 - t_3)^2/9 - (t_3 - t_1)^2/18 \\
&- (t_1 - t_2)(t_2 - t_3)(t_3 - t_1)/12.
\end{aligned}
\tag{18}
$$

satisfies (15) for $k = 1$. Moreover, these six function form a partition of unit, i.e. they are normalization in the sense $\sum_{k=1}^{6} \phi_k(t_1, t_2, t_3) = 1$.

It is easy to verify that these ϕ_k's can preserve first degree polynomials. And obviously, each ϕ_k's does not vanish on all non-A_k boundaries of Ω_0. So this is a non-conforming hexagonal finite element with normalization. We use $FE6_3^{01}$ to represent it, where '6' means hexagonal element, the lower subscript '3' indicates degree of the shape function (Note: A fractal subscript m/n will represents a rational with degree m and n), the first upper superscript '0' means non-conforming element, the second upper superscript '1' represents normalization.

4.2 Rational fraction-type polynomial interpolation

There are some different ways to get the shape functions directly. For example, if we want to get a third degree polynomial shape function ϕ_1 of vertex A_1, we try to find three straight nodal lines in Ω_0 plane, and then multiply them together. Thus, a shape function of A_1 can be taken as

$$
\phi_1(t_1, t_2, t_3) = \frac{1}{2}(t_1 - 1)(t_1 + 1)(t_2 - 1).
\tag{19}
$$

Using the same method, we can get all the six shape functions. And this type of element is non-conforming and un-normalized, named $FE6_3^{00}$.

If we divide each shape functions by their summation, we will get the normalized element $FE6_{3/2}^{01}$, which are rational fraction-type polynomials, cubic in the numerator and quadratic in the dominator.

If we elevate the degree of shape function to four, we may get some conforming elements, as an example,

$$
\phi_1(t_1, t_2, t_3) = \frac{1}{4}(t_1 - 1)(t_1 + 1)(t_2 - 1)(t_3 + 1).
\tag{20}
$$

It is un-normalized conforming element, named $FE6_4^{10}$. After normalization, it is denoted by $FE6_{4/2}^{11}$. This type of hexagonal element was originally discussed in [AT00].

4.3 Piecewise polynomial interpolation

The B-net method is another way to construct FEM shape functions. Given an element Ω_0, consider the following nineteen points: $A_1, A_2, A_3, A_4, A_5, A_6, B_1,$ $B_2, B_3, B_4, B_5, B_6, E_1, E_2, E_3, E_4, E_5, E_6, C$ on it, see Fig 6.

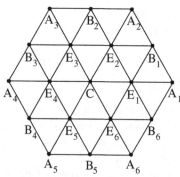

Fig. 6. The nineteen points on Δ **Fig. 7.** partial solutions of $S_2^1(\Delta)$.

Let Δ denote Ω_0 with the triangular division over the nineteen points, now we construct shape functions in $S_k^\mu(\Delta)$ with different continuity μ and different degree k.

Since $dimS_2^1 = 9$ [S01], we only need to find nine independent values out located on the nineteen points in Fig. 6 to determine a function in S_2^1 uniquely. We use $FE6S_2^1$ to denote this type of elements. After considering any essential equivalence, the number of solution is still at least 1650. But there is no solution containing all six vertices of the hexagon. If we only consider the twelve points $A_1, A_2, A_3, A_4, A_5, A_6, B_1, B_2, B_3, B_4, B_5, B_6$ on the boundaries and the center point C, the number of solution reduces to 39. If we don't even choose the center point C, the number of solution ulteriorly reduces to 16. Fig. 7 lists some interesting cases, four of the seven are symmetry along with a special line.

4.4 Other approaches

There are some other ways to construct hexagonal elements. For example, we can exclude one of the six vertices of the hexagon and add an extra point on hexagon, say, the center point C. By setting nodal lines just as in Section 4.1, the shape functions can be constructed. It can be verified that this element can preserve second degree polynomials. We name this element as $FE5_c$.

5 Numerical experiments

Case I

We consider this model problem

$$\begin{cases} -\Delta u + cu = f, \text{ in } \Omega \ (c > 0) \\ u \text{ is periodic, on } \Omega \end{cases} \tag{21}$$

where $\Omega = \{(t_1, t_2, t_3) | t_1 + t_2 + t_3 = 0; \ -1 \le t_1, t_2, t_3 \le 1\}$ is the basic hexagon domain(see Fig. 8). Assume the exact solution is

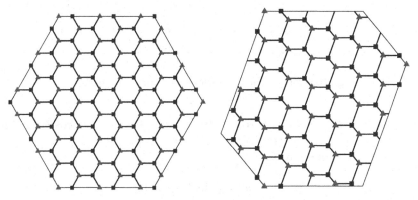

Fig. 8. a regular hexagon domain Ω with regular hexagon partition   **Fig. 9.** an irregular hexagon domain $\hat{\Omega}$ with irregular hexagon partition

$$u = sin(2\pi t_1) + sin(2\pi t_2) + sin(2\pi t_3).$$

We use $FD7p$, $FD4p$, $FE6_3^{01}$, $FE6_3^{00}$, $FE6_{3/2}^{01}$, $FE6_4^{10}$, $FE6_{4/2}^{11}$, $FE5_c$ to solve the problem mentioned above respectively.

From Table 1 and Fig. 10, we can see the coupled 4-point schemes (6a) and (6b) are second order and have almost the same accuracy with 7-point scheme (3). While, for hexagonal finite element, $FE6_3^{01}$, $FE6_{4/2}^{11}$ and $FE5_c$ are convergent, all with second order rate.

Case II

To test the generalized coupled 4-point scheme (13a) (13b), we define the periodic domain $\hat{\Omega}$ as the hexagon with these six vertices $(0,0), e_1, e_1 - e_3, e_1 +$

Table 1. scaled L_2 errors of various methods on parallel hexagon partition

$1/h$	$FD7p$	$FD4p$	$FE6_3^{01}$	$FE6_3^{00}$	$FE6_{3/2}^{01}$	$FE6_4^{10}$	$FE6_{4/2}^{11}$	$FE5_c$
16	1.56E-2	1.56E-2	3.65E-2	1.137	0.149	0.704	4.48E-2	5.59E-2
32	3.87E-3	3.87E-3	9.04E-3	1.204	0.115	1.050	1.11E-2	1.40E-2
64	9.66E-4	9.66E-4	2.25E-3	1.219	0.107	1.177	2.76E-3	3.49E-3
128	2.41E-4	2.41E-4	5.64E-4	1.223	0.105	1.212	6.89E-4	8.74E-4
256	6.03E-5	6.03E-5	1.43E-4	1.224	0.104	1.224	1.73E-4	2.19E-4

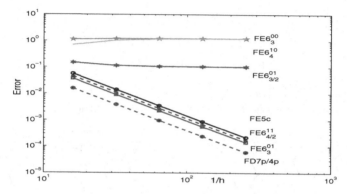

Fig. 10. Convergence behaviors for various methods.

$e_2 - e_3, e_2 - e_3, e_2$, where $e_1 = (\cos\theta_1, 0)$, $e_2 = (-\cos\theta_2\cos\theta_3, \cos\theta_2\sin\theta_3)$, $e_3 = (-\cos\theta_3\cos\theta_2, -\cos\theta_3\sin\theta_2)$, see Fig. 9.

Consider the model problem

$$\begin{cases} -\Delta u + cu = f, \text{ in } \hat{\Omega} \ (c > 0) \\ u \text{ is periodic, on } \hat{\Omega} \end{cases} \tag{22}$$

Assume the exact solution is

$$u = \sin(\frac{2\pi x}{\sin\theta_2\sin\theta_3}).$$

The numerical result shown in Table 2 verifies the second order convergence rate.

Case III

Here we use approximate eigen-decomposition via sparse approximation inverse, based on our preconditioning method [YMJ99] and software HFFT[2]. To demonstrate our preconditioning technique, consider the same model problem in Case I and assume a different exact solution

---

[2]HFFT: High-dimensional FFT on non-tenser domain, refer to [Y03].

Table 2. scaled L_2 errors for the generalized 4-point schemes with various (θ_1, θ_2)

$1/h$	$(45°, 60°)$	$(50°, 50°)$	$(40°, 65°)$	$(20°, 75°)$	$(10°, 82°)$
16	5.49E-3	5.63E-3	5.12E-3	8.34E-3	8.80E-3
32	1.36E-3	1.40E-3	1.30E-3	2.07E-3	2.19E-3
64	3.39E-4	3.48E-4	3.22E-4	5.17E-4	5.46E-4
128	8.48E-5	8.70E-5	8.05E-5	1.29E-4	1.37E-4
256	2.12E-5	2.18E-5	2.01E-5	3.23E-5	3.41E-5

$$u = exp(\frac{2}{t_1^2 - 1} + \frac{2}{t_2^2 - 1} + \frac{2}{t_3^2 - 1}).$$

We choose PETSc[3] and use CG method to solve the linear system. Use HFFT as preconditioner compared to ILU with the best fill-in level.

Table 3. comparison between ILU and HFFT as p.c. for $FD7p$ **Table 4.** comparison between ILU and HFFT as p.c. for $FD4p$

$1/h$	T-ILU	I-ILU(lv)	T-HFFT	I-HFFT	$1/h$	T-ILU	I-ILU(lv)	T-HFFT	I-HFFT
32	0.17	29(2)	0.02	1	32	0.06	25(3)	0.11	6
64	1.18	43(4)	0.17	1	64	0.52	37(5)	0.69	6
128	7.59	57(5)	0.85	1	128	3.41	45(9)	2.34	4
256	51.74	85(7)	3.84	1	256	23.86	83(10)	10.64	4
512	349.01	141(8)	17.11	1	512	158.68	140(11)	46.58	4

Table 5. comparison between ILU and HFFT as p.c. for $FE6_3^{01}$

$1/h$	T-ILU	I-ILU(lv)	T-HFFT	I-HFFT
32	0.15	25(1)	0.12	7
64	0.98	44(1)	0.60	5
128	6.49	62(2)	2.38	4
256	43.80	111(2)	8.55	3
512	291.88	166(3)	46.40	4

[3]PETSc (Portable, Extensible Toolkit for Scientific Computation), is developed by American Argonne National Laboratory, to solve PDE problems on high-performance computers. the official site: http://www-unix.mcs.anl.gov/petsc

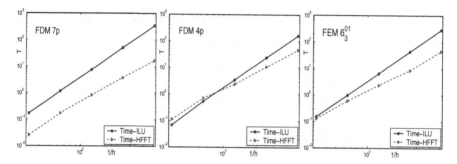

Fig. 11. comparison between ILU and HFFT as p.c. for various methods

6 Further works

The coupled 4-point scheme can be generalized into some cases of unstructured hexagon mesh. As for hexagonal finite elements, we will analyze their convergent rates theoretically in the future. Preconditioning techniques for more complicated boundary conditions will also be studied.

Acknowledgement

The authors would like to thank Dr.Li HuiYuan for valuable discussions.

References

[C76] Cialet, P.G.: The Finite Element Method for Elliptic Problems. North-Holland, New York (1976)

[I84] Ishiguro, M.: Construction of Hexagonal Basis Functions Applied in the Galerkin-Type Finite Element Method. *Journal of Information Processing* **Vol.7** No.2 (1984)

[S01] Sun, J.C.: Orthogonal piecewise polynomials basis on an arbitrary triangular domain and its applications. *Journal of Computational Mathematics* **Vol.19** No.1 (2001)

[S03] Sun, J.C.: Multivariate Fourier Series over a Class of non Tensor-product Partition Domains. *Journal of Computational Mathematics* **Vol.21** No.1 (2003)

[S04] Sun, J.C.: Approximate eigen-decomposition preconditioners for solving numerical PDE problems. (To appear in Applied Mathematics and Computation)

[Y03] Yao, J.F.: Scientific Visualization System and HFFT over Non-Tensor Product Domains. PhD Thesis, Chinese Academy of Sciences, Beijing (2003)

Efficient MPI-I/O Support in Data-Intensive Remote I/O Operations Using a Parallel Virtual File System

Yuichi Tsujita[1]

Department of Electronic Engineering and Computer Science,
Faculty of Engineering, Kinki University
1 Umenobe, Takaya, Higashi-Hiroshima, Hiroshima 739-2116, Japan
tsujita@hiro.kindai.ac.jp

Summary. A flexible intermediate library named Stampi realizes seamless MPI operations on interconnected parallel computers. In message transfer of Stampi, a vendor-supplied MPI library and TCP sockets are used selectively among MPI processes. Its router process mechanism hides a complex network configuration in message transfer among computers. Besides, dynamic process creation and MPI-I/O are also available. With the help of its flexible communication mechanism, users can execute MPI functions without awareness of underlying communication and I/O mechanisms. To realize distributed I/O operations with high performance, a Parallel Virtual File System (PVFS) has been introduced in the MPI-I/O mechanism of Stampi. Collective MPI-I/O functions have been evaluated and sufficient performance has been achieved.

Key words: MPI, MPI-I/O, Stampi, MPI-I/O process, PVFS

1 Introduction

In parallel computation, MPI [1, 2] is the de facto standard, and almost all computer vendors have implemented their own MPI libraries. Although such libraries are available inside a computer (intra-machine MPI communications), they do not support dynamic process creation and MPI communications among different computers (inter-machine MPI communications). To realize such mechanism, Stampi [3] has been developed.

Recent scientific applications handle huge amounts of data, and MPI-I/O has been proposed as a parallel I/O interface in the MPI-2 standard [2]. Although several kinds of MPI libraries have implemented it, MPI-I/O operations to a remote computer which has a different MPI library (remote MPI-I/O) have not been supported. Stampi-I/O [4] has been developed as a part of the Stampi library to realize this mechanism. Users can execute

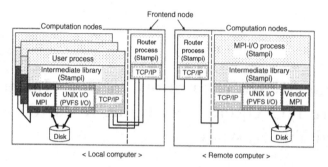

Fig. 1. Architecture of an MPI-I/O mechanism in Stampi.

remote MPI-I/O operations using a vendor-supplied MPI-I/O library. When the vendor-supplied one is not available, UNIX I/O functions are used instead of it (pseudo MPI-I/O method) [5].

Recently, PVFS [6] has been developed and available on a Linux PC cluster. It realizes distributed data management and gives users huge size of a virtual single file system. Although one of the non-commercial MPI-I/O implementations named ROMIO [7] in MPICH [8] supports the PVFS file system as one of the underlying I/O devices, it does not support remote MPI-I/O operations. To exploit high availability of the PVFS file system and realize effective distributed data management with high performance in remote MPI-I/O operations by Stampi, the UNIX I/O functions in its MPI-I/O mechanism have been replaced with native PVFS I/O functions. To evaluate this I/O mechanism, performance of primitive MPI-I/O functions has been measured on interconnected PC clusters.

In this paper, outline, architecture, and preliminary performance results of the MPI-I/O mechanism are described.

2 Implementation of PVFS in Stampi

This section describes details of the MPI-I/O mechanism of Stampi and explains a mechanism of typical MPI-I/O operations using the PVFS system.

2.1 MPI-I/O mechanism of Stampi

Architectural view of the MPI-I/O mechanism in Stampi is illustrated in Figure 1. In an interface layer to user processes, intermediate interfaces which have MPI APIs (a Stampi library) were implemented to relay messages between user processes and underlying communication and I/O systems.

In intra-machine MPI communications, a vendor-supplied MPI library is called by the Stampi library at first. Then high performance message transfer is available with a well-tuned vendor-supplied underlying communication

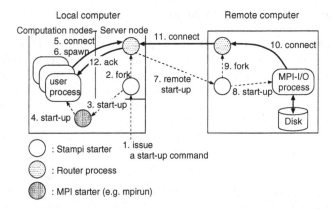

Fig. 2. Execution mechanism of MPI-I/O operations to a remote computer.

driver. On the other hand, a communication path is switched to the TCP socket connections inside the Stampi library in inter-machine MPI communications. When computation nodes are not able to communicate outside, a router process is invoked on a frontend node to relay messages from/to user processes on the computation nodes.

Remote MPI-I/O operations are carried out with the help of an MPI-I/O process which is dynamically invoked on a remote computer. I/O requests from the user processes are translated into a message data, and it is transfered to the MPI-I/O process via a communication path switched to the TCP socket connections. A bulk data is also transfered via the same communication path. The MPI-I/O process plays I/O operations using UNIX or PVFS I/O functions if a vendor-supplied MPI-I/O library is not available on a remote computer.

2.2 Execution mechanism

Stampi supports both interactive and batch modes. Figure 2 illustrates an execution mechanism of remote MPI-I/O operations with an interactive system. Firstly, an MPI starter (e.g. mpirun) and a router process are initiated by a Stampi start-up process (Stampi starter). Then the MPI starter initiates user processes. When the user processes call MPI_File_open(), the router process kicks off an additional Stampi starter on a remote computer with the help of a remote shell command. Secondly, the starter kicks off an MPI-I/O process, then a specified file is opened by it. Besides, a router process is invoked on an IP-reachable node if the MPI-I/O process is not able to communicate outside directly. After this operation, remote MPI-I/O operations are available via a communication path established in this strategy. As those complex mechanisms are capsuled in the seamless intermediate library of Stampi, users need not pay attention to the mechanisms.

2.3 Mechanism of MPI-I/O functions

Next, a mechanism of remote MPI-I/O operations using the PVFS I/O functions is explained. As an example, mechanisms of `MPI_File_read_at_all_begin()` and `MPI_File_read_at_all_end()` are illustrated in Figures 3 (a) and (b), respectively. When user processes call this function with a `begin` statement, sev-

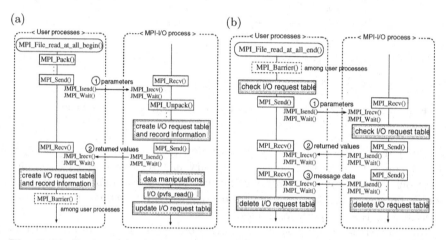

Fig. 3. Mechanisms of split collective read MPI-I/O functions with (a) `begin` and (b) `end` statements in remote MPI-I/O operations using PVFS I/O functions. MPI functions in rectangles are MPI interfaces of Stampi. Internally, Stampi-supplied functions such as `JMPI_Isend()` are called by them.

eral parameters are packed in a user buffer using `MPI_Pack()`. Then the packed data is transfered to an MPI-I/O process using `MPI_Send()` and `MPI_Recv()` of the Stampi library. Inside the functions, Stampi-supplied underlying communication functions such as `JMPI_Isend()` are used for nonblocking TCP socket communications. After the transfer, the received parameters are stored in an I/O request table which is created on the MPI-I/O process. Besides, a ticket number which is used to identify each I/O request is issued and stored in it. After this operation, the ticket number and related values which are associated with the I/O operation are sent to the user processes immediately. Once the user processes receive them, they create own I/O status table and store them in it. After synchronization by `MPI_Barrier()` among them, they are able to do next computation without waiting completion of the I/O operation by the MPI-I/O process. While the MPI-I/O process retrieves an I/O request from the table and reads data from a PVFS file system using a PVFS I/O function, `pvfs_read()`, instead of a UNIX I/O function, `read()`. The I/O operation is carried out in parallel with computation by the user processes. After the I/O operation, the information in the I/O status table is updated.

To detect completion of the I/O operation, a split collective read function with an **end** statement is used. Once this function is called, firstly synchronization by MPI_Barrier() is carried out among the user processes. Stored information in the I/O request table of the user processes is retrieved and the I/O request, ticket number, and related parameters are sent to the MPI-I/O process. After the MPI-I/O process receives them, it finds a corresponding I/O request table which has the same ticket number and reads the stored information in it. Then, several parameters and read data are sent to the user processes. Finally both I/O request tables are deleted and the I/O operation finishes.

After the I/O operation, the specified file is closed and the MPI-I/O process is terminated when MPI_File_close() is called in the user processes.

3 Performance Measurement

To evaluate the implemented MPI-I/O mechanism, performance of MPI-I/O operations was measured on interconnected PC clusters using an SCore cluster system [9]. Specifications of the clusters are summarized in Table 1. Network

Table 1. Specifications of PC clusters.

	PC cluster (I) (DELL PowerEdge 600SC × 4)	PC cluster (II) (DELL PowerEdge 1600SC × 5)
CPU	Intel Pentium-4 2.4 GHz	Intel Xeon 2.4 GHz (dual)
Chipset	ServerWorks GC-SL	ServerWorks GC-SL
Memory	1 GByte DDR SDRAM	2 GByte DDR SDRAM
Local disk	40 GByte (ATA-100 IDE)	73 GByte (Ultra320 SCSI)
Ethernet interface	Intel PRO/1000 (on-board)	Intel PRO/1000-XT (PCI-X board)
Linux kernel	2.4.19-1SCORE (all nodes)	2.4.20-20.7smp (server node) 2.4.19-1SCOREsmp (computation nodes)
Ethernet driver	Intel e1000 version 4.4.19 (all nodes)	Intel e1000 version 5.2.16 (all nodes)
MPI library	MPICH-SCore based on MPICH version 1.2.4	
Ethernet switch	NETGEAR GS108	3Com SuperStack4900

connections among PC nodes of the clusters I and II were established on 1 Gbps bandwidth network with full duplex mode via the Gigabit Ethernet switches, NETGEAR GS108 and 3Com SuperStack4900, respectively. As performance of inter-machine data transfer using on-board Ethernet interfaces (Intel PRO/1000) on the cluster II was quite lower than that using the PCI-X

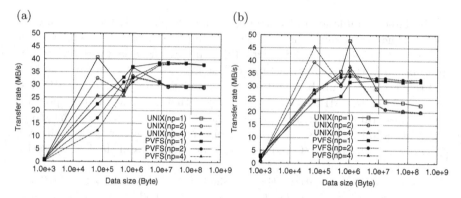

Fig. 4. Transfer rate values of (a) read and (b) write operations using blocking collective MPI-I/O functions from a PC cluster I to a PVFS file system of a PC cluster II, where *UNIX* and *PVFS* mean that MPI-I/O process played I/O operations with native UNIX and PVFS I/O functions, respectively. Besides, *np* in parentheses denotes the number of user processes.

boards (Intel PRO/1000-XT), the PCI-X boards were used during this test. Interconnection between the Ethernet switches was made with 1 Gbps bandwidth network (full duplex mode) using a category-6 unshielded twisted-pair cable.

In the cluster II, PVFS (version 1.5.8) was available on the server node. All the four computation nodes were used as I/O nodes for the file system. Size of dedicated disk on each I/O node was 45 GByte, and thus the file system with 180 GByte (4×45 GByte) was available. During this test, default stripe size (64 KByte) of it was selected.

In the both clusters, an MPICH-SCore library [10], which is a built-in MPI library of an SCore system based on MPICH version 1.2.4, was available. It was used in an MPI program on the cluster I during performance measurement of remote MPI-I/O operations.

In performance measurement of remote MPI-I/O operations from the cluster I to the cluster II, performance of blocking and split collective MPI-I/O functions with an explicit offset value was measured. In this test, TCP_NODELAY option was activated by the Stampi start-up command to optimize intermachine data transfer.

Transfer rate values of read and write operations using the blocking collective ones are shown in Figures 4 (a) and (b), respectively. In the both I/O operations, performance values in the PVFS case are better than those in the UNIX case with more than 8 MByte message data. With a 256 MByte message data, the PVFS case achieved more than 30 % and 40 % of performance improvements compared with the UNIX case in the read and write operations, respectively. It is considered that PVFS I/O functions have advantage in performance of the remote MPI-I/O operations for a large data

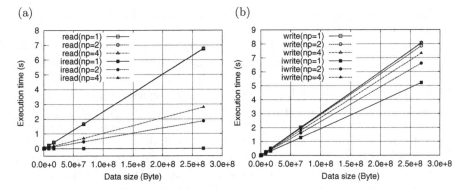

Fig. 5. Execution times of remote (a) read and (b) write MPI-I/O operations from a cluster I to a PVFS file system of a cluster II using blocking and split collective MPI-I/O functions. *read* in (a) and *write* in (b) mean execution times by MPI_File_read_at_all() and MPI_File_write_at_all(), respectively. Besides, *iread* in (a) and *iwrite* in (b) denote execution times by MPI_File_read_at_all_begin() and MPI_File_write_at_all_begin(), respectively. *np* in parentheses denotes the number of user processes.

due to its well-tuned I/O mechanism. It is also noticed that performance values are almost the same with respect the number of user processes due to a single MPI-I/O task in the pseudo MPI-I/O method. When a single collective I/O call was issued among the user processes, the I/O request was divided into individual blocking I/O requests and they were transfered to an MPI-I/O process. As the MPI-I/O process was a single process, I/O operations by it were serialized. Besides, computation by user processes and I/O operations by an MPI-I/O process on a remote computer were not overlapped due to its architectural constraints, and a required time for the inter-machine data transfer was dominant in the whole execution time. As a result, performance improvement was not observed in the case of multiple user processes compared with that in the case of a single user process.

On the other hand, such overlap of computation with I/O operations is available in nonblocking I/O functions. To examine this effect, performance of split collective MPI-I/O functions was measured. Figures 5 (a) and (b) show execution times of remote read and write MPI-I/O operations using the functions from the cluster I to the PVFS file system of the cluster II, respectively. To compare the overlap effect, execution times of the blocking collective functions are also shown in the figures.

In the read operations, execution times for the split collective function are about 0.1 %, 25 %, and 40 % of those for the blocking one with more than 1 MByte message data in the cases of a single, two, and four user processes, respectively. While the times for the split collective one in the write operations are about 60 %, 80 %, and 90 % of those for the blocking one with more than

1 MByte message data in the cases of a single, two, and four user processes, respectively.

In the split collective case, a single I/O request was divided into individual non-collective and nonblocking I/O requests. I/O operations associated with the first I/O request was carried out by the MPI-I/O process without blocking the user processes, but the remaining I/O requests were blocked until the former I/O operation finished. Due to this architectural constraints, the execution times in the cases of multiple user processes were quite longer than those in the single user process case. Besides, bulk data was also transfered when the split collective MPI-I/O function was issued in the write function case, and a required time for the transfer was dominant in the whole execution time. Therefore minimizing effect in the execution times was quite small compared with the effect in the read operations.

4 Related Work

Interface to a lower level I/O library of ROMIO is supplied by ADIO [11] which supports several kinds of I/O systems. PACX-MPI-PIO [12] was developed based on a PACX-MPI library [13], and its I/O mechanism is based on MPI_Connect I/O which supports global file I/O operations. On the other hand, Stampi provides an intermediate library which has MPI APIs for dynamic resource use in remote MPI-I/O operations. Although ROMIO supports a PVFS file system as one of the underlying I/O devices, it does not support remote MPI-I/O operations. Stampi supports the operations with the help of an MPI-I/O process.

Collective and nonblocking operations are important techniques for maximizing performance of MPI-I/O operations. ROMIO gives the nonblocking mechanism with multi-threaded method [14]. Higher overlap of computation with I/O operations has been realized on several kinds of platforms. On the other hand, Stampi provides a nonblocking remote MPI-I/O operations with the help of an MPI-I/O process and an I/O request table. In this case, once user processes send their I/O requests to an MPI-I/O process on a remote computer, they are able to do next computation without waiting the completion of I/O operations by the MPI-I/O process. The I/O operations are carried out independently with the computation.

5 Conclusions

A flexible and efficient MPI-I/O mechanism to support distributed data management in remote MPI-I/O operations has been realized in the Stampi library using PVFS I/O functions. To evaluate the effect of this implementation, performance measurement of the remote MPI-I/O operations was carried out on interconnected PC clusters. In the operations, user processes on a PC cluster invoked an MPI-I/O process dynamically on a server node of the another cluster where the PVFS file system was available, and the MPI-I/O process played I/O operations using PVFS or UNIX I/O functions on the file system. Performance of the remote MPI-I/O operations using the PVFS I/O functions was better than that using the UNIX I/O functions. Therefore implementation of the PVFS I/O functions in the MPI-I/O mechanism was effective in remote MPI-I/O operations to the PVFS file system.

Unfortunately, overlap of computation by the user processes with the I/O operations by the MPI-I/O process was not available in the blocking collective MPI-I/O functions due to its architectural constraints. Besides, inter-machine data transfer was dominant in an execution time of the remote MPI-I/O operations. Therefore performance values for multiple user processes were almost the same with those for a single user process.

While the split collective MPI-I/O functions minimized the required times to issue the functions with the help of the I/O request table which was created on both the user processes and MPI-I/O process, typically in the read operation. Once I/O requests from the user processes were stored in the table, the user processes were able to do next computation immediately. I/O operations by the MPI-I/O process were carried out in parallel with computation by the user processes.

Acknowledgments

The author would like to thank Genki Yagawa, director of Center for Promotion of Computational Science and Engineering (CCSE), Japan Atomic Energy Research Institute (JAERI), for his continuous encouragement. The author would like to thank the staff at CCSE, JAERI, especially Toshio Hirayama, Norihiro Nakajima, Kenji Higuchi, and Nobuhiro Yamagishi for providing a Stampi library and giving useful information.

This research was partially supported by the Ministry of Education, Culture, Sports, Science and Technology (MEXT), Grant-in-Aid for Young Scientists (B), 15700079.

References

1. Message Passing Interface Forum: MPI: A message-passing interface standard (1995)
2. Message Passing Interface Forum: MPI-2: Extensions to the message-passing interface standard (1997)
3. Imamura, T., Tsujita, Y., Koide, H., and Takemiya, H.: An architecture of Stampi: MPI library on a cluster of parallel computers. In *Recent Advances in Parallel Virtual Machine and Message Passing Interface*, vol. 1908 of *Lecture Notes in Computer Science*, Springer, 200–207 (2000)

4. Tsujita, Y., Imamura, T., Takemiya, H., and Yamagishi, N.: Stampi-I/O: A flexible parallel-I/O library for heterogeneous computing environment. In *Recent Advances in Parallel Virtual Machine and Message Passing Interface*, vol. 2474 of *Lecture Notes in Computer Science*, Springer, 288–295 (2002)

5. Tsujita, Y.: Flexible Intermediate Library for MPI-2 Support on an SCore Cluster System. In *Grid and Cooperative Computing*, vol. 3033 of *Lecture Notes in Computer Science*, Springer, 129–136 (2004)

6. Carns, P., Ligion III, W., Ross, R., and Thakur, R.: PVFS: A parallel file system for Linux clusters. In *Proceedings of the 4th Annual Linux Showcase and Conference*, USENIX Association, 317–327 (2000)

7. Thakur, R., Gropp, W., and Lusk, E.: On implementing MPI-IO portably and with high performance. In *Proceedings of the Sixth Workshop on Input/Output in Parallel and Distributed Systems*, 23–32 (1999)

8. Gropp, W., Lusk, E., Doss, N., and Skjellum, A.: A high-performance, portable implementation of the MPI message-passing interface standard. *Parallel Computing*, **22**(6), 789–828 (1996)

9. PC Cluster Consortium: http://www.pccluster.org/

10. Matsuda, M., Kudoh, T., and Ishikawa, Y.: Evaluation of MPI implementations on grid-connected clusters using an emulated WAN environment. In *Proceedings of the 3rd IEEE/ACM International Symposium on Cluster Computing and the Grid (CCGrid 2003)*, IEEE Computer Society, 10–17 (2003)

11. Thakur, R., Gropp, W., and Lusk, E.: An abstract-device interface for implementing portable parallel-I/O interfaces. In *Proceedings of the Sixth Symposium on the Frontiers of Massively Parallel Computation*, 180–187 (1996)

12. Fagg, G. E., Gabriel, E., Resch, M., and Dongarra, J. J.: Parallel IO support for meta-computing applications: MPI_Connect IO applied to PACX-MPI. In *Recent Advances in Parallel Virtual Machine and Message Passing Interface*, vol. 2131 of *Lecture Notes in Computer Science*, Springer, 135–147 (2001)

13. Gabiriel, E., Resch, M., Beisel, T., and Keller, R.: Distributed computing in a heterogeneous computing environment. In *Recent Advances in Parallel Virtual Machine and Message Passing Interface*, vol. 1497 of *Lecture Notes in Computer Science*, Springer, 180–188 (1998)

14. Dickens, P. and Thakur, R.: Improving collective I/O performance using threads. In *Proceedings of the Joint International Parallel Processing Symposium and IEEE Symposium on Parallel and Distributed Processing*, 38–45 (1999)

A General Approach to Creating Fortran Interface for C++ Application Libraries

Yang Wang[1], Raghurama Reddy[1], Roberto Gomez[1], Junwoo Lim[1], Sergiu Sanielevici[1], Jaideep Ray[2], James Sutherland[2], and Jackie Chen[2]

[1] Pittsburgh Supercomputing Center, Carnegie Mellon University, Pittsburgh, PA 15213, U.S.A. ywg@psc.edu
[2] Sandia National Laboratories, Livermore, CA 94550-0969, USA

Summary. Incorporating various specialty libraries in different programming languages (FORTRAN and C/C++) with the main body of the source code remains a major challenge for developing scientific and engineering application software packages. The main difficulty originates from the fact that Fortran 90/95 pointers and C/C++ pointers are structurally different. In this paper, we present a technique that allows us to circumvent this difficulty without using any nonstandard features of these programming languages. This technique has helped us to develop a FORTRAN 90/95 interface for the GrACE library, which is written in C++ for facilitating spacial grid generation, adaptive mesh refinement, and load balance maintenance. The method outlined in this presentation provides a general guideline for the creation of the FORTRAN 90/95 interface for a C/C++ library. We show that this method is system independent and has low overhead cost, and can be easily extended to situations where building the C/C++ interface for a FORTRAN 90/95 application library is required.
Key words: FORTRAN, C/C++, pointers, inter-language calling, interface

1 Introduction

For maximal portability, the best approach to scientific and engineering application software development is to use a single programming language. In many cases, it is, however, undesirable or even unfeasible to do so, mainly because of the fact that many widely used numerical or graphical software libraries are not developed in a single programming language. Consequently, incorporating various specialty libraries in different programming languages (FORTRAN and C/C++) with the main body of the source code usually becomes a major headache in the process of application software development. The primary reason for this difficulty is the differences in the conventions for argument passing and data representation. A Fortran 90/95 pointer is quite different from a pointer in C/C++. While the argument passing conventions for the basic data types has been worked out fairly well, there are significant difficulties in handling pointer data types as arguments. The primary reason for passing pointer information between C/C++ and FORTRAN is to pass references to dynamically allocated memory. In addition, the referenced memory may contain data values which are, in effect, also being exchanged and which we must therefore be able to reference from both languages. To our knowledge, no effective method is publicly known to allow FORTRAN 90/95 pointers to directly access the data space allocated in C/C++ routines. While some efforts such as Babel[DEKL04] are attempting to solve this problem in a more general purpose way, these methods are usually system dependent and involve significant overhead and learning curve.

In our recent effort to build a Terascale 3D Direct Numerical Simulation package for flow-combustion interactions, we developed a technique for building a FORTRAN interface for C++ libraries. This allows shielding the C++ implementation of the functionalities from our FORTRAN legacy code. Most importantly, the interface provides the access functions for FORTRAN 90/95 pointers, which can be one- or multi-dimensional, to access the data allocated in the C++ library. In our approach to the FORTRAN interface design, only the standard features of the programming language are used. The method outlined in this presentation provides a general guideline for the creation of the FORTRAN interface.

2 Inter-language Calling: Conventions and Challenges

It has become a widely-supported convention that the FORTRAN compilers convert the name of the FORTRAN subroutines, functions, and common blocks to lower case letters and extend the name with an underscore in the object file. Therefore, in order for C/C++ to call a FORTRAN routine, the C/C++ routine needs to declare and call the FORTRAN routine name in low case letters extended with an underscore. On the other hand, declaring the name of a C routine in lower case letters extended with an underscore makes the C routine name (without the underscore) callable from FORTRAN. Comparing FORTRAN and C/C++, the basic data types are fairly similar, e.g., REAL*8 v.s. double and INTEGER v.s. long, although the exact equivalence between these data types depends on the computer system and the compilers. Besides the similarity between their basic data types, a FORTRAN subroutine is equivalent to a C function with a return type of void, while a FORTRAN function is equivalent to a C function with a non-void return type. However, there is a major difference between FORTRAN and C in the way they handle the arguments when making routine or function calls. FORTRAN in most cases passes the arguments by reference, while C passes the arguments by value. For a variable name in the argument list of a routine call from FORTRAN, the corresponding C routine receives a pointer to that variable. When calling a FORTRAN routine, the C routine must explicitly pass addresses (pointers) in the argument list. Another noticeable difference between FORTRAN and C in handling routine calls is that FORTRAN routines supply implicitly one additional argument, the string length, for each **CHARACTER*N** argument in the argument list. Most FORTRAN compilers require that those extra arguments are all placed after the explicit arguments, in the order of the character strings appearing in the argument list. When a FORTRAN function returns a character string, the address of the space to receive the result is passed as the first implicit argument to the function, and the length of the result space is passed as the second implicit argument, preceding all explicit arguments. Whenever the FORTRAN routine involves character strings in its argument list and/or returns a character string as a function, the corresponding C routine needs to declare those extra arguments explicitly. Finally, we note that the multi-dimensional arrays in FORTRAN are stored in a column-major order, whereas in C they are stored in a row-major order.

The FORTRAN/C inter-language calling conventions summarized above have proved to be very useful in the software development for scientific and engineering applications that often requires incorporating various specialty libraries in different programming languages (FORTRAN or C/C++) with the main body of the source code. Despite the fact that the variables of basic data types passed between FORTRAN and C/C++ routines can be handled fairly well by the FORTRAN/C inter-language calling convention, one major difficulty remains to be resolved. That is we are not allowed to let a FORTRAN 90/95 pointer be passed from or to a C/C++ routine to make it associated with a C/C++ pointer, even though the pointers are of a basic data type. Unfortunately, this problem appears in many situations, and can become a major obstacle in the application software development. To give a better description of the problem, we discuss the example code shown in Table 2 which represents a fairly typical situation in reality when mixing FORTRAN code and C++ code. In this example, the FORTRAN subroutine **Application()** is a legacy code that needs to call **getGrid()** function to associate the three dimensional FORTRAN 90/95 pointer **grid** with the grid array created by a utility library, **MeshModule**, a FORTRAN 90/95 module specially designed for 3D mesh generation. The new utility library we intend to use is written in C++, and it contains a **Mesh** class that implements the functionality for 3D mesh generation and a function, **getGrid()**, for returning a C/C++ pointer associated with the memory space allocated for the grid. Unfortunately, we are required not to make any changes to the legacy FORTRAN code nor to the library code. (In many cases, the library source code is not even available.) The technical difficulty is that we are not allowed to let the FORTRAN code call the member function of **Mesh** class directly, since the FORTRAN code has no way to recognize the instance of **Mesh** class. Furthermore, even if the utility library provides a C routine **getGrid()** for the FORTRAN code to call, associating a FORTRAN 90/95 pointer directly with the C/C++ pointer returned by **getGrid()** is still not allowed, needless to mention one more difficulty that the FORTRAN 90/95 pointer here is a three dimensional array whereas the C/C++ pointer is represented by a one dimensional array.

Table 1. A problem for using FORTRAN 90/95 pointer to access the "grid" array allocated in the C++ code

FORTRAN 90/95: Application.f90	C++: Mesh.h
| ```
SUBROUTINE Application()
   USE MeshModule, ONLY :            &
      initMesh, getGrid, deleteMesh
   INTEGER :: n1, n2, n3
   REAL*8 :: geometry(3,3)
   REAL*8, pointer :: grid(:,:,:)
   ... ...
   READ(*,*)n1, n2, n3,              &
            geometry(1:3,1:3)
   CALL initMesh(n1, n2, n3,         &
                 geometry)
   grid => getGrid()
   ... ...
   CALL deleteMesh()
END SUBROUTINE Application
``` | ```
class Mesh {
public:
   void Mesh(long *,
            long *, long *,
            double *);
   void ~Mesh();
   double *getGrid() const;
   long getSizeX() const;
   long getSizeY() const;
   long getSizeZ() const;
private:
   double *grid;
   long nx, ny, nz;
   double geometry[9];
}
``` |

Generally speaking, when making a C++ numerical library usable by a FORTRAN application, we usually have to resolve these following problems: how to construct and destruct a C++ object via FORTRAN routine calls, how to access the public member functions of a C++ object via FORTRAN routine calls, how to make a multi-dimensional FORTRAN 90/95 pointer an alias of a one dimensional array allocated dynamically in the C/C++ routine, and finally how to make a memory space allocated dynamically in C/C++ routine accessible by a FORTRAN 90/95 pointer.

There are some possible ways to get around these problems. For example, we can build a simple main driver in C/C++ that calls the constructor and the destructor of the classes defined in the library. The drawback of this approach is that the life of the objects spans over the entire job process. Alternatively, we can "wrap" the C++ member function, e.g., getGrid(), with a C-style function and call the constructor and the destructor, e.g., Mesh(...) and ~Mesh() in the "wrapper". But if the member function needs to be called frequently, this approach can be very costly. As far as the FORTRAN 90/95 and the C/C++ pointer association is concerned, the CNF library[CNF] provides one of possible solutions. In the CNF library, a set of C macros are defined to help handling the difference between FORTRAN and C character strings, logical values, and pointers. For pointers, in particular, CNF does not allow FORTRAN 90/95 pointers to be associated with C pointers directly, rather, it uses FORTRAN INTEGER type to declare the C pointer passed into the FORTRAN routine so that the corresponding INTEGER variable represents the address of the array. Unfortunately, this scheme of using FORTRAN INTEGERs to hold pointer values only works cleanly if the length of an INTEGER is the same as the length of the C generic pointer type `void*`. In order to use CNF, it requires to add macros to the C++ library or to implement a C "wrapper" for the library with the macros built in. In either case, the readability of the code becomes problematic. Another approach is to use CRAY or Compaq pointers, which are nonstandard FORTRAN pointers allowed to be associated with C/C++ pointers directly. Unfortunately, only a few FORTRAN compilers support these types of pointers. Finally, we mention the Babel[DEKL04] project, that attempts to solve the problem of mixing languages in a more general way. So far, Babel only works on Linux systems, and its support for FORTRAN 90/95 pointers is still questionable.

3 Approach to the Solution

Our approach is to build a "wrapper" for the C++ library that hides the implementation details of the C++ library from the FORTRAN application. The "wrapper" handles the request from FORTRAN calls to create and destroy objects defined in the C++ library, and provides functions to return FORTRAN pointers aliased to the memory space allocated in the C++ library. It plays an intermediate role between the FORTRAN application and the C++ library and helps to minimize the changes that need to be made to the application source code.

The architecture of the "wrapper" is sketched in figure 1. The "wrapper" is made of two layers: a C-Layer and a FORTRAN-Layer, and both are written in "standard" C/C++ and FORTRAN 90/95 programming languages, together with the inter-language calling conventions summarized in the previous section. The C-Layer contains, in its global scope, a pointer to the C++ object that needs to be created or destroyed. In addition to providing a C-type interface between the FORTRAN-Layer and the public functions defined in the C++ library, the C-Layer calls the "bridge" functions, which are a set of FORTRAN routines provided by the FORTRAN-Layer, to make the FORTRAN 90/95 pointers aliased to the memory allocated dynamically in the C++ library. The FORTRAN-Layer is composed of a FORTRAN 90/95 module and the "bridge" functions, which are external to the module. The FORTRAN 90/95 module provides a set of public functions for the FORTRAN application routines to call. Each of these public functions corresponds to a public function implemented in the C++ library, and it calls the corresponding C function implemented in the C-Layer that in turn calls the corresponding public function in the C++ library. The FORTRAN 90/95 module also provides an alias function with `public` attribute for each "bridge" function to call. In its global space, the FORTRAN 90/95 module holds a set of one- or multi-dimensional FORTRAN 90/95 pointers that need to be aliased to the memory space allocated in the C++ library.

Aliasing a FORTRAN 90/95 pointer to a memory space allocated in the C++ library takes the following steps. Upon an access function call from the

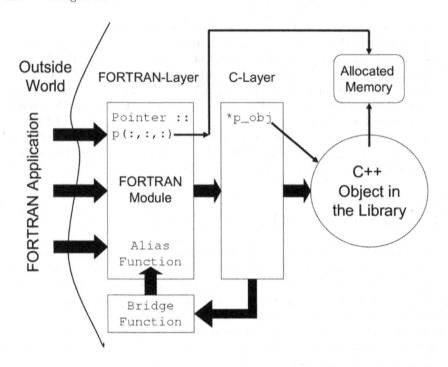

Fig. 1. The architecture of the "wrapper", a FORTRAN interface, for the C++ library. The "wrapper" is made of a C-Layer and a FORTRAN-Layer.

FORTRAN application that asks for the return of a FORTRAN 90/95 pointer, the FORTRAN 90/95 module in the FORTRAN-Layer calls the corresponding C routine in the C-Layer. The C routine calls the corresponding access function in C++ library to get the C/C++ pointer pointing to the memory space and then calls the "bridge" function to pass the C/C++ pointer into FORTRAN. The "bridge" function gets the C/C++ pointer together with its size from the caller and declares it as a FORTRAN array. The "bridge" function then calls the corresponding alias function in the FORTRAN 90/95 module to pass in the array. The alias function declares the array with appropriate dimension together with `TARGET` attribute, and then assigns the array to the FORTRAN 90/95 pointer using the operator "=>". Finally, the access function in the FORTRAN 90/95 module returns the FORTRAN 90/95 pointer to its caller (the FORTRAN application).

We again use the example in Table 2 to illustrate the method described above. The "wrapper" for `Mesh.cc`, the C++ library, is made of a FORTRAN-Layer that contains `MeshModule.f90` (see Table 3) and `grid_Bridge.f90` (see Table 4) and a C-Layer that contains `MeshWrapper.cc` (see Table 5). The private member variable `grid(:,:,:)` in the FORTRAN 90/95 module `MeshModule` is a pointer to be associated with the grid space allocated

Table 2. An example of the FORTRAN 90/95 module in the FORTRAN-Layer

MeshModule.f90

```
MODULE MeshModule
PUBLIC : initMesh, getGrid, deleteMesh, aliasGrid
PRIVATE
   INTEGER :: nx, ny, nz
   REAL*8, pointer :: grid(:,:,:)
CONTAINS
```

```
SUBROUTINE initMesh(L,M,N,g)      !==============================
   INTEGER :: L,M,N                SUBROUTINE aliasGrid(n1,n2, &
   REAL*8 :: g(9)                                         n3,a)
   CALL c_initMesh(L,M,N,g)          INTEGER :: n1,n2,n3
END SUBROUTINE initMesh              REAL*8, TARGET ::          &
   !==============================                     a(n1,n2,n3)
SUBROUTINE deleteMesh()              nx = n1
   CALL c_deleteMesh()               ny = n2
END SUBROUTINE deleteMesh            ny = n3
   !==============================
FUNCTION getGrid() result(p)         grid => a(1:nx,1:ny,1:nz)
   REAL*8, POINTER :: p(:,:,:)
   CALL c_getGrid()               END SUBROUTINE aliasGrid
   p => grid(1:nx,1:ny,1:nz)      !==============================
END FUNCTION getGrid
```

```
END MODULE MeshModule
```

Table 3. An example of the "bridge" function in the FORTRAN-Layer

grid_Bridge.f90

```
SUBROUTINE grid_Bridge(nx, ny, nz, a)
   USE MeshModule, only: aliasGrid
   INTEGER, INTENT(IN) :: nx,ny,nz
   REAL*8, INTENT(IN) :: a(nx*ny*nz)
   CALL aliasGrid(nx, ny, nz, a)
END SUBROUTINE grid_Bridge
```

by `Mesh.cc`. The member function `initMesh` in the module calls routine `c_initmesh_` in `MeshWrapper.cc` to create a `Mesh` object; function `deleteMesh` calls routine `c_deletemesh_` to destroy the `Mesh` object; function `getGrid` calls routine `c_getgrid_` and returns a pointer, aliased to `grid(:,:,:)`, to the caller; function `aliasGrid`, called by the bridge function `grid_Bridge`, makes the pointer `grid(:,:,:)` aliased to the grid array originated as a C/C++ pointer `p_mem` from routine `c_getgrid_` in `MeshWrapper.cc`. The pointer `*p_obj` in the global space of `MeshWrapper.cc` is created as an instance of class Mesh after routine c_initmesh_ is called, and

Table 4. An example of the C-Layer

```
MeshWrapper.cc

#include "Mesh.h"
Mesh *p_obj;
extern void grid_bridge_(long *, long *, long *, double *);
extern "C" {
void c_initmesh_(long *nx, long *ny, long *nz, double *geom) {
  p_obj = new Mesh(nx,ny,nz,geom);
}
void c_deletemesh_() { delete p_obj; }
void c_getgrid_() {
  double *p_mem = p_obj->getGrid();
  long nx = p_obj->getSizeX();
  long ny = p_obj->getSizeY();
  long nz = p_obj->getSizeZ();
  grid_bridge_(&nx, &ny, &nz, p_mem);    p_mem = 0;
}
}
```

is deleted after routine c_deletemesh_ is called. The routine c_getgrid_ calls
the member function getGrid of the Mesh object, to obtain the address of the
grid space and store the address in pointer *p_mem. After calling grid_bridge_,
c_getgrid_ passes the address together with the size of the grid space in all dimensions, nx, ny, and nz, into the FORTRAN routine grid_Bridge, in which
the grid space is declared as a FORTRAN array a(nx*ny*nz). This array
together with its size in each dimension is then passed into aliasGrid, where
it is declared as a three dimensional array with a TARGET attribute so that the
FORTRAN 90/95 pointer grid(:,:,:) is allowed to be aliased to it.

Obviously, the method outlined above is system independent, since the
only nonstandard features of the programming languages used here are the
widely supported inter-language conventions, summarized in Section 2. The
"wrapper" does not involve any heavy numerical computations and is only
used for passing in and out the parameters by reference without making copy
of sizeable data, and therefore it has very low overhead cost. The "wrapper" for the C++ library is able to hide all the implementation details from
the FORTRAN application, and makes the C++ library behave as a FOR-
TRAN library with interfaces provided by the FORTRAN 90/95 module in
the "wrapper". It becomes unnecessary for the FORTRAN application to be
aware of the existence of the C++ objects and the C/C++ pointers behind
the scene. In the example presented above, the FORTRAN 90/95 module in
the "wrapper" has the same module name and public functions as the module

used in `Application()` (see Table 2) so that there is no need for us to make any changes to `Application.f90`. In practice, the situation is very unlikely to be as simple as this. We can choose to build an intermediate FORTRAN-Layer between the "wrapper" and the FORTRAN application and let this intermediate FORTRAN-Layer call the FORTRAN public functions provided by the "wrapper". In this way, the FORTRAN application does not require any changes to its source code.

4 Applications

The technique presented in the previous section has been applied to the development of a FORTRAN interface for GrACE, a C++ library facilitating mesh-management and load-balancing on parallel computers. Our legacy application code, namely S3D, is written in FORTRAN 90/95 and is used for solving the Navier-Stokes equations with detailed chemistry. The FORTRAN interface for GrACE made possible to incorporate the GrACE library into S3D. The new application package can now use the functionalities implemented in GrACE for grid generation and dynamic load balance maintenance. Our ultimate goal is to develop, based on the S3D code, a terascale 3D direct numerical simulation package, capable of performing integrations at multiple refinement levels using adaptive mesh refinement algorithm while maintaining load balance, for flow-combustion interactions.

The "wrapper" for a C++ library introduced in the previous section can be turned around and viewed differently. That is, from a different perspective, we can consider that the FORTRAN-Layer plus the C-Layer is a "wrapper" for the FORTRAN application and the C++ library/application interacts with the FORTRAN application via the C-Layer acting as an interface. Therefore, the same method can also be applied to build a C/C++ interface for a FORTRAN library/module, except that the bridge functions and the alias functions should now be written in C and should belong to the C-Layer. And most importantly, it makes possible to associate a C/C++ pointer with a FORTRAN 90/95 pointer array allocated in FORTRAN. Based on this observation, we have developed a systematic way of building CCA (Common Component Architecture) [AGKKMPS99] components for FORTRAN modules, subroutines, and/or functions without having to change the FORTRAN source code. The CCA components are the basic units of software that can be combined to form applications via ports which are the abstract interfaces of the components. Instances of components are created and managed within a CCA framework, which also provides basic services for component interoperability and communication. Briefly speaking, it takes two major steps to component-wize a FORTRAN module, subroutine or function. Firstly, we use the method presented in the previous section to build a C/C++ interface for the FORTRAN code, and secondly, we follow the standard procedure to componentize the resulting C/C++ interface. A detailed description of this work will be presented in a future publication.

5 Conclusions

In summary, we have developed a general method for building a FORTRAN interface for a C++ library. This same method can also be applied to building a C/C++ interface for a FORTRAN library. The interface is made of a FORTRAN-Layer and a C-Layer. The FORTRAN 90/95 module in the FORTRAN-Layer provides an interface for the FORTRAN application and the C-Layer provides an interface for the C/C++ application. The "crux" of this method for associating a FORTRAN 90/95 pointer with a C/C++ pointer (or vise versa) is the bridge function that converts a C/C++ pointer into a FORTRAN array (or vice versa) and the alias function that makes a FORTRAN 90/95 pointer array alias to the FORTRAN array associated with the C/C++ pointer (or assigns a C/C++ pointer the address value of the FORTRAN array.)

This approach has several advantages. It is portable and has very low overhead cost. The requirement for making changes to the legacy application code is minimal and can actually be avoided. We have applied this method to build a FORTRAN interface for the GrACE library in our SciDAC project, and we have also used the same technique to componentize some of the FORTRAN 90/95 modules in our FORTRAN applications.

Acknowledgement. This work is supported in part by U.S. Department of Energy, under SciDAC Scientific Application Pilot Programs (SAPP), Contract No. DE-FC02-01ER25512.

References

[AGKKMPS99] Armstrong, B., Geist, A., Keahey, K., Kohn, S., McInnes, L., Parker, S., and Smolinski, B.: Toward a Common Component Architecture for High-Performance Scientific Computing. *8th IEEE International Symposium on High-Performance Distributed Computing*, Redondo Beach, CA (1999)
[CNF]   www.starlink.rl.ac.uk/static_www/soft_further_CNF.html
[DEKL04] Dahlgren, T., Epperly, T., Kumfert, G., and Leek, J.: Babel Users' Guide. www.llnl.gov/CASC/components/docs/users_guide.pdf (2004)
[PB00]   Parashar, M. and Browne, J.C.: Systems Engineering for High Performance Computing Software- *The HDDA/DAGH Infrastructure for Implementation of Parallel Structured Adaptive Mesh Refinement*. IMA Volume 117, Editors: S.B. Baden, N.P. Chrisochoides, D.B. Gannon, and M.L. Norman, Springer-Verlag, pp. 1-18 (2000)

A Platform for Parallel Data Mining on Cluster System

Shaochun Wu, Gengfeng Wu, Zhaochun Yu, and Hua Ban

School of Computer Engineering and Science, Shanghai University, Shanghai 200072, Chian scwu@mail.shu.edu.cn gfwu@mail.shu.edu.cn

Abstract: This paper presents a Parallel Data Mining Platform (PDMP), aiming at rapidly developing parallel data mining applications on cluster system. This platform consists of parallel data mining algorithm library, data warehouse, field knowledge base and platform middleware. The middleware is the kernel of the platform, which comprises data processing component, task manager, data manager, GUI, and so on. Taking advantage of cluster system, the middleware provides a convenient developing environment and effective bottom supports for implementation of parallel data mining algorithms. So far, the parallel data mining algorithm library has possessed several parallel algorithms, such as classification, clustering, association rule mining, sequence pattern mining and so on. With a register mechanism, the parallel data mining algorithm library is easy to be extended by users. Experiment results with the use of the PDMP show that there is a substantial performance improvement due to the cooperation of the middleware and corresponding parallel strategy.

1 Introduction

Along with more and more information available for analysis, scalability of Data Mining (DM) applications is becoming a critical factor. Implementation of Parallel Data Mining (PDM) based on high performance computer can significantly enhance the ability of data processing and the efficiency of computing [NW03] [Don02]. PDM techniques have been developed and are becoming very interesting and rewarding. However, developing, maintaining, and optimizing a parallel data mining application on today's parallel systems is an extremely time-consuming task. We believe that the following challenges still remain in effectively using parallel computing for scalable DM [CD97].
●Ease of development.
●Effectiveness of PDM algorithms.
●Ability of dealing with large data sets.
In this paper, we present a Parallel Data Mining Platform (PDMP), which

aims to enable the development of parallel data mining applications rapidly on cluster system. The PDMP is based on the observation of various parallel versions of several well-known data mining techniques. We have carefully studied parallel versions of association rule mining [AN96][HJ98], classification pattern mining [DGL89], clustering and sequential pattern mining [AGZ92][AH92]. On cluster system, which has emerged as a cost-effective and common parallel computing environment in recent years [Ort03], each of these DM methods implements parallelization by dividing the data instances among the cluster nodes. The computation on each node involves reading the data instances in an arbitrary order, processing each data instance, and performing a local reduction. After the local reduction on each node, a global reduction is performed. This similarity in structure can be exploited by the PDMP to execute the data mining tasks efficiently in parallel.

Taking advantage of cluster system, the PDMP provides a convenient PDM developing environment, a scalable PDM algorithm library, and effective bottom supports for the implementation of PDM algorithms. Based on these bottom supports, such as data storing, data fetching, data dividing and delivering,and R/W interfaces etc, we have developed several parallel versions of the well-known DM algorithms. Experimental results on ZiQiang2000, a cluster system developed by Shanghai University, reveal that there is a substantial performance improvement with the use of the PDMP.

2 Architecture of the PDMP

The architecture of the PDMP is shown in Figure 1. The PDMP is composed of two parts, one is platform middleware and the other is the organization of various resources including parallel data mining algorithm library (PDMAL), data warehouse, and field knowledge base etc.

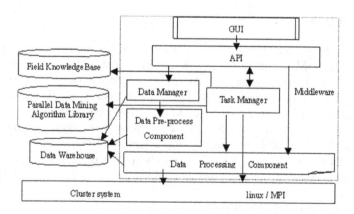

Fig. 1. Architecture of the PDMP

The middleware is the kernel of the platform. It consists of Data Processing Component (DPC), Task Manager, Data Manager, Data Pre-processing Component, GUI, and so on. By means of GUI/API, user submits a data-mining task by inputting a command or selecting data set, mining algorithm and its parameters with wizard. Task Manager accepts the task and acts as a scheduler. The DPC receives the messages from Task Manager, fetches the mining data set from data warehouse, divides the data set and delivers them to different nodes [see section 3.3]. And then Task Manager activates the selected algorithm to start corresponding mining procedure. Each correlative node operates the data subset on its own in parallel. At the end, a specified node gathers all local results from each operating node, and produces a final result. User can evaluate the result by referring to the field knowledge base or other guidelines. If necessary, they can also transfer the result into a visible style or repeat the task with new parameters. Below we will particularly describe the details of the middleware in the PDMP.

3 Platform Middleware

DM is a kind of complicated techniques integrated by many subjects. Thus, developing, maintaining, and optimizing a parallel data mining application on today's parallel systems is an extremely time-consuming task. For users, there is some limitation to well use DM techniques due to the lack of the knowledge of algorithms and relevant parameters etc. The objective of platform middleware is to simplify the use of DM platform and make the development of PDM application easier.

As described before, the middleware is the kernel of the PDMP. Through the GUI environment, users can use the Data Manager to query, import, export, or pre-processing data sets; use the Task Manager to submit a DM task. GUI also contains a wizard that allows users to select algorithms from algorithm library, input parameters and repeatedly modify parameters. While, the DPC provides bottom supports for the execution of algorithms, such as data storing, data fetching, data dividing and delivering etc. As a result, the system is in a hierarchy architecture where every component dedicates to its own function or service. The design and implementation of the platform middleware are discussed in detail as below.

3.1 Task Manager

In the PDMP, Task Manager accepts tasks and acts as a scheduler. It performs the following functions:

(1) Accepts and analyzes the DM tasks from GUI/API, converts them into a profile of selected algorithm, mining data set, and other parameters.

(2) Breaks down every task into some steps including data preparing, algorithm executing and results ingathering, and schedules them into running optimally.

(3) Does data preparing including collecting available nodes; determining how to divide the mining data set; calling the DPC to fetch the selected data set from data warehouse and execute the data division and distribute the divided data subsets to different nodes of cluster system.

(4) Calls the selected algorithm to run on each node with its own data subset.

(5) Appoints a specified node to perform the results ingathering and return the final results to GUI.

3.2 Data Manager

Data Manager is in charge of providing tools for users to query, import, export, or pre-process data sets. The data mining algorithms require sufficient and high quality data to obtain satisfied results. In many cases, the source data provided for data mining are of great quantity and complexity. They could be heterogeneous database or various data files. Therefore, tools of transforming data format and data pre-processing component are necessary for a well-designed DM platform. The primary function of the Data Manager is as follows.

(1) **Data import and export**: Data resources of DM can be of various kinds. Through transforming data format, data Manager allows users to import data from Access, SQL-Server, Oracle or various data files such as Excel files and text files, and then integrates them into data warehouse. On the other hand, Data Manager also allows users to export the data in data warehouse to various tables or files corresponding to different databases and file systems.

(2) **Data browse and query**: together with GUI, many ways of data browse and query are provided in Data Manager for users to select their mining data set expediently.

(3) **Data pre-processing**: In order to enhance the data quality, users can call the data pre-process component through Data Manager to do the data cleaning, data integration, data transformation, and data reduction etc, according to the requirements of the algorithms.

3.3 Data Processing Component (DPC)

As described above, the DPC provides bottom supports for the implementation of algorithms. Focused on cluster system, the DPC is designed for fetching, dividing and delivering data for all data mining algorithms when they were selected to run. It enables high data processing performance by adopting this hierarchical design and implementation methods. Independent of algorithm implementation, I/O, communication, and synchronization optimizations can be implemented in the DPC, enabling different parallel data mining applications to benefit from them. The architecture of the DPC is shown in Figure 2.

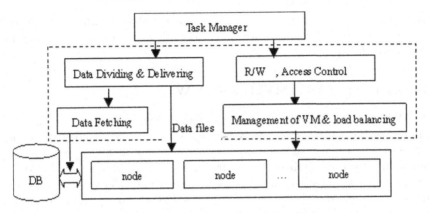

Fig. 2. Architecture of Data Processing Component

As shown in Figure 2, the DPC mainly consists of Data Fetching (DF), Data Dividing and Delivering (3D), Management of Virtual Memory and load balancing (MVM), and R/W and Access Control etc. Below we will give the details for each function module of the DPC.

Data Fetching (DF)

The DF Module is in charge of providing a unique interface for accessing data warehouse. With this interface, algorithms need not to take care of the connection of data warehouses and have high I/O performance.

When multi-algorithms are running, the data warehouse is visited so frequently that it will unavoidably become a bottleneck that results in low I/O performance. The DF Module uses asynchronous I/O operations and synchronization optimization to obtain higher I/O performance. It provides a unified method of data source selection method and allows users to fetch selected data according to the parameters inputted via GUI. The selected data sets can be stored in temporary files for further usage accordingly.

Data Division and Delivery (3D)

With responsibility for data division and delivery, the 3D module is composed of various data division methods and data distribution strategies available for different algorithms.

Obviously, how to divide data is very important. The 3D module divides data set in the ways of lines, columns, or both of them respectively. For example, division according to lines means to divide data set into q (q=n/m) blocks, where m is the number of available nodes and n is the number of the lines of data. When the Task Manager calls an algorithm, an appropriate data division method is executed.

After data set is divided into data subsets, each of them is converted into the form of files and delivered to corresponding node according to the distribute strategy of the algorithm.

Management of VM (MVM) and R/W

The MVM module provides users a transparent environment of dynamic distributed storage. In cluster system, the data transmission among the nodes may be faster than local disk access due to the use of high-speed network equipment, such as Myrinet etc [RC99]. Therefore we developed the MVM module that uses local memory as cache, and uses memory of other nodes as auxiliary storage to replace local hard disk and other storage medium. In this way, parallel data mining algorithms read their data only from local memory and the entire performance of the PDMP is significantly improved. Meanwhile, a quite large storage space is acquired.

Based on the MVM module, the R/W module encapsulates all the data accessing operations and supplies the compatibility of operations of data files upward. Parallel DM algorithms in the PDMP directly use this module to access their data.

4 Parallel DM Algorithms and Experimental Results

Effective algorithm is important to any DM task. However, selecting the most appropriate data mining techniques for a particular analysis task in a particular domain is typically done through trial and error, i.e. by applying various techniques on available datasets and viewing the results obtained. Therefore, a Parallel Data Mining Algorithm Library (PDMAL) is established in the PDMP, in perfect cooperation with platform middleware, Which provides a perfect interface for users to select proper algorithms and data set.

So far, the PDMAL contains several parallel algorithms, which are commonly used in data mining techniques such as classification, clustering, association rule mining, sequence pattern mining and so on. These algorithms are implemented on top of the platform middleware and shared a relatively similar structure. With a register mechanism, the PDMAL is easy to be extended by users.

For performance evaluation, we have conducted a series of experiments on ZiQiang2000, a cluster system developed by Shanghai University. With the peak speed of 450Gflops, ZiQiang2000 is involved with 102 SMP nodes, each with dual-CPU architecture. Experiment results show that there is a substantial performance improvement with the use of the PDMP, which profits from the cooperation of the middleware and corresponding parallel strategy. In the rest of this paper, we will give two examples of the experiments and show their results.

4.1   A Parallel Association Rule Mining Algorithm

The Parallel Frequent Pattern Tree algorithm (PFPT) is developed for association analysis based on frequent pattern tree. It generates frequent patterns instead of candidate item sets, so leads to a significant reduction of the I/O and communication time. Its implementation is on top of the platform middleware. The implementation process of the PFPT algorithm is as follows:

A) The middleware reads the mining data set from data warehouse, divides it into subsets, converts the data subsets into forms of files and delivers them to each node of the cluster according to relevant distribution strategy.

B) Each processor scans local data subset and gets counts of each 1-item, then exchanges them among all processors to get global counts of each 1-item, get the global frequent 1-item sets-Flist.

C) Each processor compresses Local data set into a Frequent Pattern Tree (LFPTree) with Flist, and gets Local Condition Pattern Base(LCPB| α) of each 1-item α in Flist from its LFPtree, passes LCPB|α to P[α](processor set to receive LCPB|α from other processors) and generate Global Condition Pattern Base (GCPB| α) in the P[α].

D) All the processors construct Condition Frequent Pattern Trees about item α (CFPTrees|α) by GCPB| α and mine all frequent item sets with the tail item α.

The PFPT algorithm has been successfully applied to the history data of elevator service to discover valuable rules that can be used to optimize the decision-making of balance between the maintenance and repair of elevators. The history data set of elevator service involves 50000 records, each with at least 12 items. For the sake of comparison, we use 4 nodes to running the PFPT and Count Distribution (CD) Algorithm, a well-know typical parallel algorithm [MS97] respectively. Figure 3 shows the running time costs of them.

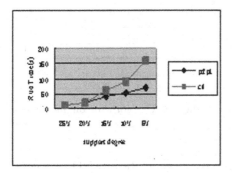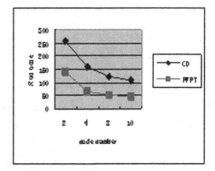

Fig. 3. Running-time cost comparison between the PFPT and Typical Parallel CD algorithm

From figure 3(a), we can find that the PFPT obtains a significant performance improvement comparing with typical Parallel CD algorithm, especially with low support degree. In figure 3(b), the running time cost decreases with the nodes increasing in both algorithms, but the PFPT always has better performance.

4.2   A Fast Scalable Parallel Classification Algorithm

Classification is a vital form of data analysis, many classification models have been proposed. In the DMPAL, Fast Scalable Parallel Classification algorithm (FSPC) is developed that can handle large data sets with lots of attributes. It introduces techniques such as partitioning data set vertically; performing the split while finding split points. We have presented the algorithm in detail in [YW04]. In the PDMP, the middleware reads the mining data set from data warehouse for the FSPC, algorithm divides it into k subsets vertically, where k is the number of attributes, then converts every subset into forms of files and delivers them to each node according to the distribution strategy. These operations are performed by the middleware with the techniques of minimizing disk seek time and using asynchronous I/O operations. With use of appropriate data partition strategy and adaptive parallelization strategy, the algorithm obtains significant improvement of performance. Figure 4 shows the experiment results of the FSPC algorithm running on the typical mushroom dataset.

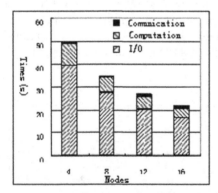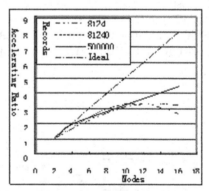

Fig. 4. Time-cost and accelerating ratio of the FSPC algorithm on mushroom dataset

Figure 4(a) shows the results comparison of the computation, I/O, and communication time cost, running on the typical mushroom dataset with 500000 records and 22 attributes. We can find that the majority of time cost is I/O and computation time. Along with the nodes number increasing, I/O and computation time will rapidly decrease, but communication time will increase appreciably. Actually, communication time cannot affect the performance, because there is a little proportion of communication time cost. In Figure 4(b), the accelerating ratio curves of the FSPC algorithm running on datasets with different records are shown. We can find that the curves close to the ideal one more and more with the data amount increasing.

5 Conclusion

Developing efficient parallel applications is a difficult task on today's parallel systems. The proposed PDMP expects to provide practitioners a convenient platform, which makes application development easier. The PDMP gives users a convenient developing environment, a scalable PDM algorithm library -PDMAL, and effective bottom supports for the implementation of algorithms.

In a hierarchical architecture, the platform middleware enables high I/O performances by minimizing disk seek time and using asynchronous I/O operations. It can be used for rapidly developing efficient parallel data mining applications that operate on large datasets. I/O, communication, and synchronization optimizations are implemented in the platform, enabling different parallel DM applications to benefit from them. Experimental results imply a substantial performance improvement with the use of PDMP, which works with the perfect cooperation of the middleware and corresponding parallel strategy.

In future works, we would like to carry on further evaluating our platform by developing parallel implementation of more effective algorithms and apply it to practical applications.

6 Acknowledgement

We are grateful to the National Joint Foundation of Earthquake (Project No.7A05709), the Nature Science Foundation of Shanghai Municipality (Project No.7A05468), the sustentation fund subject of Shanghai Municipal Education Commission (Project No. 205153), and the fourth period of key subject of Shanghai Municipal Education Commission (Project No. 205153B682) for their financial support to our work.

References

[JM94] Jiawei Han and Micheline Kamber: (1994) Data Mining: Concepts and Techniques. NY: Morgan Kaufman, pp.15.
[DZ01] Du Zhihui :(2001) High performance computing parallel programming technology-MPI parallel program design. Beijing Tsinghua University Press.

[AT99] Albert Y.Zomaya, Tarek El-Ghazawi and Ophir Frieder. 1999 "Parallel and Distributed Computing for Data Mining". IEEE Concurrency, pp.11-14, volume 1092-3063,

[JM95] J.S.Park, M.S.Chen, and P. S. Yu. : Efficient parallel data mining of association rules. 4th International Conference on information and Knowledge Management, Baltimore, Maryland,Novermber 1995.

[AJ96] Agrawal and J.Shafer. : Parallel mining of association rules. IEEE Transactions on knowledge and Data Engineering, 8(6): 962-969, June 1996

[JM97] John Darlington, Moustafa M.Ghanem, etc.: Performance models for coordinating parallel data classification. In Proceedings of the Seventh International Parallel Computing Workshop (PCW-97), Canberra, Australia,September 1997.

[KA99] Kilian Stoffel and AbdelkaderBelkoniene.: Parallel k/h-means Clustering for large datasets. In Proceeding of Europar-99, Lecture Notes in Computer Science (LNCS) Volume 1685, pages 1451 -1454. Spring Verlag, August 1999.

[TM98] Takahiko Shintani and Masaru Kitsuregawa, "Mining Algorithms for Sequential Patterns in Parallel: Hash Based Approach". Pacific-Asia Conference on Knowledge Discovery and Data Mining, pp.283-294. 1998.

[RB02] Rajkumar Buyya: High Performance ClusterComputing: Architectures and Systems Volume 1. People post press. 2002. P3838.

[RC99] R Lottiaux,C Morin.: File mapping in shared virtual memory using a parallel file system. In: Proc of Parallel Computing' 99. Delft, Netherlands, 1999. pp.606-614.

[MS97] M. J. Zaki, S. Parthasarathy, and W.Li.: A localized algorithm for parallel association mining. 9th Annual ACM Symposium on Parallel Algorithms and Architectures, Newport, Rhode Island, June 1997.

[YW04] YAN Sheng-xiang, WU Shao-chun, WU Geng-feng. : A Data-Vertical-Partitioning Based Parallel Decision Tree Classification Algorithm. Application Research of Computers. To be published in 2004.

Architecture Design of a Single-chip Multiprocessor*

Wenbin Yao[1,2], Dongsheng Wang[2], Weimin Zheng[1], and Songliu Guo[1]

[1] Department of Computer Science and Technology, Tsinghua University, P. R. China
[2] Research Institute of Information Technology, Tsinghua University, P. R. China
    {yao-wb, wds, zwm-cs}@mail.tsinghua.edu.cn;
    guosongliu01@mails.tsinghua.edu.cn

Abstract. Single-chip multiprocessor(CMP) architecture provides an important research direction for the future microprocessors. A CMP architecture(T-CMP) is put forward on the basis of the prior researches on RISC microprocessor. T-CMP integrating two MIPS-based processors is a closely coupled multiprocessor, engaging aggressive thread-level and process-level parallel techniques to improve the performance. This paper presents the key techniques of its implementation, including hardware construction and software design. A functional verification simulator is also developed to simulate the behavior of program executions in the mode of cycle-by-cycle. The results show that the design can improve the computing performance effectively.

Key words: Single-chip multiprocessor, architecture, parallelism, functional verification simulator

1 Introduction

The growth of dependency on computer systems demands the processors to provide higher and higher performance. This is especially true in the fields of microprocessor systems, which usually exploit concurrency at all levels of granularity to improve the computing performance. Currently, superscalar and VLIW general adopted in the microprocessors are two wide-spread architectures of instruction-level parallelism. However, they achieve the better performance at the cost of increasing the complexity of control logic. The static data, however, have shown that the marginal return of additional hardware was decreasing dramatically with the increased complexity[AAW96].

---

*This paper is supported by National Natural Science Foundation of China and "863" Fundamental Science Program of China and China Postdoctoral Science Foundation.

Recently, as the trend toward thread-level parallelism is getting mature, single-chip multiprocessor architecture become a promising solution to this end[SYS99, CWM01, NC01]. CMPs integrate multiple processors into a single chip, and thus combine the advantages of both microprocessor and multiprocessor architectures. From the view of construction, a standard CMP is easier to implement than other microprocessors[HNO97]. General architectures such as superscalar and VLIW are usually much difficult to design and thus results in more development and verification times. On the contrary, CMPs usually exploit very simple control logic and even reuse the existed well-defined or commercial successful CPU cores to construct the systems so as to reduce the risk of design. In addition, as the delay of interconnects between gates is becoming more significant, minimizing the length of wires and simplifying the design of critical paths are two important features. CMPs can also mitigate these design pressures greatly through making each processors take up a relatively small area on a large processor chip.

From the view of computing performance, CMPs also offer more advantages. Since the inter-processor communication latencies are lower and bandwidths are higher, CMPs are much less sensitive to data layout and communication management. As a result, they can yield higher speedup on data transaction among processors, which makes them faster than conventional multiprocessors on running parallel programs especially when threads between processors communicate frequently.

Realizing the great benefits, we design a CMP architecture named after T-CMP on the basis of the prior researches on RISC-based microprocessors (Thump-107). Through the design of this prototype system, we aim at the following goals:

(1) Verify the functionality and expandability of CMP architecture;

(2) Provide a simulation environment for the function verification of CMP and the optimization of multi-thread compiler;

(3) Assess the performance benefits of CMP, compared to other microprocessors and multi-chip multiprocessors.

The rest of the paper is organized as follows. Section 1 gives the system overview of T-CMP architecture. Section 2 shows the key techniques of T-CMP design, including hardware construction and software design. Section 3 introduces a software simulator used for verify the function of T-CMP, and give the simulation results in Section 4. Section 5 draws the conclusion.

2 System Overview

T-CMP is a closely coupled multiprocessor, which integrating two identical RISC-based microprocessors into a single chip. Its logic overview of architecture is shown in Fig. 1.

Six major functional units in T-CMP are described as follows:

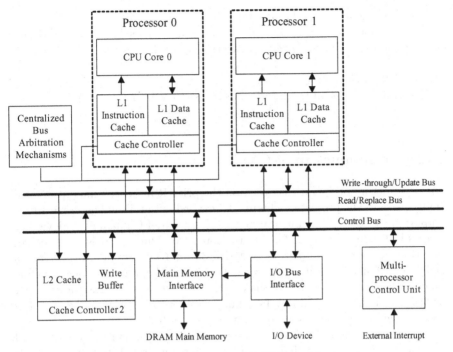

Fig. 1. Logical overview of T-CMP architecture

(1) Two identical RISC-based processors

Each processor is an independent 32-bit processor which compatible with the MIPS 4Kc instruction set, and plus four instructions for multimedia applications. Each processor also includes an 8k-bytes instruction cache and an 8k-bytes data cache and corresponding control logic.

(2) A shared secondary cache

It is a unified cache which has the capacity of 1M-bytes. The two identical processors share this cache to accelerate the access time from/to the memory.

(3) A centralized bus arbitration mechanism

This controlling mechanism is in charge of arbitrating and distributing the control privilege of the shared bus to the appropriate processor.

(4) Multi-processor control unit(MCU)

The functions of this unit are to process external interrupt distribution and inter-processor interrupts, and also to provide hardware support for synchronization of the processors.

(5) Main memory interface(MIU)

The interface handles all the interfacing transactions from/to T-CMP, including main memory accesses and external snoop processing.

(6) I/O interface unit

The unit handles all the interfacing transaction from/to T-CMP.

3 Key techniques in T-CMP

The architecture of T-CMP is based on the prior RSIC-based microprocessor (Thump-107). The project of Thump-107 carried out at Tsinghua university is supported by "863" Fundamental Science Program of China and achieves the success in the first tap-out. The instruction set of Thump-107 is the superset of MIPS-4Kc and is compatible to MIPS 32-bit RISC architecture. It has a 4k bytes instruction cache and a 4k bytes data cache, and a 7-stage pipeline which was capable of simultaneous execution of up to seven instructions per clock cycle. It has the primary frequency about more than 400Mhz, and has been applied to our net computer systems successfully.

Before the design of a CMP system, two key problems must to be solved in advance. One is how to share the information among different processors. If the lines between processors are too long or the frequency of communications among them is too high, it may decline the frequency of the microprocessor. The other is the communication latency between the microprocessor and off-chip units, demanding the pin bandwidth satisfy the requests from CPU. If these two problems are unsolved, adopting CMP architecture to raise the computing performance is no longer to be applicable.

For these two problems, we introduce a kind of software/hardware synergy approach. To the first problem, we design a shared secondary cache to reduce the communication frequency, and it also helps to control the complexity of the protocols of cache coherence. As for the secondary problem, to keep the compatibility of pin definition with our prior microprocessor, we increase the capacity of the secondary cache for more than ten times of the primary cache in the processors to decline its miss rate. Furthermore, we also adopt a software-based approach, providing the software optimization supports on the complier and cache operations to satisfy the data requirements from the memory to the processors.

3.1 Hardware design

To provide fine compatibilities, T-CMP has the same pin definitions with Thump-107. It also provides some special structures to satisfy the demands of multi-thread program executions. Here, we introduce the three key topics in the hardware design:

(1) Instruction set expansion

Every processor has the same structure as that of Thump-107, except adding two instructions specially for implementing synchronization primitives. They are load locked(LL) instruction and store conditional(SC) instruction, respectively.

(2) Cache hierarchy

A two-level cache hierarchy is designed to reduce the communication frequency between the identical processors. On the implementation, each processor has its own pair of primary instruction caches and data caches while all the processors share a single on-chip secondary cache. The secondary cache is a unified cache whose capacity is 1M-bytes. The configuration is shown in table 1.

Table 1. Cache configuration of T-CMP

Characteristic	L1 cache	L2 cache	Main memory
Configuration	Separate I SRAM and D SRAM cache, pairs for each processor	Shared, on-chip SRAM cache	Off-chip DRAM
Capacity	8K bytes each	1M bytes	128M bytes
Bus width	32-bit bus connection	64-bit read bus + 32-bit write bus	32-bit bus
Access time	1 CPU cycle	10 CPU cycles	At least 100 cycles
Associativity	2 way	4 way	N/A
Line size	32 bytes	64 bytes	4k bytes pages
Write policy	Write through, no allocate on write	Write back, allocate on writes	Write back (virtual memory)
Inclusion	N/A	Inclusion enforced by L2 on L1 caches	Includes all cached data

The program counter and register file of every processor are so separated physically as well as logically from each other as to process an instruction stream which is assigned with a thread independently.

(3) Bus arbitration mechanism

Connecting the processors and the secondary cache and other control interface together are the read and write buses, along with address and control buses. While all the buses are virtual buses, the physical wires are divided into multiple segments using repeaters and pipeline buffers. Considering the future expandability such as the addition of more processors, a bus arbitration mechanism is adopted to arbitrate the possessing privilege of shared bus during the communications of the processors.

The read bus acts as a general-purpose system bus for moving data between the processors, secondary cache, and external interface to off-chip memory. The two processors can get the required data from secondary cache via read bus simultaneously. Data from external interface can be broadcast to both of the processors while it is sending to secondary cache.

The write bus permits T-CMP to use write-update coherence protocol to maintain coherent primary caches. Data exchange between the processors is carried out via a write-through/update bus under the control of the centralized bus arbitration mechanism. When one processor modifies data shared by the other, write broadcast over the bus updates copies of the other processor while the permanent machine state is wrote back to the secondary cache.

3.2 Software design

Since each processor in T-CMP has its own program counter and register file, it is easy to implement the parallelizing executions of multiple instruction streams. Because the processors in T-CMP are single-issue processors, two independent instructions streams — generally called threads — can simultaneously issued to the different processors, respectively.

For a single application, extracting multiple threads is the premise to utilize T-CMP effectively. Because of the complexity of parallel programming, automatic parallelization technology which enables ordinary sequential programs to exploit the multiprocessor hardware without user involvement becomes the key of the implementation. Unfortunately, current compiling technology is not so mature as to fully support this goal. To solve it, T-CMP resorts to software support for achieving incremental performance improvement.

In the detailed implementation, T-CMP trades the functional distribution between the compiler and operating system to exploit both thread-level and process-level parallelisms. They are capable of dispatching multithreaded instruction streams without any hardware interference, aiming at the maximization of parallelization handling. As a result, multiple threads can be executed independently on the different processors, except synchronizations in memory references.

In the compiling level, sophisticated parallelizing compiler converses the sequence instruction queues(such as data-independent loop iterations) into parallelism instruction queues, or reorders some adjacent instructions to achieve more parallelism. In addition, T-CMP provides specific instructions to synchronize the executions of different processors, which can be exploited to explicitly divide sequence codes into threads by programmers. These instructions, which are also called synchronization points, are inserted into the sequence program to solve all possible true dependencies. Corresponding to these synchronization points, system library functions are also provided to further support the multi-threaded computing and cache optimization.

In T-CMP, a form of coarse process-level parallelism can also be scheduled by operating systems. In the multi-programmed or multi-user environments, workloads consisting of threads are generally independent each other and thus each processor can processes different programs independently. This kind of working mode avoids the problem of instruction dependence, which is the major bottleneck for the implementation of ILP in a single application. More importantly, current widespread use of multitasking operating systems makes this feasible.

4 Functional Verification Simulator

Designing a single-chip multiprocessor is a complex process, demanding the reliable evidences enough to validate the correction of functional implementation from the beginning of design phase. Therefore, a software-implemented behavior simulator described with c-language is developed to define the implementation of T-CMP and testify the characters of this architecture. It is a functional verification model, which can exactly simulate the operational behavior in the mode of cycle-by-cycle. A snapshot of the function verification simulator is shown in Fig.2.

Developing such a simulator has another benefit. It can be used to assess the function and efficiency of parallelizing complier. Software support for CMP architecture is an essential task, which requires more times and sources to parallelize the prior sequence programs. The simulator provides a studio for software designers to verify the implementations of parallelizing compiler and operating systems in the beginning phase.

5 Simulation Results

The improvement of performance can be evaluated coarsely through running some toy programs. To testify the functional features and the performance of T-CMP architectures, we designed several programs to verify of its effectiveness, which run on the microprocessors of single-processor and T-CMP respectively. The consumed cycles for these four sample programs(name as P1, P2, P3, and P4) running on the single-processor microprocessor are 50102, 141626, 61785, and 84762, respectively.

The initial state of T-CMP is just "boot-up" before the execution of programs, which means that both two levels of caches are empty. In such a mode,

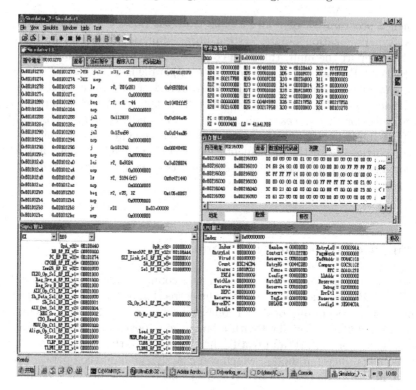

Fig. 2. A snapshot of the functional verification simulator

the programs must be filled into cache sequentially via the shared bus under the control of the centralized bus arbitration mechanism. This waste in cycles for filling up the empty caches may be released on the conditions of the execution of large and long programs. The speedup shown is not very high because the size of the sample programs stays in the toy level. With the increase of program size and the decrease of communications between threads of processors, the value of speedup may raise more quickly.

Based on the testing data obtained from the environment, it is shown that the centralized bus mechanism in the T-CMP has been the major bottleneck blocking the improvement of the performance. So we further refine the bus mechanism by expending the number of read/write ports, and at the same time add the bypass paths to the caches. The succeeding coarse testing results show that the refined T-CMP improve the average performance about a factor of 1.0~1.2, compared with the former version. The results of comparisons between different running modes are shown in Table 2 and Fig.3.

Table 2. Running results under the different running modes

Groups	CMP Mode	Sequent Mode	Speedup	CMP Mode (Refined)	Speedup
P1/P2	162,328	191,728	1.18	155,342	1.234
P2/P3	178,425	203,411	1.14	164,986	1.232
P3/P4	103,879	146,547	1.41	99,465	1.473

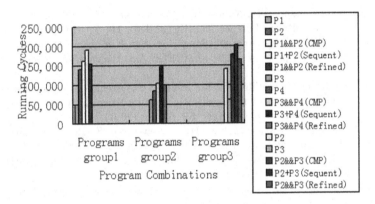

Fig. 3. Comparison between different modes on the aspect of running cycles

6 Conclusion

The trend of development in semiconductor technology shows that within 5 years it is possible to integrate a billion transistors on a reasonably sized silicon chip. At this integration level, it is necessary to find parallelism to effectively utilize the transistors. CMP is such a technology due to its merit of relatively simple design process, compared to other complex architectures[HHS00].

This paper presented the design of T-CMP, based on prior microprocessor Thump-107. We first describe the system overview from a micro-architecture perspective. Some key techniques including hardware and software design used for raising the performance are presented in detail. To verify the functionality and effectiveness, we develop a simulator to simulate its operational behaviors in the mode of cycle-by-cycle. To raise the efficiency of verification and shrink the design time, we also have designed a random test generator which could generate test programs automatically. With the help of these two tools, we can improve our design more quickly and effectively.

The further performance evaluation of T-CMP based on its function implementation is in progress now. The coarse results show that T-CMP improves the average performance about a factor of 1.2~1.7, compared with microprocessor Thump-107. These results show that compared with other architectures a fine-designed CMP architecture can improve the performance of microprocessor more greatly and easily, especially for some parallel applications. Our further work will focus on the architecture optimization from the different aspects such as low-power and reliability.

References

[AAW96]   Amarasinghe, S.P.; Anderson, J.M.; Wilson, C.S.; et al: Multiprocessors from a Software Perspective. IEEE Micro, **16**, 52–61.(1996)

[SYS99]   Sang-Won Lee; Yun-Seob Song; Soo-Won Kim; et al: Raptor: A Single Chip Multiprocessor. The First IEEE Asia Pacific Conference on ASIC, 217–220.(1999)

[CWM01]   Codrescu, L.; Wills, D. S.; Meindl, J.: Architecture of Atlas Chip-Multiprocessor: Dynamically Parallelizing Irregular Applications. IEEE Transactions on Computers, **50**, 67–82. (2001)

[NC01]   Nickolls, J.; Calisto L. J. M. III.: A Low-Power Single-Chip Multiporcessor Communications Platform. IEEE Micro, **23**, 29–43. (2003)

[HNO97]   Hammond, L.; Nayfeh, B. A.; Olukotun K.: A Single-Chip Multiprocessor. IEEE Computer, **30**, 79–85. (1997)

[HHS00]   Hammond, L.; Hubbert, B. A.; Siu, M.; et al: The Stanford Hydra CMP. IEEE Micro, **20**, 71–84. (2000)

Metadata Management in Global Distributed Storage System*

Chuanjiang Yi, Hai Jin, and Yongjie Jia

Cluster and Grid Computing Lab
Huazhong University of Science and Technology, Wuhan, 430074, China
hjin@hust.edu.cn

Abstract. Data-intensive applications require the efficient management and transfer of terabytes or petabytes of information in wide-area. Data management system in distributed computing environment could not meet this requirement efficiently. Storage visualization is an efficient way to solve these problems. It integrates all kinds of high performance storage system to a unit one. It can not only share resources and make fully use of resources, but efficiently avoid conflict of data explosion and limited storage ability. To resolve these conflicts, we propose a *Global Distributed Storage System* (GDSS).

With data explosion, metadata becomes more and more enormous. Exinting methods could not keep up with the need of scalability, availability and efficiency to manage metadata. To solve these problems, MDC (*MetaData Controller*), a novel metadata management scheme based on MatchTable, is introduced in GDSS. In MDC, MatchTable is responsible for communication between SSP (*Storage Service Point*) and DS (*Directory Server*), and cache module is responsible for increasing of metadata access efficiency and as an assistant to maintain the replica coherence of metadata. MDC locates metadata only once by matching MatchTable, which avoids searching it from root directory server step by step. Experiment results indicate that MDC can locate metadata fast and supply fault tolerance, and also has better availability and scalability.

1 Introduction

Data-intensive, high-performance computing applications require the efficient management and transferring of terabytes or petabytes of information in wide-area, distributed computing environments. As there are large self-governed

---

*This paper is supported by National Science Foundation under grant 60125208 and 60273076, ChinaGrid project from Ministry of Education, and National 973 Key Basic Research Program under grant No.2003CB317003

data in internet, resource sharing is becoming more and more important. Storage visualization is an efficient way to achieve this goal. It integrates all kinds of high performance storage system to a unit one. It can not only share resources and make fully use of resources, but efficiently avoid conflict of data explosion and limited storage ability. *Global Distributed Storage System* (GDSS) [JRW03] is our attempt to storage visualization.

In GDSS, metadata is very important. It describes data in the system. The functions of metadata are as the follows. (1) Implementing "single global user space". "Single" means GDSS can supply users their own view, and "global" means each user has a unique id (user_name + domain_name) in GDSS. (2) Recording information of data to implement data location, data search, data register, and access control. (3) Managing user and group, such as adding and deleting users in a certain group.

A flexible and scalable metadata management is needed for conveniently and availably accessing the data set consisting of all types of heterogeneous data. Many systems, such as SRB (*Storage Resource Broker*), manage metadata in the hierarchy directory structure. But with data explosion, metadata becomes more and more enormous. This method could not keep up with the need of scalability, availability and efficiency to manage metadata.

To solve these problems, MDC (*MetaData Controller*), a novel metadata management scheme based on MatchTable, is introduced in GDSS. MDC locates metadata only by matching MatchTable once, which could avoid searching it from root directory server step by step. Experiment results indicate that this management can fast locate metadata and provide fault tolerance, and also it has better availability and scalability.

2 Related Works

CSFS (*Campus Star File System*) [HGD01] from Peking University describes single user information by using XML and achieves transparent accessing to storage resource. But it provides neither "group" (only "single user"), nor file sharing among users.

Globus [EKK03][VTF01] from Argonne National Laboratory is a middleware, which focuses on resource management, security, information service, data management and so on in grid computing. Under standard protocol, Globus implements several basic mechanisms such as data file moving, remote accessing GASS [BFK99], and high speed transferring Gridftp [TFG03]. Based on those functions it implements metadata replica management. In Globus, MDC manages metadata, however, which could not transparently access storage resources.

SRB [RWM02][WRM03] from San Diego Supercomputer Center is a middleware accessing heterogeneous storage system in a uniform way in the distributed environment, which can supply transparent data service for the higher users/applications. MCAT is adopted to manage metadata in SRB. MCAT records metadata in the way like UNIX file. This method could not keep up with the need in scalability, availability and efficiency. First, this central scheme will lead to single point of failure in the whole system. Second, when metadata becomes enormous the system employs several metadata servers, and these servers must cooperate and the result will return from the root node that adds the overload to the root server. Third, when the root server is out of service the whole metadata server will be out of service. It is difficult to provide high availability. It is also difficult to scale up the metadata servers.

3 Metadata Controller—MDC

To solve these problems, fast lightweighted metadata is required, improving metadata accessing efficiency and guaranteeing better metadata replica coherence. We propose a novel metadata management method, called MDC, in GDSS.

3.1 Metadata Structure

GDSS manages the whole system by dividing it into several logic domains. All of metadata in a certain domain form a metadata logical tree, which is just like the UNIX file system. Fig. 1 is an example of the logical tree. In this figure, DS-A, DS-B, DS-C, and DS-D is DS in this domain; "root" is the root directory of this tree. A1, A2, . . . is sub directories of the tree.

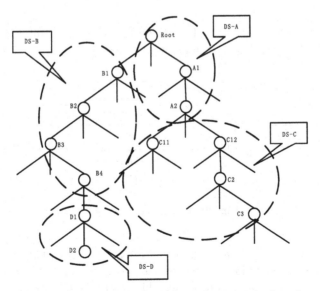

Fig. 1. Example of the logical tree in a certain domain

Each user belongs to a certain domain. User's root directory is bonded to one of sub directories in this domain. While a user logins, system automatically directs it to the user's root.

3.2 Synopsis of MDC

To maintain self-organization in each domain and high performance, metadata service is built on LDAP protocol, which is a distributed and consolidated protocol. In each domain metadata is stored in DS (*Directory Server*), and a GNS (*Global Naming Server*) is created to manage all the DS in a domain. Fig.2 is the sketch of GNS and DS in GDSS.

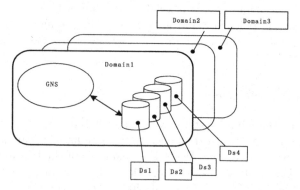

Fig. 2. Sketch of GNS and DS in GDSS

Fig.3 describes architecture of GNS. There are three modules in GNS, and function of each module is as the follows:

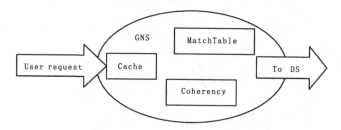

Fig. 3. Architecture of GNS

MatchTable. It records name and root directory of each DS. When "read" operation does not hit in cache, find out a DS that the file/directory user requested is located from MatchTable, which can take full use of replicas and reduce load of DS in this domain.

Cache. It stores the hot metadata and metadata. Write operation is firstly done only in cache, and then is sent to DS, which can avoid conflict arousing by accessing varying replicas of one metadata at the same time.

Coherency. While write operation comes, the update information will be recorded in cache, and then synchronize all of the DS that store this metadata.

3.3 MatchTable

The search efficiency of metadata affects the whole efficiency of the system. With the increasing of resources and information of the files and the directories, root DS will become a bottleneck with respect to expansibility and availability. MatchTable in GDSS is an efficient way to locate metadata in these DS fast and accurately.

MatchTable preserves each DS's name and root directory, and it is a structure always in memory to improve the locate efficiency. MatchTable management module is set on GNS.

When the system receives user's request, it firstly queries MatchTable to find out which DS the file/directory user requests exists, and then directly access this DS, which avoids to search from root DS step by step and reduce the load of root DS.

For example, the tree in Fig.1 is a file/directory structure in a domain with four DS. The root directory of tree exists in DS-A, directory "/root/B1" in DS-B, directory "/root/A1/A2/C11" and "/root/A1/A2/C12" in DS-C, and directory "/root/B1/B2/B3/B4/D1" in DS-D. Table 1 is the corresponding MatchTable.

Table 1. MatchTable corresponding to Fig.1

root directory of DS	name of DS
/root	DS-A
/root/B1	DS-B
/root/A1/A2/C11	DS-C
/root/A1/A2/C12	DS-C
/root/B1/B2/B3/B4/D1	DS-D

Suppose that a user wants to access directory "/root/B1/B2/B3/B4/D1/D2". If using the present accessing scheme, it will start from DS-A, the root DS, passing through DS-C to DS-D, finally return the metadata to user, shown in Fig.4. But if use MatchTable, system first looks up the MatchTable and makes the furthest matching, and DS-D will be quickly found out, shown in Fig.5.

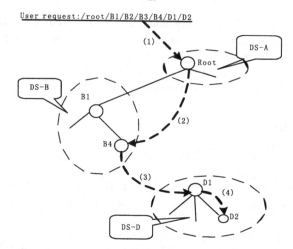

Fig. 4. The present accessing scheme to access metadata

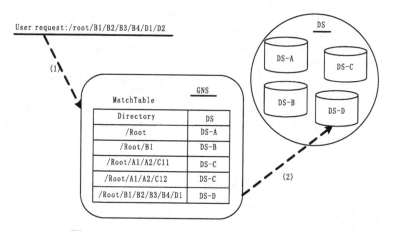

Fig. 5. MatchTable scheme to access metadata

3.4 Cache

While a user's request is received, GNS firstly searches it in its cache. If hit, read/write cache directly, the metadata returned to user is the latest. If not hit and is a "read" operation, system reads a DS. If it is "write" operation, system records these update information in its cache, and then update the related DS.

Hash map is used to implement the cache module. Each object in cache is described by "Attributes". There are two kinds of metadata in cache, one read object, and the other write object. Two single links are used to maintain those two objects. Read objects are organized by single link "R_lastAccessedList", and is sorted by the last accessing time. Write objects are organized by single link "W_ageList", and is sorted by the creating time. If a write object has been updated to every related DS, it is moved from "W_ageList" to "R_lastAccessedList". If cache is full, the front of "R_lastAccessedList" is deleted and only object in "R_ lastAccessedList" can be deleted.

3.5 Metadata Replica

Metadata is so important that replica is necessary. Based on metadata characteristics in GDSS, directory is the smallest unit to set up replica for metadata. All the replicas to a certain metadata are equal. Different users have different rights to decide the number of metadata replicas.

If a metadata has replicas, name and root directory of the DS that stores these replicas also will be recorded in MatchTable. When the request comes, MatchTable provides a DS list to system. If something wrong with one DS, replica on other DS will be accessed smoothly.

Fig.6 is an example of metadata tree of a domain, and directory "/root/B1" is stored in "DS-B" and "DS-C". Table 2 is the corresponding MatchTable to this tree. If "DS-B" goes down, "DS-C" could be accessed.

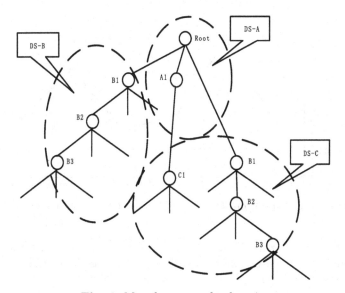

Fig. 6. Metadata tree of a domain

4 Performance Evaluation

The test environment of GDSS consists of several servers in Cluster and Grid Computing Laboratory and Storage System Key Laboratory. The former

Table 2. MatchTable corresponding to metadata tree in Fig.6

Root directory of DS	Name of DS
/root	DS-A
/root/B1	DS-B
/root/A1/C1	DS-C
/root/B1	DS-C

laboratory has fifty host computers, a cluster with sixteen nodes and a gateway, and the later laboratory has several disk arrays, SAN, and NAS. There are two domains in the system. One is iccc.hust.edu.cn, the other is storage.hust.edu.cn.

Response time to directory browsing. Login GDSS and test the system response time to directory browsing. Test each group for fifty times and record the average response time. The result is shown in Fig.7.

Response time to user login. Response time to user login is the time from user beginning to login GDSS to user seeing his root directory. The result is shown in Fig.8.

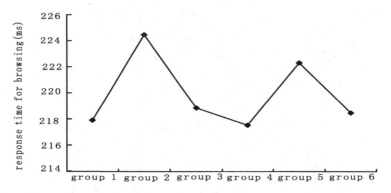

Fig. 7. Test result on the system response time for user browsing

Response time to read/write file. Response time to directory browsing represents the time from user submitting request to file beginning to transfer. The result is shown in Fig.9.

Response time to metadata searching. Totally there are 10694 files, including 9904 files and 790 directories, are tested. To describe these files, DS builds 33900 metadata items, which costs space 6M. Averagely, size of metadata to each file is 185Byte. Fig.10 is the the result of response time to metadata searching.

From these three test results we can see that using this metadata management scheme, the response time to user's operations is always in a stable time range. From the last test result, size of metadata information is very

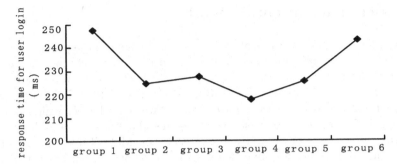

Fig. 8. Response time for user login

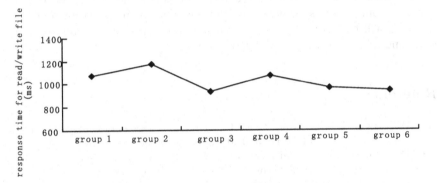

Fig. 9. Result of response time for read/write files

small so that GDSS has the ability to manage massive storage device. With the increasing of search range and precision, system response time decreases sharply.

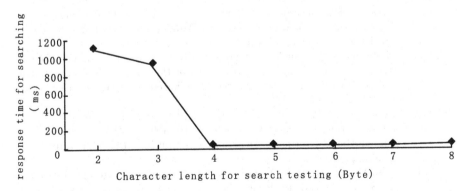

Fig. 10. Response time for metadata searching

5 Conclusion and Future Work

This paper analyzes shortages of metadata management in the present storage visualization systems, and then brings forward a novel metadata hierarchy management scheme. In this scheme, metadata is stored in all of the DS in this domain, and a GNS is created to manage all the DS in the domain. All metadata of a certain domain form a metadata logical tree, which is just like the UNIX file system. System locates metadata by MatchTable, which records the root directory and name of each DS. MatchTable can not only locate metadata fast and reduce load of root directory server, but get better system scalability and availability.

In our future work, we will strengthen function of metadata fault tolerant and bring out a more suitable scheme for metadata replica creation and selection. We will bring out a searching scheme based on attribute, and optimize searching scheme to make MDC run better.

References

[JRW03]   Jin H., Ran L., Wang Z., et al.: Architecture Design of Global Distributed Storage System for Data Grid. High Technology Lett., **9**(4), 1–4 (2003)

[HGD01]   Han H., Guo C., Dai Y., et al.: A scheme to construct global file system. In: Proceedings of the Second International Conference on Web Information Systems Engineering. IEEE, Inc. Kyoto (2001),pp.206–212

[EKK03]   Ellert M., Konstantinov A., Konya B., et al.: The NorduGrid project: Using Globus toolkit for building GRID infrastructure. Nuclear Instruments and Methods in Physics Research, Section A. Elsevier **502**, 407–410 (2003)

[VTF01]   Vazhkudai S., Tuecke S., Foster I.: Replica Selection in the Globus Data Grid. In: Proceedings of the 1st International Symposium on Cluster Computing and the Grid. IEEE, Inc. Brisbane (2001),pp.106–113

[BFK99]   Bester J., Foster I., Kesselman C., et al.: GASS: A data movement and access service for wide area computing systems. In: Proceedings of the Sixth Workshop on Input/Output in Parallel and Distributed Systems. ACM Press Atlanta (1999),pp.78–88

[TFG03]   Thulasidasan S., Feng W.C., Gardner M.K: Optimizing GridFTP through Dynamic Right-Sizing. In: Proceedings of the 12th IEEE International Symposium on High Performance Distributed Computing. IEEE, Inc. Seattle (2003),pp.14–23

[RWM02]   Rajasekar A., Wan M., Moore R.: MySRB and SRB—components of a Data Grid. In: Proceedings of the 11th IEEE International Symposium on High Performance Distributed Computing. IEEE, Inc. Edinburgh (2002),pp.301–310

[WRM03]   Wan M., Rajasekar A., Moore R., et al.: A simple mass storage system for the SRB data grid. In: Proceedings of 20th IEEE/11th NASA Goddard Conference on Mass Storage Systems and Technologies. IEEE, Inc. San Diego (2003),pp.20–25

Adaptive Parallel Wavelet Method for the Neutron Transport Equations

Heng Zhang[1,2] and Wu Zhang[1]

[1] School of Computer Science and Engineering, Shanghai University Shanghai 200072, China wzhang@mail.shu.edu.cn
[2] Department of Mathematics, Shihezi University, Shihezi 832000, China zhheng01@163.com

Abstract The numerical solution of the time-dependent neutron transport problem in 2-D Cartesian geometry is considered. The problem is described by a coupled system of hyperbolic partial differential equations with the parameters of multiple groups. The system, for given group parameter, is discretized by a discrete ordinate method (SN) in angular direction and adaptive spline wavelet method with alternative direction implicit scheme (SW-ADI) [CZ98] in space-time domain. A parallel two-level hybrid method [SZ04] is used for solving the large-scale tridiagonal systems arising from the SW-ADI-SN discretization of the problem. The numerical results are coincided with theoretical analysis well.

Key Words: Neutron transport equations, Parallel computing, Adaptive spline wavelet

1 Introduction

Wavelet approximations have drawn great attention as a potentially efficient numerical technique for solving PDEs. To list a few, the pioneering work by Beylkin et al.[BCR91] has started several works in deriving sparse matrix representations of integral operators with Daubechies' wavelet basis[Dau88] or wavelet -like basis [Alp93]. In the area of differential operators, most of the attempts of using wavelets are to generate an adaptive mashing structure (select a partial set of full wavelet basis functions) upon which differential operations can be carried out, thus reducing the amount of mesh points (number of wavelet basis functions) needed to resolve detailed structures in the solutions of PDEs[CW96][Jam94][Jam95][MPR91][FS97].

Because of their advantageous properties of localizations in both space and frequency domains, wavelets seem to be a great candidate for adaptive and multi-resolution schemes to obtain solutions which vary dramatically both in space and time and develop singularities. However, to take advantage of the nice properties of wavelet approximations, we have to find an efficient way to deal with the nonlinearity and general boundary conditions in the PDEs. This is made possible by the localization properties of the wavelet basis both in space (like traditional finite element basis functions) and in frequency(like the Fourier basis functions).

In this paper, we will extend our studies of using a specially designed spline wavelet basis in $H^2(I)$ on a closed interval I to the resolution of the initial boundary value problems, in this case, two-dimensional neutron transport equations. There are two equally important issues in the design of adaptive methods for the efficient solution of time-dependent PDEs. The first one is the ease of generating locally refined adaptive meshes to resolve the gradients of solutions, the other is an efficient solver for the resulting algebraic system (linear or nonlinear). The second issue should be paid more attention because it is the bottleneck to achieve efficiency of a time-dependent solver of many problems. The wavelet methods turn out to be a good candidate to address both issues. We will resolve them by using an efficient adaptive SW-ADI method which combines adaptive spline wavelet methods and the traditional alternative direction implicit (ADI) schemes proposed in [CZ98].

The discrete ordinate method (SN) or SN-finite element method is frequently used for solving the neutron transport problem. In this paper, we will use the SN method for discretization of the phase space and only discuss the problems with single group transport problem as a instance of multiple groups in order to propose a highly efficient and precise method–adaptive SW-ADI-SN algorithms.

Parallel computing is a highly efficient way. For the tridiagonal systems arising from the adaptive SW-ADI-SN scheme, we will use an more efficient and accurate parallel computing algorithms to solve them. Finally, we will propose the adaptive parallel SW-ADI-SN algorithms.

The paper is organized into following sections. In Section2, we describe the two-dimensional neutron transport equations. The wavelet methods, a fast discrete wavelet transform between function values and wavelet interpolative expansion coefficients proposed [CZ98]and the SW-ADI method which combines spline wavelet methods and ADI schemes is discussed briefly in Section3. Parallel SW-ADI method are introduced in Section4. In Section5, we develop an adaptive parallel SW-ADI-SN method and consider their implementation. Numerical experiments have been performed on the cluster of workstation. The conclusion are given in Section6.

2 The 2D Neutron Transport Equations

We will consider the 2D time-dependent transport equations under the 2-D Cartesian coordinates as following:

$$\frac{1}{\nu_g}\frac{\partial \phi_g}{\partial t} + \xi\frac{\partial \phi_g}{\partial x} + \eta\frac{\partial \phi_g}{\partial y} + \sum_g \phi_g = Q_{fg} + Q_{sg} \tag{1}$$

where $g = 1, 2, ..., G_0, \mathbf{\Omega} = \xi \mathbf{e}_x + \eta \mathbf{e}_y$ is the independent variable for moving directions of neutron, and
   fusion source

$$Q_{fg} = \chi_g \sum_{g'=0}^{G_0} \nu \sum_{f}^{g'} \psi_{g'}(x, y, t) \tag{2}$$

scatter source

$$Q_{sg} = \sum_{g'=0}^{g} \sum_{s}^{g' \to g} \psi_{g'}(x, y, t) \tag{3}$$

$$\psi_g(x, y, t) = \frac{1}{2\pi} \int_0^1 d\mu \int_0^{2\pi} \phi_g(x, y, \mu, \omega, t) d\omega \tag{4}$$

initial condition

$$\phi_g(x, y, \mu, \omega, t) \mid_{t=0} = \phi_g^0(x, y, \mu, \omega) \tag{5}$$

boundary condition

$$\phi_g(x, y, \mu, \omega, t) = 0, (x, y, \mu, \omega, t) \in \Gamma_\Omega = \{(x, y) \in \Gamma, \mathbf{\Omega} \cdot \mathbf{n}_\Gamma < 0\} \tag{6}$$

Obviously, equation (1) involves spatial, temporal, direction, energy, and density variables.

3 SW-ADI method

In this section, we introduce SW-ADI methods for one-dimensional initial-boundary value problems of the PDEs , which can be extended to two-dimensional or three-dimensional problems in a tensor product form. The SW-ADI methods are given in[CZ98]for detail.

3.1 Wavelet spaces and Wavelet approximations

Let I denote a close interval $[0, L]$ and let $H^2(I)$ and $H_0^2(I)$ denote the following two Sobolev spaces with finite L^2-norm up to the second derivative, respectively[CZ98]:

$$H^2(I) = \{f(x) \mid x \in I, \|f^{(i)}\|_2 < \infty, i = 0, 1, 2\} \tag{7}$$

$$H_0^2(I) = \{f(x) \in H^2(I) \mid f(0) = f(L) = f'(0) = f'(L) = 0\} \tag{8}$$

For a given integer $J \geq 0$, let V_J denote the following subspace

$$V_J = V_0 \oplus W_0 \oplus W_1 \oplus ... \oplus W_{J-1} \tag{9}$$

where

$$V_0 = span\{\varphi_{0,k}, -3 \le k \le L-1\} \tag{10}$$

$$W_j = span\{\psi_{j,k}, -1 \le k \le n_j - 2, \}, n_j = 2^j L, 0 \le j < L \tag{11}$$

and

$$H^2(I) = V_0 \oplus_{j=0}^{\infty} W_j \tag{12}$$

It can be checked that $\dim V_0 = L + 3$, $\dim W_j = n_j$. The collocation points

$$X^{(-1)} = \{0, \frac{1}{2}, 1, 2, ..., L-1, L-\frac{1}{2}, L\} = \{x_k^{(-1)}\}_{k=1}^{L+3} \tag{13}$$

for V_0 are chosen and

$$X^{(j)} = \{\frac{1}{2^{j+2}}, \frac{3}{2^{j+1}}, \frac{5}{2^{j+1}}, ..., L-\frac{5}{2^{j+1}}, L-\frac{3}{2^{j+1}}, L\frac{1}{2^{j+2}}\} = \{x_k^{(j)}\}_{k=-1}^{n_j-2} \tag{14}$$

for $W_j, j \ge 0$ and the number of collocation points in V_0 is $L+3$ and that in W_j is n_j which both match the dimensions of spaces V_0 and $W_j, j \ge 0$.

The point vanishing property holds, i.e., for $j > i$,

$$\psi_{j,k}(x_l^{(i)}) = \delta_{kl} \tag{15}$$

where $j \ge i, -1 \le k \le n_j - 2$, and $1 \le l \le L-1$ if $i = -1; -1 \le l \le n_j - 2$ if $i \ge 0$. The point vanishing property of (15) assures that $\{\psi_{j,k}(x)\}$ forms a hierarchical basis function over the collocation points (13)– (14).

For any function $u(x) \in H^2(I)$ we have an approximate function $u_J(x)$ in the form of

$$u_J(x) = u_{-1}(x) + \sum_{j=0}^{J} u_j(x) \in V_J = V_0 \oplus W_0 \oplus W_1 \oplus ... \oplus W_{J-1} \tag{16}$$

where

$$u_{-1}(x) = \sum_{k=-3}^{l} u_{-1,k}\phi_{0,k}(x) \in V_0 \tag{17}$$

$$u_j(x) = \sum_{k=0}^{n_j-2} u_{j,k}\psi_{j,k}(x) \in W_j, 0 \le j \le J \tag{18}$$

and the approximation of derivation $u^{(i)}(x)$

$$u_J^{(i)}(x) = u_{-1}^{(i)}(x) + \sum_{j=0}^{J} u_j^{(i)}(x) \quad i = 1, 2 \tag{19}$$

where

$$u_{-1}^{(i)}(x) = \sum_{k=-3}^{l} u_{-1,k}\phi_{0,k}^{(i)}(x) \tag{20}$$

$$u_j^{(i)}(x) = \sum_{k=0}^{n_j-2} u_{j,k}\psi_{j,k}^{(i)}(x), 0 \le j \le J \tag{21}$$

Let us denote the expansion coefficient of (17) or (18) in a vector form: $U = (u^{(-1)}, u^{(0)}, ..., u^{(J)})^T$, $u^{(-1)} = (u_{-1,-3}, u_{-1,-2}, ..., u_{-1,L-1})^T$, $u^{(j)} = (u_{j,-1}, u_{j,0}, ..., u_{j,n_j-2})^T$. Denote the values of $u(x)$ at collocation points by $\widehat{U} = (\hat{u}^{(-1)}, \hat{u}^{(0)}, ..., \hat{u}^{(J)})^T$, $\hat{u}^{(-1)} = (\hat{u}_{-3}^{(-1)}, \hat{u}_{-2}^{(-1)}, ..., \hat{u}_{L-1}^{(-1)})$, and $\hat{u}^{(j)} = (\hat{u}_{-1}^{(j)}, \hat{u}_0^{(j)}, ..., \hat{u}_{n_j-2}^{(j)})^T$. The coefficient $u^{(j)}$ will be determined by satisfying interpolation conditions at the collocation points defined in (13) and (14): $\hat{u}_k^{(-1)} = u(x_k^{(-1)})$, $\hat{u}_k^{(j)} = u(x_k^{(j)})$.

According to the point vanishing property (31), we have

$$Pu^{(-1)} =: \hat{u}^{(-1)}, M_j u^{(j)} =: \hat{u}^{(j)} - u_{j-1}(x_k^{(j)})_{k=-1}^{n_j-2} \tag{22}$$

where P and M_j are the $(L+3) \times (L+3)$ and $n_j \times n_j$ transform matrices, respectively. Then the discrete wavelet transform (DWT) maps from $\widehat{U}$ to U (from $u(x)$ to U).

3.2 ADI methods for The 2D Neutron Transport Equations

To achieve unconditional stability, we may resort to a fully implicit method for time discretization of equation(1). Unfortunately, this will result in a sparse system of algebraic equations, which, however, may require a large amount of computation. One remedy is to use ADI methods, which only require solving one-dimensional implicit problems for each time step.

We discuss the SW-ADI schemes which will be used in our numerical simulations for the 2D neutron transport equations. Based on the ADI formulae the SW-ADI schemes with non-uniform schemes are developed, which are important in implementing wavelet approximation and adaptive procedure.

3.2.1 1st-order homogeneous hyperbolic equation

We consider 1st-order homogeneous hyperbolic equation

$$\begin{cases} u_t + \eta u_x + \xi u_y = 0 \\ u\,|_{t=0} = u_0, u\,|_{\partial\Omega} = g \end{cases} \tag{23}$$

where $u = u(t,x,y), \eta = \eta(t,x,y), \xi = \xi(t,x,y), u_0 = u_0(x,y), g = g(t)$, $(x,y) \in \Omega, t \in [0,T]$, $\partial\Omega$ is the boundary of domain, $\Omega = [x_2, x_2] \times [y_1, y_2]$.

In the first time step from t_n to $t_n + \frac{\Delta t}{2}$, an implicit difference is used for u_x and an explicit difference is used for u_y. In the second step from $t_n + \frac{\Delta t}{2}$ to t_{n+1}, a reversed procedure is used. The corresponding ADI format is as following:

$$\begin{cases} u^{n+\frac{1}{2}} - u^n + \frac{\Delta t}{2}(\eta D_x u^{n+\frac{1}{2}} + \xi D_y u^n) = 0 \\ u^{n+1} - u^{n+\frac{1}{2}} + \frac{\Delta t}{2}(\eta D_x u^{n+\frac{1}{2}} + \xi D_y u^{n+1}) = 0 \end{cases} \tag{24}$$

i.e

$$\begin{cases} (1 + \frac{\eta \Delta t}{2} D_x)u^{n+\frac{1}{2}} = (1 - \frac{\xi \Delta t}{2} D_y)u^n \\ (1 + \frac{\xi \Delta t}{2} D_y)u^{n+1} = (1 - \frac{\eta \Delta t}{2} D_x)u^{n+\frac{1}{2}} \end{cases} \tag{25}$$

where $u^{n+\frac{1}{2}} = u(t_n + \frac{\Delta t}{2}, x, y), u^n = u(t_n, x, y)$ is an intermediate value. The (25) is consistent with (23), and truncation error is $O(\Delta t^2 + h^2)$, where the step size $h = \Delta x = \Delta y$. The approximate solution must satisfy the initial and boundary conditions, i.e.,

$u \mid_{t=0} = u_0$, at all mesh points;

$u^n \mid_{\partial \Omega} = g \mid_{t=t_n}, n = 0, 1, ..., N$, on the boundary $\partial \Omega$.

The intermediate value $u^{n+\frac{1}{2}}$ introduced in each ADI schemes above is not necessarily an approximate to the solution at any time levels. As a result, particularly with the higher order methods, the boundary values at intermediate level must be obtained, if possible, in terms of boundary values t_n and t_{n+1}. The following formulae give $u^{n+\frac{1}{2}}$ explicitly in terms of the central difference of g^n and g^{n+1}

$$\begin{aligned} u^{n+\frac{1}{2}} &= \tfrac{1}{2}(1 - \tfrac{\xi \Delta t}{2} D_y)u^n + \tfrac{1}{2}(1 + \tfrac{\xi \Delta t}{2} D_y)u^{n+1} \\ &= \tfrac{1}{2}(1 - \tfrac{\xi \Delta t}{2} D_y)g^n + \tfrac{1}{2}(1 + \tfrac{\xi \Delta t}{2} D_y)g^{n+1} \end{aligned} \tag{26}$$

where $g^n = g(t_n, x, y)$.

If the boundary conditions are independent of the time, the formulae giving $u^{n+\frac{1}{2}}$ on the boundary $\partial \Omega$ reduce to

$$u^{n+\frac{1}{2}} = g \tag{27}$$

3.2.2 1st-order non-homogeneous hyperbolic equation

We consider 1st-order non-homogeneous hyperbolic equation

$$\begin{cases} u_t + \eta u_x + \xi u_y = S \\ u \mid_{t=0} = u_0, u \mid_{\partial \Omega} = g \end{cases} \tag{28}$$

where $u = u(t, x, y), \eta = \eta(t, x, y), \xi = \xi(t, x, y), S = S(t, x, y), u_0 = u_0(x, y), g = g(t), (x, y) \in \Omega, t \in [0, T], \Omega = [x_2, x_2] \times [y_1, y_2], \partial \Omega$ is the boundary of domain.

We have the ADI scheme as following

$$\begin{cases} (1 + \frac{\eta \Delta t}{2} D_x)u^{n+\frac{1}{2}} = (1 - \frac{\xi \Delta t}{2} D_y)u^n + \frac{\Delta t}{2} S^{n+\frac{1}{2}} \\ (1 + \frac{\xi \Delta t}{2} D_y)u^{n+1} = (1 - \frac{\eta \Delta t}{2} D_x)u^{n+\frac{1}{2}} + \frac{\Delta t}{2} S^{n+\frac{1}{2}} \end{cases} \tag{29}$$

or

$$\begin{cases} (1 + \frac{\eta \Delta t}{2} D_x)u^{n+\frac{1}{2}} = (1 - \frac{\xi \Delta t}{2} D_y)u^n + \frac{\Delta t}{2} S^n \\ (1 + \frac{\xi \Delta t}{2} D_y)u^{n+1} = (1 - \frac{\eta \Delta t}{2} D_x)u^{n+\frac{1}{2}} + \frac{\Delta t}{2} S^{n+1} \end{cases} \tag{30}$$

where $S^{n+\frac{1}{2}} = S(t_n + \frac{\Delta t}{2}, x, y), S^n = S(t_n, x, y)$.

If the source term is dependent on the unknown function,i.e.,

$S = S(t, x, y, u)$, we will use the Runge-Kutta scheme as following

$$\begin{cases} u^{n+1} = u^n + \frac{1}{6}(k_1 + k_2 + k_3 + k_4) \\ k_1 = \Delta t[-\eta D_x u^n - \xi D_y u^n + S(t_n, x, y, u^n)] \\ k_2 = \Delta t[-\eta D_x(u^n + \frac{k_1}{2}) - \xi D_y(u^n + \frac{k_1}{2}) + S(t_n, x, y, (u^n + \frac{k_1}{2}))] \\ k_3 = \Delta t[-\eta D_x(u^n + \frac{k_2}{2}) - \xi D_y(u^n + \frac{k_2}{2}) + S(t_n, x, y, (u^n + \frac{k_2}{2}))] \\ k_4 = \Delta t[-\eta D_x(u^n + k_3) - \xi D_y(u^n + k_3) + S(t_n, x, y, (u^n + k_3))] \end{cases} \quad (31)$$

Then We have the ADI scheme for equation (28) as following

$$\begin{cases} (1 + \frac{\eta \Delta t}{2} D_x)u^{n+\frac{1}{2}} = (1 - \frac{\xi \Delta t}{2} D_y)u^n + \frac{\Delta t}{2} S^n \\ (1 + \frac{\xi \Delta t}{2} D_y)u^{n+1} = (1 - \frac{\eta \Delta t}{2} D_x)u^{n+\frac{1}{2}} + \frac{\Delta t}{2} S^{n+1} \end{cases} \quad (32)$$

where $S^n = S(t_n, x, y, u^n)$, the Runge-Kutta scheme(31) gives u^{n+1} in S^{n+1}.

Here the derivatives matrices D_x, D_y be computed by the derivative matrices method given in [CZ98].

4 Parallel Implementation of SW-ADI methods

Parallel computing have been used for solving large scale problems in science and engineering. Message Possing Interface (MPI) is the primary parallel programming models, and is used for the parallel implementation in this paper.

4.1 Parallelizing of sequential algorithm and program of ADI method

By analyzing the source program we can choose parts of the sequential algorithm which can be parallelized. The structures of the source program are shown in Figure 1. According to the analyzed results, the computing time percentage of ADI subroutine is 90.8%. Where the subroutine adi1 is used to solve the system of algebra equations, and hence we will mainly parallelize adi1 .

The further analysis of the subroutine adi1 for a time step is shown in Figure 2. The subroutines x_loc and y_loc of coordinate values don't need to be changed. We only amend the subroutines der2d, adi_rk2, transpose, sweep, and their subroutine .

Solving the equation (25), (30) and (32) are equivalent to solving tridiagonal systems. Replacing the LU decomposition method, we use parallel 2-level hybrid PPD algorithm [SZ04] for solving the tridiagonal systems. The corresponding subroutine ADI is modified to achieve parallelization. Considering data communication, the whole source codes are parallelized with MPI.

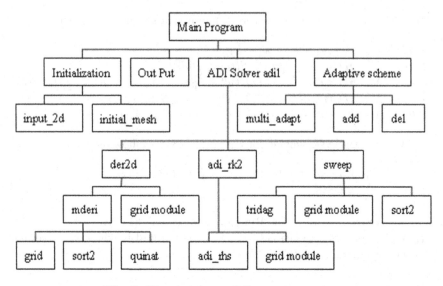

Fig. 1. The structures of the source program

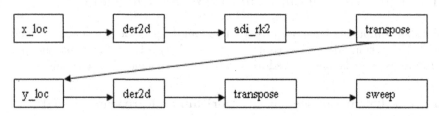

Fig. 2. The function flows of adi

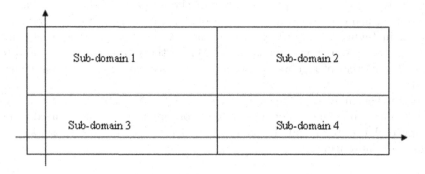

Fig. 3. The rectangle domain decomposition

4.2 The domain decomposition method for ADI scheme

A domain decomposition procedure is considered for the ADI scheme. An example of rectangle domain is given in Figure 3.

For esch time step of ADI method, the computation on a node only depends on itself and four adjacent nodes . To compute the value of the function on the nodes on the boundary of sub-domain, we need the data of the nodes on the boundary of adjacent sub-domain, which results in the communications between the processors. Two instances of the communications corresponding to computation on the boundary are shown in Figure 4.

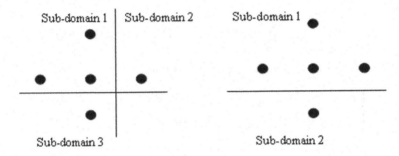

Fig. 4. The instances of the communications between three sub-region and between two

When the data in some sub-domains are stored, one extra row and column, called the pseudo boundary, are needed to stored data for communication. Given a time step, as long as the computation is fulfilled, each processor exchanges the latest data with adjacent processors. After the communications are achieved, the processor will begin new computation at the next time step till the final time.

5 Parallel SW-ADI-SN Algorithms

We now discuss the parallel SW-ADI-SN algorithms used to solve 2D transport equations. The two-dimensional data structure and adaptive procedure described in Section5.1 of [CZ98] are used. For the sake of simplicity, we only consider Eq. (28) and ADI scheme (30). The treatment of other situations of the ADI schemes are similar to it.

5.1 Parallel SW-ADI-SN algorithm of solving transport equations

First of all, direction variables are discretized by using SN method

$$\frac{1}{\nu_g}\frac{\partial \phi_{g,m}}{\partial t} + \xi_m \frac{\partial \phi_{g,m}}{\partial x} + \eta_m \frac{\partial \phi_{g,m}}{\partial y} + \sum_g (x,y)\phi_{g,m}(x,y,\mathbf{\Omega}_m,t) = Q_{fg} + Q_{sg}$$

(33)

where $m = 1, 2, ..., M, \ g = 1, 2, ..., G_0$

$$\psi_g(x,y,t) = \frac{1}{2\pi}\sum_{m=1}^{M}\phi_{g,m}(x,y,\mathbf{\Omega}_m,t)P_m$$

(34)

Then, rewrite the equations into

$$\frac{\partial \phi_{g,m}}{\partial t} + \nu_g \xi_m \frac{\partial \phi_{g,m}}{\partial x} + \nu_g \eta_m \frac{\partial \phi_{g,m}}{\partial y} = S(\phi_{g,m},x,y)$$

(35)

$$S(\phi_{g,m},x,y) = -\nu_g \sum_g (x,y)\phi_{g,m}(x,y,\mathbf{\Omega}_m,t) + \nu_g(Q_{fg} + Q_{sg})$$

(36)

And, using the ADI scheme of hyperbolic equation

$$\begin{cases} (1 + \frac{\nu_g \xi_m \Delta t}{2} D_x)\phi_{g,m}^{n+\frac{1}{2}} = (1 - \frac{\nu_g \eta_m \Delta t}{2} D_y)\phi_{g,m}^{n} + \frac{\Delta t}{2} S^n \\ (1 + \frac{\nu_g \eta_m \Delta t}{2} D_y)\phi_{g,m}^{n+1} = (1 - \frac{\nu_g \xi_m \Delta t}{2} D_x)\phi_{g,m}^{n+\frac{1}{2}} + \frac{\Delta t}{2} S^{n+1} \end{cases}$$

(37)

where $S^n = S(x,y,\phi_{g,m}^n), S^{n+1} = S(x,y,\phi_{g,m}^{n+1})$.

Finally, the unknown functions $\phi_{g,m}^n, \phi_{g,m}^{n+\frac{1}{2}}, \phi_{g,m}^{n+1}$, and source term S^n, S^{n+1}, in discrete transport equations are interpolated in both x and y directions by using cubic spline. The derivative matrices D_x, D_y are derived with the same step as SW-ADI. The solutions are obtained by solving the tridiagonal systems repeatedly by parallel 2-level hybrid PPD algorithm.

5.2 The general Steps of the Implementation

Given

J_x, J_y-wavelet levels in x and y direction

L_x, L_y-interval lengths of x and y direction

Step 1. Generate spline wavelet mesh; Store x_r, x_c, and solutions ϕ_r, ϕ_c in the row-compressed and column-compressed way.

Step 2. Given initial time t_0 in each discrete direction $\{\xi_m, \eta_m\}$. Given initial values of $\{\phi_{r0}, \phi_{c0}\}$, compute the initial source term

$$S_0 = \nu(\frac{\chi\nu\sum_f + \sum_s}{2\pi})\sum_{m=1}^{M}\phi_m^0 P_m$$

(38)

Step 3. Solve the linear systems of SW-ADI

$$(1 - \frac{\nu_g \eta_m \Delta t}{2} D_y)\phi_{g,m}^{n} + \frac{\Delta t}{2} S^n \Rightarrow \phi_m^{*}$$

(39)

$$(1 + \frac{\nu_g \xi_m \Delta t}{2} D_x)\phi_{g,m}^{n+\frac{1}{2}} = \phi_m^* \tag{40}$$

$$(1 - \frac{\nu_g \xi_m \Delta t}{2} D_x)\phi_{g,m}^{n+\frac{1}{2}} + \frac{\Delta t}{2} S^{n+1} \Rightarrow \phi_m^{**} \tag{41}$$

$$(1 + \frac{\nu_g \eta_m \Delta t}{2} D_y)\phi_{g,m}^{n+1} = \phi_m^{**} \tag{42}$$

by parallel 2-level hybrid PPD algorithm.[SZ04]
Step 4. By using ϕ_m^{n+1} compute new source term

$$S_{n+1} = \nu(\frac{\chi\nu \sum_f + \sum_s}{2\pi}) \sum_{m=1}^{M} \phi_m^{n+1} P_m \tag{43}$$

When $t < 6_N, t = t + \Delta t$ and repeat Steps 3-4.
Otherwise stop and computation ends.

5.3 Numerical Experiments

The MPI codes for the SW-ADI-SN method are run on the supercomputer
"ZQ-2000"of Shanghai University. The numerical experiments were performed
for the ADI method of hyperbolic PDEs (see Figure 5), then parallel SW-ADI
algorithm and parallel SW-ADI-SN method on the "ZQ-2000" finally.

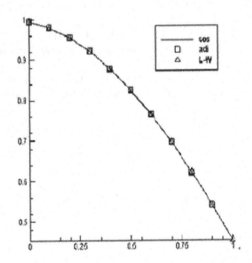

Fig. 5. The numerical experiment for the ADI method of hyperbolic PDEs

The elementary numerical results coincide with the theoretical analysis,
and parallel computing shows much higher efficiency than sequential computing.

Acknowledgements. This paper is supported by the fourth key subject construction of Shanghai, the Natural Science Funding project of Shanghai Educational Committee (No.205155) and the key project of Shanghai Educational Committee (No.03az03).

References

[CZ98] W. Cai and W. Zhang (1998), An adaptive SW-ADI method for 2-D reaction diffusion equations, J. Comput. Phys., Vol.139, No.1, 92-126.

[SZ04] X. Sun and W. Zhang (2004), A parallel two-level hybrid method for Tridiagonal systems and its application to fast Poisson solvers, IEEE Trans. Parallel and Distributed Systems, Vol.15, 2, 97-107.

[BCR91] G. Beylkin, R. Coifman and V. Rokhlin (1991), Fast wavelet transforms and numerical algorithms I, comm. Pure Appl. math.44. 141.

[Dau88] I. Daubechies (1988), Orthogonal bases of compactly, comm. Pure Appl. Math. 41. 909.

[Alp93] B. Alpert (1993), A class of bases in L^2 for the Sparse representation of intefral operators, SIAM J. Math. Anal.24(1, 246).

[CW96] W. Cai and J. Wang (1996), Adaptive multiresolution collocation methods for initial boundary value problems of nonlinear PDEs, SIAM J. Math. Anal. 33(3), 937.

[Jam94] L.Jameson (1994), On the Wavelet-Optimized Finite Difference Method, ICASE Report No. 94-9, NASN CR-191601.

[Jam95] L.Jameson (1995), On the spline-based Wavelet differentiationmatrix, Appl.numer.Math. 17(33), 33.

[MPR91] Y. Maday, V. Perrier, and J. C. Ravel (1991), Adaptive dynamique sur bases d'ondeelettes pour l'approximation d'equatioms aus derivees partielles, C, R. Acad. Sci. Paris 312,405.

[FS97] J. Froehlich and K. Schneider (1997), An Adaptive wavelet-Vagvelette algorithm for the solution of PDEs, J. Comput. Phy. 130, 174.

Ternary Interpolatory Subdivision Schemes for the Triangular Mesh*

Hongchan Zheng[1] and Zhenglin Ye[2]

Department of Applied Mathematics, Northwestern Polytechnical University, Xi'an, Shaanxi 710072, P.R. China

Abstract. A ternary interpolatory subdivision scheme based on interpolatory $\sqrt{3}$-subdivision is proposed first. The limit surface is C^1-continuous. To improve its property, a kind of ternary interpolatory subdivision scheme with two shape parameters is constructed and analyzed. It is shown that for a certain range of the parameters the resulting surface can be C^1-continuous.

Keywords. interpolation, $\sqrt{3}$-subdivision, ternary subdivision, C^k-continuity

1 Introduction

Subdivision curves and surfaces are valued in geometric modelling applications for their convenience and flexibility. They permit the representation of objects of arbitrary topological type in a form that is easy to design, render, and manipulate. Subdivision schemes can be classified in approximating and interpolating schemes. In general, approximating schemes produce smoother surfaces. But interpolation is an attractive feature because the original control vertices defining the curve or surface are also vertices of the limit curve or surface, which allows one to control it in a more intuitive manner. On the other hand, for many applications it is mandatory that the position of the vertices in the control mesh are not changed. Therefore interpolatory schemes have to be analyzed and used.

Binary and ternary 4-point interpolatory curve subdivision schemes have been derived by [1] and [2] respectively. The former has been extended to the case of bivariate binary subdivision. The butterfly subdivision scheme [3], which is the most famous interpolatory surface subdivision scheme, is its extension to the case of the triangular meshes, and was subsequently improved by [4] and [5]. The former's extension to the case of the quadrilateral meshes with arbitrary topology is presented by [6]. The latter is improved by [7].

---

*Supported by the Doctorate Foundation of Northwestern Polytechnical University(CX200328)

All these interpolatory surface subdivision schemes have considered 1-to-4 split operation, namely, for triangular meshes, a new vertex for every edge of the original mesh is computed and the vertices are triangulated so that one triangle of the mesh is split into four triangles of the refined mesh.

A special surface subdivision scheme using a different split operation is introduced by [8]. It is so-called $\sqrt{3}$-subdivision scheme, which use the vertex insertion and edge flipping operator for the refinement. [9] improved the $\sqrt{3}$-subdivision scheme to an interpolatory $\sqrt{3}$-subdivision scheme, which can be looked upon as a 1-to 9 ternary surface scheme after two subdivision steps.

Ternary subdivision scheme performs quicker topological refinement than the usual dyadic split operation. We can get a quicker approximation of a subdivision surface after the same subdivision steps by using a ternary subdivision scheme than by using a binary one.

Motivated by these observations, we consider the construction of a ternary smooth interpolatory surface subdivision scheme which works on the regular triangular mesh where all vertices have valence 6. We perform a 1-to-9 triangle split for every triangular face: we leave all the old vertices unchanged, tri-sect all the edges by inserting two new edge-vertices between every adjacent pair of old ones, and introduce one new face-vertex corresponding to a triangular face in the old control net.

In this paper we first present a ternary interpolatory surface subdivision scheme based on interpolatory $\sqrt{3}$-subdivision scheme. But the stencils for new vertices are big and the shape of the subdivision surface is not modifiable once the initial control mesh is given. To overcome the two shortcomings, we further our discuss on the construction and the analysis of a kind of ternary interpolatory subdivision scheme with two shape parameters, which is an extension of the ternary 4-point interpolatory curve subdivision scheme [7] to the case of the regular triangular meshes. By analyzing the eigenstructure of the subdivision matrix we put forward the necessary conditions on the C^1-continuity and C^2-continuity of the limit surface. By analyzing the contractibility of certain difference subdivision schemes we perform the actual C^1-continuity analysis on the scheme.

2 Interpolatory $\sqrt{3}$-subdivision scheme

Interpolatory $\sqrt{3}$-subdivision scheme works on triangle meshes. Instead of splitting each edge and performing a 1-to-4 split for every triangle, it uses the vertex insertion and edge flipping operator for the refinement of the mesh. Since the subdivision scheme is interpolatory, only the positions of new vertices have to be computed. In the regular case a new face-vertex F for an interior triangle is computed by using a 12-neighborhood stencil shown in Figure 1, where the new face-vertex F is marked by circular solid dot. The rule have the form:

$$F = a(P_1 + P_2 + P_3) + b(P_4 + P_5 + P_6) + c(P_7 + P_8 + P_9 + P_{10} + P_{11} + P_{12}),$$

where $a = \frac{32}{81}, b = -\frac{1}{81}, c = -\frac{2}{81}$. To achieve a regular mesh structure, the edges between old vertices are removed and the edges between the new vertices are inserted. This process is illustrated in Figure 2.

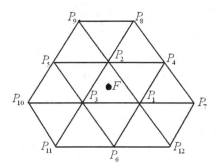

Fig. 1. Stencil for the computation of a new vertex

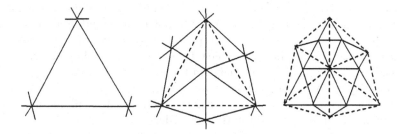

Fig. 2. Interpolatory $\sqrt{3}$ refinement of the mesh

For the computation of new face-vertices of triangles near the boundary of the mesh, some stencil vertices for the subdivision rule may not exist, in this case virtual vertices are introduced by reflecting vertices across the boundary of the mesh to apply the normal new face-vertex subdivision rule. When a boundary triangle is subdivided, special method is introduced. A boundary edge is divided in every second refinement step by inserting two new edge-vertices [2] and split into three parts. Thus the boundary triangle is split into nine parts. This process is illustrated in Figure 3, where the vertices of the boundary edge are marked by circular hollow dots and the new edge-vertices are marked by circular solid dots.

We can see from Figure 2 and Figure 3 that after two refinement steps every triangle is split into nine new triangles. So interpolatory $\sqrt{3}$-subdivision

---

[2] The stencil of the two new edge-vertices are depicted in Fig 5.

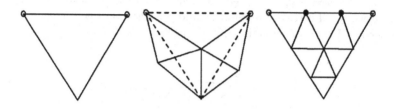

Fig. 3. Subdividing boundary triangles

scheme reproduces a ternary scheme after two subdivision steps. In next section we consider the construction of a ternary smooth interpolatory surface subdivision scheme directly.

3 A ternary interpolatory subdivision scheme for the triangular mesh

In this section we propose a ternary interpolatory subdivision scheme for the regular triangular mesh based on interpolatory $\sqrt{3}$-subdivision scheme. Since the subdivision scheme is interpolatory, we leave all the old control vertices unchanged, introduce one new face-vertex for a triangular face, and two new edge-vertices for an old edge in the old control mesh.

The topological connecting rules are as follows:

(t1) The new face-vertex of a triangle is connected to all the new edge-vertices of the three edges of the triangle orderly.

(t2) The new edge-vertices of the edges of a triangle are connected to each other orderly.

By using the rules t1 and t2,a triangle is split into nine smaller triangles and all vertices inserted into the mesh have valence 6 except the boundary ones. So a regular triangular mesh will be refined into a new regular one. Using this splitting operation the number of triangles in the mesh grows by factor 9 in every refinement step instead of factor 4 with the normal 1-to-4 split. The quicker growth of the mesh size allows a faster generation of a subdivision surface.

Since we want to develop rules for a smooth ternary interpolatory surface subdivision scheme, the rules for the computation of new face-vertices and edge-vertices are needed. The geometric smoothing rules are as follows.

(g1) The new face-vertex is computed by using the same 12-neighborhood stencil depicted in Figure 1. For the triangles near the boundary of the mesh, virtual vertices are introduced.

(g2) New edge-vertices for an interior edge are computed by a 2-ring neighborhood stencil. Figure 4 shows the stencil of one of the new edge-vertices with

13 weights, where $\alpha = 2a^2 + a + 2ab + 2bc + 2ac$, $\beta = 2a^2 + b + 2b^2 + 2ac + 2c^2$, $\varepsilon = 4ac + 2b^2$, $\nu = 2ac + 2bc$, $\kappa = 2c^2$, $\gamma = a^2 + 2ab + 2bc + 2c^2 + ac + c$, $\delta = 2ab + 2ac + 2bc + c^2$, $\xi = ab + 2ac + bc + c^2$, $\theta = 2bc + c^2$, $\eta = c^2$, $\rho = bc + c^2$, $\lambda = ac + b^2 + 2c^2$, $\tau = ac + 2bc$ and a, b, c are defined as before. The stencil of the other can be obtained symmetrically. Similar to (g1), if it is necessary, virtual vertices are introduced.

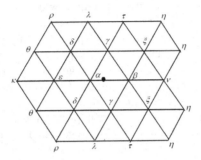

Fig. 4. The stencil of an interior new edge-point

(g3) New edge-vertices for a boundary edge are computed by four consecutive boundary vertices in every subdivision step, which is the same as Labsik' method. The stencil is illustrated in Figure 5.

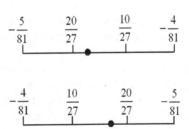

Fig. 5. The stencils of boundary new edge-vertices

The subdivision surface generated by applying the above ternary scheme n steps can be looked as a surface gotten by applying the interpolatory $\sqrt{3}$-subdivision scheme $2n$ steps. According to the C^1-continuity of the interpolatory $\sqrt{3}$-subdivision scheme, we know that the subdivision scheme presented in this section is C^1-continuous.

4 Ternary interpolatory subdivision scheme with parameters

The ternary interpolatory subdivision scheme presented in Section 3 can be used to model a C^1-continuous surface. But the stencil for a new edge-vertex is complicated and has a bad locality. On the other hand, the shape of the subdivision surface is determined completely by the initial control mesh and can not be adjusted once it is specified, which is not convenient in the application. To overcome the two shortcomings, in this section we discuss the construction of an improved ternary interpolatory subdivision scheme.

In [7] we proposed a ternary curve interpolatory subdivision scheme with two shape parameters which have distinct geometric characteristics and can be used to control and modify the shapes of the subdivision curves. Now we extend it to the use for surface modelling.

For regular triangular mesh, we still perform the above 1-to-9 triangle split operation and connecting operation to get the topology of the refinement mesh. To improve the locality of the ternary subdivision scheme in Section 3 and adjust the shape of the limit surface, we present a ternary interpolatory subdivision scheme with two shape parameters. The subdivision rules are:

(1) New face-vertex F for a triangle is still computed by its 1-ring neighborhood stencil depicted in Figure 6(a), but we introduce parameters in weights as follows:

$$F = \eta(P_1 + P_2 + P_3) + \kappa(P_4 + P_5 + P_6) + \theta(P_7 + P_8 + P_9 + P_{10} + P_{11} + P_{12}),$$

where $\eta = \frac{2}{3} - \mu - 2\omega, \kappa = -\frac{1}{3} + \mu + \omega, \theta = \frac{1}{2}\omega$ and μ, ω are two shape parameters [7].

(2) New edge-pvertices for an interior edge are computed by a butterfly like stencil. The stencil of one of the two new edge-vertices is depicted in Figure 6(b), where $\alpha = \frac{4}{3} - 2\mu - 2\omega, \beta = -\frac{1}{3} + 2\mu + 2\omega, \gamma = \frac{1}{3} - \mu - 3\omega, \delta = -\frac{1}{3} + \mu + 2\omega$ and $\xi = \omega$. The stencil of the other is similar to this one.

(3) New edge-vertices for boundary edge are computed by the old vertices of the old edge and their own immediate boundary neighbor. Here we use the 4-point ternary interpolatory subdivision rule in [7]. The corresponding stencil of a new edge-vertex is shown in Figure 6(c). Similar to this we can get a symmetric result about the stencil of the other new edge-vertex.

In practice we often consider the case of one parameter.

Case 1: For the case of $\mu = \frac{2}{9} - 3\omega$ [7], we get a ternary interpolatory subdivision scheme with one shape parameter ω which is generally implemented. Figure 7 shows the results of applying ternary interpolatory subdivision scheme with one shape parameter after four subdivision steps. For the same initial control mesh the left figure in Figure 7 shows the result obtained with $\omega = -\frac{1}{60}$, while the right figure describes the result obtained with $\omega = -\frac{1}{20}$. From Fig 7 we may conclude that we can adjust the shape of the subdivision surface by choosing the shape parameter ω appropriately. Furthermore, due to the ternary property of the scheme, the implementation can

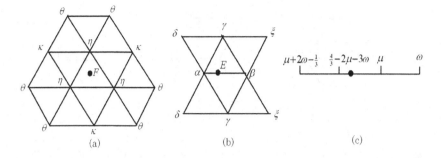

Fig. 6. The stencils of the ternary interpolatory scheme with two shape parameters

speed up the generation and display of the subdivision surface. In practice we can get a "good" approximation to the limit surface only after 4–5 subdivision steps. Hence the presented algorithm is effective.

Fig. 7. Examples of ternary interpolatory subdivision

Case 2: To make the stencils as small as possible, we may set $\omega = 0$ while allowing μ to be free. In this case we get a ternary interpolatory subdivision scheme with one shape parameter μ. For the case $\omega = 0$, we have $\theta = 0, \xi = 0, \eta = \frac{2}{3} - \mu, \kappa = -\frac{1}{3} + \mu, \alpha = \frac{4}{3} - 2\mu, \beta = -\frac{1}{3} + 2\mu, \gamma = \frac{1}{3} - \mu$, and $\delta = -\frac{1}{3} + \mu$. The corresponding stencils of new vertices are depicted in Figure 8. These may be the minimum size that the stencils can be while keeping the scheme be C^1.

5 Continuity analysis of the ternary interpolatory subdivision scheme with shape parameters

To analyze the convergence and the continuity of the ternary interpolatory subdivision scheme with shape parameters, we consider a subnet M_0 consisting

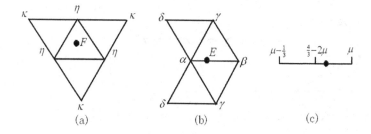

Fig. 8. The stencils of the ternary interpolatory scheme with shape parameter μ

of a point P_0 (marked by •) surrounded by 2 rings of points at an arbitrary subdivision level as illustrated in Figure 9.

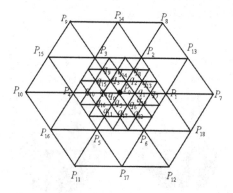

Fig. 9. A 2-ring neighborhood of P_0

Let N_1 be a refinement net whose vertices depend only on M_0. Further let M_1 be a subnet of N_1 and be with the same size and connectedness as M_0. Let $\mathbf{p} = [P_0, P_1, P_2, \cdots, P_{18}]$ and $\mathbf{q} = [q_0, q_1, q_2, \cdots, q_{18}]$ denote the vertices of M_0 and M_1, respectively. Since the vertices q_i are affine combinations of the vertices P_i, we have

$$\mathbf{q} = \mathbf{Sp},$$

where

$$\mathbf{S} = \begin{pmatrix}
1 & 0 & 0 & 0 & 0 & 0 & 0 & 0 & 0 & 0 & 0 & 0 & 0 & 0 & 0 & 0 & 0 & 0 & 0 \\
\alpha & \beta & \gamma & \delta & 0 & \delta & \gamma & 0 & 0 & 0 & 0 & 0 & 0 & \xi & 0 & 0 & 0 & 0 & \xi \\
\alpha & \gamma & \beta & \gamma & \delta & 0 & \delta & 0 & 0 & 0 & 0 & 0 & 0 & \xi & \xi & 0 & 0 & 0 & 0 \\
\alpha & \delta & \gamma & \beta & \gamma & \delta & 0 & 0 & 0 & 0 & 0 & 0 & 0 & 0 & \xi & \xi & 0 & 0 & 0 \\
\alpha & 0 & \delta & \gamma & \beta & \gamma & \delta & 0 & 0 & 0 & 0 & 0 & 0 & 0 & 0 & \xi & \xi & 0 & 0 \\
\alpha & \delta & 0 & \delta & \gamma & \beta & \gamma & 0 & 0 & 0 & 0 & 0 & 0 & 0 & 0 & 0 & \xi & \xi & 0 \\
\alpha & \gamma & \delta & 0 & \delta & \gamma & \beta & 0 & 0 & 0 & 0 & 0 & 0 & 0 & 0 & 0 & 0 & \xi & \xi \\
\beta & \alpha & \gamma & \xi & 0 & \xi & \gamma & 0 & 0 & 0 & 0 & 0 & 0 & \delta & 0 & 0 & 0 & 0 & \delta \\
\beta & \gamma & \alpha & \gamma & \xi & 0 & \xi & 0 & 0 & 0 & 0 & 0 & 0 & \delta & \delta & 0 & 0 & 0 & 0 \\
\beta & \xi & \gamma & \alpha & \gamma & \xi & 0 & 0 & 0 & 0 & 0 & 0 & 0 & \delta & \delta & 0 & 0 & 0 & 0 \\
\beta & 0 & \xi & \gamma & \alpha & \gamma & \xi & 0 & 0 & 0 & 0 & 0 & 0 & 0 & \delta & \delta & 0 & 0 & 0 \\
\beta & \xi & 0 & \xi & \gamma & \alpha & \gamma & 0 & 0 & 0 & 0 & 0 & 0 & 0 & 0 & \delta & \delta & 0 & 0 \\
\beta & \gamma & \xi & 0 & \xi & \gamma & \alpha & 0 & 0 & 0 & 0 & 0 & 0 & 0 & 0 & 0 & \delta & \delta & 0 \\
\eta & \eta & \eta & \kappa & \theta & \theta & \kappa & \theta & \theta & 0 & 0 & 0 & 0 & \kappa & \theta & 0 & 0 & 0 & \theta \\
\eta & \kappa & \eta & \eta & \kappa & \theta & \theta & 0 & \theta & \theta & 0 & 0 & 0 & \theta & \kappa & \theta & 0 & 0 & 0 \\
\eta & \theta & \kappa & \eta & \eta & \kappa & \theta & 0 & 0 & \theta & \theta & 0 & 0 & 0 & \theta & \kappa & \theta & 0 & 0 \\
\eta & \theta & \theta & \kappa & \eta & \eta & \kappa & 0 & 0 & 0 & \theta & \theta & 0 & 0 & 0 & \theta & \kappa & \theta & 0 \\
\eta & \kappa & \theta & \theta & \kappa & \eta & \eta & 0 & 0 & 0 & 0 & \theta & \theta & 0 & 0 & 0 & \theta & \kappa & \theta \\
\eta & \eta & \kappa & \theta & \theta & \kappa & \eta & \theta & 0 & 0 & 0 & 0 & \theta & \theta & 0 & 0 & 0 & \theta & \kappa
\end{pmatrix}$$

is a 19×19 subdivision matrix. By analyzing the eigenvalues of the subdivision matrix $\mathbf{S}$ we can get the necessary conditions on the continuity of the limit surface [10].

The eigenvalues of $\mathbf{S}$ are:

$$1, \frac{1}{3}, \frac{1}{3}, -\frac{5}{3} + 6\mu + 12\omega, -2\omega, -\frac{3}{2}\omega, -\frac{3}{2}\omega, -\frac{1}{2}\omega, -\frac{1}{2}\omega, -\frac{1}{3} + \mu, -\frac{1}{3} + \mu, -\frac{1}{3} + \mu, -\frac{1}{3} + \frac{3}{2}\mu + 2\omega \pm \frac{1}{6}\sqrt{9\mu^2 - 72\mu\omega - 72\omega^2 + 48\omega}.$$

Analyzing these eigenvalues could be tricky when μ and ω vary simultaneously and independently. So we just consider the effect of one parameter on the continuity of the limit surface.

First we consider the case $\mu = \frac{2}{9} - 3\omega$. In this case the ternary interpolatory subdivision scheme contains one shape parameter ω. The corresponding eigenvalues of $\mathbf{S}$ are:

$$1, \frac{1}{3}, \frac{1}{3}, \frac{1}{9}, \frac{1}{9}, \frac{1}{9}, -\frac{1}{3} - 6\omega, -\frac{1}{9} - 5\omega, -\frac{1}{9} - 5\omega, -\frac{1}{9} - 5\omega,$$
$$-\frac{1}{9} - 3\omega, -\frac{1}{9} - 3\omega, -\frac{1}{9} - 3\omega, -2\omega, -2\omega, -\frac{3}{2}\omega, -\frac{3}{2}\omega, -\frac{1}{2}\omega, 0.$$

By calculation it is easy to see that the limit surface is C^1 only if when ω is kept in $(-\frac{4}{45}, 0)$, and it is C^2 only if when ω is kept in $(-\frac{2}{45}, -\frac{1}{27})$. By analyzing the contractibility of certain difference subdivision schemes [8, 11] we can prove that for the range $-\frac{1}{18} < \omega < 0$, the limit surface is C^1. Here we will not give the details of the proof owing to the limitation of space.

Then we consider the behaviour of the free parameter μ. Let $\omega = 0$ and μ be free, the ternary interpolatory subdivision scheme contains one shape parameter μ. The corresponding eigenvalues of $\mathbf{S}$ are:

$$1, \frac{1}{3}, \frac{1}{3}, -\frac{5}{3} + 6\mu, -\frac{1}{3} + 2\mu (three\ times), -\frac{1}{3} + \mu (six\ times), 0(six\ times),$$

which indicates that the limit surface could be C^1 only for the range $\frac{2}{9} < \omega < \frac{1}{3}$. We can prove that for this range, the limit surface is at most C^1.

6 Conclusions

We have presented a few of ternary interpolatory surface subdivision schemes for regular triangle meshes. They can produce C^1 interpolatory subdivision surfaces efficiently and quickly.

Next work should aim at the generalization of the presented scheme for the application to the case of arbitrary control nets. Special rules have to be constructed and applied for the subdivision around extraordinary points. The limit behavior and shape characterization of subdivision surface near extraordinary points should be analyzed.

References

1. Dyn, N., Levin, D., Gregory, J.A.: A 4-point interpolatory subdivision scheme for curve design. Computer Aided Geometric Design (1987) 257-268
2. Hassan, M.F., Ivrissimitzis, I.P., Dodgson, N.A.,Sabin, M.A.: An interpolating 4-point C^2 ternary stationary subdivision scheme. Computer Aided Geometric Design (2002) 1-18
3. Dyn, N., Gregory, J.A., Levin, D.: A butterfly subdivision scheme for surface interpolation with tension control. ACM Transactions on Graphics (1990) 160-169
4. Zorin, D., Schroder, P., Sweldens, W.: Interpolating subdivision for meshes with arbitrary topology. In Computer Graphics Proceedings, ACM SIGGRPH (1996) 189-192
5. Prautzsch, H., Umlauf, G.: Improved triangular subdivision schemes. In: Wolter, F.E.,Patrikalakis, N.M.(ed.): Proceedings of the Computer Graphics, Hannover,Germany (1998) 626-632
6. Kobbelt, L.: Interpolatory subdivision on open quadrilateral nets with arbitrary topology. Computer Graphics Forum (Proceedings of EUROGRAPHICS 1996) 409-420
7. Zheng, H., Ye, Z.: A class of C^2 interpolatory subdivision scheme with two parameters. Journal of Northwestern Polytechnical University (2004) 329-332
8. Kobbelt, L.: $\sqrt{3}$-subdivision. In Computer Graphics Proceedings, Annual Conference Series, ACM SIGGRAPH (2000) 103-112
9. Labsik, U., Greiner, G.: Interpolatory $\sqrt{3}$-subdivision. Computer Graphics Forum (Proceedings of EUROGRAPHICS 2000) 131-138
10. Sabin, M.A.: Eigenanalysis and artifacts of subdivision curves and surfaces. In: Iske, A., Quak, E., Floater, M.S. (eds.): Tutorials on Multiresolution in Geometric Modelling. Springer (2002) 69-92
11. Dyn, N.: Subdivision schemes in computer-aided geometric design. In: Light, W. (ed.): Advances in numerical analysis,Vol. 2, Clarendon Press (1992) 36-104

Part II

Contributed Papers

Modeling Gene Expression Network with PCA-NN on Continuous Inputs and Outputs Basis

Sio-Iong Ao, Michael K. Ng, and Waiki Ching

Department of Mathematics, The University of Hong Kong, Hong Kong, China
siao@hkusua.hku.hk, mng@maths.hku.hk and wkc@maths.hku.hk

In this work, the numerical modeling of the gene expression time series has been constructed with the principal component (PCA) and the neural network (NN). Our approach is different from other studies in bioinformatics. It can be considered as an extension of the linear network inference modeling to a nonlinear continuous model. The predictions have been compared with other popular continuous prediction methods and the PCA-NN method outperforms the others. In order to avoid over-training of the network, we have adopted the AIC tests and cross-validation. Our contribution is that we can know more about the cell development pathways, which is useful for drug discovery, etc.

1 Introduction

The first step of our PCA-NN algorithm is to form the input vectors for the time series analysis. They are the expression levels of the time points in the previous stages of the cell cycle. Then, these input vectors are processed by the PCA. Thirdly, we use these post-processed vectors to feed the neural network. To our best knowledge, our paper is the first one to employ the principal component and neural network for modeling the gene expression with continuous input and output values.

Previous studies have mainly focused on the clustering and the classification of the gene expressions. The principal component analysis has been applied to reduce the dimensionality of the gene expression data in studies like [HHW03]. Neural network has been reported for its successful applications in the gene expression analysis. Khan et al. [Kha01] have applied the PCA with the neural network for the classification of cancers using gene expression profiling. The neural network is trained as a classifier for discrete outputs EWS, RMS, BL, and NB of the cancer types, with discrete input values like 0, 1, etc., for representing the genes' co-expression level. Here, we are going to study the removal of this discrete output constraint.

The outline of this paper is as follows. In Section 2, we give the description of the microarray data. In Section 3, we talk about the details of the methodology of our method. In Section 4, we discuss about the results and comparison. Lastly, we will explore the further works that can be done.

2 Microarray Data

The DNA microarray technology [COB03] has enabled us to have the gene expression levels at different time points of a cell cycle. A microarray is a small chip that can measure thousands of DNA molecules. In the experiment, the time series data of a biological process can be collected. New algorithms have been developing to analysis the results from this high-throughput technology. A difficulty of these studies is that the number of genes are usually much larger than the number of conditions or time points. Another difficulty is that the measurement error is still quite large.

Our first microarray dataset is from the experiment of Spellman et al. [Spe98]. It contains the yeast's gene expression levels at different time points of the cell cycle (18 data points in one cell cycle). From the Spellman's data set, there are totally 613 gene time series data that do not have missing values and show positive cell cycle regulation by periodicity and correlation algorithms. While the number of variables is large, the number of observations per variable is small (18 time points for each gene). We will also test the dataset of Cho et al. [Cho98]. There are 17 time points for a total of 384 genes in this data set.

3 The PCA-NN Algorithm

The typical three-layer network architecture is employed, with the input layer, the hidden layer and the output layer. The mathematical structure for the neural network structure can be expressed as follows [PEL00]:

$$y = g(\sum_{j=1}^{J} w_j^{(2)} f(\sum_{i=1}^{I} w_{ji}^{(1)} x_i))$$ (1)

where I denotes the number of inputs, J the number of hidden neurons, x_i the ith input, w^1 the weights between the input and hidden layers, w^2 the weights between the hidden and output layers. And for our study, we have used the tansig activation function:

$$f(x) = \frac{2}{1 + \exp^{-2x}} - 1$$ (2)

This is mathematically equivalent to $\tanh(x)$, but runs faster in Matlab than the implementation of tanh [VMR88]. And for the output activation function, we have simply used the linear combination of the inputs.

Our neural network results show that the prediction error is better than other methods compared here but it is still high. It may be due to the lack of enough training data and also to the nature of the gene expression levels The absolute percentage error for the second dataset is 14.76%. We have studied the efforts of different network architectures and the number of lag terms on the network. And, to avoid over-training, we have adopted the AIC tests and cross-validation.

Applying PCA for Feature Extraction

The PCA is applied to see if it can assist the neural network for more accurate prediction. Its basic idea is to find the directions in the multidimensional vector space that contribute most to the variability of the data. The representation of data by PCA consists of projecting the data onto the k-dimensional subspace according to

$$x' = F(x) = A^t x \qquad (3)$$

where x' is the vectors in the projected space, A^t is the transformation matrix which is formed by the k largest eigenvectors of the data matrix, x is the input data matrix. Let $x_1, x_2, ..., x_n$ be the samples of the input matrix x. Then, the principal components and the transformation matrix can be obtained by minimizing the following sum of squared error:

$$J_k(a, x') = \sum_{k=1}^{n} \left\| (m + \sum_{i=1}^{k} a_{ki} x_i') - x_k \right\|^2 \qquad (4)$$

where m is the sample mean, x_i' the i-th largest eigen-vector of the co-variance matrix, and a_{ki} the projection of x_k to x_i'.

We have applied the PCA to the whole spectrum of the genes in the two datasets separately. With the PCA obtained by Singular Value Decomposition (SVD) method, the gene expression matrix of dimension 613×18 has been reduced to 17×18. There exists a more or less clear trend for the neural network to make the prediction. And our purpose of dimension reduction is successful.

4 Results and Comparison

The results from the Naïve prediction, moving average prediction (MA), which takes the average of past three expression values for prediction, autoregression (AR(1)), neural network prediction, ICA-NN method and PCA-NN method are listed in Table 1. Naive method has simply used the previous expression value as the prediction value. Moving average prediction has used the average of a certain number of previous expression values as a predictor. In fact, the

Naive method can be regarded as the moving average of one lag term only. The first-order autoregression AR(1) is of the form:

$$x_t = \rho x_{t-1} + \varepsilon_t \tag{5}$$

where x_t is the expression level at time t, ρ the coefficient of x_{t-1}. ε_t is the white noise time series with $E[\varepsilon_t] = 0$, $[\varepsilon_t^2] = \sigma_t^2$, and $Cov[\varepsilon_t, \varepsilon_s] = 0$ for all $s \neq t$. These three methods are popular for continuous numerical predictions and their corresponding errors are in-samples errors. It can be observed that the NN and PCA-NN models are better than these methods.

Table 1. Prediction Results from the Different Methods

Results	Naïve	MA	AR	NN	ICA-NN	PCA-NN
Abs. Err (1st set)	94.92%	125.87%	80.34%	75%	83.93%	51.31%
Abs. Err (2nd set)	28.52%	39.16%	27.31%	22.52%	27.67%	12.91%

Discussions on the NN Structure

We have tested different combinations of the lag lengths for the training of the neural network with one hidden layer and ten hidden neurons. Different network architectures have been checked for their performance.

Table 2. Prediction Results for NN Method with Different Input Lag Lengths

Results	NN1	NN2	NN3	NN4	NN5
Abs. Err (1st set)	66.23%	63.89%	59.68%	57.82%	55.85%
Abs. Err (2nd set)	19.07%	16.68%	14.76%	13.42%	11.77%
AIC (1st set)	189.39	208.31	225.86	245.45	263.63
AIC (2nd set)	375.13	391.73	409.38	425.41	443.31

In Table 2, the results of feeding the neural network of 10 hidden neurons with different input lag terms are shown. NN1 represents network with the input lag term t-1, NN2 with two input terms t-1 and t-2, NN3 with inputs t-1, t-2 and t-3, NN4 with input t-1, t-2, t-3 and t-4, NN5 with input t-1, t-2, t-3, t-4 and t-5.

The AIC results are also listed in Tables 2 and 3. It has been shown that Akaike's criterion is asymptotically equivalent to the use of cross-validation [PEL00]. Akaike's information criterion (AIC) is defined as:

$$AIC = T \ln(residual\ sum\ of\ squares) + 2n \tag{6}$$

where n is the number of parameters estimated, and T the number of usable observations [End95].

From the AIC results, the architecture of the t-1 input with 5 hidden neurons is suggested. And, feeding the neural network of 5 hidden neurons with input t-1, the error in the 1st data set is found to be 75% while that in the 2nd data set is 22.52%, which are better than the Naïve, MA, and AR methods.

Table 3 shows us the prediction results and AIC values for the PCA-NN method with different network structures. T1 is for feeding the network with input t-1 term only, T2 with input t-1 and t-2 terms, T3 with input t-1, t-1 and t-3 terms. N5 refers to the network of 5 hidden neurons and N10 of 10 neurons. While for data set 1, the AIC values of TIN5 and T2N5 are more or less the same, the AIC values of the second data set suggests clearly that model T2N5 should be employed. This AIC result here is slightly different from that of stand-alone neural network.

Table 3. Prediction Results for PCA-NN Method with Different Neural Network Structures

Results	T1N5	T2N5	T3N5	T1N10	T2N10	T3N10
Abs. Err (1st set)	67.63%	51.31%	46.85%	57.97%	44.22%	30.99%
Abs. Err (2nd set)	22.02%	12.91%	14.65%	19.01%	8.68%	9.47%
AIC (1st set)	160.61	160.87	168.77	184.81	195.75	204.74
AIC (2nd set)	348.91	338.61	351.59	372.42	367.08	389.33

In our cross-validation testing, the algorithm's performance in the out-of-sample results are better than other methods' in-sample prediction. Furthermore, we have tested the possibility of replacing the PCA component with the ICA component for the modeling. However, the results suggest that ICA does not supplement well with neural network for the gene expression time series modeling here.

5 Concluding Remarks

The relationship of the gene expression level has been studied with PCA-NN. A main difficulty in our numerical prediction is that the time points in one cell cycle are short. We have found that the NN performs better than the other methods like Naive and AR for this problem. The PCA-NN method can give us more accurate model of the genome network.

There are still much room for improvement on the PCA-NN algorithm. First of all, we have employed the standard neural network. There have been studies like [HSP02] that have shown how to improve the predictive capability of the NN. The genetic programming can also be applied for the optimization of the NN structure [RWP03]. Furthermore, the genetic algorithm is useful for weighting the genes' importance in the model. As pointed by Keedwell [KN02], GA is a promising tool for the optimization of the gene weightings, which itself is a NP-hard problem. Similarly, we can regard our NN numerical prediction as the fitness function of the GA component. And we are going to select the most influential genes for each gene's development in its cell cycle with GA. The methodology will be the GA-PCA-NN hybrid system.

References

[Kha01]   Javed Khan, et al.: Classification and Diagnostic Prediction of Cancers Using Gene Expression Profiling and Artificial Neural Networks. Nature Medicine, 7(6), 2001, Pages 673-679.

[COB03]   Helen C. Causton, John Quackenbush and Alvis Brazma: Microarray Gene Expression Data Analysis: a Beginner's Guide. Blackwell Publishing, 2003.

[HHW03]   Michael Hornquist, John Hertz and Mattias Wahde: Effective Dimensionality of Large-Scale Expression Data Using Principal Component Analysis. BioSystem, 65, 2003, Pages 147-156.

[KN02]   E. C. Keedwell and A. Narayanan: Genetic Algorithms for Gene Expression Analysis. First European Workshop on Evolutionary Bioinformatics (2002), Pages 76-86.

[Spe98]   Paul T. Spellman, et al.: Comprehensive Identification of Cell Cycle-regulated genes of the Yeast Saccharomyces cerevisiae by Microarray Hybridization. Molecular Biology of the Cell, 9, December 1998, Pages 3273-3297.

[Cho98]   Raymond J. Cho, et al: A Genome-Wide Transcriptional Analysis of the Mitotic Cell Cycle. Molecular Cell, 2, July 1998, Pages 65-73.

[PEL00]   Jose C. Principe, Neil R. Euliano and W. Curt Lefebvre: Neural and Adaptive Systems: Fundamentals through Simulations. John Wiley & Sons, 2000.

[VMR88]   Vogl, T. P., J.K. Mangis, A.K. Rigler, W.T. Zink, and D.L. Alkon: Accelerating the convergence of the backpropagation method. Biological Cybernetics, vol. 59, pp. 257-263, 1988.

[End95]   Walter Enders: Applied Econometric Time Series. Wiley, 1995.

[HSP02]   Hoo, K.A., Sinzinger E.D., and Piovoso, M.J.: Improvements in the predictive capability of neural networks. Journal of Process Control, vol. 12, pp. 193-202, 2002.

[RWP03]   Ritchie, M.D., White, B.C., Parker, J.S., Hahn, L.W, and Moore, J.H.: Optimization of neural network architecture using genetic programming improves detection and modeling of gene-gene interactions in studies of human diseases. BMC Bioinformatics, 4:28, 2003.

A Reuse Approach of Description for Component-based Distributed Software Architecture

Min Cao[1], Gengfeng Wu[1], and Yanyan Wang[2]

[1] School of Computer Science & Engineering, Shanghai University, Shanghai 200072, P.R. China mcao@mmail.shu.edu.cn
[2] Information & Scientific Computation Groups, Zhongyuan Technical College, Zhengzhou 475004 P.R. China

Abstract Software reuse concerns the systematic development of reusable components and the systematic reuse of these components as building blocks to create new systems. The latter attracts more researchers. This paper presents a novel approach for Component-Based Software architecture description with Graph-Oriented Programming model (CBSGOP). The components of distributed software are configured as a logical graph and implemented using a set of operations defined with the graph in CBSGOP. In this way, users can specialize the type of the graph to represent a particular style of architectures for software development. The practice shows that CBSGOP provides not only build-in support for describing flexible and dynamic architectures but also reuse of software architecture itself.

Keyword Software Reuse, Software Architecture, Component-Based Software, Graph-Oriented Programming, Software Architecture Reuse

1 Introduction

The reuse of software has become a topic of much interest in the software community due to its potential benefits, which include increased product quality and decreased product cost and schedule. Software reuse is the process of implementing or updating software systems using existing software assets, or say, component [MK93]. It has two sides: the systematic development of reusable components and the systematic reuse of these components as building blocks to create new systems. The latter attracts much research.

In this paper, we present a novel approach for Component-Based Software architecture description with Graph-Oriented Programming model (CBSGOP), which tries to narrow the gap between design and implementation of distributed software. It aims at the description of component-based distributed software architecture, as well as the reuse of software architecture.

2 CBSGOP Approach

Graphs have been considered to be intuitive for the expression of software architectures [LK01]. CBSGOP is also based on this intuition, and directly supports users to model architectures as graphs. The graphs in CBSGOP are not only an abstract description of architecture but also a concrete construction for system implementation and components programming in final working systems. CBSGOP has its advantage in the modeling, implementation and reuse of software architectures.

2.1 CBSGOP Model

CBSGOP is defined as a four-tuple (G, f_c, f_p, L), where G is a directed or undirected logical graph, whose nodes are associated with components, and whose edges define the interaction between components. f_c is a component-to-node mapping, which allows users to bind components with Interface Description Language (IDL) description and implementation information to specific nodes. f_p is an optional node-to-processor mapping, which allows users to explicitly specify the mapping of the logical graph to the underlying network of processors. When the mapping specification is omitted, a default mapping will be performed. L is a library of language-level graph-oriented programming primitives. Each element of CBSGOP is defined as follows:

Definition 1: The graph G = $\langle$N, R, E$\rangle$, where N is a limited set of vertices (nodes), R is a limited set of labels, which represents relations between nodes, and E $\subseteq$ N×R×N is a limit set of edges. Each edge e $\in$ E, and e = $\langle n_1, rel, n_2 \rangle$, where $n_1, n_2 \in$ N, rel $\in$ R. n_2 is the directed succeeder of n_1, and n_1 is the directed precedent of n_2.

Definition 2: f_c: C $\longrightarrow$ N is a mapping. C is a limited set of components that compose component-based distributed software. These components contain their IDL description and implementation information. N is a limited set of nodes in a graph g $\in$ G.

Definition 3: f_p: N $\longrightarrow$ P is a mapping. N is a limited set of nodes in a graph g $\in$ G. P is a limited set of processors on which component-based distributed software is distributed.

The L is a limited set of primitives predefined on the graph. The predefined primitives are divided into four classes: Communication and Synchronization, Sub-graph generation, Update and Query.

Software designed by CBSGOP consists of a group of distributed components. Each component is bound to a node of the logical graph. For simplicity, each node currently can host one component at one moment. The relations among the components are presented by the edges between the nodes. That is, we can reasonably limit the edges to communication links. Different kinds of communication relationships, e.g., data flows and control signals, can coexist in a graph, and they are distinguished by associating different labels to the edges. The *Graph* is distributed over the physical network by binding the nodes to the *processors*. A *processor* can be any computing device with CBSGOP Runtime installed.

Once a logical graph that represents component-based distributed software is distributed to the physical network, the components in the software may use graph-oriented primitives for inter-object communication. The basis for the graph-oriented communication is the graph-oriented reference of nodes and edges. So some build-in references, such as *precedents, succeeds, closure of precedents, closure of succeeds*, are defined. Users can derive new references by specify computations, such as union, on existing reference.

2.2 Description of Software Architecture with CBSGOP

Traditionally, architectures are represented informally as "box and line" drawings. But such drawings are mainly visual aids for understanding. With CB-SGOP, users are required to specify the graph more strictly. The topology of the graph should be precisely specified as a set of nodes and the directed edges linking these nodes up. CBSGOP provides its API for graph construction.

By using API of CBSGOP, a graph depicting Master-Slave architecture (Figure 1), can be expressed as:

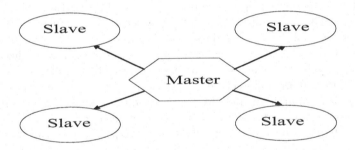

Fig. 1. A Simple M/S Architecture

Graph mygraph=new Graph ("exampleGraph",{"master","slave1","slave2", "slave3", "slave4"}, {{"slave1","master"}, {"slave2","master"}, {"slave3","master"}, {"slave4","master"}}).

In mygraph, the master node sits in the center, and slaves are distributed around the master. There is an edge linking each slave to the master.

Now the components are prepared for each nodes of this graph. Suppose we have an *average-value computing* component as the master, and four local *calculation* servers as slaves. The components may be bound to the nodes:

mygraph.bindCOToNode("averageComputing", "Master").

The components should use graph-oriented primitives for inter-object communication. In order to communicate with the components behind the build-in references, primitives with style of method invocation are provided. Thus for our example, the *average-value computing* may invoke a method named `result`, which is defined in *calculation*: `succeedNodes.result(parameters);`

With this programming style, it is possible to specify the dynamism in software architecture. For instance, if users want to add a local calculation server as slave while the application is running, they may call the following CBSGOP API to finish the task: `mygraph.addNode(new Node("Slave5"));` `mygraph.addEdge("slave5","master");` `mygraph.bindCOToNode("calculation5", "Slave5");`

Then the build-in reference `succeedNodes` is updated automatically, and the *average-value computing* still invokes method `result` through `succeedNodes.result (parameters);` which needs no modification to source code.

3 Reuse of Software Architecture Supported by CBSGOP

From the example discussed above, we see that CBSGOP software has little to do with architecture except the topology during its execution. Thus we may extend CBSGOP for software architecture reuse. The solution is based on the inheritance of the general class Graph. Users derive a new type from the general Graph type, and implement their architecture as a specialized graph such as a "tree" or a "star" with predefined CBSGOP primitives. Specialized graphs contain more architecture information and provide more specific functionality. For example, a node in a tree reasonably knows that there is one and only one root node, and one and only one center node in a star. More specific reference, such as *parent*, *children*, *siblings*, and *root* can be provided for tree nodes. Behaviors in specialized architecture are defined in this new class, and grouped as different interfaces for different nodes of the graph.

Along with more and more data joining the calculation to improve the performance, the node Master in Figure 1 waits longer and longer for the result. This leads the new requirement of adding more slaves in the system. Suppose another six slaves are needed, and the current software architecture is changed as shown in Figure 2. This situation is not foreseen by the system designer and cannot be directly solved by the system itself.

According to the previous discussion, the dynamism in software architecture may be implemented by using CBSGOP API. This is one of the methods to realize the dynamism in software architecture. Now we can implement it by using another kind of method based on the inheritance of the general class Graph. For above example, a new graph type, named MasterSlave, is defined as a subclass of Graph:

class MasterSlave extends Graph implements I_Master, I_Slave {......};

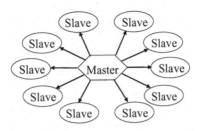

Fig. 2. The Changed MasterSlave Architecture

where I_Master and I_Slave are the interfaces of this MasterSlave type for the master node and slave nodes respectively. This type of graph will restrict the topology to MasterSlave consisting of one master node, a number of slave nodes. Each slave node is linked by an edge to the master. Then the instantiation of the MasterSlave with ten slaves can be simplified as

MasterSlave myMSGraph = new MasterSlave(10);

And the binding of components can be expressed as

myMSGraph.bindMaster("averageComputing");

myMSGraph.bindSlaves({"calculation1", "calculation2", "calculation3", "calculation4", "calculation5", "calculation6", "calculation7", "calculation8", "calculation9", "calculation10"});

Now components can directly use those more convenient and intuitive communication operations defined in their specific interfaces of the graph object. In the example, the master node uses the operations declared in interface I_Maser:

double ResultBuffer [ ] = Slaves.Result(parameters);

Average(ResultBuffer);

In this way, users may construct a graph to generate an application by using MasterSlave myMSGraph = new MasterSlave(n) (where n represents the number of slaves in the application), no matter it is a new application or the one dynamically changed from an existed one.

We can benefit more from the mechanism of the graph class inheritance. For instance, if we define a new class, named MulM, as follows:

class MulM extends MasterSlave implements I_MultiMaster { ...... }

And a new behavior adding/deleting master is defined and provided through interface I_ MultiMaster: addMaster(masterName) or removeMaster (masterName). Then an application, whose software architecture is shown in Figure 3, can be generated.

Therefore, the expressiveness of a graph type in CBSGOP is somewhat determined by how the specific graph type is. Users may derive subclass, which represents specific software architecture, from the general class Graph.

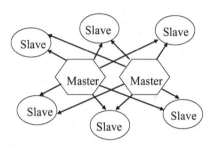

Fig. 3. The Extended Multiple Master Architecture

Users generate an instance of the subclass either for creating a new application with software architecture represented by the subclass, or for dynamically building an application with similar architecture from an existing one. Thus the software architecture reuse is performed in CBSGOP.

4 Acknowledgements

This research is partially supported by Shanghai Municipal Education Commission Research Grant 01A02, as well as the fourth period of key subject of Shanghai Municipal Education Commission (Project No. 205153 and B682). We are also grateful to Professor Jiannong Cao at Hong Kong Polytechnic University for his wonderful support, advice and encouragement.

References

[CWW03] Cao, M., Wu, G.F, Wang, Y.Y.: Architecting CORBA-Based Distributed Application. In: Proceedings of GCC 2003, Volume LNCS 3032, pp266-268. Springer, Shanghai (2003)

[DD96] Department of Defense. : Software Reuse Executive Primer. Falls Church, VA, April(1996)

[Met98] Metayer, Danie.L.:Describing Software Architecture Styles Using Graph Grammars.IEEE Transactions on Software Engineering, Vol., **24**, No.7, 521–533, July(1998)

[SDK95] Shaw,Mary.,DeLine,Robert.,Klein,Daniel.V.,et, al.:Abstractions for Software Architecture and Tools to Support Them.IEEE Transaction on Software Engineering,Vol., **21**, No. 4, 314–335, April(1995)

[LVM00] Luckham,David C.,Vera,Luckham.,Meldal,Sigurd.: Key Concepts in Architecture Definition Languages. In:Foundations of Component-Based Systems, Cambridge University Press(2000)

An Operation Semantics Exchange Model for Distributed Design

Chun Chen[1,2], Shensheng Zhang[1], Zheru Chi[2], Jingyi Zhang[1], and Lei Li[1]

[1] Dept. of CSE, Shanghai Jiaotong University, China chenchun@cs.sjtu.edu.cn
[2] CMSP, Dept. of EIE, PolyU., Hong Kong, China enzheru@inet.polyu.edu.hk

Abstract. This paper studies on operation semantics which is closely associated with design activities. Operation semantics describes designer's operations, operational parameters and the meanings. An operation semantics exchange model is proposed to illustrate how operation semantics can be used to support distributed product design. The key issues involved in the model are also discussed, such as the operation semantics generation process, operation semantics retrieval methods and operation semantics representation by XML. A prototype system based on this model is developed to demonstrate collaborative process.

Key words: interoperability, operation semantics, data exchange, ontology

1 Introduction

Conventional design tools such as AutoCAD usually define their own data formats. This makes product data difficult to be communicated and exchanged between these tools in distributed design environment. The common solution is to establish product data exchange standard such as STEP to enable the data communication and exchange between heterogeneous design tools. However individual STEP translators for each tool have to be developed [1]. Further, the STEP is a neutral data representation format which explicitly represents forms in great detail, but does not explicitly represent functions and behavior [2], and it still does not provide all the resources needed to achieve a high-level, semantic data communication between engineering processes [3]. Actually in distributed design, designers are more concerned about cooperative information of design activities. Therefore, if the data shared and exchanged carry more semantic information of design process such as design intent and content, the collaborative design process will proceed in a way that makes design more effective and interactive [4][5]. This paper presents an operation semantics exchange model which is used to describe and exchange the cooperative information of design activities.

2 Operation semantics exchange model

When designers discuss a design task in a face-to-face way, they are operating on the same object and everyone knows what is going on. But in distributed environment the collaborative situation has been changed. The designers' operation processes as well as design results must be shared to each participant's local machine. This section first describes the representation of operation semantics, and then presents the operation semantics exchange model.

2.1 Operation semantics

Operation semantics is a primitive concept which represents information used for semantic interoperability. Operation semantics describes designer's operations, operational parameters and the meanings. All of the semantics of operation-related information can be retrieved when a designer performs a set of operations on the CAD tools. For example, if the designer draws a circle, the operation semantics with related operation sets may looks like:

Operation Semantics := create a circle with the center point and radius.
Operation1 := invoke a "create circle" command.
Operation2 := designate the center point of the circle with the point coordinates.
Operation3 := designate the radius of the circle with a parameter of float value.

With different design tool the set of operations may not be the same as above. So when operation semantics is shared, precise interpretation and mapping of semantics in each design tool is essential to achieve the exchange.

2.2 The operation semantics exchange model

Each design tool always defines its own rules including the command format of operations, principle of object naming, mechanism of data access and storage. So they are incompatible with each other for data exchange. In summary, there exist two problems in operation semantics exchange:

(1) Due to the difference of operational interface and software framework of tools, the set of operations may be different when designer tries to achieves the same semantics. Some tools even provide several sets of operations to serve the same purpose. This may increases the difficulty in semantics retrieval.

(2) Mismatching of operation semantics may occur between design tools. Design tools developed to meet diverse requirements may provide some special functions. So operation semantics domain should be predefined by ontology.

In order to solve the above-mentioned problems, interpretation and mapping mechanism based on ontology is introduced to the operation semantics exchange model. Ontology is a term commonly used in the fields of AI and knowledge management [6]. Ontology can be used to abstract and describe the concepts of domain knowledge and its objective is to share and reuse domain knowledge [7]. The interpretation mechanism must be established as the linking of operation semantics and domain knowledge and also the mapping mechanism for domain knowledge. Fig. 1 describes the operation semantics exchange model based on ontology. In this model, local operation instructions are mapped into semantic meta-unit which is an independent semantic unit. By using the data transform engine and data representation engine a semantic meta-unit is translated into a meta-aim unit which is independent to the design tool. All the mapping rules and data transformation and representation methods are specified by ontology.

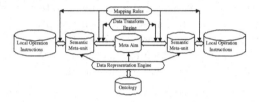

Fig. 1. An operation semantics exchange model based on ontology

3 Implementation

3.1 Operation semantics generation process

In distributed environment, it is important to generate and manipulate data in a minimum time to meet the real-time and synchronization requirements. The generation process of operation semantics consists of two stages: recognition and representation. In recognition stage, the high layer semantics of product design is converted into low layer operations of CAD tool commands. By using CAD API tools or message mechanism, operations are optimized and transformed into basic operation instructions. In representation stage, basic operation instructions can be mapped into sharable operation semantics by referring to ontology definition rules and ontology concepts database.

3.2 Operation semantics retrieval methods

There are two methods to retrieve the operation semantics: using CAD API and using message mechanism of the operation system.

Semantics retrieval using CAD API. Usually most conventional CAD tools provide API for further development or customization to meet personal requirements. Via the CAD API, what the designer has done can be captured and some manual operation instructions can be executed automatically.

Semantics retrieval using message mechanism. All the user inputs must be first processed by the operation system, then the operation system recombines the user inputs into messages and sends the messages into the CAD message queue. So if we intercept and capture the messages before they are sent to the CAD message queue, we can get the user's original operations and translate them into operation semantics.

3.3 Operation semantics representation by XML

XML (eXtensible Markup Language) is a basic data modeling language providing a DTD document to make up contents of domain knowledge. XML/DTD documents are well structured and suitable for data storage and retrieval, and the abstract concepts in ontology are used for communications. DTD is a collection of tags, and its meaning is interpreted by associated ontology. When a set of operation semantics are captured, an related XML document is then generated to encode the operations, parameters and meanings referring to relevant DTD documents. Then the XML document is sent to other sites to be parsed into local executable instructions.

4 An experiment

A prototype platform for distributed product design is built to illustrate how collaborative design proceeds. The socket communication, hook technology and CAD API have been used in the development of the platform with a client/server framework. Here a demonstration of collaborative design with

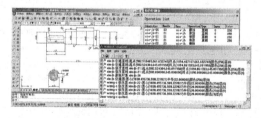

Fig. 2. Collaborative design sample based on operation semantics exchange

two distributed sites is given(shown in Fig. 2). Two users at geographically different locations begin the collaborative design to confirm the design of an axle. An AutoCAD2002 software interface and an operation list window are displayed in the client site. The users do their design with their familiar design tools and all of the operation semantics are captured by the system. Then in the server site which can be located at the place available to all the client sites, all of operation semantics which are captured by every client site are stored and can be used for further processing. It is shown that two users can simultaneously discuss with each other and conveniently manipulate the product design drawing to do the modifications in distributed environment. As the server stores the logs of operation semantics, it can be effectively used to review the collaborative work in the future.

5 Conclusions and Acknowledgements

In this paper an operation semantics exchange model is proposed to facilitate information sharing in distributed product design. Different from the STEP based data exchange, operation semantics exchange is to extract the semantics of the designer's operations as the exchanged data to achieve collaborative design. The operation semantics are defined to enable the interpretation and mapping among heterogeneous design tools, and XML/DTD is well used to describe the contents of semantics and facilitate the content retrieval and execution. The demonstration shows that geographically distributed designers proceed with their designs successfully, and all the information of operation semantics are stored to show what they have done in the design.

This research was sponsored by the China 863 Program (No. 2002AA411420) and the NSFC grant No. 60374071. The work is also partially supported by a research grant (No. G-T851) from Hong Kong Polytechnic University.

References

1. Zhang Y.P., Zhang C., Wang H.P.: An Internet Based STEP Data Exchange Framework for Virtual Enterprises. Computers in Industry, 41 (2000) 51-63.
2. Teresa D.M., Bianca F., Stefan H.: Design and Engineering Process Integration through a Multiple View Intermediate Modeler in a Distributed Object-Oriented System Environment. Computer-Aided Design, 30 (1998) 437-452.
3. Mark J.C., John C.K., Martin A.F.: Rapid Conceptual Design Evaluation Using a Virtual Product Model. Engng Applic. Artif. Intell, 9 (1996) 439-451.
4. Seitamaa-Hakkarainen P., Raunio A.M., Raami A.: Computer Support for Collaborative Designing. Intern. J. Technology and Design Education, 11 (2001) 181-202.
5. Baek S., Liebowitz J., Lewis M.: An Exploratory Study: Supporting Collaborative Multimedia Systems Design. Multimedia Tools Appl, 12 (2000) 189-208.
6. Ram S., Park J.: Semantic Conflict Resolution Ontology(SCROL): An Ontology for Detecting and Resolving Data and Schema-Level Semantic Conflicts. IEEE TKDE, 16 (2004).
7. Zhan C., Paul O.: Domain Ontology Management Environment. IEEE Proc. of 33rd HICSS, 8 (2000), Jan. 04-07.

OGSA-based Product Lifecycle Management Model Research

Dekun Chen[1], Donghua Luo, Fenghui Lv, and Fengwei Shi

Shanghai University and CIMS Center, Shanghai University, Shanghai, P.R. China. dkchen@mail.shu.edu.cn

Abstract. PLM application in manufacturing industry is becoming more and more popular. At the same time, integration in different enterprises with respective PLM system is also becoming more difficult, which have really affected and deterred the cooperation and product coordinate design and relative manufacturing activities and so on. The successful application of computing Grid provides a new approach for the development of PLM. The article puts forward an OGSA-based PLM architecture model (GPLM), which use key Grid technologies to resolve cooperative problems between manufacturing enterprises. GPLM doesn't surround an enterprise's management but focuses on interoperability during the dynamic processes of development and manufacturing between enterprises. Moreover, it focuses on the data exchange standard with STEP/XML that is internationally confirmed.

Keywords. OGSA PLM XML data exchange

1 Disparity between Ideal and Reality of PLM

PLM is playing an important role in network manufacturing. The situation about building enterprise information system, however, is not as optimistic as we expected. In inner-enterprise, it is very difficult to realize a senior level integration by dint of simple integration methods in different application systems such as PDM, ERP, CAD, and CAPP etc, which have differences among platforms, development languages and structure models etc. In extended enterprises, especially in machine manufacturing, it is involved with many enterprises, organizations, different users and global information to realize distributed product collaborative development and different place manufacturing, which make product data management in extended enterprises and collaboration of application systems in inter-organization and inter-enterprise more difficult. To realize information integration would need an expensive and complicated application information system if the system is developed by traditional technologies which are not only make middle and small enterprises unaffordable but also objectively lead to an invisible constrains to the development of manufacturing units. Therefore the coming true of PLM ideal is looking forward a new breakthrough in network and computing environment.

2 Characters of OGSA

OGSA is a service structure focusing on service. In early Grid various computing capacities were abstracted as various resources such as computing resources, storage resources, and communication resources etc which simplified system structures. Further combining with characters of web service these resources are encapsulated by Globustoolkits3 as service in OGSA. Grid service is web service, which provides a set of interfaces. The definition of those nails down and complies with specified routines, resolving problems about service discovery, dynamically service development, life cycle management and notification etc. In OGSA each resource is regarded as Grid service. As a result Grid is regarded as an aggregation of extendable Grid service. Grid service may aggregate by various formats to meet request of different virtual organizations.[YCL02] OGSA includes two supporting technologies that are Grid technology such as Globus and web service.

3 OGSA based GPLM Model with Fractal Character

According to viewpoint above, combining with manufacturing request for product data management, we propose a product life-cycle management model based on OGSA, which is named for short as GPLM. It acts as middle layer model in Grid application and provides basic product information service for top layer manufacturing application layer. GPLM Deals with distributed, collaborative and intelligent product data. GPLM's aims are as follows: (1) Connectivity and consistency of product data based on collaborative enterprises. (2) Open system structure. Layer format shaped by a lot fractal and seamless integration of software show dynamic and diversity. (3) Product data standard management. GPLM is a new integration model based on STEP standard that is internationally confirmed. (4) Low complexity and cost of system.[And02].

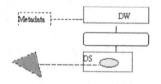

Fig. 1. GPLM Basic Unit

From these aims above we can get a conclusion that GPLM actually is product data Grid in manufacturing field. GPLM bales heterogeneous product data sources as a normative Grid service or web service, which is abstracted as basic product data unit (Figure 1). Various complex application systems with respective use that are relative to product data in manufacturing all can be regarded as extension of basic product data unit and aggregation of a lot connection. GPLM model is divided into four layers that are product data application layer, data window service management layer, data service node fabrication layer, and product data resources layer. Information service fabrication layer and information management layer are the core layers of GPLM from top to down, which can connect bottom various information sources and top various application. Details as follows:

3.1 Product Information Resources Layer

As showed in figure 2, GPLM defines various information system of enterprise as information sources that are heterogeneous, open or semi-open system. We suggest that product data can be divided into four kinds in management' s view:

- Engineering structure data being stored by relative database.
- Standard document such as Electronic table-excel.
- Various vector drawing generated by CAD.
- Else document such as electronic document.

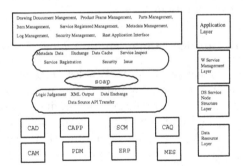

Fig. 2. GPLM Structure

The first step of GPLM is inner-enterprise information integration. We propose information resources to abstract various information sources such as ERP, CAD, and CAPP etc., construct enterprise standard, realize enterprise information Grid service encapsulated normatively, and shield differences of information structure between departments according to Grid principle so that we can make real-time and dynamic information in department becoming really shared resources; The second is information collaboration between upriver and downriver collaborative enterprises. The third is sharing information of virtual organization.

3.2 Data Service Node Fabrication Layer

The layer designs class of data collection service, product data release service, local buffer storage service, product data exchange service, and security authentication service. Agent combined with services mentioned above makes specific information sources become preconcerted normative output of standard product information sources having nothing to do with origin. The main functions of GPLM consist of services. The layer logical relations are defined as follows:

- Resources belong to enterprises. Every enterprise has the only standard serial number and name. An enterprise can join a lot of virtual organizations and has a lot of resources. The enterprises and virtual enterprises can be resources according to fractal theory.
- Each resource can be baled as a Grid service or web service by agent.
- Every Grid service has the only standard serial number and name. A Grid service may have a lot of functions.
- The interface of the Grid service is divided into a common part and special-purpose part. Various Grid services have the unity of common interface. Every kind of Grid service has one's own unified special purpose interfaces.
- Every function has the only standard serial number and name. The same function in different Grid services must use standard serial number and name, in order to facilitate user's search.

3.3 Data Window Service Management Layer

This layer is service management layer. Besides finishing the essential functions of management, it obtains the standard form products data of the lower layer through Grid service, then changes the products data for the display form of the data window ,print, upper layer XML data to output and products data buffer storage.[FKT01] This layer's functions include as follows:

- Service registry management.
- Service supervision management.
- Metadata management.
- Products data buffering storage management.
- Products data exchange management.

3.4 Product Data Application Layer

This layer includes the functions in every stage of product life cycle: drawing document management, product structure management, part clan management, project management, service registry manage, metadata maintenance management, daily record management, and security management etc. Applying open interfaces of GPLM can integrate more applications.

4 GPLM Basic Unit Prototype Design and Realization

Taking the complex circumstances of product life cycle phases in simulated manufacture domain into account, we have designed various exploitation platformvarious programming toolsvarious service modes and various data encapsulations, therefore we can make a experiment on GPLM Core Technology through Prototype System. We chose GT3 and the two most popular exploitation modes: J2EE and .NET while adopting two encapsulation methods: Grid Service and Web Service .The network environment is LAN that is in Shanghai University Engineering Training Center (one District) and Shanghai University CIMS Center (another District). The two LAN are connected by VPN. Three information fountains are three independent GPLM basic units in experiment circumstance. In order to compare effect of different service encapsulation tools, we use uniform DW module to transfer three different DS module as the concentrative test project.

Prototype system designs the uniform service interface. Work flow: When the service provider is transferred, the query conditions (CompanyID, ResourceID, FunID, PartsID and so on) which are firstly parsed by XML, then take the corresponding information out of the Meta Data Frame according to the requirement. Following the information, it takes the data out of database then switch to standard format. At last it returns the same criterion of product information through service interface. When DW transfer product datawith uniform product data middle layer of XML format, the local format has switched to standard format. DW module can conveniently take data out as its database is in local area by the intersect service transfers. Then after it has been parsed, the browser will see the page. Figure3 depicts the experiment list interface.

Fig. 3. Interface of Corresponding Table

5 Conclusion

A common PLM application system emphasizes particularly on the management of product data, neglecting management of manufacturing resources generating product data. However, GPLM combining with Grid techniques strengthens different information's relation and is much more close to the ideal of PLM. With Grid techniques it makes breakthrough in application of manufacturing field and creates favorable condition for establishing Manufacturing Information Grid in manufacturing enterprise.

References

[YCL02]   Yu Zhihui,Chen Yu and Liu Peng: Computing Grid. Publishing of Ts-
          inghua University. (2002)
[And02]   Andreas Schreiber. OpenPDM -Online PDM Integration based on Stan-
          dards. www.prostep.com (2002)
[FKT01]   Foster, I., Kesselman, C. and Tuecke, S. :The Anatomy of the Grid: En-
          abling Scalable Virtual Organizations. In:International Journal of High
          Performance Computing Applications, 15 (3). 200-222. (2001)

Study of Manufacturing Resource Interface Based on OGSA

Dekun Chen[1], Ting Su[2], Qian Luo[3], and Zhiyun Xu[4]

Shanghai University and CIMS Center, Shanghai University, Shanghai, P.R. China. dkchen@mail.shu.edu.cn

Abstract. Traditional models of manufacturing resource can't cover all the manufacturing information, including resource information, process information and product information, which cannot make enterprise function in a whole. From the experience of the Grid Computing and the summary of manufacturing feature, the paper introduces a model named GPPR, integrating three information parts of the manufacturing resources based on OGSA (Grid-based Product Process Resource). An example is presented to prove the feasibility and the possibility of wide spread of GPPR model by using globus3.0 is to create the prototype system of GPPR. Make an interface model including all the manufacturing information.

Keywords. OGSA; manufacturing resource; service encapsulation

1 From the Computing Grid to the Manufacturing Grid

Network manufacturing needs to integrate manufacturing resources distributed in different fields and areas in a dynamic optimization. As a great technology progress after the coming of the Internet, the Grid can provide such solution. But when we lucubrate to the key technologies of the MG, such as enterprise resource scheduling, supply chain process optimization and product lifecycle management [YCL02], we can't avoid facing the problem how to encapsulate the various manufacturing resources with normative means. These various manufacturing resources include enterprise product data, manufacturing knowledge, experts, machine tools and so on.

T hese years, the success of the CG(Computing Grid) can be a good example to the MG when encapsulating resources. No matter how powerful and how far the computer is, it can be encapsulated to the computing resources and can be found and rented as a reliable normative computing capability [YCL02]. For such reason, the CG has appealing a great many of experts and a large deal of investment. But researches on how to encapsulate manufacturing resources in the Grid are obviously less. So we can conclude:

- The encapsulation of manufacturing resources in traditional network is of ripe. But its limitation cannot meet the requirement of cooperative design and manufacturing on the Internet, largely hindering the developing of advanced manufacturing.
- MG is the trend of network manufacturing, but it is difficult to import the Grid technologies to the manufacturing field. One obstacle is the encapsulation of manufacturing resources in the Grid.

2 The Characteristic of Resources in Rapid Manufacturing

The encapsulation of manufacturing resources is the basis of network manufacturing. One typical application is RM(Rapid Manufacturing). There are many types of RM resources, including software, rapid processing equipment, rapid modeling equipment, co-ordinate measuring machine, manufacturing information database, manufacturing knowledge database and so on. We can classify these resources into three parts according to the way of operation. One is the automatic resources, including computers, information databases, knowledge databases, and NC equipment. The second part is the semi-automatic resources, such as semi-automatic equipment and the interface software. The third part is the resources manipulated by the human beings, involving the common machine tools and the experts. In the light of this division, the encapsulation will be a very tough work because we need to control these distributed resources in a dynamic way.

R -MG (Rapid MG) can be divided into SG (Service Grid), DG (Data Grid) and KG (Knowledge Grid) according to its application. CAD and rapid processing will be available in the SG. The DG will provide the service about product data and resource data, such as storage, transmission and format change. The KG will supply the description of manufacturing knowledge in a clear or unclear way. But, there will be a large difference between the encapsulation of these applications.

3 Manufacturing Resource Interface Model Based on OGSA - GPPR Node Model

In manufacturing field, PLM (Product Lifecycle Management) is a set of solution about the manufacturing information management and service architecture, which allows the enterprise to design and manufacture the product together with its partners, thus making all the product information managed well [LLG02]. It is believed that the control of design and manufacturing process is very difficult when the product management data has been changed. When PDM/PLM has becoming popular in manufacturing industry,experts are trying to replace the traditional information integration based on the product with the dynamic process integration based on the supply chain. To achieve this goal, we propose a new dynamic pattern integrated the product, the process and the resource in the light of the theory of the Grid and PLM. We abstract manufacturing resources into a standard node with Grid Interface, which constitute the MG(Fig 1).

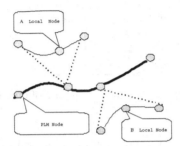

Fig. 1. The Principle of GPPR

The node is wrapped with a standard interface based on the Grid. Inside the node, agent is used to transform the given manufacturing resources. We call such a node as GPPR, which is a MG node integrating the product, the process and the resource information with the standard interface. According to the similarity theory in the Grid, the local GPPR node is similar to the whole nodes. So if we encapsulate every stage of manufacturing as a GPPR node, we can eliminate the heterogeneousness among manufacturing resources, thus greatly reducing the complex of the system . Furthermore, we think the MG is just the system in which GPPR node can optionally group.

- GPPR is designed not only for one enterprise's management, but also for the cooperation when the enterprises dynamically develop products.
- GPPR is available to the SG, DG, and KG.
- In GPPR, the information exchange between the enterprises follows the International Standardization such as STEP/XML
- GPPR makes the integration of enterprises application systems in the security of Grid Service.

S o GPPR is the basis to realize Rapid SG, Rapid DG and Rapid KG in the future.

3.1 The Prototype of the GPPR and Its Design

3.2 The Prototype and its Logic Structure

We have discussed in the Section 2 that a lot of manufacturing resources can't be encapsulated directly while they are absolutely necessary in manufacturing process. So we propose a combination way to encapsulate the resources (Fig 2). as follows:

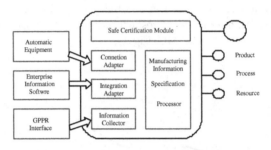

Fig. 2. The Logic Structure of GPPR

- Direct Encapsulation. We can develop a general module that can collect the real-time data from those automated equipments such as PCs, PLCs by its communication ports, then encapsulate it as a standard Grid Service. We call this module Connection Adapter.
- Indirect Encapsulation. It's known that other types of resource information can be found in the enterprise' MISs, including PDM, ERP and so on.
- We also design an Information Collector to get the data user input, which can totally meet the needs of manufacturing.

3.3 An Example for GPPR

In this chapter, we will show a classic example to describe how to create a GPPR node.

R M involves lots of detail manufacturing technologies. The transformation of CAD drawing format is the common one. Due to the history, there are several formats in CAD drawing, so that an operative project often needs many times to change the format of the drawings. We have abstracted this action as a manufacturing resource, then encapsulated it as a Grid Service.

A ccording to the drawing's complexity, we can divide the drawings into three types:

- Auto-transformation, including parts made by standard geometries. This type can be transformed with some software.
- Non-Auto- transformation, including parts of complex curve. Experts can change these drawings with their domain knowledge.
- Semi-Auto- transformation. Some drawings should be examined by the experts when they have been changed.

J ust as analyzed in the Section 4.1, we have encapsulated the Auto-transformation with the first way, and the others with the second way (Fig 3).

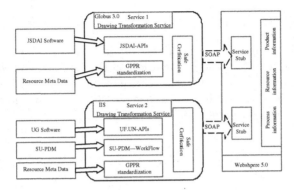

Fig. 3. The Drawing Exchange Node of GPPR

T he work flow of this GPPR is:

- When a user wants to change the STEP file to the XML file, he/she can inquire the description and the state of the GPPR. When the situation is OK and the system has got the file, GPPR will automatically invoke Service 1. After finishing, the system will return the transformation information and the changed file to the user. This service is totally automation.
- Service 2 is an indirect type. That means this transformation needs human beings. Like Service1, the system automatically invokes the service. The process will be separated to transformation, examination, amend, re-examination, check and accept so that the user can get the latest transformation state without knowing the details of the workflow. When the process ends, the system will return the result.

3.4 The Implementation of the Example

We develop these services based on the tools in Table 1.

Table 1. The Implementation Tools of Two services

Tools	Service 1(Grid Service)	Service 2(Web Service)	Server
Hardware	PC	PC	PC
OS	Windows 2000 server	Windows 2000 server	Windows2000 server
Software	j2sdk1.4.101	UFUN API	IBM websphere 5.0
	JSDAI3.6 API	SUPDM	Globus 3.0
	Globus 3.0 API	Microsoft Visual	
		Studio.Net	

4 Conclusion

Having tested from many aspects, the GPPR node has proven its practicability
in enterprises integration and cooperation on the Internet, the correctness of
encapsulating resources based on the service technology. Though this research
is focused on the RM, but it is sure that GPPR will be of significant to the
whole discrete manufacturing field.

References

[LLG02]   Li Shanping,Liu Nainuo and Guo Ming:Product Data Standard and
          PDM.ublishing of Tsinghua University.(2002)
[YCL02]   Yu Zhihui,Chen Yu and Liu Peng: Computing Grid. Publishing of Ts-
          inghua University. (2002)
[And02]   Andreas Schreiber. OpenPDM -Online PDM Integration based on Stan-
          dards. www.prostep.com (2002)
[TCF03]   Tuecke S , Czajkowski K, Foster I: Grid Service Specification [EBP OL]
          . http ://www. grid forum. orgPogsi2wg , (2003)
[FDT01]   Foster I ,Desselman C , Tuecke S. The Anatomy of the Grid : Enabling
          Scalable Virtual Organizations[J] . International J Supercomputer Appli2
          cations ,(2001)
[YSW03]   Yang G G, Shi S M, Wang D X, et al. DSI : Distributed Service Integration
          for Service Grid. Journal of Computer Science and Technology, 18(4) :
          474483 (2003)

Hybrid Architecture for Smart Sharing Document Searching *

Haitao Chen[1], Zunguo Huang[1], Xufeng Li[1], and Zhenghu Gong[1]

School of Computer, National University of Defense Technology, Changsha 410073, China nchrist@163.com

The file sharing application is one mainstream application of P2P technology, in which users can share files in their local hard disc. There are many research projects devoting to solve the problems of current file sharing applications from different angles, such as Gnutella[G01], NeuroGrid[JH02] and so on. In this paper we propose a novel searching architecture for sharing documents-SmartMiner. The storage and search of raw documents are based on DHT network, but the storage and search of meta-data are based on unstructured P2P network in SmartMiner. SmartMiner combines the deterministic searching capacity of DHT network and the fuzzy searching capacity of unstructured P2P network. It implements the intelligent query forwarding based on the overlay network of file relation to discovery adequately the relations between the users and files. It also considers the trust metrics of nodes and implements the efficient ranking of searching results.

Key words: peer-to-peer, file sharing, search, file relation, SmartMiner

1 Introduction

Peer-to-peer network is a novel Internet application architecture which has been applied in many fields. The file sharing application is one of mainstream applications of P2P technology. The number of users who use Kazza[KA04] and Morpheus[MO04] which are very popular P2P file sharing software exceeds one hundred millions. That is more than half of the number of all network users. From the view of network management, sixty percent of main bandwidth of many ISP is consumed by P2P file sharing applications, and the proportion is increasing. Searching is key problem of P2P file sharing systems which has decisive influence on the performance of system. This paper

---
*This project is supported by the National 863 High-Tech Research and Development Plan of China under Grant No.2003AA142080, the National Grand Fundamental Research 973 Program of China under Grant No.2003CB314802.

proposes a novel searching architecture for sharing documents-SmartMiner. SmartMiner adopts hybrid architecture in which the storage and search of raw documents are based on DHT network, but the storage and search of meta-data are based on unstructured P2P network.

2 Related Work

There are many research projects devoting to solve the problems of current file sharing applications from different angles. Napster[NA04] is a typical centralized searching model. The directory server of centralized search model is a single point of failure and performance bottleneck, so the scalability of this model is limited. This model's advantage is that it supports for semantic partial-match queries as well as high efficiency and security assure. Gnutella[G01] is typical broadcast search model. The main disadvantage of broadcast model is high bandwidth consumption which leads to bad scalability. Furthermore, the searching results of this model is uncertain which means the documents existing somewhere in the network maybe can not be located. NeuroGrid[JH02] learns the distribution of documents from the results of the searching. NeuroGrid makes routing decision based on the keywords of the documents stored in other nodes. The disadvantages of NeuroGrid are that the information updating speed is slow; the result of query is uncertain; the survivability under attack is low; it doesn't support for ranking of searching results.

3 Research on File Sharing Architecture

Existing P2P file sharing architectures can be coarsely partitioned into three groups: structured, unstructured and hybrid. Structured P2P network supports well for deterministic search, but it is hard to carry out fuzzy search because of its simple structure. On the contrary, unstructured P2P network supports well for fuzzy search, but it is hard to carry out deterministic search because of its unstable and erase structure. There are some attempts to support fuzzy search in structured networks such as Semplesh[SE04] which maps the RDF Triples to hash and faces problems like low efficiency, PSearch[TXD03] which compresses the document vector space to DHT space and faces problems like single searching capacity for keywords query. Broadcast is a basic method used in unstructured networks to implement deterministic search.

We present a hybrid architecture in which storage and search of raw documents are based on structured network while the storage and search of meta-data are based on unstructured network. The architecture of SmartMiner is shown in Figure 1. The unstructured network can present the complex relations among the file meta-data and it can meet the users' diverse query needs. DHT network can effective solve the problem such as file moving, replicating and downloading, and support the discovery of documents relations.

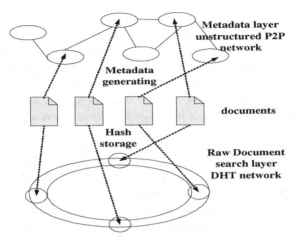

Fig. 1. Sketch Map of SmartMiner Architecture

4 Research on Overlay Network of Document Relation

We propose the notion of overlay network of documents relation to discovery adequately the relations between the users and files and to optimize the traditional query routing algorithms. The overlay network of documents relation includes five layers: raw documents layer, meta-data layer, documents relation layer, role relation layer, user relation layer. The architecture is shown in Figure 2. SmartMiner adopts multi-attribute meta-data to present file meta-data.

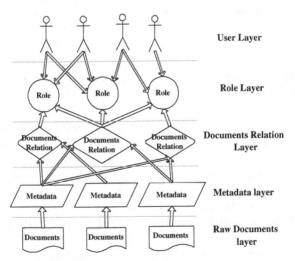

Fig. 2. Sketch Map of Overlay Network of Documents Relation

Documents relation is the relations among the files which satisfy special rule such as same relation, quote relation, similar relation and so on. Files can form a multi-dimension file relation space according to multi-dimension attributions such as time, space, theme, application, keywords. The evaluation of relations among files can use different multi-dimension formula according to different search requirements. Similar content can aggregate to link each other, and the same content can link together through some kind of discovery method.

A user can belong to several roles. Every role represents a file set which the user is interested in. The role relation is used to evaluate the similar interest of users. Users with similar interests can link each other and gradually form a group through discovery of role relations.

User relation is used to evaluate relations among correlative users, which is used for route probing and ranking of searching result. User can distinguish the users in the same role interest group from different angles such as probability of success for historical searching, download speed, trust degree and so on. The node with higher service quality will get higher score. The score will influence the route probing and ranking of searching result.

Different file relations network can be constructed according to different rules of file relations in the file relation layer, such as follows: (1) Same File Link means to find the same files and link them each other. (2) Similar File Link means to find the similar files and link them each other. (3) User Recommendatory Link means that the users can recommend files and attach the level of recommendation. (4) User Reading Link means to link files according to users' reading and downloading behavior. Users' downloading and searching are indication of the interests of the users.

Those networks of meta-data relation are dynamic and have mechanism of feedback. Users' searching and visiting behavior can change the meta-data relations.

5 Searching Algorithm Based on Overlay Network of Documents Relation

The overlay network of documents relation can be used to carry out effective searching. The process of searching is as follow: (1) The user sends out search requirements. Local resources will firstly be searched and then the query is forwarded to other nodes if local search fails. The query forwarding is based on overlay network of documents relation. If the search component can confirm the role of the query, the query will be forward according to the role relation tables in the local, and forward prior to the nodes that get high score. Otherwise, it will decompose the query to the documents relation layer which confirms the destination nodes according to the matching degree of file attributions and searching requirements. (2) The node receiving the query will match its local resources with the query. It will return the files if the match successes, otherwise it will judge whether forwards the query according to some policy. (3) The user ranks the searching results and selects files which he wants to download. There are many ways of ranking. One is that the user scores the results according to the correlative degree of matching condition and trust degree of the user. The second one is that the user retrieves the detailed meta-data in the DHT network, and then the user gives a integrate score according to the historical evaluation such as times of download, times of recommendation. (4) User begins to download simultaneously files from multiple nodes, and the download process will register the status of download. (5) The user gives the evaluation after downloading the files. (6) At last the feedback process begins. There are many kinds of feedback. The routing table updates its content according to the downloading records. Trust feedback updates its trust table according to the users' evaluation on the files. It will reduce the reputations of users who usually are off-line. It will identify the users who provide malicious files as enemy. It will increase reputations of online users who provide files with high quality. With the help of feedback, the whole searching network will get evolution.

6 Advantages and Problems

SmartMiner has advantages as follow: combining the deterministic searching capacity of DHT network and the fuzzy searching capacity of unstructured P2P network; owning good searching capacity and supports for many kinds of searching methods such as theme searching, keywords searching and so on through multi-attribute meta-data; introducing the concept of overlay network of documents relation which can discovery adequately the relations between the users and files; making full use of feedback to optimize searching process; supporting effective ranking of searching results.

There are still some problems in SmartMiner. It will generate overhead when the nodes join and quit in the bottom DHT network. Further researches are needed on quantitative analyzing and testing.

References

[G01]   Kan. G.: Gnutella. Peer-to-Peer: Harnessing the Benefits of Disruptive Technologies , Ed Oram, A. O'Reilly & Associates , http://gnutella.wego.com/, 94-122 (2001).

[JH02]   Sam joseph, S.R.H: NeuroGrid: Semantically Routing Queries in Peer-to-Peer Networks. International Workshop on Peer-to-Peer Computing, Pisa , (2002).

[MCPD02] Cuenca-Acuna, F.M., Peery, C., Martin R.P. and Nguyen, T.D.: PlanetP: Using Gossiping to Build Content Addressable Peer-to-Peer Information Sharing Communities. International Workshop on Peer-to-Peer Computing, Pisa , (2002).

[TXD03] Chunqiang TangZhichen Xu Sandhya Dwarkada, Peer-to-Peer Information Retrieval Using Self-Organizing Semantic Overlay Networks , sigcom 2003.

[OHA01] Babaoglu, O., Meling, H. and Montresor, A.: Anthill: a Framework for the Development of Agent-Based Peer-to-Peer Systems. Technical Report UBLCS-2001-09 , (2001).

[KA04] Kazaa, www.kazaa.com.

[MO04] Morpheus, www.morpheus.com.

[NA04] Napster, www.napster.com.

[SE04] Semplesh, www.plesh.net.

ID-based Secure Group Communication in Grid Computing *

Lin Chen, Xiaoqin Huang, and Jinyuan You

Department of Computer Science and Engineering, Shanghai Jiao Tong University,
No.1954, HuaShan Road, Shanghai 200030, China chenlin@sjtu.edu.cn

The security requirements especially for secure group communication are discussed in grid circumstance. When several processes generated by a user want to communicate securely, they have to share a secret key to communicate. We use the identity-based cryptography to implement the secure group communication. Finally, we describe the advantages of our scheme.

1 Introduction

Grid systems and applications manage and access resources and services distributed across multiple control domains [FK99]. Grid computing research has produced security technologies based not on direct interorganizational trust relationships but rather on the use of the VO (virtual organization) as a bridge among the entities participating in a particular community or function [WSF03]. The results of this research have been incorporated into a widely used software system called the Globus Toolkit [BFK00] that uses public key technologies to address issues of single single-on, delegation [GM90], and identity mapping, while supporting standardized APIs such as GSS-API [Lin97]. The Grid Security Infrastructure (GSI) is the name given to the portion of the Globus Toolkit that implements security functionality. The user population is large and dynamic. The resource pool is large and dynamic. A computation may acquire, start processes on, and release resources dynamically during its execution. Resources may require different authentication and authorization mechanisms and policies, which we will have limited ability to change. An individual user will be associated with different local name spaces, credentials, or accounts, at different sites, for the purposes of accounting and access control. The processes constituting a computation may communicate by using a

---

*This paper is supported by the National Natural Science Foundation of China under Grant No.60173033.

variety of mechanisms, including unicast and multicast. Group context management is needed to support secure communication within a dynamic group of processes belonging to the same computation [FKT98]. In recent years, the identity-based cryptography is developed rapidly. The advantages and convenience are obvious. In this paper, we discuss the secure group communication mechanism based on identity cryptography in grid circumstance.

2 Security Requirements

Grid systems may require any standard security functions, including authentication, access control, integrity, privacy, and nonrepudiation [FKT98]. In order to develop security grid architecture, we should satisfy the following characteristics:

- Single sign-on: A user should be able to authenticate once.
- Protection of credentials: User credentials must be protected.
- Interoperability with local security solutions: Security solutions may provide interdomain access mechanisms.
- Exportability: We require that the code be (a) exportable and (b) executable in multinational testbeds.
- Uniform credentials/certification infrastructure: Interdomain access requires a common way of expressing the identity of a security principle such as an actual user or a resource.
- Support for secure group communication.

A computation can comprise a number of processes that will need to coordinate their activities as a group. The composition of a process group can and will change during the lifetime of a computation. Therefore, secure communication for dynamic groups is needed. So far, no current security solution supports this feature; even GSS-API has no provisions for group security contexts [FKT98].

3 Related Work

3.1 ID-based Non-Interaction Secret Sharing Protocol

Ryuichi Sakai et al. [SOK00] mentioned the non-interaction secret sharing protocol. It can be proved easily using $\hat{e}$ map properties. ID_A and ID_B are respectively the identity information of Alice and Bob. $S_{IDA} = sH(ID_A)$ and $S_{IDB} = sH(ID_B)$ are respectively their secret key. s is the system secret key, $P_{IDA} = H(ID_A)$, $P_{IDB} = H(ID_B)$. The protocol is as follows:

- Alice computes the K_{AB}, $K_{AB} = \hat{e}(S_{IDA}, P_{IDB})$
- Bob computes the K_{BA}, $K_{BA} = \hat{e}(S_{IDB}, P_{IDA})$.

- Using map $\hat{e}$ properties, we can verify $K_{AB} = K_{BA}$. $K_{AB} = \hat{e}(S_{IDA}, P_{IDB}) = \hat{e}(sP_{IDA}, P_{IDB}) = \hat{e}(P_{IDA}, P_{IDB})^s = \hat{e}(P_{IDA}, sP_{IDB}) = \hat{e}(P_{IDA}, S_{IDB}) = K_{BA}$

The protocol doesn't need interaction process.

3.2 One Round Tripartite Diffie-Hellman Protocol

Antoine Joux [Jou00, Jou04] presented a new efficient one-round tripartite key agreement protocols. A pairing is a bilinear map $\hat{e} : G_1 \times G_1 \rightarrow G_2$ with group G_1 and G_2 of a large prime order q , which satisfies the bilinear, non-degenerate and computable properties [Jou00]. If Alice, Bob and Charlie want to get a secret share secret, the protocol requires each party to transmit only a single broadcast message to establish an agreed session key among three parties.

$$A \rightarrow B, C : aP \tag{1}$$
$$B \rightarrow A, C : bP \tag{2}$$
$$C \rightarrow A, B : cP \tag{3}$$

After the session, A computes $K_A = (bP, cP)^a$. B computes $K_B = (aP, cP)^b$ and C computes $K_C = (aP, bP)^c$. The established session key is $K = K_A = K_B = K_C = (P, P)^{abc}$.

4 Secure Group Communication Based on Identity Cryptograph

In Grid computing circumstance, a group member often join/leave the group dynamically. When a user joins the group, the group key should change to a new value. Also when a user leaves a group, the original key should become invalid and the remaining member should have a new group key. So our group communication scheme should support the dynamic circumstance. Our algorithm consists of three steps as follows:

Setup: Suppose our system have two members initially. From Section 4 ID-based Non-Interaction secret sharing protocol, the user u_1 computes: $K_{AB} = \hat{e}(S_{IDA}, P_{IDB})$ and u_2 computes: $K_{BA} = \hat{e}(S_{IDB}, P_{IDA})$. u_1 and u_2 hold the sharing secret key $K_{AB} = K_{BA}$. The structure is as Fig.1.

Join: As in Fig.1, u_3 wants to join the group. We use the A. Joux's one round tripartite Diffie-Hellman protocol. A bilinear map $\hat{e} : G_1 \times G_1 \rightarrow G_2$ with group G_1 and G_2, $P \in G_1$, which has the following properties:

1. Bilinear: For all $P, Q, R, S \in G_1$,
   $\hat{e}(P + Q, R + S) = \hat{e}(P, R)\hat{e}(P, S)\hat{e}(Q, R)\hat{e}(Q, S)$.

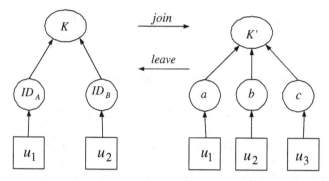

Fig. 1. Share key graphs before and after a join (leave)

2. Non-Degenerate: For a given point $Q \in G_1, \hat{e}(Q, R) = 1_{G_2}$ for all $R \in G_1$ if and only if $Q = 0_{G_1}$.
3. Computable: There is an efficient algorithm to compute $\hat{e}(P, Q)$ for any $P, Q \in G_1$.

$P \in G_1$, to achieve the sharing secret key, the process is as follows:

- u_1 randomly chooses a and computes $P_A = aP$.
- u_2 and u_3 randomly chooses b, c and respectively compute the values:$P_B = bP, P_c = cP$.
- $\hat{e}(P_B, P_C)^a = \hat{e}(P_A, P_C)^b = \hat{e}(P_A, P_B)^c = \hat{e}(P, P)^{abc}$.

So u_1, u_2, u_3 holds the sharing secret key $\hat{e}(P, P)^{abc}$.

Remove: Suppose u_2 want to be deleted in Fig.3. The sharing secret key $\hat{e}(P, P)^{abc}$ is revoked. u_1 and u_2 get their sharing secret key using Sakai's Non-Interaction Secret Sharing Protocol. It is very simple.

When there exists sub-processes, for example, a parent process generates three sub-processes or there is a lot of users which want to share their secret key, we can use the following methods. If a user generates three separate processes in order to complete a task, he generates three processes and each process generates three sub-processes as in Fig.2. The three sub-processes and the parent process can get their shared secret key as above. The u_1, u_2, u_3 and their parent process can get the sharing key K_{123}. The u_4, u_5, u_6 and their parent process can get the sharing key K_{456}. The u_7, u_8, u_9 and their parent process can get the sharing key K_{789}. All users can get their sharing key K_{1-9}.

5 Discussion and Comparison

In grid computing circumstances, a user's process can be created at a remote site, and the site may belong to another domain. Before these processes get the shared Diffie-Hellman key, these processes need to authenticate each

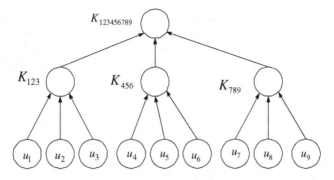

Fig. 2. Three processes each with three processes secret key exchange

other. We should need to use x.509 certificates to authenticate before these
processes begin to one round tripartite Diffie-Hellman secret key exchange.
If the sites belong to different domains, we also should consider the local security policies. When the one round tripartite Diffie-Hellman protocol is put
into the grid application, a lot of related problems should be considered. Our
secure group communication scheme can also be used in different entities that
need to communicate securely in grid circumstance. In this paper, we use the
identity-based cryptography scheme to construct the secure group communication; there is a lot of advantages than the authenticated Diffie-Hellman
secret exchange. In the ID-based system, the public key can be randomly selected, for example, the e-mail address. The public key can be changed easily,
for example, using the public key: bob@hotmail.com ∥ current-year. In doing
so the user can use his private key during the current year only. Thus, we
get the effect of annual private key expiration. In grid security communication,the Id-based scheme is very suitable for dynamic circumstance, because
it is very convenient to get the sharing secret key. In ID-based secure group
communication system, two users don't need to be on line at same time, but
in the authenticated Diffie-Hellman protocol several users need to be on line
at the same time to get their sharing secret key. In ID-based system, for two
users to get the sharing secret key, there is no need for interaction. For three
users to get their sharing secret key, there is just one round for Tripartite
Diffie-Hellman key exchange.

6 Conclusion

In this paper, we have discussed some security requirements, especially for
the secure group communication in grid circumstance. Because secure group
communication is seldom considered before, we propose the identity-based

Diffie-Hellman secure group communication protocol. The protocol also consider the dynamic situation when the members join/leave the group. The situation that processes have sub-processes is also being considered. When the protocol put into real application, some related problems should be considered.

References

[FK99] Foster, I. and Kesselman, C.: The Grid: Blueprint for a New Computing Infrastructure. Morgan Kaufmann 2-48 (1999)

[WSF03] Welch, V., Siebenlist, F. and Foster, I. et al.: Security for grid services. Proceedings of the 12th IEEE International Symposium on High Performance Distributed Computing (HPDC'03), (2003)

[BFK00] Butler, R., Foster, D. and Kesselman, I. et al.: A National-Scale Authentication Infrastructure. IEEE Computer, $33(12)$, 60-66 (2000)

[GM90] Gasser, M. and McDermott, E.: An architecture for practical delegation in a distributed system. Proc. 1990 IEEE Symposium on Research in Security and Privacy, IEEE Press, 20-30 (1990)

[Lin97] Linn, J.: Generic Security Service Application Program Interface. Version 2. INTERNET RFC 2078, (1997)

[FKT98] Foster, I., Kesselman, C. and Tsudik, G.: A Security Architecture for Computational Grids. ACM Conference on Computers and Security, 83-91 (1998)

[SOK00] Sakai, R., Ohgishi, K. and Kasahara, M.: Cryptosystems based on pairing. In SCIS, Okinawa, Japan, (2000)

[Jou00] Joux, A.: A one-round protocol for tripartite Diffie-Hellman. Algorithm Number Theory Symposium -ANTS-IV, Lecture Notes in Computer Science 1838, Springer-Verlag 385-394 (2000)

[Jou04] Joux, A.: A One Round Protocol for Tripartite Diffie-Hellman. Journal of Cryptology. $\mathbf{17}$, 263-276 (2004)

Performance Prediction in Grid Network Environments Based on NetSolve

Ningyu Chen, Wu Zhang, and Yuanbao Li

School of Computer Engineering and Science, Shanghai University, Shanghai 200072, PRC cny314@sohu.com,wzhang@mail.shu.edu.cn,leon_lyb@163.com

Abstract. The remarkable growth of computer network technology has spurred a variety of resources accessible through Internet. The important feature of these resources is location transparency and obtainable easily. NetSolve is a project that investigates the use of distributed computational resources connected by computer networks to efficiently solve complex scientific problems. However, the fastest supercomputers today are not powerful enough to solve many very complex problems with NetSolve. The emergence of innovative resource environments like Grids satisfies this need for computational power. In this paper, we focus on two types of Grid: Internet-connected collection of supercomputer and megacomputer, and explore the performance in these grid network environments.

Key word: performance prediction, NetSolve, Grid network environments

1 Introduction

More than fifty years of innovation has increased the raw speed of individual supercomputers by a factor of around one million, but they are still far too slow for many scientific problems. The emergence of innovative resource environments like Grid computing satisfies this need for computational power. It is now widely in practice in industry, research labs and supercomputer centers. However, not all complex problems, at the first sight, can be solved in the new environments. For example, many tightly coupled problems are very sensitive to the environments.

We evaluate the performance of that kind of tightly coupled scientific application problems on two Grids: one is a set of geographically distributed supercomputers[NW03], and these supercomputers are connected by a Grid middleware infrastructure like the NetSolve toolkit, as well as a collection of the scale of one million Internet-connected workstations[Don02]. We want to know the factors that limit the performance of the grid environments, and to design the performance prediction model.

2 the NetSolve System

The NetSolve system[CD97] is a representative implementation of Grid. It allows users to access computational resources, such as hardware and software, distributed across the network. The NetSolve system is comprised of a set of *loosely* connected machines [AN96]. It consists of three major components: the NetSolve client, the NetSolve agent and the NetSolve server (computational resource). Figure Fig.1 shows the working mechanism of NetSolve: a NetSolve client sends a request to the NetSolve agent. The agent chooses the "best" NetSolve resource according to the size and nature of the problem to be solved. Several instances of the NetSolve agent can exist on the network.

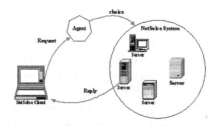

Fig. 1. The NetSolve System

3 Analysis on the Performance in Grid Environments

Knowing the performance of an application in a Grid environment is an important issue in application development and for scheduling decisions. We will deal with two different computational Grids: a set of geographically distributed supercomputers and a collection of the scale of one million Internet-connected workstations.

3.1 Performance on a Supercomputers Cluster

Computational Grids are collections of shared resources customized to the needs of their users. However, because of the characteristics of this new environment, many tightly coupled, synchronous applications, at first sight, don't be capable of exploiting computational Grids fully. In order to better exploiting computational Grids to solve these applications, they must implement unigrid domain decomposition. It decomposes the global domain over processors and places an overlap region (referred to as *ghost-zone*) on each processor. This reduces the number of messages (and hence the communication latency costs) while the total amount of data exchanged remains constant. We predict the performance on a collection of Internet-connected supercomputers and study the environment characteristics that limit performance. We also investigate ways to increase performance by tuning application parameters and improving the code. Let's make the following assumptions: Firstly, all of these supercomputers are identical, and they are connected through identical network linking to the Internet. Secondly, the problem space is decomposed with an 1D decomposition among supercomputers and a 3D decomposition among the processors of each supercomputer, and each supercomputer is regular assigned a grid space of size $X * Y * Z$. Finally, we assume there exist M supercomputers, each having N processors; meanwhile, we assume ghost-zone depth D for inter-supercomputer communication and ghost-zone depth d for intra-supercomputer communication. To build the performance model for the architecture described above, we model a supercomputer as a 'faster' processor with a 'big' associated memory[HJ98].

Time on Useful Work. Each supercomputer spends its time on *useful work*, communicating, and on redundant work. We ignore idle time due to synchronization, assuming perfect load balance (identical supercomputers and identical work load per supercomputer). Each supercomputer is assigned a grid space of size $X * Y * Z$. Since we assume a 3D regular partition at supercomputer level, each processor will be assigned a grid space of size $\frac{X}{\sqrt[3]{N}} * \frac{Y}{\sqrt[3]{N}} * \frac{Z}{\sqrt[3]{N}}$ The execution time of an application is always proportional to the size of the problem in despite that the execution model is sequential or parallel. So the time spent for useful work on a problem of size $X * Y * Z$ on a supercomputer with N processors is: $T_1 = kXYZ$ (k is the average time a grid is executed over I iterations on each supercomputer). At the same time, we know that the same amount of time is spent by each processor solving its part of the problem $\frac{X}{\sqrt[3]{N}} * \frac{Y}{\sqrt[3]{N}} * \frac{Z}{\sqrt[3]{N}}$. but working in parallel with efficiency E, so we can get:

$$T_1 = \frac{l}{E} * \frac{X}{\sqrt[3]{N}} * \frac{Y}{\sqrt[3]{N}} * \frac{Z}{\sqrt[3]{N}} \tag{1}$$

(l is the average time a grid is executed over I iterations on each supercomputer by the sequential execution model). We can now compute k the time it takes to process a grid-point: $k = \frac{l}{EN}$.

Redundant work Time. Consider that each processor has ghost-zones of depth d and each supercomputer has ghost-zones of depth D. This is meant to accommodate the variation in communication costs (inter-supercomputer vs. intra-supercomputer). Since replicated time on the processor ghost-zones (of size d) is already included in the model through the efficiency value E, the time spent by each supercomputer on doing replicated work is a function of $(D-d)$. In each iteration replicated work is done on $(D-d)XY$ grid points. Each supercomputer has at most two ghost-zones. The total time spent doing replicated work over I iterations is therefore:

$$T_2 = 2IK(D - d)XY \tag{2}$$

Communication Costs. Now we begin to compute the communication costs. The values corresponding to each grid point in the problem space are updated with each iteration based on values of the neighboring grid points. To reduce the number of messages exchanged, larger chunks of data can be sent at once. For each message sent from a supercomputer to another, communication time is: $t = t_1 + t_2 * L$, where t_1 is the message start-up time and t_2 is the cost for sending one byte of data. Over I iterations there are I/D communication phases, in which each supercomputer sends two messages of $L=DXYh$ bytes each. Incoming and outgoing messages share the communication link. Therefore, the time spent communicating is:

$$T_3 = \frac{I}{D} * 2(t_1 + 2t_2DXYh) = \frac{2It_1}{D} + 4It_2XYh \qquad (3)$$

Execution Time . So for M identical supercomputers with N processors each and a problem space of a $X * Y * Z$ grid points, each supercomputer has to solve a $X * Y * Z$ grid point problem, we can get that total execution time for I iterations is:

$$T = IT_1 + T_2 + T_3 = IkXYZ + 2Ik(D - d)XY + \frac{I}{D} * 2(t_2DXYh) \qquad (4)$$

Optimal Ghost-Zone Size and Predicted Efficiency. Through computing the minimum of the T from equation (4), we can get optimal ghost-zone size for:

$$D_{opt} = \frac{1}{X}\sqrt{\frac{t_1}{k}} \qquad (5)$$

From equation (5), we can know that larger inter-supercomputer communication latency requires larger ghost-zones while slower processors or larger problems would require smaller ghost-zones. At the same time, we can compute the value of overall efficiency easily.

3.2 Performance on a megacomputer

There are over 500 million PCs around the world, with many as powerful as early 1990s supercomputers. Internet Computing is motivated by the observation that at any moment millions of these computers are idle. The available middleware infrastructure and the soon-to-be-available network connectivity bring the vision of a general-purpose 1 million-processor system (*megacomputer*) closer to reality. However, for applications like NetSolve, increased computational power may be less significant than the aggregated memory of millions of computers which will allow solving problems of unprecedented scale and resolution.

We consider the processor space divided in 'clusters':groups of computers on the same Gigabit local area or campus network and assume there are 1000 clusters with 1000 machines for each. We assume that communication within a cluster is low delay, high bandwidth. We imagine a cluster to have 100s to 1000s machines. We analyzed the performance of a megacomputer using the same model as in previous sections. We just summarize our conclusions. Efficiency of 15-25 percent can be obtained even without modifying NetSolve' tightly coupled computational pattern. To conclude, we estimate that Net-Solve could run at an execution rate of 20 TFLOPS on a megacomputer. This is about 30 times faster than the best execution rate achievable now on a supercomputer[DGL89].

4 Summary and Future Work

We provided a detailed performance prediction model in two grid network environments —-an Internet-connected collection of supercomputers and a megacomputer, and we get the optimal ghost-zone size and predicted efficiency through computing. The prediction model shows that using computational Grids, scientists will shortly have a powerful computational platform at a low cost. Our future work is to validate our model with experiments executed in the two kinds of grid environments.

Acknowledgements. This paper are supported by "SEC E-Institute: Shanghai High Institutions Grid" project, the fourth key subject construction of Shanghai, the Natural Science Funding of Shanghai Educational Committee (No.205155).

References

[BNSS99] Benger, W., Novotny, J., Seidel, E., Shalf, J.: Numerical Relativity in a Distributed Environment, Proceedings of the Ninth SIAM Conference on Parallel Processing for Scientific Computing, March 1999. 808

[F00] Foster, I.: Internet Computing and the Emerging Grid, Nature 408(6815), 2000. 808, 814

[ABD00] Arnold, D.C., Bachmann, D. and Dongarra, J.: Request Sequencing: Optimizing Communication for the Grid. [J].In Euro-Par 2000-Parallel Processing, August 2000.

[AD03] Arnold, D.C. and Dongarra, J.: The NetSolve Environment: Progressing Towards the Seamless Grid [M]. http://icl.cs.utk.edu/netsolve

[RI01] Ripeanu, M., Iamnitchi, A.: Performance Predictions for a Numerical Relativity Package in Grid Environments, Technical Report TR-2001-23, U. of Chicago, 2001. 808, 809

[MFK03] Makino, J., Fukushige, T., Koga, M.: A 1.349 TFLOPS simulation ofbl ack holes in a galactic center in GRAPE-6, In Supercomputing 2003. 815

A Hybrid Numerical Algorithm for Computing Page Rank

Waiki Ching[1], Michael K. Ng[1], and Waion Yuen[1]

Department of Mathematics, The University of Hong Kong, Pokfulam Road, Hong Kong, China {wkc,mng}@maths.hku.hk, h9923304@hkusua.hku.hk

Summary. The computation of PageRank is an important issue in Internet. Especially one has to handle a huge size of web with its size growing rapidly. In this paper, we present an adaptive numerical method for solving the PageRank problem. The numerical method combines the Jacobi Over-Relaxation (JOR) method with the evolutionary algorithm. Numerical examples based on simulations are given to demonstrate the efficiency of the proposed method.

1 Introduction

In surfing the Internet, surfers usually use search engines to find the related webpages satisfying their queries. Unfortunately, very often there can be thousands of webpages which are relevant to the queries. Therefore a proper list of the webpages in certain order of importance is necessary. The list should also be updated regularly and frequently. Thus it is important to seek for fast algorithm for the computing the PageRank so as to reduce the time lag of updating. It turns out that this problem is difficult. The reason is not just because of the huge size of the webpages in the Internet but also the size keeps on growing rapidly.

PageRank has been proposed by Page et al. (1998) [12] to reflect the importance of each webpage, see also [8]. In fact, a similar idea has been proposed by Garfield (1955, 1972) [5, 6] as a measure of standing for journals, which is called the *impact factor*. The impact factor of a journal is defined as the average number of citations per recently published papers in that journal. By regarding each webpage as a journal, this idea was then extended to measure the importance of the webpage in the PageRank Algorithm.

The PageRank is defined as follows. Let N be the total number of webpages in the web and we define a matrix Q called the *hyperlink matrix*. Here

$$Q_{ij} = \begin{cases} 1/k & \text{if webpage } i \text{ is an outgoing link of webpage } j; \\ 0 & \text{otherwise;} \end{cases}$$

and k is the total number of outgoing links of webpage j. For simplicity of discussion, here we assume that $Q_{ii} > 0$ for all i. This means for each webpage, there is a link pointing to itself. Hence H can be regarded as a transition probability matrix of a Markov chain of a random walk. The analogy is that one may regard a surfer as a random walker and the webpages as the states of the Markov chain. Assuming that this underlying Markov chain is irreducible, then the steady-state probability distribution $(p_1, p_2, \ldots, p_N)^T$ of the states (webpages) exists. Here p_i is the proportion of time that the random walker (surfer) visiting state (webpage) i. The higher the value of p_i is, the more important webpage i will be. Thus the PageRank of webpage i is then defined as p_i. If the Markov chain is not irreducible then one can still follow the treatment in Section 2.

Since the size of the Markov chain is huge, direct method for solving the steady-state probability is not desirable. Iterative methods Baldi et ak. (2003) [3] and decomposition methods Avrachenkov and Litvak (2004) [1] have been proposed to solve the problem. Another pressing issue is that the size of the webpages grows rapidly, and the PageRank of each webpage has to be updated regularly. Here we seek for adaptive and parallelizable numerical algorithms for solving the PageRank problem. One potential method is the hybrid iterative method proposed in Yuen et al. (2004) [14]. The hybrid iterative method was first proposed by He et al. [10] for solving the numerical solutions of PDEs and it has been also successfully applied to solving the steady-state probability distributions of queueing networks [14]. The hybrid iterative method combines the evolutionary algorithm and the Successive Over-Relaxation (SOR) method. The evolutionary algorithm allows the relaxation parameter w to be adaptive in the SOR method. Since the cost of SOR method per iteration is more expansive and less efficient in parallel computing for our problem (as the matrix system is huge), here we replace the role of SOR method by the Jacobi Over-Relaxation (JOR) method. We develop a hybrid iterative method based on JOR and evolutionary algorithm. The hybrid method allows the relaxation parameter w to be adaptive in the JOR method. The remainder of this paper is organized as follows. In Section 2, we give a brief mathematical discussion on the PageRank approach. In Section 3, we present the adaptive numerical algorithm for solving the PageRank problem. In Section 4, we give a convergence analysis of the proposed numerical algorithm. In Section 5, simulated numerical examples are given to illustrate the effectiveness of the proposed method.

2 The PageRank Algorithm

The PageRank Algorithm has been used successfully in ranking the importance of web-pages by Google [8]. Consider a web of N webpages with Q being the hyperlink matrix. Since the matrix Q can be reducible, to tackle this problem, one can consider the revised matrix P:

$$P = \alpha \begin{pmatrix} Q_{11} & Q_{12} & \cdots & Q_{1N} \\ Q_{21} & Q_{22} & \cdots & Q_{2N} \\ \vdots & \vdots & \vdots & \vdots \\ Q_{N1} & Q_{N2} & \cdots & Q_{NN} \end{pmatrix} + \frac{(1-\alpha)}{N} \begin{pmatrix} 1 & 1 & \cdots & 1 \\ 1 & 1 & \cdots & 1 \\ \vdots & \vdots & \vdots & \vdots \\ 1 & 1 & \cdots & 1 \end{pmatrix} \tag{1}$$

where $0 < \alpha < 1$. In this case, the matrix P is irreducible and aperiodic, therefore the steady state probability distribution exists and is unique [13]. Typical values for α are 0.85 and $(1 - 1/N)$, see for instance [3, 8, 9]. The value $\alpha = 0.85$ is a popular one because power method works very well for this problem [9]. However, this value can be considered to be too small and may distort the original ranking of the webpages, see the example in Section 3.

One can interpret (1) as follows. The idea of the algorithm is that, for a network of N webpages, each webpage has an inherent importance of $(1-\alpha)/N$. If a page P_i has an importance of p_i, then it will contribute an importance of αp_i which is shared among the webpages that it points to. The importance of webpage P_i can be obtained by solving the following linear system of equations subject to the normalization constraint:

$$\begin{pmatrix} p_1 \\ p_2 \\ \vdots \\ p_N \end{pmatrix} = \alpha \begin{pmatrix} Q_{11} & Q_{12} & \cdots & Q_{1N} \\ Q_{21} & Q_{22} & \cdots & Q_{2N} \\ \vdots & \vdots & \vdots & \vdots \\ Q_{N1} & Q_{N2} & \cdots & Q_{NN} \end{pmatrix} \begin{pmatrix} p_1 \\ p_2 \\ \vdots \\ p_N \end{pmatrix} + \frac{(1-\alpha)}{N} \begin{pmatrix} 1 \\ 1 \\ \vdots \\ 1 \end{pmatrix}. \tag{2}$$

Since $\sum_{i=1}^{N} p_i = 1$, (2) can be rewritten: $(p_1, \ldots, p_N)^T = P(p_1, \ldots, p_N)^T$.

3 The JOR Method and the Hybrid Method

In this section, we first give a review on the JOR method for solving linear system, in particular solving the steady state probability distribution of a finite Markov chain. We then introduce the hybrid algorithm based on the JOR method and the evolutionary algorithm. We consider a non-singular linear system $Bx = b$, the JOR method is a classical iterative method. The idea of JOR method can be explained as follows. We write $B = D - (D - B)$ where D is the diagonal part of the matrix B. Given an initial guess of the solution, x_0, the JOR iteration scheme reads:

$$x_{n+1} = (I - wD^{-1}B)x_n + wD^{-1}b \equiv B_w x_n + wD^{-1}b. \tag{3}$$

The parameter w is called the relaxation parameter and it lies between 0 and 1 [2]. Clearly if the scheme converges, the limit will be the solution of $Bx = b$. The choice of the relaxation parameter w affects the convergence rate of the JOR method very much, see for instance the numerical results in Section 5. In general, the optimal value of w is unknown. For more details about the JOR method and its property, we refer readers to Axelsson [2].

We recall that the generator matrix P of an irreducible Markov chain is singular and has a null space of dimension one (the null vector corresponds to the steady-state probability distribution). One possible way to solve the steady-state probability distribution is to consider the following revised system:

$$A\mathbf{x} = (P + \mathbf{e}_n^t \mathbf{e}_n)\mathbf{x} = \mathbf{e}_n^t \tag{4}$$

where $\mathbf{e}_n = (0, 0, \ldots, 0, 1)$ is a unit vector and $\mathbf{e}_n^t$ is the transpose of $\mathbf{e}_n$. The steady-state probability distribution is then obtained by normalizing the solution $\mathbf{x}$, see for instance Ching [4]. We remark that the linear system (4) is irreducibly diagonal dominant. The hybrid method based on He et al. [10] and Yuen et al. [14] consists of four major steps: *initialization, mutation, evaluation and adaptation.*

In the initialization step, we define the size of the population k of the approximate steady-state probability distribution. This means that we also define k approximates to initialize the algorithm. Then use the JOR iteration in (4) as the "mutation step". In the evaluation step, we evaluate how "good" each member in the population is by measuring their residuals. In this case, it is clear that the smaller the residual the better the approximate and therefore the better the member in the population. In the adaptation step, the relaxation parameters of the "weak" members are migrated (with certain probability) towards the best relaxation parameter. The hybrid algorithm reads:

Step 1: Initialization: We first generate an initial population of k ($2 \leq k \leq n$) identical steady-state probability distributions as follows: $\{\mathbf{e}_i : i = 1, 2, \ldots, k\}$ where $\mathbf{e}_i = (1, 1, \ldots, 1)$. We then compute $r_i = ||B\mathbf{e}_i - \mathbf{b}||_2$ and define a set of relaxation parameters $\{w_1, w_2, \ldots, w_k\}$ such that

$$w_i = \tau + \frac{(1 - 2\tau)(k - i)}{k - 1}, \quad i = 1, 2, \ldots, k.$$

Here $\tau \in (0, 1)$ and therefore $w_i \in [\tau, 1 - \tau]$. We set $\tau = 0.01$ in our numerical experiments. We then obtain a set of ordered triples $\{(\mathbf{e}_i, w_i, r_i) : i = 1, 2, \ldots, k\}$.

Step 2: Mutation: The mutation step is carried out by doing a JOR iteration on each member $\mathbf{x}_i$ ($\mathbf{x}_i$ is used as the initial in the JOR) of the population with their corresponding w_i. We then get a new set of approximate steady-state probability distributions: $\mathbf{x}_i$ for $i = 1, 2, \ldots, k$. Hence we have a new set of $\{(\mathbf{x}_i, w_i, r_i) : i = 1, 2, \ldots, k\}$. Goto Step 3.

Step 3: Evaluation: For each $\mathbf{x}_i$, we compute and update its residual $r_i = ||B\mathbf{x}_i - \mathbf{b}||_2$. This is used to measure how "good" an approximate $\mathbf{x}_i$

is. If $r_j < tol$ for some j then stop and output the approximate steady state probability distribution $\mathbf{x}_j$. Otherwise we update r_i of the ordered triples $\{(\mathbf{x}_i, w_i, r_i) : i = 1, 2, \ldots, k\}$ and goto Step 4.

Step 4: Adaptation: In this step, the relaxation factors w_k of the weak members (relatively large r_i) in the population are moving towards the best one with certain probability. This process is carried out by first performing a linear search on $\{r_i\}$ to find the best relaxation factor, w_j. We then adjust all the other w_k as follows:

$$w_k = \begin{cases} (0.5 + \delta_1) * (w_k + w_j) \text{ if } (0.5 + \delta_1) * (w_k + w_j) \in [\tau, 1 - \tau] \\ w_k \qquad\qquad\qquad\qquad \text{otherwise,} \end{cases}$$

where δ_1 is a random number in $[-0.01, 0.01]$. Finally the best w_j is also adjusted by

$$w_j = \delta_2 * w_j + (1 - \delta_2) * \frac{(w_1 + w_2 + \ldots + w_{j-1} + w_{j+1} + \ldots + w_k)}{k - 1}$$

where δ_2 is a random number in $[0.99, 1]$. A new set of $\{w_i\}$ is then obtained and hence $\{(\mathbf{x}_i, w_i, r_i) : i = 1, 2, \ldots, k\}$. Goto Step 2.

Proposition 1. *The cost per JOR iteration is $O(kN^2)$ when the population size is k.*

4 Convergence Analysis

In this section, we consider the linear system $B\mathbf{x} = \mathbf{b}$ such that B is strictly diagonal dominant, i.e.

$$|B_{ii}| > \sum_{j=1, j \neq i}^{N} |B_{ij}| \quad \text{for} \quad i = 1, 2, \ldots, N$$

where N is the size of the matrix. We prove that the hybrid algorithm converges for a range of w. We first have

Lemma 1. *Let B be a strictly diagonal dominant square matrix and*

$$K = \max_i \left\{ \sum_{j=1, j \neq i}^{N} \frac{|B_{ji}|}{|B_{ii}|} \right\} < 1,$$

then

$$\|B_w\|_1 \leq 1 - (1 - K)w < 1 \quad \text{for} \quad \tau < w < 1 - \tau$$

where B_w is defined in (1).

By using the similar approach in [14], one can prove that

Proposition 2. *The hybrid iterative method converges for $w \in [\tau, 1 - \tau]$.*

Proof. We observe that

$$f(\tau) = \max_{w \in [\tau, 1-\tau]} \{||B_w||_1\}$$

exists and less than one and let us denote it by $0 \leq f(\tau) < 1$. Therefore in each iteration of the hybrid method, the matrix norm ($||.||_1$) of the residual is decreased by a fraction not less than $f(\tau)$. By using the fact that $||ST||_1 \leq ||S||_1||T||_1$, the hybrid algorithm is convergent.

We note that the matrix A in (2) is irreducibly diagonal dominant only but not strictly diagonal dominant. Therefore the condition in Lemma 1 is not satisfied. However, one can always consider a regularized linear system as follows:

$$(A + \epsilon I)\mathbf{x} = \mathbf{b}.$$

Here I is the identity matrix and $\epsilon > 0$ can be chosen as small as possible. In our numerical examples, however, we find that there is no need to add this regularization.

5 Numerical Experiment

We demonstrate the fast convergence rate of the hybrid algorithm by some simulated examples. We also compare the numerical results with the hybrid iterative method with SOR [14]. The stopping criteria in all the examples are set to be $||(P + \mathbf{e}_n^t \mathbf{e}_n)\mathbf{x} - \mathbf{e}_n||_2 < 10^{-6}$ where P is the revised matrix of the webpages. The initial guess for all the methods are the zero vector for all the numerical examples.

We simulate three cases of webs of maximum size $N = 400$ by using the functions round() and rand() with "seed=1" in MATLAB. The three cases considered here are: Case 1: "round(rand(400,400))", Case 2: "round(rand(400,400)+0.2)", and Case 3: "round(rand(400, 400)-0.2)". For each case, we assume the web size grows from 100 to 200, then to 300 and then to 400. The linkage matrix for the case of 300 is obtained by taking the first 300×300 principal matrix of the linkage matrix for the case of 400 with normalization. Similarly the linkage matrix for the case of 200 is obtained by taking the first 200×200 principal matrix of the linkage matrix for the case of 300 with normalization. Finally the linkage matrix for the case of 100 is obtained in a similar way. In the following tables, we report the numerical results for two typical cases of α.

Table 1: $\alpha = 1 - 1/N$.

JOR	Case	1			Case	2			Case	3		
N	100	200	300	400	100	200	300	400	100	200	300	400
$k=2$	41	56	42	42	57	95	58	70	31	26	32	25
$k=3$	56	60	42	42	56	75	57	61	31	35	43	25
$k=4$	46	59	42	42	55	72	58	62	31	32	38	25
$k=5$	56	60	43	43	56	68	57	60	32	30	36	26
SOR	Case	1			Case	2			Case	3		
N	100	200	300	400	100	200	300	400	100	200	300	400
$k=2$	20	18	17	17	16	15	16	15	18	14	19	15
$k=3$	30	27	17	25	16	23	16	23	18	21	29	15
$k=4$	25	24	19	22	17	21	16	21	18	19	26	18
$k=5$	30	28	19	23	17	21	16	20	20	20	25	17

Table 2: $\alpha = 0.85$.

JOR	Case	1			Case	2			Case	3		
N	100	200	300	400	100	200	300	400	100	200	300	400
$k=2$	42	56	44	47	61	82	66	64	18	28	32	26
$k=3$	55	60	45	52	62	81	63	62	18	36	42	26
$k=4$	53	59	45	49	58	71	62	62	18	33	38	26
$k=5$	53	65	45	49	61	70	64	62	18	32	37	26
SOR	Case	1			Case	2			Case	3		
N	100	200	300	400	100	200	300	400	100	200	300	400
$k=2$	19	17	17	16	16	14	15	15	15	14	19	16
$k=3$	28	26	17	24	16	22	15	23	15	23	29	16
$k=4$	24	23	19	21	16	20	16	21	17	20	25	16
$k=5$	28	26	19	21	17	21	16	20	16	20	23	16

First, we observe that for each N, the hybrid algorithm terminates with a relaxation parameter close to the "optimal relaxation" in both JOR and SOR parameters. Second, in the tested examples, we find that the optimal values of k is 2. Third, in the hybrid algorithm, the main computational cost comes from the mutation which is a JOR (SOR) iteration and the cost is of $O(N^2)$. It is clear that the hybrid iterative method with SOR is faster but the hybrid iterative method with JOR is more efficient in parallel implementation. This can therefore compensate the overhead.

Acknowledgment: The authors would like to thank the anonymous reviewer for the helpful comments. Research supported in part by RGC Grant No. HKU 7126/02P and HKU CRCG Grant Nos. 10204436 and 10205105.

References

1. Avrachenkov, L. and Litvak, N.: Decomposition of the Google PageRank and Optimal Linking Strategy, Research Report, INRIA, Sophia Antipolis (2004)
2. Axelsson, O.: Iterative Solution Methods, Cambridge University Press, Cambridge (1996)
3. Baldi, P., Frasconi, P. and Smith, P.: Modeling the Internet and the Web, Wiley, England (2003)
4. Ching, W.: Iterative Methods for Queuing and Manufacturing Systems, Springer Monograph in Mathematics, London (2001)
5. Garfield, E.: Citation Indexes for Science: A New Dimension in Documentation Through Association of Ideas, Scicence **122** (1955) 108-111
6. Garfield, E.: Citation Analysis as a Tool in Journal Evaluation, Scicence **178** (1972) 471-479
7. Golub, G. and van Loan, C.: Matrix Computations, The John Hopkins University Press, Baltimore (1993)
8. www.search-engine-marketing-sem.com/Google/GooglePageRank.html
9. Haveliwala, T. and Kamvar, S.: The Second Eigenvalue of the Google Matrix, Stanford University, Technical Report (2003)
10. He, J., Xu, J. and Yao, X.: Solving Equations by Hybrid Evolutionary Computation Techniques, IEEE Trans. Evol. Comput. **4** (2000) 295-304.
11. Kamvar, S., Haveliwala, T. and Golub, G.: Adaptive Methods for the Computation of PageRank, Linear Algebra and Its Applications **386** (2004) 51-65
12. Page, L., Brin, S., Motwani, R. and Winograd, T.: The PageRank Citation Ranking: Bring Order to the Web, Technical Report, Stanford University (1998)
13. Ross, S.: Introduction to Probability Models, (7th Edition) San Diego, Calif.: Academic Press (1997)
14. Yuen, W., Ching, W. and Ng, M.: A Hybrid Algorithm for Queueing Systems, CALCOLO **41** (2004) 139-151

A Parallel Approach Based on Hierarchical Decomposition and VQ for Medical Image Coding

Guangtai Ding[1], Anping Song[1], and Wei Huang[1]

School of Computer Engineering and Science, Shanghai University, Yanchang Rd. 149, Shanghai 200072, China {dgt, apsong, planewalker}@mail.shu.edu.cn

Abstract. A scheme of medical image coding is proposed. The method involves the following two steps: 1) Using hierarchical Cosine transform, a medical image is decomposed into a set of so-called compressed-units. Each unit can be transformed and computed parallely in a medical image processing system. 2) Based on the set of compressed-units, generating functions for the codebook used in the vector quantization coding method are constructed.

Key words: Hierarchical Cosine transformation, compressed-units, parallel implementation, Vector Quantization, medical image coding.

1 Introduction

VQ(Vector Quantization) has been widely used in image coding by many researchers. It is well-known that the main problem of VQ is that the compression parameters depend on an image training set, the appropriate codebook may vary from image to image[SHB99, PCT98], and the codebook may be transmitted together with the coded data in a certain way. How to construct the codebook is always a difficult task. In [C02], a pyramidal lattice VQ coding scheme is introduced. Several fast search algorithms are presented to speed up the VQ closest codeword search process in encoding in [BSG94, SR02]. Two new design techniques for adaptive orthogonal block transforms based on VQ codebooks are presented by [CGS98]. Many researchers indicated that VQ and DCT are two algorithms of lossy compression, which are suited to medical images(CT,MRI etc) [SHB99].

A compression scheme based on VQ and hierarchical DCT (HDCT for short) is proposed in this paper. The HDCT is similar to that of Adaptive Block Size Discrete Cosine Transform (ABSDCT) [CGS98], but is easer to operate. Using HDCT, the DCT of a higher order image can be easily calculated from DCTs of its lower order subimages. We create codebook database(*codebooks* for short) by decomposing standard images into 4×4 blocks, and calculating their DCTs as codewords. Codebooks do not need to be transmitted together with the coded data, because codebooks used in different machines are the same as others.

The notation (m, δ)-similarity is defined and discussed in this paper also. Using (m, δ)-similarity, it is easy to search a proper codeword in codebooks for the data. Not only the codewords are used, but also the difference images between standard medical images and images of a human body are used in coding.

2 Hierarchical orthogonal transform

Lemma. *Let $f(x, y)$ be an $N \times N$ image, $N = 2n$, f_{2n} be the matrix corresponds to it, and F_{2n} be the Cosine Transform matrix of f_{2n}. Dividing f_{2n} into 4 $n \times n$ blocks (we denote the (I,J)-th block as $f_{n,IJ}$, and $F_{n,IJ}$ as its Cosine transform, where $I, J = \{0, 1\}$), then the following two formula hold:*

$$F_{2n} = \begin{pmatrix} A_{2n}^{11} A_n^T & A_{2n}^{12} A_n^T \\ A_{2n}^{21} A_n^T & A_{2n}^{22} A_n^T \end{pmatrix} \begin{pmatrix} F_{n,11} & F_{n,12} \\ F_{n,21} & F_{n,22} \end{pmatrix} \begin{pmatrix} A_n(A_{2n}^{11})^T & A_n(A_{2n}^{21})^T \\ A_n(A_{2n}^{12})^T & A_n(A_{2n}^{22})^T \end{pmatrix} \quad (1)$$

$$\begin{pmatrix} F_{n,11} & F_{n,12} \\ F_{n,21} & F_{n,22} \end{pmatrix} = \begin{pmatrix} A_n(A_{2n}^{11})^T & A_n(A_{2n}^{21})^T \\ A_n(A_{2n}^{12})^T & A_n(A_{2n}^{22})^T \end{pmatrix} F_{2n} \begin{pmatrix} A_{2n}^{11} A_n^T & A_{2n}^{12} A_n^T \\ A_{2n}^{21} A_n^T & A_{2n}^{22} A_n^T \end{pmatrix} \quad (2)$$

The proof is very simple and is omitted. The formula (1) is called *hierarchical orthogonal(orthonomal) transform*, (2) is called the *inverse* of it. In this paper, Cosine transform is used as the orthogonal transform to compress images.

3 Linear similarity of images

Let $m \geq 2$ be an integer, $\delta > 0$ be a real number. f is called (m, δ) -similar to g at (x_0, y_0), if for every pixel of $f(x, y)$, there exists a linear transform T, a real number d, and a convolution operator $\mathbf{k}$, such that

$$\|(f(x, y) - \mathbf{k} * g(x', y') - d)\chi_{(x_0, y_0), m}(x, y)\|_2 < \delta, \quad \textbf{(I)}$$

where

$$T : \begin{pmatrix} x' \\ y' \end{pmatrix} = \begin{pmatrix} a_1 & b_1 \\ a_2 & b_2 \end{pmatrix} \begin{pmatrix} x \\ y \end{pmatrix} + \begin{pmatrix} c_1 \\ c_2 \end{pmatrix} \quad \textbf{(II)}$$

$$\chi_{(x_0, y_0), m}(x, y) = \begin{cases} 1 & if \quad x_0 \leq x < x_0 + m, y_0 \leq y < y_0 + m \\ 0 & otherwise \end{cases},$$

$\| \cdot \|_2$ is the Euclid norm. Let

$$\begin{pmatrix} x_0' \\ y_0' \end{pmatrix} = \begin{pmatrix} a_1 & b_1 \\ a_2 & b_2 \end{pmatrix} \begin{pmatrix} x_0 \\ y_0 \end{pmatrix} + \begin{pmatrix} c_1 \\ c_2 \end{pmatrix}.$$

Then (x_0', y_0') is said to be the similar pixel of g corresponds to (x_0, y_0). If f is (m, δ)-similar to g at every pixel site (x, y), then we call f is fully (m, δ)-similar to g (*fully similar* or *fully linear similar* for short). (m, δ)-similarity describes how similar or closer to each other the two images are, under a certain co-ordinate transformation, a shift of pixel values, and a convolution restoration. (m, δ)-similarity assumption is the presupposition in this paper.

4 The compression scheme

The main method in our scheme is to calculate the difference image blocks between standard medical image and that of a human body, and then extend each blocks as possible large as we can. Because we represent each so-called compressed unit by few number of parameters, and the codebook is not needed to transfer with the data together, thus the transmission problem can be solved in this way. The scheme includes the following 5 steps.

(1)Construct local codebooks

Let $I = \{I_i\}$, $i = \overline{1, N}$, (the integer set $\{1, 2, ..., N\}$ is defined as $\overline{1, N}$) be a set of radiological slice images (standard images for short) of a certain organic part of a human body(e.g. CT slice images set). The acquiring of medical images are always normative in certain parameters, thus the images of different patients(source images for short) are similar to each other while the parameters are similar. Divide I_i into 4×4 subimage-blocks $I_i^{k,l}$, where $(k, l) \in \overline{1, P} \times \overline{1, Q}$, $P \times Q$ is the number of the blocks. Let $J = \{J_i^{k,l}\}$, where $J_i^{k,l}$ is the Cosine transform of $I_i^{k,l}$, the (k, l)-th subimage of I_i. Put all these Cosine transform into so-called codebook database.

(2)Near to code-book-image estimate

Suppose f and I_i are acquired in the same system with similar parameters. For convenience, the assumption that the image discussed is linear similar to a certain standard image, is always satisfied in this paper.

(3)Decomposite source image

With the same way to divide I_i, we divide f into 4×4 subimages. Let $f = \{f^{r,s}\}$, $(r, s) \in \overline{1, P} \times \overline{1, Q}$, and $F = \{F^{r,s}\}$, where $F^{r,s}$ is the Cosine transform of the (k,l)-th subimage $f^{r,s}$ of f.

(4)Construct compressed unit

Divide source image into subimages in the way of to divide standard image. Let $f^{r,s}$ be the (r,s)-th subimage of f, we construct compressed unit starting from (r,s) in the following steps:

1) Calculate $f = \{f^{r,s}\}$, $F = \{F^{r,s}\}$, $(r, s) \in \overline{1, P} \times \overline{1, Q}$.

2) Let $\sigma > 0$ be a control parameter, find (k, l, z, p), such that

$$||J_i^{k,l} - zF^{r,s} - p\Gamma||_2 = \min, |(k, l) - (r, s)| < \sigma,$$

where Γ is the DCT of the matrix $(1)_{4 \times 4}$, which elements are 1. Let

$$g_{r,s}^{k,l} = J_i^{k,l} - zF^{r,s} - p\Gamma$$

be the Cosine transform of (k,l)-th difference image.

3) Let $\delta_4 > 0$ be an error. If there exist (z_1, z_2, z_3, z_4, p), satisfying

$$g_{r,s}^{k,l} = J_i^{k,l} - z_1 F^{r,s} - p_1 \Gamma, g_{r,s}^{k+1,l} = J_i^{k+1,l} - z_2 F^{r+1,s} - p_2 \Gamma,$$
$$g_{r,s+1}^{k,l+1} = J_i^{k,l+1} - z_3 F^{r,s+1} - p_3 \Gamma, g_{r+1,s+1}^{k+1,l+1} = J_i^{k+1,l+1} - z_4 F^{r+1,s+1} - p_4 \Gamma,$$

$$||g_{r,s}^{k,l}||_2 + ||g_{r,s}^{k+1,l}||_2 + ||g_{r,s+1}^{k,l+1}||_2 + ||g_{r+1,s+1}^{k+1,l+1}||_2 < 4\delta_4,$$

then, denote

$$\Delta_3^8 = \begin{pmatrix} A_8^{11} A_4^T & A_8^{12} A_4^T \\ A_8^{21} A_4^T & A_8^{22} A_4^T \end{pmatrix} \begin{pmatrix} g_{r,s}^{k,l} & g_{r,s+1}^{k,l+1} \\ g_{r+1,s}^{k+1,l} & g_{r+1,s+1}^{k+1,l+1} \end{pmatrix} \begin{pmatrix} A_4(A_8^{11})^T & A_4(A_8^{21})^T \\ A_4(A_8^{12})^T & A_4(A_8^{22})^T \end{pmatrix},$$

we call f can be extended, Δ_3^8 is called *3-direction 8-extension difference image*. In the same way, 0,1,2-direction 8-extension Δ^8, and 16-extension Δ^{16}, 32-extension image Δ^{32} can be defined also.

We call $U = \{k, l, r, s, o, \Delta^o, \mathbf{d}[i], z_0, \mathbf{z}[i,j]\}$ to be a compressed unit, where (k, l) is the starting site of the corresponded codeword in the codebook, (r, s) is the starting site of the subimage in source image, o is the extension order(may be 8,16,32), Δ^o is the difference image, $d[i]$ is the i-th extension direction, z_0 is the z-parameter of the starting unit, $z[i,j]$ is the j-th direction of the i-th extension,where $i = \overline{1, \log o - 2}, j = \overline{0, 3}$.

(5)Parallel algorithm

PRAM(Parallel Random Access Machine) parallel model is suitable to create compressed units and to reconstruct the image. In this paper, we adopt the PRAM(CREW) as the model. The maim step of the algorithm is to determine p point(s) in image field (p is the number of processors in using), to process the same procedure to create the compressed units.

5 Experimental results

We got a sequence of MRI images (about 58 slices), and a sequence of CT images(about 124 slices) in our experiment. We only used 4 CT images. The distance of CT slices was 1.7mm. For simplicity, we define δ as the average error of pixel values. Simulations were carried out with 256×256 CT images in a microcomputer (256M Memory, P4 2.1GHz Processor). We did programming in Matlab 6.5. The decomposing and extending time were calculated, while different parameters were tried. The total processing time was also estimated.

6 Conclusions

Using HDCT, we proximate a medical image with images combined by codewords in codebooks. The main contribution in our paper is that we adopt the

Table 1. 256×256 CT Images, $\delta = 16$, $\sigma = 6$

Processor number	4×4	8×8	16×16	32×32	Decomposing time(seconds)	Total processing time(seconds)
1	1407+1	296	50	11	7.687	8.172
4	1403+1	265	46	14	7.641	8.063
8	1443	259	41	15	8.087	8.235

DCTs of standard image-bloks as codewords of codebooks to construct the images to be compressed. Almost every steps of the scheme can be executed parallely. Table 1 shows the distributions of numbers of $4 \times 4, 8 \times 8, 16 \times 16, 32 \times 32$ subimages of a image to be divided. We find that the amount of pixels of $8 \times 8, 16 \times 16, 32 \times 32$ blocks are 50% more over the amount of all pixels. Because the higher order blocks have a superior performance in coding, the HDCT have superior performance.

Acknowledgments We acknowledge Professor Wu Zhang for giving us many useful helps in this work. We also acknowledge Radiology Department, University of Iowa Hospitals & Clinics. We download the CT&MR medical images as raw materials in our experiment from their web site www.radiology.uiowa.edu/download. This work was supported partly by SEC E-institute: Shanghai High Institution Grid Project and the Fourth Key Subject Construction of Shanghai.

References

[SHB99] Sonka, M., Hlavac, V., Boyle, R.: Image Processing, Analysis, and Machine Vision, Second Edition. Thomson Learning and PT Press(1999)

[CGS98] Caglar,H., Gunturk,S., Sankur,B., et al.: VQ-Adaptive block transform coding of images. IEEE Trans. Image Processing, **7**, 110–115 (1998)

[C02] Chang, H.T.: Gradient match and side match fractal Vecter Quantizers for images. IEEE Trans. Image Processing, **11**, 1–9(2002)

[SR02] Song, B.C., Ra, J.B.: A fast search algorithm for Vecter Quantization using L_2 pyramid of codewords. IEEE Trans. Image Processing, **11**, 10–15(2002)

[BSG94] Barland, M., Sole, P., Gaidon, T., Antonini, M., Mathieu, P.: Pyramidal lattice Vecter Quantization for multiscale image coding. IEEE Trans. Image Processing, **3**, 367–381(1994)

[PCT98] Perlmutter, S.M., Cosman, P.C., Tseng, C.W., Olshen, R.A., et al.: Medical image compression and Vector Quantization. Statistical Science, **13**, 30–53(1998)

A High-resolution Scheme Based on the Normalized Flux Formulation in Pressure-based Algorithm

M. H. Djavareshkian[1] and S. Baheri Islami[1]

Faculty of Mechanical Engineering, University of Tabriz, Tabriz, Iran
Djavaresh@tabrizu.ac.ir, baheri@tabrizu.ac.ir

Abstract. The paper presents an implementation of NVD scheme into an implicit finite volumes procedure, which uses pressure as a working variable. The newly developed algorithm has two new features: (i) the use of the normalized flux and space formulation (NFSF) methodology to bound the convective fluxes and (ii) the use of a high-resolution scheme in calculating interface density values to enhance the shock-capturing property of the algorithm. The virtues of the newly developed method are demonstrated by solving a wide range of flows spanning the subsonic, transonic and supersonic spectrum. Results obtained indicate higher accuracy when calculating interface density values using a high-resolution scheme.

Keywords: High resolution scheme, Pressure-based algorithm, Normalized flux formulation

1 Introduction

In computational fluid dynamics (CFD), great research efforts have been devoted to the development of accurate and efficient numerical algorithms suitable for solving flows in the various Reynolds and Mach number regimes. Leonard [1] has generalized the formulation of the high-resolution flux limiter schemes using what is called the normalized variable formulation (NVF). Examples of schemes based on NVF can be found in [2]. Most of NVf methods use different differencing schemes through the solution domain. This procedure includes some kind of switching between the differencing schemes. Switching introduced additional non-linearity and instability in to the computation. The worst case is that instead of a single solution for steady state problem, the differencing scheme creates two or more unconverged solution with the cyclic switching between them. In that case it is impossible to obtain a converged solution and convergence stalls at some level. The SBIC scheme was introduced by Javareshkian [3]; it is integrated from central interpolation and first and second order interpolation procedures. In all of these methods, the NVD scheme is applied to the primitive variables of convected quantities. Alvec et al [4] developed one of the procedures and used for normalized flux for boundedness. in a density- based algorithm. Djavareshkian and Reza-zadeh [5] were among the first to implement a normalized flux formulation (NFD) scheme into a pressure-based finite volume method. The NVD scheme is applied to the fluxes of convected quantities, excluding mass flux by them. In this paper a high-resolution scheme based on normalized fluxes is developed to calculate the fluxes of convected quantities, including mass flux and this scheme is also used to calculate interface density values in correction step.

2 Finite Volume Discretization

The basic equations, which describe conservation of mass, momentum and scalar quantities, can be expressed in the following vector form, which is independent of used coordinate system.

$$\frac{\partial \rho}{\partial t} + \nabla \cdot (\rho \mathbf{V}) = S_m \tag{1}$$

$$\frac{\partial (\rho \mathbf{V})}{\partial t} + \nabla \cdot (\rho \mathbf{V} \otimes \mathbf{V} - \mathbf{T}) = \mathbf{S_v} \tag{2}$$

$$\frac{\partial (\rho \phi)}{\partial t} + \nabla \cdot (\rho \mathbf{V} \phi - \mathbf{q}) = \mathbf{S_\phi} \tag{3}$$

where ρ , $\mathbf{V}$ and ϕ are density , velocity vector and scalar quantity respectively, $\mathbf{T}$ is the stress tensor and $\mathbf{q}$ is the scalar flux vector. The latter two are usually expressed in terms of basic dependent variables. The stress tensor for a Newtonian fluid is:

$$\mathbf{T} = -(P + \frac{2}{3}\mu \nabla \cdot \mathbf{V})\mathbf{I} + 2\mu \mathbf{D} \tag{4}$$

and the Fourier-type law usually gives the scalar flux vector:

$$\mathbf{q} = \Gamma_\phi \nabla \phi \tag{5}$$

Integration of (3) over a finite volume (see e.g. Fig. 1) and application of the Gauss divergence Theorem yield a balance involving the rate of change in ϕ, face fluxes and volume-integrated net source. The diffusion flux is approximated by central differences and can be written for cell-face east (e) of the control volume in Fig. 1 as an example as:

$$I_e^D = D_e(\phi_P - \phi_E) - S_e^\phi \tag{6}$$

The discretization of the convective flux, however, requires special attention and is the subject of the various schemes developed. A representation of the convective flux is:

$$I_e^C = (\rho v A)_e \phi_e = F_e \phi_e \tag{7}$$

The normalization can be applied on the either primitive variable ϕ_e or convective flux,I_e . The details of the application NVD technique for primitive variable with SBIC scheme can be found in Djavareshkian[3]. The expression for the I_e by the SBIC scheme is given in [5]. Details of a high resolution scheme in calculating interface density values is dealt with later. The discretised equations resulting from each approximations take the form:

$$A_P \cdot \phi_P = \sum_{m=E,W,N,S} A_m \phi_m + S_\phi^{\circledcirc} \tag{8}$$

where the $A's$ are coefficients [5].

3 High Resolution Scheme in Correction Step

The compressible version of SIMPLE algorithm is used in this paper. The complete method of obtaining linearized mass fluxes in this algorithm [7] is

$$\rho^* u^{**} \approx \overset{\circ}{\rho} u^* - (\rho^\circ)^{\mathrm{HR}} D \nabla \delta p + u^* (\frac{d\rho}{dp}) \delta \rho \tag{9}$$

Substitution of (20) into continuity equation yields

$$A_P \cdot \delta p_P^* = A_E \cdot \delta p_E^* + A_W \cdot \delta p_W^* + A_N \cdot \delta p_N^* + A_S \cdot \delta p_S^* + S_P \tag{10}$$

where S_P is the finite difference analog of $\nabla(\rho^\circ u^*)$, which vanishes when the solution is converged. The A coefficients in (21) are

$$A_E = (\rho_e^\circ)^{HR}(\tilde{a}D)_e - \lambda_e(\tilde{a}u^*)_e \cdot (\frac{d\rho}{dp})_e \tag{11}$$

$$(\rho_e^\circ)^{HR} = \rho_W^\circ + \tilde{\rho}_e^\circ(\rho_E^\circ - \rho_W^\circ) \tag{12}$$

λ in (22)is a factor whose significance is explained subsequently:

$$
\begin{aligned}
\lambda &= 1 \; ; & F_e &> 0 & and & \quad \tilde{\rho}_P^\circ \leq 0, \tilde{\rho}_P^\circ \geq 1 \\
\lambda &= 0 \; ; & F_e &< 0 & and & \quad \tilde{\rho}_P^\circ \leq 0, \tilde{\rho}_P^\circ \geq 1 \\
\lambda &= 1 - \frac{(1-R_e)\tilde{\rho}_P}{k} \; ; & F_e &< 0 & and & \quad 0 \leq \tilde{\rho}_P^\circ \leq k \\
\lambda &= \frac{R_e \times \tilde{\rho}_\eta \mathrm{P}}{k} ; & F_e &> 0 & and & \quad 0 \leq \tilde{\rho}_P^\circ \leq k \\
\lambda &= R_e \; ; & & & & \quad k \leq \tilde{\rho}_P^\circ \leq 1 \quad (13)
\end{aligned}
$$

The structure of the coefficients in (21) simulates the hyperbolic nature of the equation system. Indeed, a closer inspection of expression (22) would reveal an upstream bias of the coefficients (A decreases as u increases), and this bias is proportional to the square of the Mach number. Also note that the coefficients reduce identically to their incompressible form in the limit of zero Mach number. The overall solution procedure follows the same steps as in the standard SIMPLE algorithm, with the exception of solving the hyperbolic-like pressure-correction (21).

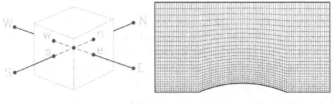

Fig. 1. Finite volume (left) and bump geometry (right).

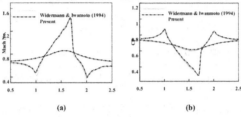

(a)   (b)

Fig. 2. Transonic (M=0.675): Mach number; pressure coefficient distribution.

4 Results

The results of inviscid transonic and supersonic flow calculations over a bump in a channel are presented. For this test, at the inlet of the domain all flow variables are specified if supersonic flow is considered. For subsonic inlet flow, stagnation pressure P_o , stagnation temperature T_o and the inlet angle are specified. At the outlet, all the flow variables are given by extrapolation for supersonic velocity. The static pressure is fixed for subsonic outlet flows. Slip boundary conditions are used on the upper and lower walls. A non-uniform grid of 98 × 26 in which the grid lines are closely packed in and near the bump region is shown in Fig. 1. The results of transonic flow with inlet Mach number equal to 0.675 over a 10 percent thick bump for two schemes on gird 98 × 26 are presented in Fig. 2. The value of in SBIC scheme for this case is 0.5. The Mach number and pressure ratio distributions on the walls are compared in Figs. 2(a) and 2(b) respectively. The results are compared with those of Widermann and Iwanmoto [6] on a same grid, 98 × 26 nodes, which were carried out with a TVD scheme based on characteristic variable. The agreement between the two solutions is close, thus verifying the validity of the present high-resolution scheme. The second case is supersonic flow over 4 percent thick bumps on a channel wall. The computations were performed on a grid 90 × 30 . The value of the k in present scheme for this case is 0.5. Figures 3(a) and 3(b) show the Mach number distribution on the upper and lower surfaces and Mach contour distributions respectively. These results are compared with the results of TVD schemes based on characteristic variable [7]. This comparison shows that the resolution of the leading edge shock, the reflection of leading edge shock at the upper wall and trailing edge shock for two schemes are nearly the same, thus verifying the validity of the present high-resolution scheme for supersonic flow.The results of the third case are presented for a planar convergent-divergent nozzle reported by Mason et al.[8]. The geometrical details for this test case are given in Fig. 4. Calculations have been performed for inviscid flow and using a grid of 24 × 7 nodal points packed near the throat in the axial direction. The treatment of boundary conditions for the inviscid case is similar to the inviscid test cases presented earlier. The stagnation pressure and temperature were presented at inlet and a static-to-stagnation pressure ratio of 0.1135 is defined at the outlet. The agreement between the numerical results and the experimental data is seen to be very good along both the nozzle wall and centerline. In this case, a compression wave can be observed downstream of the throat (Fig. 4). This compression wave starts in the upper part of the throat and touches the center line downstream; blow it. The Mach number is almost constant in the cross-section; above it, the Mach number suddenly drops.

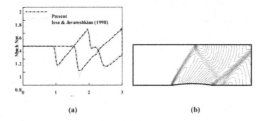

(a)   (b)

Fig. 3. Supersonic (M=1.4): (a) Mach number distribution; (b) Mach contoures.

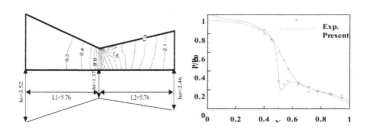

Fig. 4. Nozzle geometry and Iso Mach lines; pressure ratio (supersonic nozzle).

References

1. Leonard, B.P.: Simple High-Accuracy Resolution Program for convective modeling of discontinuities. Intern. J. Numer. Meth. Fluids **8** (1988), 1291–1318.
2. Darwish, M.S.: A New High-Resolution Scheme based on the normalize variable Formulation. Numer. Heat Trans. **B,24** (1993), 353–371.

3. Djavareshkian, M.H.: A New NVD Scheme in Pressure-Based Finite-Volume Methods, 14th Australasian Fluid Mechanics Conf., Adelaide University, Adelaide, Australia, 2001, 339–342.
4. Alves, M.A. *et al.*: Adaptive multiresolution approach for solution of hyperbolic PDEs. Comput. Meth. Appl. Mech. Engrg. **191** (2002), 3909–3928.
5. Djavareshkian, M.H., Reza Zadeh S.: Comparison Between Normalized Flux and Normalized Primitive Variable in Pressure-Based Algorithm. 12th annual conf. of fluid dynamics, Adelaide University, Ottawa, Canada, May 2004.
6. Widermann, A., Iwanmoto, J.: A Multigrid TVD Type scheme for computing inviscid and viscous flows. Computers Fluids **23** (1994), 771–735.
7. Issa, R.I.: Solution of Implicit discretised fluid flow equations by operator splitting. J. Comput. Phys. **62** (1986), 40–46.
8. Mason, M.L., Putnam, l.E. and Re, R.J.: The effect of throat contouring in two-dimensional converging-diverging nozzles at static conditions. NASA TP 1704, 1982.

Aerodynamic Calculation of Multi-element High-lift Airfoil with a Pressure-based Algorithm and NVD Technique

M. H. Djavareshkian[1] and S. Memarpour[1]

Faculty of Mechanical Engineering, University of Tabriz, Tabriz, Iran
Djavaresh@tabrizu.ac.ir, shmemarpour@yahoo.com

Abstract. An improved pressure-based procedure to solving Navier Stokes equations on a nonorthogonal mesh with collocated finite volume formulation is developed for incompressible flows. The boundedness criteria for this procedure are determined from NVD scheme. The procedure incorporates the $K - \varepsilon$ eddy-viscosity turbulence model. The aim of the present study is to develop above scheme in incompressible flow and investigate the calculation of this flow around the multi-element airfoils. The results of this simulation around the multi-element airfoil MDA 30P-30N are compared with numerical data. Computation visualization of the external flows show that the algorithm presented in this paper is successful in incompressible regime for coarse meshes.

Keywords: High resolution scheme, Pressure-based algorithm, Normalized Variable Diagram, High Lift, Multi-element

1 Introduction

Now the high-resolution schemes that have boundedness criteria have removed the defects of first and high order differential scheme. Leonard [1] has generalized the formulation of the high-resolution flux limiter schemes using what is called the normalized variable formulation (NVF). The NVF methodology has provided a good framework for development of high-resolution schemes that combine simplicity of implementation with high accuracy and boundedness. Gaskell and Lau [2] introduced SMART scheme. The SFCD scheme was represented by Ziman [3]. Zhu and Rodi [4] introduced SOUCUP scheme and etc. Javareshkian [5] has recently developed the SBIC scheme with the minimum number of adjustable parameters. One advantage of this scheme in comparison with all other differencing schemes is some kind of switching only two differencing schemes; central differencing and blending between up-wind and central differencing are included that blending factor is determined automatically. This approach has previously shown a comparable ability of predicting compressible aerodynamic flow.

The aim of the present study is to develop above scheme in incompressible flow and investigate the calculation of this flow around the multi-element airfoils. The results of this simulation around the multi-element airfoil MDA 30P-30N are compared with numerical data. Computation visualization of the external flows show that the algorithm presented in this paper is successful in incompressible regime with coarse meshes.

2 Finite Volume Discretization

The basic equations, which describe conservation of mass, momentum and scalar quantities, can be expressed in the following vector form, which is independent of used coordinate system:

$$\frac{\partial(\rho)}{\partial t} + \nabla \cdot \rho \mathbf{V} = S_m \tag{1}$$

$$\frac{\partial(\rho \mathbf{V})}{\partial t} + \nabla \cdot (\rho \mathbf{V} \otimes \mathbf{V} - \mathbf{T}) = \mathbf{S_v} \tag{2}$$

$$\frac{\partial(\rho \phi)}{\partial t} + \nabla \cdot (\rho \mathbf{V} \phi - \mathbf{q}) = \mathbf{S_\phi}, \tag{3}$$

where $\rho, \mathbf{V}$ and ϕ are density, velocity vector and scalar quantity respectively, $\mathbf{T}$ is the stress tensor and $\mathbf{q}$ is the scalar flux vector. The stress tensor for a Newtonian fluid is

$$\mathbf{T} = -(\mathrm{P} + \frac{2}{3}\mu \nabla \cdot \mathbf{V})\mathbf{I} + 2\mu \mathbf{D} \tag{4}$$

and the Fourier-type law usually gives the scalar flux vector:

$$\mathbf{q} = \Gamma_\phi \nabla(\phi) \tag{5}$$

.Integration of (3) over a finite volume and application of the Gauss divergence Theorem yield a balance involving the rate of change in ϕ , face fluxes and volume-integrated net source. The diffusion flux is approximated by central differences and can be written for cell-face east (e) of the control volume as an example as:

$$I_e^D = D_e(\phi_P - \phi_E) - S_e^\phi \tag{6}$$

A representation of the convective flux is

$$I_e^c = (\rho \cdot \mathrm{V} \cdot \mathrm{A})_f \phi_f = F_f \phi_f \tag{7}$$

,the value of the dependent variable ϕ_e is not known and should be estimated using an interpolation procedure, from the values at neighboring grid points. ϕ_e is determined by the SBIC scheme, that it is based on the NVD technique, used for interpolation from the nodes E, P and W. The expression can be written as

$$\phi_e = \phi_w + (\phi_E - \phi_W) \cdot \widetilde{\phi}_e \tag{8}$$

The functional relationship used in SBIC scheme for $\widetilde{\phi}_e$ is given by:

$$
\begin{aligned}
\widetilde{\phi}_e &= \widetilde{\phi}_P & \phi_P > 0, \quad \phi_P < 1 \\
\widetilde{\phi}_e &= -\frac{\widetilde{X}_P - \widetilde{X}_e}{k(\widetilde{X}_P - 1)} \widetilde{\phi}_P^{\,2} + \left(1 + \frac{\widetilde{X}_P - \widetilde{X}_e}{k(\widetilde{X}_P - 1)}\right) \widetilde{\phi}_P & 0 \le \phi_P > k \\
\widetilde{\phi}_e &= -\frac{\widetilde{X}_P - \widetilde{X}_e}{\widetilde{X}_P - 1} + \frac{\widetilde{X}_e - 1}{\widetilde{X}_P - 1} & k \le \phi_P \ge 1
\end{aligned}
$$

$$\tag{9}$$

, where

$$\widetilde{\phi}_p = \frac{\phi_P - \phi_W}{\phi_E - \phi_W}, \quad \widetilde{\phi}_e = \frac{\phi_e - \phi_W}{\phi_E - \phi_W}, \quad \widetilde{X}_e = \frac{x_e - x_w}{x_E - x_W}, \quad \widetilde{X}_p = \frac{x_P - x_w}{x_E - x_W}$$

The details of how the interpolation is made is dealt with Ref. [5]; it suffices to say that the discretized equations resulting from each approximations take the form:

$$A_P \cdot \phi_P = \sum_{n=E,W,N,S} A_n \cdot \phi_n + \acute{S}_\phi \tag{10}$$

3 Turbulence Model

In order to consider the turbulent effects, the Navier-Stokes equation is considered the average value for variable flow and turbulence model k-ε . In other hand turbulent viscosity (Boussinesq) is used which relates shear stress with the average of velocity gradient for incompressible flow(See Ref.[6]).

4 Solution Algorithm

The present work employs the SIMPLE technique in which the implicitly discretised equations are solved at each time step by a sequence of predictor and corrector steps.

5 Boundary Condition

A C-type structured grid was used with free stream conditions on the far-field boundaries, in inlet boundary, the values of velocities u and v are known as the physical boundary condition and the pressure p is extrapolated from domain solution as numerical boundary condition. In outlet boundary, it is supposed that the static pressure is constant. For top and bottom boundaries of computational domain, slip boundary condition is used. No-slip wall boundary conditions were applied at the airfoil surfaces. the far-field boundary placed at 15-chord length away from the airfoil surface.

6 Results

In this paper the main emphasis is concentrated on the implementation of an improved pressure based method in incompressible flow around the multi-element airfoil cases. Computational results are shown in followed figures for a baseline series of test cases. Fig. 1 close-up three-element MDA 30P-30N airfoil grids. Figs. 1-2 also show the far-field boundary placed at 15-chord length away from the airfoil surface. The value of K in SBIC method for all cases is 0.3. Three cases considered are incompressible flow around an MDA 30P-30N airfoil at a Reynolds number of $Re_c = 9.0 \times 10^6$ and angles of attack $\alpha=4\ deg$,$\alpha=16\ deg$ and $\alpha=19\ deg$. The distribution of pressure coefficients on the upper and lower surfaces of airfoil for three cases and contours of pressure coefficients, Cp, are shown in Figs.3-8. It can be seen that two computed results show good agreement.

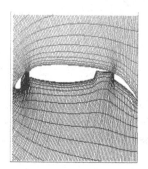
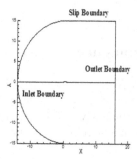

Figure 1. Part of grid used for MDA 30P-30N airfoil.   **Figure 2.** Far-field boundary (right).

7 Conclusion

A high resolution scheme has been implemented in a pressure-based, finite-volume procedure which uses an implicit solution algorithm. The mention scheme based on normalized variables is developed to calculate the fluxes of convected quantities. The method is applied to incompressible flow and the results have been compared with the results of the fine mesh. These comparisons show, the results of the fine and coarse meshes are nearly the same.

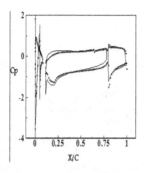

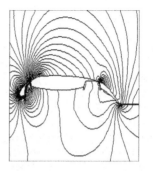

Figure 3. Pressure coefficient distribution for base line configuration at $\alpha = 4°$:$(-)$ 167×103 (o) 167×203.

Figure 4. Pressure coefficient contours.

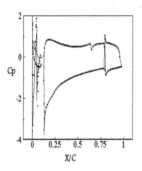

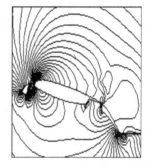

Figure 5. Pressure coefficient distribution for base line configuration at $\alpha = 16°$:$(-)$ 167×103 (o) 167×203.

Figure 6. Pressure coefficient contours.

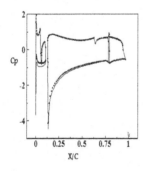

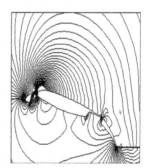

Figure 7. Pressure coefficient distribution for base line configuration at $\alpha = 19°$:$(-)$167×103 (o) 167×203.

Figure 8. Pressure coefficient contours.

References

1. Leonard, B.P.: Simple High-Accuracy Resolution Program for convective modeling of discontinuities. International Journal for Numerical Methods in Fluids,8,1291-1318 (1988)
2. Gaskell, P.H., Lau, A.K.C.: Curvature-Compensated convective Transport: SMART, a new boundedness-preserving transport algorithm. International Journal for Numerical Methods in Fluid, 8, 617–640 (1988)
3. Ziman, H.:A computer Prediction of Chemically Reacting Flows in Stirred Tanks. PhD thesis, University of London (1990)
4. Zhu, J., Rodi, W.: Low Dispersion and Bounded Convection Scheme. Computer Methods in Applied Mechanics and Engineering,92, 87–96 (1991)
5. Djavareshkian, M.H.: A New NVD Scheme in Pressure-Based Finite-Volume Methods. 14th Australasian Fluid Mechanics conference, Adelaide University, Adelaide, Australia, 10-14 December 2001, 339-342 (2001)
6. Launder, B. E. and Spalding, D. B. The Numerical Computation of Turbulent Flows , Computer Methods in Applied Mechanics and Engineering, Vol.3, pp.269, 1974.

Active Learning with Ensembles for DOE

Tao Du and Shensheng Zhang

Dept. of Computer Science of Shanghai Jiaotong University, Shanghai 200030, China {dutao,sszhang}@sjtu.edu.cn

Summary. In this paper, a novel active learning algorithm for design of experiments (DOE) is presented. In this algorithm, a boosting method for regression is firstly used to generate ensemble of learners from existing data. And then the average ensemble ambiguity among the element learners in the ensemble is proposed to determine which data point would be labeled by executing experiments. The results of simulations have shown that when the number of experiment is limited, the algorithm is better compared with traditional passive learning algorithms.

Key words: Machine learning, active learning, design of experiment, query by boosting

1 Introduction

In many aspects of applications, we need predict the performance of a system or product. To solve this problem, the basic idea is to construct surrogate models which are sufficiently accurate to approximate the original system or products. And those surrogate models are then used to facilitate design space exploration, optimization and reliability analysis.

In many circumstances, the data used for training the surrogate models could only be achieved by doing experiments. And it is usually very costly and time consuming. As a result, the size of data set can not be very large. But limited size of data set would make constructing an accurate model from observed data very hard. To relieve this dilemma, in this paper, we introduce active learning into the context of DOE, and a novel active learning by boosting algorithm for regression is proposed.

The remainder of this paper is organized as following. Section 2 would offer an introduction to 'active learning'. In section 3, an active learning algorithm by boosting for regression is detailed out. In section 4, the simulation results of the algorithm proposed in this paper would be given and analyzed. Finally, in section 5, we summarize and conclude the paper.

2 Active learning for DOE

Active learning [NA01] is a framework in which the learner has the freedom to select which data points are added to its training set. The active learning problem for design of experiments is formulated as below.

Define the underlying function to be simulated in experiment is $g(x)$, the goal of learning is to construct a surrogate model $f(x)$ to approximate $g(x)$ well enough. In the beginning, we have an initial training set with limited size D_0. In the context of active learning, during the training, the learner could add samples to the training set iteratively. For example, in the ith step , the learner presents a new sample x^* to request its label y^*, and add labeled sample (x^*, y^*) to the training set D_i, in the step $i+1$, the learner would be training again using the training set $D_{i+1} = D_i \cup (x^*, y^*)$, and then presents the new sample , gets its label and add the labeled samples to the training set. Those processes would iterate until the termination conditions are reached.

Among approaches for active learning, 'query by committee' [SOS92] is an attractive and general active learning strategy with theoretical performance guarantee. A drawback of 'query by committee' is that it can't be applied on a deterministic component learning algorithm. To overcome this drawback, several contributions [AM98, VCT00, PR04] have presented another active learning strategy similar to 'query by committee' which is called 'query by boosting'. In 'query by boosting', the boosting approach is used to generate the multiple learners for selecting the next sample to be labeled.

As we know, all the contributions on active learning by boosting are only concerned with the classification problem. But in the context of design of experiments, most input and output variables are continuous real-valued. Therefore, we need take the regression problems into consideration and extending existed active learning by boosting algorithms to the regression applications.

3 Description of our method

To extend the active learning by boosting algorithm from classification to regression, the ensemble ambiguity of regressors is used. The rigorous definition of the ensemble ambiguity A of unlabeled sample i is given as below:

$$A_i = \sqrt{\frac{\sum_{j=1}^{N} (y_{i,j}^{(p)} - \bar{y}_i^{(p)})^2}{N}}$$

Where N is the number of the component learners in the ensemble generated by boosting, $y_{i,j}^{(p)}$ is the prediction for the $i'th$ unlabeled sample by the $j'th$ component learner , $\bar{y}_{i,j}^{(p)}$ is the mean of the prediction for the $i'th$ unlabeled sample among all component learners.

It is easy to see that samples with greater ensemble ambiguity also have greater uncertainty according to the ensembles of regressors. Hence selecting samples with greater ensemble ambiguity to query its labels would make the information gain on each query maximized and make the learner's performance improved with training samples as fewer as possible.

In this paper, the boosting algorithm for improving regressors proposed in [Dru97] is adopted to generate the ensemble of regressors. For regression machines, those samples whose predicted values differ most from their observed values are defined to be "most" in error, and the sampling probabilities would be adjust to bias those samples with great error. The regressors in the ensemble are combined using the weighted median to give the final output, whereby those predictors that are more "confident" about their predictions are weighted more heavily.

Detailed description of our active learning algorithm for DOE is given as below, the details of the boosting algorithm for regression could be found in [Dru97].

Algorithm: Query by boosting for Regression (QBR)

Input: Number of trials: T

Component learning algorithm: C

Number of time re-sampling is done: N

Number of query candidates: R

Number of samples in initial training set D_0: l

Initialization: set $D_0 = \{(x_1, y_1), \cdots (x_l, y_l)\}$ using uniform design or random sampling.

For i $= 1, \ldots, T$

1. Run boosting algorithm for regression [Dru97] on input (D_i, C, N) and get an ensemble of regressors E_i.

2. Generate the samples set Q with the size R as the query candidates.

3. Pick the sample x^* with greatest ensemble ambiguity A as the query point from the sample set Q.

4. Query the label of sample x^*, and obtain y^*.

5. Update the training set as follow:

$$D_{i+1} = D_i \cup (x^*, y^*)$$

End For Output: Output the training set D_T and ensemble of regressors E_T

In the step 2 of the algorithm QBR, there are several methods to generate the query candidates set Q depending on the context the algorithm is used. If the algorithm is used to construct a simulation model to approximate the underlying function, the random sampling or uniform design could be used. If the algorithm is used to find the optimum of the underlying function, the method similar to Q2 [MS98] could be used, in which the region of interest is truncate into smaller region step by step.

When constructing the simulation model from D, QBR could be used to generate training set and construct simulation model simultaneously.

4 Experiments and discussions

To evaluate the performance of the active learning algorithm QBR proposed in this paper, we apply the algorithm on both artificial and real dataset.

The artificial data is from the [Fri91], which is generated according to the equation $y = \tan^{-1}\left[\dfrac{x_2^2 - (^1/_{x_2 x_4})}{x_1}\right] + noise$. The real data used in experiments is 'Boston Housing' which is obtaining from UCI database.

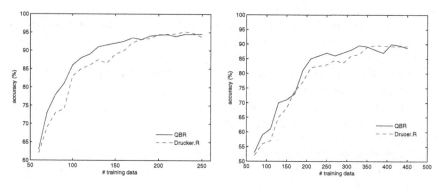

Fig. 1. artificial data   **Fig. 2.** 'Boston housing'

The initial training set and test set is selected randomly. During the process of training, for artificial data, the equation used for generating data is used as an oracle to give any query x^* its value of label y^*. And for real data, since there is no oracle to give any arbitrary query its label, the query candidates would be chosen in remain samples not in training set.

In the experiments, an RBF network [Gun01] is used as the component learning algorithm for constructing regressors. In the two cases, number of query candidates R is both set to 20, and Number of times re-sampling T is set to 100.

The results of the experiments are shown by Fig.1 and Fig.2 respectively. Those figures show the prediction accuracy on test set of the algorithm QBR proposed in this paper and of the passive boosting learning algorithm Drucker.R.

It is obviously from Fig.1 and Fig.2 that the prediction accuracy of QBR is better than the regression algorithm Drucker.R in [Dru97] when the size of training set is limited. When the number of training samples is small, the training set generated by QBR is more informative and could improve the learner's performance greatly. And as the training set is increasing, the training set generated by QBR and by random sampling is tent to be equal, therefore, the prediction accuracy of both algorithms is becoming identical.

5 Conclusions

In the context of Design of Experiments, it needs to construct regression model from data for simulation. In this paper, an active learning algorithm for regression problems is proposed, by incorporating the ideas of boosting regressions and query by committee. Experiments on both artificial data and real data show that the proposed active learning algorithm QBR achieve much improvement in data efficiency as compared to the algorithm of boosting regressions.

References

[AM98]   Abe, N., Mamitsuka, H.: Query learning strategies using boosting and bagging. International conference on Machine Learning (ICML).

[Dru97] Drucker, H. 1997 . Improving regressors using boosting techniques. In Machine Learning: Proceedings of the Fourteenth International Conference 107-115. Morgan Kauffman, San Francisco.

[FS95]   Y. Freund, R. Schapire: A decision theoretic generalization of on-line learning and an application to boosting. In Proceedings of the Second European Conference on Computational Learning Theory (Euro COLT'95), pages 148-156, 1995.

[Fri91] Freedman, J.H.: Multivariate Adaptive Regression Splines, The Annals of Statistics, 19.

[Gun01] Gunnar, R.; A Documentation of RBF and AdaBoost Software Package. 2001.

[MS98]   Moore, A.W., Schneider, J. G.: Q2: Memory-based active learning for optimizing noisy continuous functions. In Proceedings of the Fifteenth International Conference on Machine Learning(pp.386-394).

[NA01]   N. Roy and A. McCallum, "Toward optimal active learning through sampling estimation of error reduction," in Proc. 18th International Conference on Machine Learning(ICML01), 2001, pp. 441–448.

[PR04]   Prem Melville, Raymond J. Mooney: Diverse Ensembles for Active Learning. In Proc of the 21st International Conference on Machine Learning.

[SOS92] H. S. Seung, M. Opper, and H. Sompolinsky.: Query by committee. In Proc. 5th Annu. Workshop on Comput. Learning Theory, pages 287-294. ACM Press, New York, NY, 1992.

[VCT00] Vijay S., Chidanand Apte, Tong Zhang: Active Learning using Adaptive Resampling. Proceedings of the sixth ACM SIGKDD international conference on Knowledge discovery and data mining, pages 91-98, 2000.

Parallel Computing for Linear Systems of Equations on Workstation Clusters

Chaojiang Fu[1,2], Wu Zhang[1] and Linfeng Yang[1]

[1] School of Computer Engineering and Science, Shanghai University, Shanghai
200072, PRC cjfu@163.com, zhang@staff.shu.edu.cn, bennvie@sina.com
[2] School of Architectural Engineering, Nanchang University, Nanchang 330029,
PRC sdfcj@eyou.com

Abstract In this paper the parallel algorithm of preconditioned conjugate
gradient method (PCGM) is presented and implemented on DELL workstation cluster. Optimization techniques for the sparse matrix vector multiplication are adopted in programming. The storage schemes are analyzed in detail.
The numerical results show that the designed parallel algorithm has good
parallel performance on the high performance workstation cluster. This illustrates the power of parallel computing in solving large-scale problems much
faster than on a single processor.
Key words: parallel computing; preconditioned conjugate gradient method;
network; workstation cluster

1 Introduction

In recent years, parallel computing has become an important field in science
and engineering computing. There are two main reasons which make it reasonable. First, the use of parallel computing can handle large scale problems.
Second, networks of computers under LINUX result in a workstation cluster.
On such hardware platforms, one can gain first experiences in parallel computing. So a workstation cluster is poised to become the primary computing
infrastructure for science and engineering computing. It may achieve cheap,
highly available, and provide multiple CPUs for parallel computing.

Solving linear systems of equations is one of the most important problems
in science and engineering computing. For example, the application of finite
difference or finite element method to engineering problems requires the solution of a system of linear equations. In general, the systems of equations can
be expressed as, $AX = b$. Where A is a $n \times n$ sparse matrix, X is the vector
of unknowns and b is n-dimensional vector. In most case, this yields a linear
system with a symmetric positive definite sparse system matrix.

Unfortunately, the direct methods in elementary linear algebra classes (e.g., Gaussian elimination) are difficult to adapt for distributed memory systems, especially if the coefficient matrix is sparse. So most of the sparse linear system solvers for distributed memory systems use iterative methods[NW03, Don02]. Since the finite difference or finite element method generally yields a linear system with a symmetric positive definite sparse system matrix. This property of the system matrix allows the use of PCGM as the linear solver.

2 Parallel Implementation of PCG Method

The parallelization of the PCG method can be implemented since matrix and vector operations of this algorithm can be parallelized [CD97]. The parallelism comes from storing and operating on main sections of the working vectors and the system matrix. The parallel implementation is presented below.

Initialize: choose $X_0 \Longrightarrow r_0 = b - AX_0$

solve $M\tilde{r}_0 = r_0 \Longrightarrow p_0 = \tilde{r}_0$

$$c = dot(\tilde{r}_0, r_0)$$

M is the preconditioning matrix, using $M = diag(a_{11}, a_{22}, \ldots, a_{nn})$

Main iterate over:

$(1) z = Ap$ $\longleftarrow$ *Parallel Matrix − Vector Product*

$(2) \alpha = c/dot(p, z)$ $\longleftarrow$ *Parallel Vector Dot Product*

$(3) X = X + \alpha * \beta$ $\longleftarrow$ *BLAS daxpy Operation*

$(4) r = r - \alpha * z$ $\longleftarrow$ *BLAS daxpy Opreation*

$(5) solve M\tilde{r} = r$ *for* $\tilde{r}$ $\longleftarrow$ *Solve Matrix System*

$(6) d = dot(r, r)$ $\longleftarrow$ *Parallel Vector Dot Product*

$(7) if(\sqrt{d} < tolerance)$ break;

$(8) \beta = d/c$

$(9) p = \tilde{r} + \beta * p$ $\longleftarrow$ *BLAS daxpy Operation*

$(10) gather(p)$ $\longleftarrow$ *Parallel Gather*

$(11) c = d$

(12) go to (1)

The parallel algorithm is implemented in the programming language C++ using the Message Passing Interface (MPI) standard for communications between processors[AN96]. This algorithm involves 1 matrix-vector product, 2 dot products, and 3 vector updates (daxpy operations) per iteration. This implementation is optimized with respect to memory usage by storing exactly 4 vectors of length n, and by using a function that implements the matrix-vector multiplication in matrix-free form[HJ98, DGL89]. The vectors stored are the approximation to the solution X, the search direction p, the residual r := b - AX, and the auxiliary vector q.

The parallel algorithm requires 4 communication operations per iteration. To compute the matrix-vector product, processors need to interchange one data with their neighboring processors. These communications are implemented with nonblocking MPI_ISEND/

MPI_IRECV commands. Since the dot products apply to vectors split across the processors and since the results are needed on all processors, an MPI_Allreduce operation is required in each of the 2 dot products. Optimization techniques for the sparse matrix vector multiplication are adopted in programming[AGZ92, AH92], i.e., the serial part of the dot products, local to each processor, is implemented in C++ inner function dot. The overlap of the communications with the computations is implemented by using asynchronous messages in the message-passing model in order to hide the latency of the communications[Ort03].

3 Storage scheme

The aim of the matrix storage is to achieve a better performance. There are several possibilities to store a matrix. The key is not to store unnecessary data. The emphasis of the storage is on developing an algorithm which will solve problems too large to be solved efficiently on a single CPU.

Storing all n^2 elements of $A \in R^{n \times n}$ explicitly is known as "dense storage mode". For dense matrices this is good technique. For sparse matrix it is not, because all zero elements are stored. This is wasteful both in memory and in runtime, as many multiplications involving zeros are needlessly computed.

A common idea is to take advantage of the sparsity of the matrix. In this "sparse storage mode", only non-zero elements are stored . This reduces the memory requirements. This also improves performance, as only multiplications with those non-zero elements are computed.

To reduce the memory usage further, the constant coefficients in predetermined positions of the system matrix are taken. A function is provided that accepts a vector p as input and returns the vector $z = Ap$ as output; each component z_k is computed as summation of the appropriate components of p multiplied with hard-coded coefficients[HJ98]. This technique is known as a "matrix-free implementation" because no elements of the system matrix are explicitly stored at all[DGL89]. Such a matrix-free method dramatically reduces the memory requirements. It is the most efficient approach to memory usage. The scheme is adopted in this paper.Table 1 shows memory usage for three kinds of storage scheme in case of pentadiagonal matrix, when double precision arithmetic is used.

Table 1. Predicted memory usage for three storage methods in MB

n	dense	sparse	Matrix-free
256	< 1	< 1	< 1
1024	8	< 1	< 1
4096	134	< 1	< 1
16384	2184	1	< 1

4 Numerical Test and Analysis

The numerical and performance tests of the developed parallel PCG algorithm are performed on DELL workstation cluster in School of Computer Engineering and Science, Shanghai University. It is a cluster with 12 processors arranged in 1 four-processor node with 2.0GHz Intel Xeon chips (1MB cache) and 4 dual-processor nodes with 2.4GHz Intel Xeon chips (512KB cache) and 1.0GB of memory per node. These nodes are connected with a 100.0Mbps Ethernet interconnect. Using the Message Passing Interface (MPI) paradigm.

The numerical problem is a classical prototype problem, the Poisson equation with a homogeneous Dirichlet boundary condition, using finite differences in two dimensions. This yields a linear system with a symmetric positive definite system matrix. The results of computation are shown in Fig. 1.

Memory is an important limitation to the size of the system to be solved. A close look at Table 1 indicates that less than 1 MB of memory is used for a sparse or matrix-free storage compared to 134 MB for a dense storage when dimension (n) of the system matrix is equal to 4096. Adding the dimension (n) to 16384, it takes over 2 GB using dense storage. This is not possible to compute this problem for single processor of the workstations because only 1 GB of memory is available for the single processor, so n equals to 11180 that is the largest system for memory of the single processor. Clearly, a matrix-free implementation is the best storage method. With a matrix-free implementation, the method is optimal with respect to memory usage, and we are able to solve problems that are much too large for single processor computers.

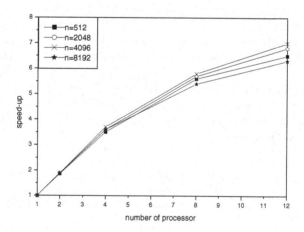

Fig. 1. Plot of the speedup for different n on the 12-processor cluster

Fig. 1 shows speedups for this cluster, using up to 12 processors. It is readily apparent from Fig. 1 that the slopes of the speedup curve continue to descend all the way up to 12 processors. When the number of processor is greater than 4, an apparent descend is observed because the communication overhead between processors increases significantly. Within the 4 processors, the computation is implemented on 4-processor node. The communication overhead between processors is less; thus its performance is better.

5 Conclusions

Parallel computing is a useful method for solving large-scale problems faster. PCG algorithm is parallelized and this parallel algorithm is implemented on workstation cluster. Its performance is analyzed. The reduction in memory due to the matrix-free implementation not only allows for the solution of much larger problems but also decreases computing time. The parallel performance studies show that a high-performance parallel platform is necessary to obtain excellent speedup on a cluster.

Acknowledgements. Our works are supported by "SEC E-Institute: Shanghai High Institutions Grid" project, the fourth key subject construction of Shanghai, the Key Project of Shanghai Educational Committee (No.03AZ03).

References

[Bar94] Barret, R., et al.: Templates for the Solution of Linear Systems, Building Blocks for Iterative Methods. SIAM, Philadelphia (1994)

[Law86] Law, K.H.: A Parallel Finite Element Solution Method. Computers & Structures, 23(6),845-858(1986)

[Geo03] George, E.K., et al.: Parallel Scientific Computing in C++ and MPI. Cambridge University Press (2003)

[Pac97] Pacheco, P.S.: Parallel Programming with MPI. Morgan Kaufmann (1997)

[All03] Allen, K.P.: Efficient Parallel Computing for Solving Linear Systems of Equations. Graduate Student Seminar, University of Maryland, Baltimore County (2003)

[BH86] Brown, P.N., Hindmarsh, A.C.: Matrix-free methods for stiff systems of ODE's. SIAM J. Numer, Anal, 23,610-638(1986)

[Sor97] Sorin, G.N., et al.: Load-Balanced Sparse Matrix-Vector Multiplication on Parallel Computers. Parallel and Distributed Computing, 46,180-193 (1997)

[AH92] Aliaga, J.I., Hernandez, V.: Symmetric sparse matrix-vector product on distributed memory multiprocessors. Conference on Parallel Computing and Transputer Applications,Barcelona,Spain (1992)

[Ort03] Ortigosa, E.M., et al.: Parallel scheduling of the PCG method for banded matrices rising from FDM/FEM. J. Parallel Distrib. Comput., 63,1243-1256 (2003)

A Service Model Based on User Performances in Unstructured P2P *

Jianming Fu[1,2], Weinan Li[1], Yi Xian[1], and Huangguo Zhang[1]

[1] School of Computer, Wuhan University, Wuhan 430072, China
[2] State Key Laboratory of Software Engineering, Wuhan University, Wuhan
  430072, China `fujms@public.wh.hb.cn`

Abstract. This paper presents a service model based on user performances
to deal with free riding in unstructured P2P, which exploits historical logs. In
our model, after receiving a request, the server peer calculates the user performances of the client peer to itself, that of the client peer to the server's group,
and that of the client's group to the server's group, and then provides the
different quality of services according to these user performances. Moreover,
a peer with high user performance can obtain high quality of service, whereas
the peer with low user performance may get service with certain delay or be
rejected in a certain probability.

Key words: peer to peer; free riding; user performance; service model

1 Introduction

Peer-to-Peer (P2P) systems are typically designed on the assumption that all
peers will voluntarily contribute resources to a global pool. However, the behavior of peers is quite different, some peers might be benevolent in providing
services, whereas another peers might be malicious or stingy and not provide
services to others. The behavior of participants that consume many resources
while contribute little is regarded as free riding and such participants are
called freeloaders [1].

This paper proposes a service model based on user performances, which
represents the user contribution. And a server peer provides different quality
of services according to the performances of a client or the client's group.
Meanwhile the user performance is computed by the server not by the client
peer, so it is impossible for the client to fake its performance. Our model is

---
*Foundation item: This work is supported by the National Natural Science Foundation of China with the Grant No.90104005 & No.60373089 and the Natural Science
Foundation of Hubei in China No.2002ABB036.

different from SLIC [2]. We use three kinds of performances, one is of peer vs. peer, the next is of peer vs. peer's group, and the last is of peer's group vs. peer's group. However, SLIC only uses performance of peer vs. peer, and these peers are neighbors.

2 Our Service Model

Our service model includes two parts, service process and responding policy. The details are given below.

2.1 Service Process

First of all, we describe basic organization about a P2P system. A peer has a peer identifier PID, and this peer must belong to a group. Meanwhile, a group has a group identifier GID.

Preliminary information each peer keeps is showed in the following. For each peer, $MI = \{PID, Timestamp\}$ contains all members of its group. $GT=\{PID, GID, FID, FS, Timestamp\}$ stores all records when a peer has downloaded any file, and GID is the group identifier for any server peer with the PID. $ST=\{PID, GID, FID, FS, Timestamp\}$ stores records for uploaded files, and GID is the group identifier for any client peer with the PID. $Timestamp$ is used to denote the time inserted a record, FID is the file identifier, and FS is the length of a file.

The service process is showed below.

Step 1 A peer N_r knows that another peer N_p holds a file FID via searching mechanism, and sends request (F_REQ message) to N_p.

Step 2 The peer N_p responds the request by its own service policy mentioned-below, then sends the file with FID to N_r (F_RES message), then inserts this responding record in its ST.

Step 3 After N_r receives the response from N_p, it adds a record about this request in its GT

2.2 Responding Policy

After the peer N_p receives a request from N_r, N_p will deal with the request by certain policy. We use user performances to choose responding policy. The definition of three performances is shown below.

$Definition1$: Performance of Peer vs. Peer (PPP) is defined as:

$$PPP(N_r, N_p) = \frac{\sum\limits_{i \in N_p.GT and i.PID=N_r.PID} i.FS}{\sum\limits_{i \in N_p.ST and i.PID=N_r.PID} i.FS}$$

PPP is a ratio of the service level to the consumption level of N_r vs. N_p.

$Definition2$: Performance of Peer vs. peer's Group (PGP) is defined as:

$$PGP(N_r, N_p) = \frac{\sum\limits_{j \in N_p.MI} (\sum\limits_{i \in j.GT and i.PID=N_r.PID} i.FS)}{\sum\limits_{j \in N_p.MI} (\sum\limits_{i \in j.ST and i.PID=N_r.PID} i.FS)}$$

PGP is a ratio of the service level to the consumption level of N_r vs. N_p'group.

$Definition3$:Performance of peer'Group vs. peer's Group (GGP) is defined as:

$$GGP(N_r, N_p) = \frac{\sum\limits_{j \in N_p.MI} (\sum\limits_{i \in j.GT and i.GID=N_r.GID} i.FS)}{\sum\limits_{j \in N_p.MI} (\sum\limits_{i \in j.ST and i.GID=N_r.GID} i.FS)}$$

GGP is a ratio of the service level to the consumption level of N_r'group vs. N_p'group.

Now we introduce threshold α and a constant $DELAY\_CONST$. On the basis of user performances, we can choose a response policy according to the following steps.

Step 1 respond immediately if $N_p.GID = N_r.GID$ or $PPP(N_r, N_p) \geq \alpha$.

Step 2 delay the response within $DELAY\_CONST$ if $PGP(N_r, N_p) \geq \alpha$.

Step 3 delay the response within $2*DELAY\_CONST$ if $GGP(N_r, N_p) \geq \alpha$.

Step 4 delay the response within $3*DELAY\_CONST$ in probability ρ.In this case, there is either no interactions between two groups, or mentioned-above performances maybe too low, so the server peer can reject the request in $1-\rho$.

3 Conclusion

The user performance is computed by the server, so juggling performance for clients is avoided. If only the performance of the client to the server is used to choose policy, the number of response is little, for this kind of performance maybe lower. Therefore, three kinds of performances is provided to improve the rate of successful responses.

References

1. E. Adar and B. A. Huberman. Free riding in Gnutella, Xerox PARC report, Oct 2000, available at www.parc.xerox.com/istl/groups/iea/papers/gnutella.
2. Qixiang Sun, Hector Garcia-Molina. SLIC: A Selfish Link-Based Incentive Mechanism for Unstructured Peer-to-Peer Networks. 24th International Conference on Distributed Computing Systems (ICDCS'04), March, 2004. Hachioji, Tokyo, Japan.

DMR: A Novel Routing Algorithm in 3D Torus Switching Fabric

Jianbo Guan, Xicheng Lu, and Zhigang Sun

School of Computer Science, National University of Defense Technology, Changsha 410073, China guanjb@nudt.edu.cn

Summary. In recent years 3-D Torus network is widely considered in implementing large scale switching fabrics. Based on analysis of several problems faced in packets routing in 3-D Torus switching fabrics, a novel routing algorithm called DMR (Dimension-order-based Multi-path Routing) is proposed. The DMR algorithm is discussed in detail and it is proved that DMR can achieve high throughput by balancing traffic loads on multiple equivalent paths while maintaining packets order in one TCP flow. The performance of DMR routing algorithm is evaluated using a simulation approach and it is compared with two other representative routing algorithms: e-cube routing and random routing. The simulation results and the analysis are shown afterwards. At last a conclusion is given.

Key words: torus, switch, routing algorithm

1 Introduction

Internet traffic is growing at rates of 70 to 150 percent per annum while the performance of core routers evolves much slower[CA01]. Service providers want to deploy routers supporting higher speed interfaces and larger scale switching fabrics than before when they upgrade their backbone networks. Switching fabrics heavily affects the overall performance and size of routers. Traditional single-stage switching fabrics cannot satisfy the increasing demands for speed and scale because of their inherent poor scalability so novel multi-stage architectures should be considered[ZL02].

Nowadays 3-D Torus network is widely recommended for implementing large scale switching fabrics because of its obvious advantages over other multi-stage architectures[WIL02][ZL02]. Packets routing algorithms are very important for 3-D Torus switching fabrics since they are pivotal for the overall performance.

We will propose a novel packet routing algorithm called DMR (Dimension-order-based Multi-path Routing) and show that it can solve two conflicting problems in 3-D Torus switching fabrics: balancing traffic loads on multiple paths and maintaining packets order in one TCP flow. Performance evaluations in comparison with other two routing algorithms are also given.

2 3-D Torus Switching Fabric

3-D Torus is a direct interconnection network, which has been widely used in MPP supercomputers. It is considered as the most appropriate interconnection network to implement large scale switching fabrics for high performance routers. 3-D Torus is a kind of k-ary n-cube network which has 3 dimensions, with the first and the last nodes directly connected on each dimension.

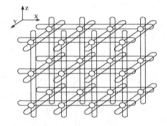

Fig. 1. A 4x2x3 3-D Torus network

Fig. 1 shows a 4x2x3 3-D Torus network. 3-D Torus network has following advantages for implementing large switching fabrics:

- The degree of interconnection for each node is deterministic 6, which decreases complexity for practical implementation.
- This architecture can be scaled easily and cost efficiently. 3-D Torus network can increase node one by one and the cost increases linearly with number of nodes.
- Diameter of the network is low which facilitates to reduce the packets delay to traverse the network.
- 3-D Torus network offers a high-degree of path diversity. That means there are many different paths from node A to node B. This property facilitates traffic loads balancing and fault tolerance.

3 DMR routing algorithm

3.1 Routing in 3-D Torus Architecture

Routing style determines how packets are forwarded through the switch fabric and heavily influences the overall performance. In practice we can implement a 3-D Torus switching fabric as follows: each node is made up with a line

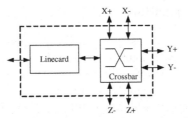

Fig. 2. A node in 3-D Torus switch fabric

card and a 7x7 crossbar(one port for line card and the other six ports for connection to six neighbors), as Fig. 2 shows.

Packets may pass several hops to traverse the switch fabric. There exist two routing styles in 3-D Torus switching fabrics:

Source routing: the path of a packet is determined at its source node.
Distributed routing: the next hop of a packet is determined at each node it passes.

Distributed routing is complex for implementation since lookup operation should be taken on each node a packet passes by. So source routing is usually selected in practice. Two important issues should be considered in designing routing algorithms for 3-D Torus switching fabrics:

- Multi-path for load balance and fault tolerance: taking good use of path diversity of 3-D Torus and distributing the traffic over multiple paths, help to reduce the probability of congestion in the switch fabric under non-uniform traffic and provide more system availability as well.
- Maintain the packets order: disorder of packets in a certain flow may cause performance degradation in current TCP implementation. So it is recommended that routers should maintain packets order in one TCP flow[JCN99].

E-cube routing and random routing algorithms are both representative source routing algorithms which are widely used. In e-cube routing all packets from A to B will traverse the switch through a deterministic path which follows certain dimension order such as XYZ. E-cube routing will not cause packets disorder but cannot carry out loads balancing because only one path among the available paths is utilized to transmit packets, so congestion may occur on certain hot path when heavy traffic arrives. On the contrast, random routing will randomly select a path from all paths to the destination for each packet. Random routing can take good use of the path diversity of 3-D Torus network and balance the traffic well, but it cannot maintain the packets order in one flow which may cause heavy performance decrease in TCP applications.

3.2 DMR routing algorithm

As we described above it seems that the two problems, loads balancing and maintaining packets order, conflict with each other . Here we propose a novel hybrid routing algorithm called DMR (Dimension-order-based Multipath Routing) can solve both problems well.

The essence of DMR routing algorithm is : We need not maintain order among all packets through the switch fabric but only the packets in one flow, so packets that belong to different flows can be transmitted on different paths and would not cause any problem. Select a path for each flow is too complex and too costly in implementation. So we hash all packet flows into six groups and give each group a unique path. The six different paths used to forward packets are determined by the six permutations of the three dimensions respectively (XYZ, YZX, ZXY, XZY, YXZ, ZYX). The information in the packets header can uniquely identify a TCP flow, so they can be used as the key for the hash function.

It is not difficult to design a hash algorithm using packets header information that can divide all incoming traffic into six approximately even groups. In practice the hash function is implemented on each line card after the lookup operation. Each packet is assigned with one of six available paths based on the hash result. A packet is attached with a route label and then injected into the switching fabric. At each of the following nodes the packet pass by, it would be switched only based on the route label information and no more lookup operation is needed.

DMR routing algorithm use flow label as key to spread traffic flows over six equivalent paths based on dimension orders, so traffic loads can be balanced well if hash algorithm is designed properly. At the same time, packets order in a certain flow is preserved because they traverse the switch fabric along the same path.

3.3 Performance simulation and analysis

We have evaluated the performance of the proposed algorithm in comparison with two widely used routing algorithm, e-cube routing and random routing using a simulation approach. We carry out simulation using the SIM simulator developed by Stanford University. Original SIM simulator is designed for single-stage switching fabric simulation. We modified it to adapt the 3D Torus switch environment. The simulation is based on a 128 (8x4x4) nodes 3D Torus switching fabric and under burst traffic with burst length equal to 50 and 100. We think that burst traffic is more similar with practical network traffic than uniform i.i.d. traffic. The simulation has been run under traffic loads of 50%, 60%,70%,80% and 90% respectively. The performance is measured by an average packet delay. The simulation is shown in Fig. 3.

The simulation results show that the performance of DMR routing algorithm is approximate to random routing algorithm and much better than

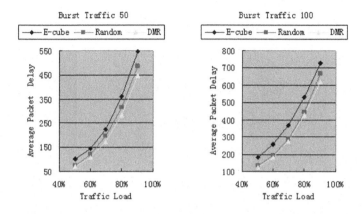

Fig. 3. Simulation result in 3-D Torus

e-cube routing algorithm. The performance promotion of DMR and random routing is up to utilizing multiple paths for loads balancing. At the same time DMR routing algorithm is better than random routing algorithm because it can maintain packets order in one TCP flow while random routing algorithm cannot.

The DMR routing algorithm proposed to be used in 3-D Torus switching fabrics can balance traffic loads over multiple paths to achieve high throughput and system availability. At the same time it can maintain packets order in one TCP flow. The DMR routing algorithm makes 3D Torus switching fabrics practical and efficient.

Acknowledgements The work has been co-supported by National Natural Science Foundation of China Under grant NO. 90104001.

References

[CA01] Coffman, A.G., and A.M. Odlyzko. "Growth of the Internet." Optical Fiber Telecommunications IV Journal. July 2001

[WIL02] William J. Dally, "Scalable Switching Fabrics for Internet Routers", http://www.avici.com/technology/whitepapers/TSRfabric-WhitePaper.pdf

[ZL02] Zhigang Sun and Lei Jia "Using MPP Interconnection Network In terabit RouterDesign", Computer Engineering and Science, June. 2002

[JCN99] J. Bennett, C. Partridge, and N. Shectman. "Packet Reordering is not Pathological Network Behavior", IEEE/ACM Trans. on Networking, 1999, vol 7, No.6: 789-798

Multiagent-based Web Services Environment for E-commerce *

Tianqi Huang[1], Xiaoqin Huang[1], Lin Chen[1], and Linpeng Huang[1]

Department of Computer Science and Engineering, Shanghai Jiao Tong University, Shanghai 200030, China huangxq@sjtu.edu.cn, huang-lp@cs.sjtu.edu.cn

The use of Web services on the World Wide Web is expanding rapidly, as the need for application-to-application communication and interoperability grows on distributed computing systems. It is essential to find the mechanism that allowing customers and providers to communicate and perform e-commerce transactions in a fluent and precise manner, minimizing the time normally spent in finding useful information, which tends to be quite high. The agent-based computing introduces an important new paradigm for the implementation of distributed applications in an open and dynamically changing environment. We present how agent technology can be used to improve electronic commerce environment based on Web Services. Compare to other electric commerce platform, it has the advantages of stability, efficiency and personality.

1 Introduction

The use of Web services on the World Wide Web is expanding rapidly, as the need for application-to-application communication and interoperability grows on distributed computing systems. The agent based computing introduces an important new paradigm for the implementation of distributed applications in an open and dynamically changing environment. One very interesting application for agent is Internet computing such as web-based systems and electronic commerce.

One of the most important things in actual electronic commerce is to achieve an accurate information interchange between participating entities (typically customers and providers), using any available communication infrastructure [FKB03]. The current networking environment with all the Internet and Intranets communication services make possible the access to a great

---

*This paper is supported by Shanghai Information Grid Project, National 863 Program 2001AA113160 and SEC E-Institute: Shanghai High Institutions Grid project 200308.

amount of information [IM02]. Nowadays, it is essential to find a mechanism that allow customers and providers to find each other and perform e-commerce transactions in a fluent and precise manner, minimizing the time normally spent in finding useful information, which tends to be quite high. This kind of mechanism is provided by the so called mediation platform [MA99] that typically includes the set of duties which are supported to rule every transaction associated with e-commerce, and are oriented to decrease the distance between customers and providers from a commercial point of view.

Currently, available systems that provide integration between multiagent and Web servers are few [FKB03]. Research efforts, different mechanisms for the close integration of agent platforms in existing infrastructures of web services based primarily on Servlet and JavaBeans Specifications. In contrast, our implementation is based on the web services model and particular on the SOAP protocol, UDDI, WSDL as used on the World-Wide Web. Again, this approach departs from the currently available systems because it adds multi-agents functionalities on top of already up and running web servers. The rest of the paper is organized as follows: Section 2 describes the overview of web services. Section 3 discusses agents and multi-agent. Section 4 is the framework of Tele-portal. Section 5 is about using multi-agent ideas to develop Tele-portal system and Section 6 gives the conclusion.

2 Overview of Web Services

A web service is a platform and implementation independent software component that can be: described using a service description language, published in a registry of services, discovered through a standard mechanism, invoked through a declared API, and compound with other services. The three roles of the Web service model are as follows:

- The service provider: The entity that hosts one or more Web services. The service providers publish their services in the service registry.
- The service requester: The entity that invokes the Web services using SOAP. A service requester seeks the services in the service registry.
- The service registry: The entity where the service definitions and binding information are kept and made available for searching. The service provider publishes its services in the service registries.

A web service has special behavioral characteristics: XML-based, Loosely coupled, coarse-grained, ability to be synchronous or asynchronous, supports remote procedure call (RPC), supports documents exchange. Figure 1 shows service-oriented architecture.

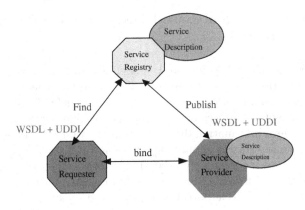

Fig. 1. A service-oriented architecture

3 Agents and Multiagent

Agent based systems facilitate the deployment of a widely distributed architecture, with high capabilities for communication and negotiation among all the components. From a not formal point of view, an agent is a piece of software that receives and store information from the environment, is allowed to make decision, may act in a proactive way and is able to communicate with other agents for, on one hand, asking them to do any activity they can perform or, on the other hand, doing one of its own activities for other agents.

The notion of agent-based software interoperability is based on the idea of a loosely coupled collection of agents that can cooperate to achieve a common goal. Each individual agent is presumed to be a specialist for a particular task, and the expectation is that, just as is in the sphere of human engineering, complex projects can be undertaken by a collection of agents, no one of which has the capability of performing all the required tasks of the project.

The development of agent-based systems offers a new and existing paradigm for production of sophisticated programs in dynamic and open environment, particularly in distributed domains such as web-based systems and electronic commerce. The multi-agent paradigm is an extension of distributed objects, exhibiting features such as active threading, run-time code mobility with autonomous navigation, and knowledge-based inter-agent communication. These characteristics favor a uniform implementation of essential mobile computing services such as multicast communication, intelligent fault-tolerant routing, proxy server/client handling, pessimistic and optimistic data replication management, and multi-level security models.

4 Framework of Agent-based Distributed Information System Tele-portal

Tele-portal is an agent-oriented and service-oriented environment for deploying dynamic distributed system. It realizes a service-agent programming model, which is a combination of two concepts in the field of distributed computing: the concept of services and multiagent. In Tele-portal, mobile agents are used to retrieve information distributed across a large heterogeneous network such as the Internet to support applications, and service interface agent are used to wrap services. Closely related to interface agents are information agents. The main function of these agents is to manage information from many different sources [And]. Tele-portal is implemented by such software support environment: operation system is Sun Solaris8 or up version, OLTP, Apache SOAP Tomcat, J2EE platform, oracle8i database and Bea Weblogic server. The platform can be easily deployed on Apache Tomcat web servers using the Apache SOAP library and compromise a mobile agent Application Programming Interface (API) and a collection of Java Server Pages (JSPs) and Enterprise JavaBeans (EJB) for its management. The platform was designed having in minded the extensibility, interoperability and flexibility.

5 Using Multi-agents Ideas for Tele-portal System

The notion of agent-based software interoperability is based on the idea of a loosely coupled collection of agents that can cooperate to achieve a common goal. Each individual agent is presumed to be a specialist for a particular task, and the expectation is that, just as in the sphere of human being engineering, complex projects can be undertaken by a collection of agents, no one of which has the capability of performing all the required tasks of the project [Sun]. In addition, if the system has open agent architecture, then improved models can replace individual agents, thereby enabling the system to improve gradually, grow in scope, and generally adapt to changing circumstance.

In our portal system, portal reverse proxy is a multiagents includes a lot of agents to perform the tasks. As showing in the above component architecture, each function model is an agent to realize the function.

The core part is also composed of many multi-agents, such as presentation function module, logic module and entity module. Each module is composed of a lot of software agents. For example, the maintenance manager is composed a lot of software agents, including operation management, system management, service management and operation data management. Each agent realizes different functions separately.

The approach taken by the Tele-portal project is to harness the potential to be derived from combining various information sources by employing a distributed collection of collaborating software agents. Certainly at the abstract modeling level, the notion of using cooperating agents has several attractive features [FKB03]:

- Using a collection of problem-solvers makes it easier to employ divide-and-conquer strategies in order to solve complex, distributed problems. Each agent only needs to process the capabilities and resources to solve an individual, local problem.
- The mental image of autonomous, human-like agents facilitates the mapping of real-world problems into a computational domain.
- The idea of several agents cooperating to solve a problem that none could solve individually is a powerful metaphor for thinking about various ways that individual elements can be combined to solve complex problems.

In a word, our project Tele-portal deals with complex cooperating systems based on the agent concept. Agents are components that aim at conveying models inspired from the real life. Such an approach is put forward as a new generation model for the engineering of complex cooperating systems.

6 Conclusion

In this paper we present an agent-base complex system. We also provide the ideas to develop complex software system by integrate a lot of agents to implements some function. This main idea is decomposing the problem in terms of autonomous agents that can engage in flexible, high level interactions. The main contribution of this work is that it provides a methodology and prototype implementation of an electronic commerce platform. We suggest that this approach has the flexibility necessary to provide.

References

[FKB03] Foukarakis,I.E., Kostaridis,A.I., Biniaris,C.G., et.al.: Implementation of a Mobile Agent Platform Based on Web Services. MATA 2003. Lecture Notes in Computer Science, Vol.2881. Springer-Verlag, Berlin Heidelberg New York 190-199 (2003)
[IM02] Irene Sygkouna, Maria Strimpakou, et.al.: Seamless Incorporation of Agents in an E-Commerce Intermediation Platform. MATA 2002. Lecture Notes in Computer Science, Vol.2521. Springer-Verlag, Berlin Heidelberg New York 292-301 (2002)
[MA99] Martin Bichler, Arie Segev.: A Brokerage Framework for Internet Commerce. Distributed and Parallel Databased, vol.7, Number 2. 133-148 (1999)
[Sun] Sun Microstems Inc.: The Servlet Specification 2.3. http://www.javasoft.com/servlet
[And] Andrew S.T.: Distributed Systems Principles and Paradigms. 173-180

Linux Cluster Based Parallel Simulation System of Power System*

Ying Huang[1] and Kai Jiang[2]

[1] Zhejiang University, Hanzou, P.R. China
[2] Shanghai Supercomputer Center, Shanghai P.R. China kjiang@ssc.net.cn

Summary. In research on power system relay protection, a lot of computer simulations must be done for various system models, running conditions and fault conditions. Because of the huge computation requirements, the simulations would cost quite much time. The paper describes a Linux cluster based parallel simulation system of power system developed by the authors, and shows good performance-cost ratio through using existing PC hardware and network resources. The parallel software is based on MPI parallel programming model. Computation examples indicate that the parallel simulation system improves the computation speed significantly.

keyword: Power system, EMTP, Linux, cluster, MPI

1 Introduction

In research on power system relay protection, a lot of computer transient simulations must be done. For example, in research on filter algorithm, the filter performance of algorithm must be inspected of various system models, running conditions and fault conditions; in research on fault phase selection, the verification of phase selection principle must be done through computer simulations of various conditions. Traditional power system computer transient simulation program can only perform simulation for one running condition and fault condition, and can't repeat performing simulation for various conditions automatically. If various simulations are done through modifying the input file of traditional simulation program by hand, it will result in too much work, and cannot include all conditions. A program could be developed to modify the input file and then call the traditional simulation program, which would make the computer simulations for various conditions automatically. But if all

*This work is supported by The National Natural Science Foundation of China (Grant number:90412010) and The Science & Technology Committee of Shanghai Municipality Key Technologies R&D Project (Grant number:03dz15026)

the computation is done on one PC, the computation capability is too limited to satisfy the huge computation requirements. To resolve the problem above, a Linux cluster based parallel simulation system of power system is developed.

2 Linux Cluster Configuration

The Linux cluster can be constructed by purchasing commercial products, but the cost is expensive, so, we construct the cluster with existing PC hardware and network device. Figure 1 shows the configuration of Linux cluster.

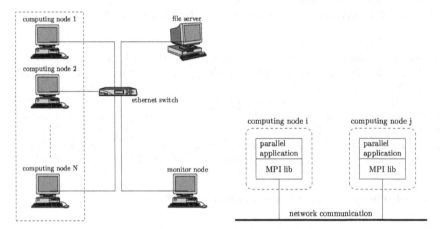

Fig. 1. Cluster system configuration   **Fig. 2.** MPI programming model

In figure 1, the Linux cluster is composed of several computation nodes, a file server and a monitor node, they are connected through high speed ether-net, so, the ethernet switch or hub is required too. The function and config-uration of various node is as follow: *1. Computing node.* Computing node is the computation component of cluster. It uses the existing PC hardware and Linux operating system. Linux is the excellent server and computing platform, but now has some shortcoming in its usability and software supporting. So most PCs of academic institute in China are installed with "Windows". It is impractical to install Linux operating system on all these PCs. The solution proposed by the paper is not installing Linux on hard disk of these PC, but us-ing "Linux operating system on CD (Linux Live CD)" to perform computing using CPU, memory and network device on PC hardware. "Linux Live CD" means, creating a bootable Linux CD, using it boot PC, so that, the Linux operating system in CD is used instead of the Windows operating system in hard disk. By that, any PC with network and CD drive can be a cluster com-puting node without any specific configuration and software installing. *2. File server.* The function of file server in cluster is storing the result of computing. Because there are only large capacity read-only storage devices (CD) but no large capacity read-write storage devices in computing nodes, the result of computing must be stored in file server. File server is a PC with Linux installed, and it is the only PC in cluster in which Linux must be installed. It provides storage and file service through NFS server software. *3. monitor node.* The function of monitor node is monitoring the conditions of every computing node and file server, so that the emergency (such as hardware and software fault) of cluster can be dealt with in time. In monitor node, Linux or Windows can be installed. It uses telnet tool to login file server and every computing node, performs the monitoring though running "top" software in telnet session, which can display the usage of various computing resources such as CPU, memory etc dynamically, and the dispatching of these resources among every computing task. *4. network device* Network performance is crucial to cluster system, so 100M ethernet with ethernet switch is chosen.

3 Parallel Simulation Software

3.1 Introduction to MPI Programming Model

The parallel simulation software is based on MPI (Message Passing Interface) parallel programming model, which performs simulation computing for various running and fault conditions automatically. The model expresses the network communication with "Message", and exchanges information, synchronizes each other, controls execution through passing message among every parallel executing component. The model provides flexible control and expresses method on parallel programming. The implementation of MPI is to call to a message library. The parallel software developed by the paper uses one of the MPI implementation: MPICH. The structure of the parallel simulation software with MPI parallel programming model is as figure 2.

With MPI, the parallel software is separated from details of network communication, so that, the complexity of parallel programming is reduced and the efficiency of programming is increased.

3.2 Parallel Simulation Software

The main modules of parallel computing software is as figure 3.

Software is composed of electric magnetic simulation module (EMTP), system model, running conditions, fault conditions set module, EMTP input file analysis and control module, EMTP result file analysis and control module, storage access module and parallel control module. The function and implementation of every module is as follows: *1. Electric magnetic simulation module (EMTP).* The module performs power system transient simulation of

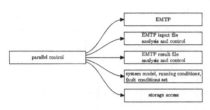

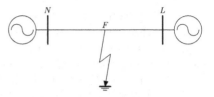

Fig. 3. Modules of software system

Fig. 4. The model of emulation calculation

one system model, running condition and fault condition. There are many existing programs that can perform the function. The module chooses popular ATP-EMTP program. *2. System model, running conditions, fault conditions set module.* The module records a set of power system models, and defines a set of running conditions and fault conditions for every model. All combinations of system models, running conditions and fault conditions construct a condition set of simulation. The main function of the module is managing the condition set, which includes the input files of system models and various conditions; and combining the conditions automatically, which is feed to EMTP input file analysis and control module. *3. EMTP input file analysis and control module.* To perform power transient simulation for various conditions, the input file of EMTP program must be modified according to condition automatically. That is just the function of this module. *4. EMTP result file analysis and control module.* The computing result of EMTP is analyzed by the module, to extract useful information or convert it to other formats. *5. Storage access module.* After being analyzed and processed by EMTP result file analysis and control module, the result of EMTP is stored to file system or database through this module and is used by research on power system relay protection. *6. Parallel control module.* Parallel control module is the core of software parallelization. After parallel software starts running on cluster, many copies of parallel software (so called "parallel process") start running on every computing node of cluster, the parallel control module of every parallel process communicates each other through MPI, performs the simulation cooperatively.

Parallel control module uses master slave programming module, the process with number 0 is the master process, other processes are computing process. The flowchart of parallel control module of master process and computing processes is as figure 5, the message flowchart is as figure 6.

Process is master or computing process, which is decided according whether the process number of process is 0. The master process gets simulation condition from system model, running conditions, fault conditions set module, sends the simulation condition to idle computing processes, and sends "end" message after all of the simulation conditions are computed. After performing simulation of one condition, computing process sends "computing finished" to the master process to indicate that it is idle, then waits for the message coming from the master process, if message received is "computing condition", it

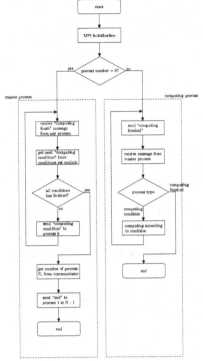

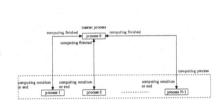

Fig. 5. Flowchart   **Fig. 6.** Message flowchart

performs the simulation continued, if message received is "end", it exits. Performing simulation includes 4 steps: modifying input file of EMTP through calling EMTP input file analysis and control module according to the condition, performing real simulation through calling EMTP, calling EMTP result file analysis and processing module, storing simulation result through calling storage access module.

4 Computation Examples

Reference [HY2003] does some research on optimal model of power system least error square algorithm. In order to compare the filter performance of various models of power system least error square algorithm, computer transient simulation for various running conditions and fault conditions must be done. The power system model used for transient simulation is as figure 4.

There are totally 368640 combinations of conditions. In research, the parallel software developed by the paper is used. Computing with one Pentium IV 1.8GHz PC on all of the conditions costs 20 hours and 5 minutes, computing with 5 Pentium IV 1.8GHz costs just 4 hours and 21 minutes. The computing speed almost improves linearly, the time of computing is reduced greatly.

Conclusions: (1) Linux cluster built with existing resources is good performance/cost and choice for many institutes. (2) Power system parallel simulation improves the computing speed greatly. (3) Linux cluster and MPI will make the research methods depending on large scale scientific computing be applied widely.

References

[HY2003]   Huang Ying, He Ben-teng, Study on the optimal model of Least Error Square Algorithm, Proceedings of 19th work shop on power system automation major of Chinese universities, Chengdu, China, **10**, 2366-2371 (2003)

[LYL2003]   Li Ya-lou, Zhou Xiao-xin, Wu Zhong-xi, Personal computer cluster based parallel algorithms for power system electromechanical transient stability simulation, Power System Technology, **27**, 6-12 (2003)

[MJL1999]   Ma Ji-lan, Fen Xiu-fang, Operating system principal and Linux, Posts and Telecom Press, (1999)

[KNOPPIX]   http://www.knoppix.org

[DZH2001]   Du Zhihui, Parallel computing technology of high performance computing, MPI parallel programming, Tsinghua University Press, (2001)

Pipeline Optimization in an HPF Compiler [*]

Kai Jiang[1], Yanhua Wen[2], Hongmei Wei[2], and Yadong Gui[3]

[1] Shanghai Supercomputer Center, Shanghai, P.R. China kjiang@ssc.net.cn
[2] Jiangnan Computing Technology Institute, Jiangnan, P.R. China
[3] Shanghai Supercomputer Center, Shanghai, P.R. China

Summary. Generally, in scientific applications, parallelism is extracted from loops because they spend the most of execution time. Iterations of such loops are executed in parallel to achieve speedup over the sequential program. However, a number of these scientific applications exhibit recurrences that give rise to data dependencies across processors or nodes. These dependencies tend to slow down parallel execution and sometimes even serialize the loop. This paper proposes pipeline technology to resolve such problems, which breaks the computation into blocks. Each processor performs the computation of a block, which enables the next processor in the pipeline to compute its corresponding block. Once the pipeline is filled, the computation of blocks on different processors proceeds in parallel. We describe the design and implementation of the pipelining in an HPF compiler and show that the computation achieves better parallel performance using our method.

Key word: pipeline, HPF, data-parallel, data dependency

1 Introduction

There are several parallel programming models in common use, such as Shared Memory, Threads, Message Passing, Data Parallel, Hybrid etc. The data parallel is one of the most popular models.

High Performance Fortran (HPF) is the first standard data-parallel language. It hides most low-level programming details and allows programmers to concentrate on the high-level exploitation of data parallelism. The compiler automatically manages data distribution and generates the explicit interprocessor communications and synchronizations needed to coordinate the parallel execution.

---

[*]This work is supported by The National Natural Science Foundation of China (Grant number:90412010) and The Science & Technology Committee of Shanghai Municipality Key Technologies R&D Project (Grant number:03dz15026)

Unfortunately, many important parallel applications do not fit a pure data-parallel model and exhibit recurrences in their problem formulations. These recurrences give rise to cross-processor dependencies. These dependencies tend to slow down parallel execution and sometimes even serialize the loop. These types of dependencies are quite common in scientific codes.

This paper presents pipeline method to resolve the data dependency problem, and implements it in an HPF compiler. The pipeline optimization breaks the computation into blocks, each processor performs the computation of a block, which enables the next processor in the pipeline to compute its corresponding block. Once the pipeline is filled, the computation of blocks on different processors proceeds in parallel.

The rest of this paper is organized as follows. Section 2 discusses the data dependency problem and presents the pipeline method . Section 3 describes the implementation of pipeline optimization in an HPF compiler. Section 4 gives the performance result and Section 5 concludes.

2 Pipeline Optimization for Cross-processor Data Dependency

2.1 Data Dependency Analysis

Typically, in scientific applications, parallelism is extracted form loops because that is where most of the time is spent. Interactions of such loops are executed in parallel to achieve speedup over the sequential program. However, a number of these scientific applications exhibit recurrences that give rise to data dependencies, that means, an update to a data element on one node depends on the value of a data element on another node. These dependencies enforce an ordering on the execution of the iterations in the loop where they occur, sometimes even forcing sequentialization of the execution of the entire loop.

As in figure 1,this is a piece of typical program to describe the data dependency problem

```
DO  k = lbk,ubk
DO  j = lbj,ubj
DO  i = lbi,ubi
Ai, j, k= A(i, j-1, k) + A(i-1, j, k) + A(i-1, j-1, k)
ENDDO
ENDDO
ENDDO
```

Fig. 1. program model of cross-processor data dependency

Assume the data (array A) is tow-dimensional distributed (block, block, *). So there have a cross-processor dependency because A(i,j,k) depends on A(i,j-1,k), A(i-1,j,k) and A(i-1,j-1,k); the distribution of A by I-dimension and j-dimension results in some nodes requiring access to the last row or the last column or the last element of some other nodes in order to update its own fist row or first column or first element. Figure 2 shows the data distribution and dependency of the above loop, which run on the 16 nodes (from node0 to node15).

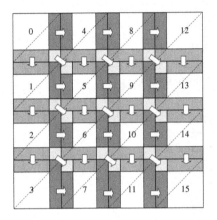

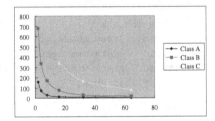

Fig. 2. data dependency of the loop

Fig. 3. Performance of pipeline optimization

From the figure 2, we can find that not all nodes have data dependencies each other. There is no dependency between some nodes, while these nodes that have no data dependency get the data they need, they can execute in parallel.

2.2 Program Transformation with Pipeline Method

Through analyzing the data dependency, partial parallelism is found. If let 16 processors execute the above loop as in figure 2. First it can be computed on node0, after finishing the computation on node0, node0 sends its last row to node1, last column to node4, last element to node5, then node1 and node4 can compute together; after they finish their computation, they send the corresponding data to their neighbor nodes which need the data, then those nodes can execute together.Thus, the partial parallelism is extracted from this loop. This method is called pipeline.

Using pipeline method, this piece of program can be transformed to the pipelined parallel program. According to the different positions where the message statement could be added, there are two kinds of transformation programs, as in figure 4 and 5.

```
if(myid...)recv message(I direction )
if(myid...)recv message(J direction )
if(myid...)recv message(I&J direction )
DO  k = lbk,ubk
DO  j = lbjn,ubjn
DO  i = lbin,ubin
A(i, j, k)= A(i, j-1, k) +
       A(i-1, j, k) + A(i-1, j-1, k)
ENDDO
ENDDO
ENDDO
if(myid...)send message(I direction )
if(myid...)send message(J direction )
if(myid...)send message(I&J direction )
```

Fig. 4. Program Transformation I

```
DO  k = lbk,ubk
if(myid...)recv message(I direction )
if(myid...)recv message(J direction )
if(myid...)recv message(I&J direction )
DO  j = lbjn,ubjn
DO  i = lbin,ubin
A(i, j, k)= A(i, j-1, k) +
        A(i-1, j, k) + A(i-1, j-1, k)
ENDDO
ENDDO
if(myid...)send message(I direction )
if(myid...)send message(J direction )
if(myid...)send message(I&J direction )
ENDDO
```

Fig. 5. Program Transformation II

2.3 New Problem of the Two Transformation

One important parameter in a pipelined program is the amount of work performed by each processor before communicating. This is usually referred to as the block size. In Transformation I, the block size is the whole length of k-dimension, first, the initial node should compute all k-dimension data, while the other nodes are all waiting for it. So there is a long pipeline fill time. In Transformation II, the block size is only 1, pipeline fill time decreases, but message passing is frequent, which increases the communication overhead. Two solutions are both inefficient.

2.4 Pipeline in Block

An efficient solution to the new problem is pipeline in block. That means, splitting up the size of k-dimension (take example for the above loop) into blocks, and let the initial node just compute the block size of k-dimension , and then begin to send messages to its neighbor nodes, meanwhile compute the next block of k-dimension. The other nodes do the same as the initial node.Thus, message passing and computing execute at the same time, which reduces more time than the transform I and the transform II.

According to this method, the loop can be transformed to as figure 6.

```
DO  lk = lbk, ubk, message_length
uk = min( lk+message_length-1, ubk)
if(myid...)recv message(I direction )
if(myid...)recv message(J direction )
if(myid...)recv message(I&J direction )
DO  k = lk, uk
DO  j = lbjn,ubjn
DO  i = lbin,ubin
Ai, j, k= A(i, j-1, k) +
    A(i-1, j, k) + A(i-1, j-1, k)
ENDDO
ENDDO
ENDDO
if(myid...)send message(I direction )
if(myid...)send message(J direction )
if(myid...)send message(I&J direction )
ENDDO
```

Fig. 6. Program Transformation III

```
do 10 k = 2, nz0-1
do 10 j = 2, ny0-1
do 10 i = 2, nx0-1
do 10 m = 1, 5
do 10 l = 1, 5
v(m,i,j,k) = v(m,i,j,k) -
omega* (a(m,l,i,j,k) * v(l,i,j,k-1)
+ b(m,l,i,j,k) * v(l,i,j-1,k)+
c(m,l,i,j,k) * v(l,i-1,j,k)
10 continue
```

Fig. 7. a piece of program of LU

The block size is crucial to the performance of a pipelined parallel program. A large block size decreases the communication overhead but increases node idle time, while a small block size decreases node idle time but increases the communication overhead. It should be be chosen cautiously according to the scale of array and the processors number.

3 Implementing Pipeline in an HPF Compiler

High Performance Fortran (HPF) is an informal standard for extensions to Fortran 90 to assist its implementation on parallel computers, particularly for data-parallel computations.Foremost among these extensions are directives for telling the compiler how to distribute the data and providing assertions that improve the optimization of the generated code. All directives of HPF are of the form !HPF$. Some important directives are PROCESSORS, DIS-TRIBUTE, ALIGN, REDISTRIBUTE, INDEPENDENT, etc.

To deal with the loops which contain the cross-processor dependency, a new compile directive "pipeline" is developed. Its format is as follows:

```
!hpf$  pipeline(i, j) on A(i, j, k)
!hpf$  pipeline(i, j) on A(i, j, k) strip (k,message_length)
```

While the compiler detect the pipeline directive, it is told that the loop behind the directive perhaps contains the data dependency. It passes the loop to the pipeline control program to be dealt with, then the pipeline control program does everything.

First, all information is gathered, such as all arrays in the loop, iteration node, the distribution dimensions, index variables, etc, which are used to make all kinds of examinations later.

Second, analyze data dependency.It has two steps. One is dependency examination, which decide that whether there really has data dependency in the loop. The other is direction examination.

Finally, after making sure that the loop should be pipeline optimized, transform it to the pipelined parallel program.

4 Performance

This section reports the performance of pipeline optimization on one program, LU decomposition. The program has several loops that has data dependencies. Figure 7 shows one of them.

In this loop, the array v is distributed (block, block, *). To optimized it with pipeline method, add the pipeline directive as the following in front of the loop:

```
!HPF$ pipeline(j, i) on v(m, i, j, k) strip(k, blocksize)
```

Our experiment was run on a Massively Parallel Processor system (MPP) ,which is comprised of 96 compute nodes, 16 I/O nodes and 6 access nodes. Each compute node has 4 CPUs for computation and 1 CPU for communication. Its peak performance is 384Gflops. All tests were compiled with the HPF compiler. We choose various block size, different sizes result in different execute time. Figure 3 shows the best test result of this piece of program.

5 Conclusion

The paper analyzes the data dependency problem which is quite common in scientific applications, and presents pipeline method to solve it. And in order to get better efficiency, we improve the pipeline method to pipeline in block. The block size is proved to be very important to the performance, and it is necessary to do more research about it.Furthermore, we implement pipeline optimization in an HPF compiler through developing a new compile directive. Using this compiler, the program of LU decomposition is tested with pipeline optimization, and gets good performance. Finally, we think that the pipeline method can also be used in other parallel program model.

References

[HPFF1993] High Performance Fortran Forum,High Performance Fortran Language Specification, Sci. Prog. special issue.2 1-2 (1993)
[CHK1994]   C.H. Koelbel, D.B. Loveman, R.S. Schreiber, G.L. Steele Jr, M.E. Zosel, The High Performance Fortran Handbook, MIT Press, (1994)
[GRA2000]   Gregory R. Andrews. Foundations of Multithreaded, Parallel, and Distributed Programming. Addison/Wesley, (2000)
[DP1986]   David Padua, Michael Wolfe, Advanced compiler optimizations for super computers, Communications of the ACM, **29** 1184-1201 (1986)
[MHPCC]   http://www.mhpcc.edu/training/workshop/hpf
[LLNL]   http://www.llnl.gov/computing/tutorials/parallel_comp/

Clustering Approach on Core-based and Energy-based Vibrating

Shardrom Johnson[1], Daniel Hsu[1], Gengfeng Wu[1], Shenjie Jin[2], and Wu Zhang[1]

[1] School of Computer Engineering and Science, Shanghai University, Shanghai 200072, China. jshardrom@staff.shu.edu.cn

[2] Global Delivery China Center, Hewlett-Packard Development Company, Shanghai 201206, China.

Abstract. Clustering, focused mainly on *distance-based clustering*, has been studied extensively for many years. To discover clusters with arbitrary shapes, *density-based clustering* methods have been developed. The *Core-based and Energy-based Vibrating Method*, presented in this issue, is a clustering method which improves the *density-based clustering* methods of data mining in some fields. *Density-based clustering* is firstly used to find the *core object*. Then, the *core object* is described by the conception of *energy*. Based on the energy analysis, the peculiar objects is *vibrating* by the interfering energy to weaken their dissimilarity. Therefore, *Vibrating Method*, as one of the *cluster reducing strategy*, can reduce the number of clustering and highlight the correlation among clusters.

Keywords Core-based and Energy-based Vibrating Method, Cluster Reducing Strategy, Density-based Clustering, Distance-based Clustering.

1 Introduction

Clustering, as an unsupervised method, receives a series of unlabelled examples in the space defined by the attribute variable, and organizes them into clusters according to some rules. The target is to make a collection of data objects that are *similar* to one another within the same cluster and are *dissimilar* to the objects in other clusters[FK99][FK98].

Some of the current classical clustering methods calculate the difference between attributes corresponding to different objects, map the difference to a *distance*, and cluster the data objects according to the distances. These *distance-based clustering* methods rarely involve the connection and influence between the data objects and their neighbor[DCL02].

Another group of clustering methods is called *density-based clustering* methods, which have the following philosophy: if the data objects are mapped into an attribute space, they are in high density in the center and low density in the boundary of a cluster[MHX01]. Therefore the clusters and their boundaries can be found up by calculating the variation of the density of the data objects. *Density-based clustering* methods can deal with the arbitrary shape of clusters, and efficiently identify the *noise* data objects, but they will probably create too many clusters and are weak in dealing with the connections between the data objects in the boundary of clusters[EKSX96].

To reduce the number of clusters created, we propose the *Core-based and Energy-based Vibrating Method* which synthesizes the merits of the *distance-based clustering* and the *density-based clustering*.

2 Preliminary Principle

The basic ideas of the proposed *Method* involve a number of new definitions.

Definition 1. *The neighborhood within a radius ε of a given object is called the ε-**neighborhood** of the object.*

Definition 2. *If the ε-neighborhood of an object contains at least a minimum number σ of objects, then the object is called a σ-**core object**.*

Definition 3. *Given a set of data objects, D, we say that an object p is **directly density-reachable** from object q if p is within the ε-neighborhood of q, and q is a σ-**core object**.*

Definition 4. *An object p is **density-reachable** from object q with respect to ε and σ in a given set of data objects, D, if there is a chain of objects $p_1, \ldots, p_n, (p_1 = q) and (p_n = p)$ such that p_{i+1} is directly density-reachable from p_i with respect to ε and σ, for $1 \leq i \leq n, p_i \in D$.*

Definition 5. *An object p is **density-connected** from object q with respect to ε and σ in a given set of data objects, D, if there is an object $o \in D$ such that both p and q are density-reachable from o with respect to ε and σ.*

According to the above definitions, it only needs to find out all the maximal density-connected spaces to cluster the data objects in an attribute space. And these density-connected spaces are the clusters. Every object, which is not contained in any clusters, is considered to be *noise* and can be ignored.

Definition 6. *Suppose x and y be two data objects in a d-dimensional feature space, F^d. The **influence function** of data object y on x is a function $f_B^y : F^d \rightarrow R_0^+$, which is defined in terms of a basic influence function f_B.*

$$f_B^y(x) = f_B(x, y) \tag{1}$$

Theoretically, the influence function can be an arbitrary function that can determine the distance between two data objects in a neighborhood, only if the influence function has the reflexive property and the symmetric property[DCL02], such as a *Gaussian Influence Function*

$$f_{Gauss}(x,y) = e^{-\frac{d(x,y)^2}{2\sigma^2}}. \tag{2}$$

Definition 7. *Given a d-dimensional feature space, F^d. The **density function** at a data object $x \in F^d$ is defined as the sum of all the influence to x from the rest of data objects in F^d.*

$$f_B^D(x) = \sum_{i=1}^{n} f_B^{x_i}(x). \tag{3}$$

For example, the density function that results from the Gaussian influence function (2) is

$$f_{Gaussian}^D(x) = \sum_{i=1}^{n} e^{-\frac{d(x,y)^2}{2\sigma^2}}. \tag{4}$$

According to **Definition 6** and **Definition 7**, it is easy to convert the task of clustering the data objects by finding out the maximal density connection spaces in a complex feature space to the influence function and density function model.

3 Core-based and Energy-based Vibrating Method

Given a d-dimensional feature space, F^d. The data objects set $X = \{X_1, X_2, \ldots, X_n\} \subset F^d$ has n data objects, and each data object $X_i = \{x_{i1}, x_{i2}, \ldots, x_{it}\}$ has t attributes at most, for $1 \leq i \leq n$. And some definitions are presented as follows.

Definition 8. *Core, containing all of the core object, is a model set to describe the given cluster's data objects, $C = \{c_1, c_2, \ldots, c_k\}$.*

In principle, we can choose the arbitrary *core object* as the *core*, such as the local maxims of the cluster's data objects:

$$C = \max_{X_j \in Cluster} x_{ji}(1 \leq i \leq d). \tag{5}$$

Definition 9. *Energy represents the difference among the data objects, which is based on the core. The relative energy $E_{ij} = d(X_i, X_j)$ is the difference between the data object X_i and X_j. The absolute energy for data object X_i, $E_i = d(X_i, C)$, is the relative energy between the data object X_i and the Core of its cluster.*

If the initial clusters is given by the *density-based clustering* methods, we can take over the influence function (**see Definition 6**) and density function (**see Definition 7**) to calculate the *energy* of the data objects by subtracting the value of their density function to the value of the **Core**'s :

$$E_i = |f_B^D(X_i) - f_B^D(C)| = f_B^D(C) - f_B^D(X_i). \tag{6}$$

3.1 The Process of the Method

Core-based and Energy-based Vibrating Method selects the **Core(see Definition 8)**, which is the representative of the cluster's data objects, as the reference to calculate each data objects' **Energy(see Definition 9)**. And then, according to some specific rules, the peculiar data objects is *vibrating* by the interfering energy to separate the data objects, merge similar clusters and highlight the connections between the data objects. We will describe the process of the method as follows: **Step 1: Initial Clustering**. The purpose of initial clustering is to get an initial understanding of the data objects and find out the *core* of the clusters. Firstly, we compute the value of each data object's density function by *Equation (3)*. Secondly, base on the *Gradient* of the data objects' density function, we divide the data objects into several parts (clusters). Among each part (cluster), we choose the data objects, which have the maximal density function value, as the core of the cluster.

Fig. 1. The Distribution of The Data Objects

Figure 2(L) and Figure 3(L) show the distribution of the data objects(as Figure 1) when using the *Square Wave Density Function* and the *Gassian Density Function* respectively. The position of the data object is determined by its attributes and the height of the data object is determined by its density function or its energy.

Step 2: Vibrating. In this step, we calculate the energy of the data objects at first. As described in **Definition 9**, the energy of each data object is the difference between the value of its density function and the value of the *Core* by *Equation (6)*. Next, we select some peculiar data objects as the vibrated ones, based on the energy analysis. And then, we adapt *the vibrating value (Q)* to make all of the peculiar data objects vibrating, and attain the new distribution of the data objects.

Step 3: Evaluation. After the vibrating, we should give the evaluation about the candidate result. The evaluation method could be classical support or confidence calculation[DCL02]. For the unsatisfied result, we will redo the process by step 1 and step 2, or give up.

Evaluation is the Key in this step. On one hand, the result, submitted with large support and confidence, will probably be a fact that is known to everyone. On the other hand, after too much vibration, some data objects with low connection will be in the same cluster, and the result without evaluation will never be meet to the requirement of the end user.

To summarize, the *Core-based and Energy-based Vibrating Method* can be described as below:

Algorithm of Core-based and Energy-based Vibrating Method:
(1) Calculate the density of all the data objects. (2) Cluster the data objects according to their gradient. (3) Find out the core of each cluster. (4) Calculate the energy of each data object. (5) Select the peculiar data objects. (6) Vibrate the peculiar data objects. (7) Broadcast and spread the interfering energy. (8) Evaluate the result and make the decision.

3.2 Experiments and Performance Evaluation

Suppose that we use the *Square Wave Density Function* and the *Gassian Density Function* respectively. The comparison of the result is shown in Figure 2-3.

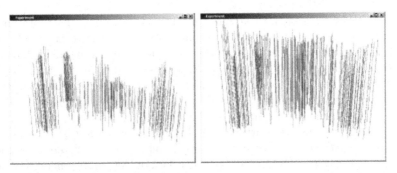

Fig. 2. Before(L) and After(R) Vibration Using the *Square Wave Density Function*

It shows in Figure 2-3 that the clusters with clear boundaries have somewhat merged after 3 levels of vibration. From the comparison of Figure 2(L) to Figure 2(R) and Figure 3(L) to Figure 3(R), we can figure out that those clusters, which are close to each other but are not density connected (refer to

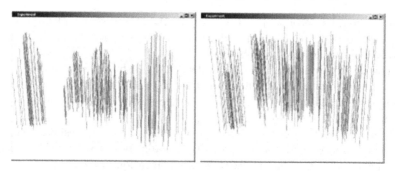

Fig. 3. Before(L) and After(R) Vibration Using the *Gassian Density Function*

Definition 5), have somewhat mixed together. And the gradient of the data objects' density function in the boundaries has been decreased.

4 Conclusion

The proposed *Method* makes up the shortage of the *Aggregate Hierarchical Cluster Method*[Zhe97]. It can be applied only to some parts of the data space (especially the boundaries of the clusters) so as to gather together the data objects in the boundary and merge the similar clusters. And, it can find out the *bridges* between the data objects and clusters by recording the path of the spreading of the vibration. Moreover, the proposed *Method* is reversible. We could change the *direction* of the vibration (to the core or away from the core) to recall the steps at any time, in order to change its *developing direction* after changing of the parameters.

Since the proposed *Method* relies mostly on spreading the vibration to adapt the energy of the data objects, and assimilate the data objects, the result is apparent in the space area with high data density, while in that with low data density, its affect is weak. And these results in the high density space domain (close to the *Core* of the cluster) have been different adoption in energy from that in the low data density space domain. Therefore, the data objects in the boundaries of cluster are easy to get closer, and spur the mergence of the clusters. Furthermore, the vibrating value (Q) is the key of the proposed *Method*. Since every data object has effect on its ε-neighborhood, the vibration will be spread to all the other data objects in its ε-neighborhood. But the vibration will be decaying with the increasing of distance.

In the current definitions, the energy of the data objects (**Definition 9**) only rejects the different between the data objects and the core of the cluster. And it is only a static rejection. While in data warehouses, there always exists such data objects that they change with time frequently. It can be defined as "motion energy" in the future so as to find out more instructive information.

References

[Bha99] Bhavani, M.T.: Data Mining: Technologies, Techniques, Tools and Trends. CRC Press LLC (1999)

[FSAS98] Fei, C., Stephen, F., Andreas S.W., Steven R. W.: Modeling financial data using clustering and tree-based approaches. In: Ebecken, N. F. F. (ed) Data Mining. WIT Press, COPPE/UFRJ Federal University of Rio de Janeiro (1998)

[JM00] Jiawei, H., Micheline, K.: ata Mining: Concepts and Techniques. Morgan Kaufmann Publishers (2000)

[MHX01] Manoranjan, D., Huan, L., Xiaowei, X.: '1 + 1 > 2': Merging distance and density based clustering. In: Proc. The 7th International Conference on Database Systems for Advanced Applications. DASFAA'2001, HongKong (2001)

[AD98] Alexander, H., Daniel, A.K.: An efficient approach to clustering in large multimedia databases with noise. In: Proc. 1998 Int. Conf. Knowledge Discovery and Data Mining. KDD'98, New York (1998)

[Zhe97] Zhexue, H.: Clustering large data sets with mixed numeric and categorical value. In: Proc. The 1st Pacific-Asia Conference on Knowledge Discovery and Data Mining. World Scientific, Singapore (1997)

[EKSX96] Ester, M., Kriegel, H.P., Sander, J., Xu, X.: A density-based algorithm for discovering clusters in large spatial databases. In: Proc. 1996 Int. Conf. Knowledge Discovery and Data Mining. KDD'96, Portland (1996)

Design and Implementation of Detection Engine Against IDS Evasion with Unicode

Dongho Kang, Jintae Oh, Kiyoung Kim, and Jongsoo Jang

Security Gateway System Team, Electronics and Telecommunications Research Institute, 161 Gajeong-Dong, Yuseoung-Gu, Daejeon, 305-350, Korea dhkang {showme, kykim, jsjang}@etri.re.kr

Summary. The fast extension of inexpensive computer networks has increased the problem of unauthorized access and tampering with data. As a response to increased threats, many Signature-based Intrusion Detection Systems have been developed. Current NIDSs are barely capable of real-time traffic analysis and detecting IDS evasion techniques on Fast Ethernet links. Gigabit Ethernet has become the actual standard for large network installations. Therefore, there is an emerging need for enhanced security analysis techniques that can keep up with the increased network throughput. This paper introduces the whole architecture of our system designed to perform intrusion detection on high-speed links and proposes the efficient Detection Engine against IDS evasion techniques that is run by FPGA logic.

keywords : Unicode Attack, IDS Evasion technique, Gigabit IDS

1 IDS Evasion Techniques

Although there are various categories of intrusion detection systems, evasion techniques have also become sophisticated. The basic idea behind evasion is to fool the IDS into seeing different data than what the target host will see, thus allowing the attacker to slip through undetected. Some IDS evasion techniques are:

1.1 Path obfuscation

An attacker can use a Web browser's URL to enter a path statement in order to ac-cess a file on the Web server with the intention of causing damage, or to retrieve sensitive information. Normally, the path statement would be incorporated into the attack signature, and the attack could be recognized. However, the attacker could alter the URL's path statement to appear different to evade detection and cause harm. For ex-ample, "/winnt/. /. /. /test" is the same as "/winnt/test," but the signatures don't match, and so an IDS trained to alert on "/winnt/test" will miss this attack.

1.2 Unicode Encoding

Hex encoding can be used to represent characters in URLs. This type of encoding has been heard about by most people who deal with security. The famous IIS exploit that used this encoding method is an example of what a Unicode request looks like.

//127.0.0.1/scripts/..%c0%af../winnt/system32/cmd.exe?+c+dir+c:

Directory traversal exploits use strings like ".. /.. /.. / ". Most IDSs have signatures to detect this, but attackers replace the "/" with the Unicode equivalent, "In fact, many variations of the same string could be created.

2 Gigabit-IDS Architecture

Hardware based and high performance Security Gateway System (SGS) provide security functions such as firewall, IDS, Rate-limiting, and traffic metering which are implemented on two FPGA (Xilinx Vertex II Pro) chips in each security board module. Security board also has embedded CPU MPC860 that embedded Linux OS operating in. Total five security boards can be installed to SGS. Internal modules for detection operation is consists of Unicode conversion and Unicode attack detection function, pattern matching function, Rule management function.

-   Unicode conversion and Unicode attack detection function is to decode and decide whether attack is or not against Unicode string in packet.
-   Pattern Matching module matches the incoming packet with fixed field patterns based on packet header information that is easily examined by fixed size and fixed offset. Briefly, it performs pattern matching for detecting intrusions.
-   Rule Manager manages the ruleset that is required for intrusion detection.

Through the interoperability of these components, SGS analyzes data packets as they travel across the network for signs of external or internal attack.The major functionality of SGS is to perform the real-time traffic analysis and intrusion detection on high-speed links. Therefore, we focus on effective detection strategies applied FPGA logic and kernel logic. Fig 1. shows Intrusion detection chip architecture.

3 Unicode Attack Detection Unit

Packet data is passed to the units through a 16-bit bus. The header information of each packet is compared with the predefined header data. If the header information matches the rule, the payload is sent to Unicode pattern match

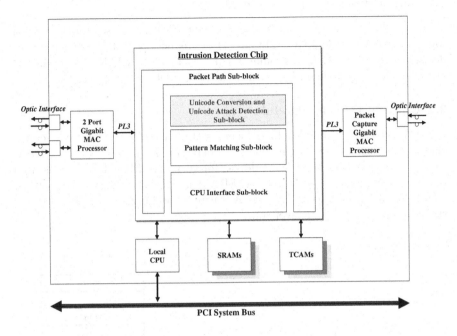

Fig. 1. Gigabit-IDS Architecture

units included detection algorithms against IDS evasion techniques. However, unlike the software implementation, all the rule chains are matched in parallel to achieve predictable high performance. Fig.2 is the block diagram of our architecture.

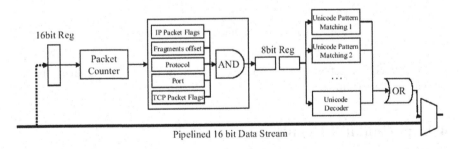

Fig. 2. Paralled Datapath of Unicode Attack Detection Unit in Reconfigurable Hardware

3.1 Path obfuscation Detection Module

To detect Path Obfuscation technique, PODM(Path Obfuscation Detection Module) extracts URI field's value in HTTP request packet. And then makes decision whether URI field's value is valid or invalid. If URI field's value is out of value range defined URI syntax in RFC 2396.

3.2 Unicode Decoder Module

UDM (Unicode Decoder Module) inspects the packet's payload for Escaped characters and converts them back to their ASCII Values. At first, UDM receive input packets and make address for searching SRAM. If **** is used as address for SRAM searching key. SRAM is consists of ASCII characters. UDM get ASCII characters from SRAM using matching address. Fig.3 shows the process for Unicode conversion.

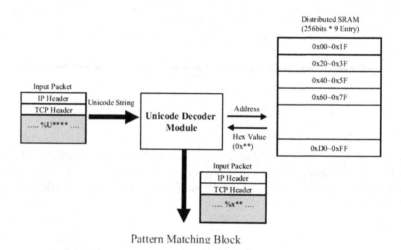

Fig. 3. The precess for Unicode Conversion

Converted Packet is sent to Pattern Matching Block for Signature based intrusion Detection.

4 Experimental Result

We have used IXIA Traffic Generator/Analyzer to generate and transmit packet to SGS. Security board of SGS has two gigabit interface fiber ports and two ports are connected to IXIA Traffic Generator. One port is used to receive packets from IXIA and the other port is used to send packets to IXIA again after processing. We send IDS evasion packets with background traffic to SGS. Table 1 is the experimental result where D/D=Decode/Detect.

Table 1. The Result of IDS Evasion Test Under Load

Attack Type	0Gbps	1Gbps	1.5Gbps	2Gbps
URL Encoding	D/D	D/D	D/D	D/D
/./ Directory Insertion	Decode	Decode	Decode	Decode
Premature URL Ending	Decode	Decode	Decode	Decode
Fake/TAB Parameter	Decode	Decode	Decode	Decode
Case Sensitivity	Decode	Decode	Decode	Decode
Windows / delimiter	Decode	Decode	Decode	Decode
Session Splicing	Decode	Decode	Decode	Decode

5 Conclusion and Future work

In this paper, we have presented hardware based Detection Engine against IDS evasion with Unicode. One of important requirements of security system is on high performance. If not satisfied with network speed, those may not be used. We have developed security board which provides Intrusion Detection function that support total 2gigabit throughput (two gigabit ports). The key idea is to apply unicode conversion and detection program to Intrusion detection module to detect IDS evasion techniques in wire speed. Our future work includes developing De-fragmentation and TCP Reassembly functions. Those functions are one of detection functions against IDS evasion techniques.

References

[KIM04] B.-K. Kim, Ik-K. Kim, K.-Y. Kim, J.-S.Jang: Design and Implementation of High Performance Intrusion Detection System, ICCSA (2004).

[PTA98] T. Ptacek and T. Newsham: Insertion, Evasion, and Denial of Service: Eluding Network Intrusion Detection, Secure Networks Inc (1998).

[Roe99] M.Roesch: Snort-Lightweight Intrusion Detection for Networks, USENIX LISA '99 (1999).

[PAX99] V.Paxson, Bro: a system for detecting network intruders in real-time, Computer Networks31, 23-24 (1999).

[Den87] D.Denning: An Intrusion Detection Model, IEEE Trans.on Software Engineering (1987).

[Kem02] R. Kemmerer and V. Giovanni: Intrusion Detection: A Brief History and overview, IEEE Security and Privacy, 27-30 (2002).

[Val00] A. Valdes and K. Skinner: Adaptive, Model-based Monitoring for Cyber Attack Detection, RAID (2000).

[Eric01] Eric Hacker: IDS Evasion with Unicode, SecurityFocus Infocus(Online), Available: http://www.securityfocus.com (2001).

Derivative Based vs. Derivative Free Optimization Methods for Nonlinear Optimum Experimental Design

Stefan Körkel[1], Huiqin Qu[2], Gerd Rücker[3], and Sebastian Sager[1]

[1] Interdisciplinary Center for Scientific Computing, University of Heidelberg, Im Neuenheimer Feld 368, D-69120 Heidelberg, Germany
[2] Intelligent Information Processing Laboratory, Fudan University, Shanghai, China
[3] Deutsche Börse AG, 60485 Frankfurt am Main, Germany

1 Introduction

An important task in the procedure of the validation of dynamic process models is nonlinear optimum experimental design. It aims at computing experimental layouts, setups and controls in order to optimize the statistical reliability of parameter estimates from the resulting experimental data. The models we consider usually arise from applications in chemistry or chemical engineering and consist of nonlinear systems of differential equations, e.g. ordinary differential equations (ODE) or differential algebraic equations (DAE). Here we sketch our numerical approach (implemented in our software package VPLAN) which is based on sequential quadratic programming with a tailored derivative computation and compare it to the easy to implement but much less powerful derivative free approach.

2 Problem Statement

In many practical and industrial applications dynamic processes play an important role. To simulate, understand, control and optimize these processes, they are described by dynamic mathematical models, usually systems of ordinary or partial differential equations. We concentrate on differential algebraic equations (DAE):

$$\dot{y} = f(t, y, z, p, q), \quad 0 = g(t, y, z, p, q)$$

where $t \in [t_0; t_{end}] \subset \mathbb{R}$ is the time and $x = (y, z) : [t_0; t_{end}] \to \mathbb{R}^{n_x}$ are the system states. The values of the quantities $p \in \mathbb{R}^{n_p}$, the parameters, are known only roughly. The controls q describe the layout, the setup and the processing of experiments, we distinguish between time independent control variables $q_1 \in \mathbb{R}^{n_{q_1}}$ and time dependent control functions $q_2 : [t_0; t_{end}] \rightarrow \mathbb{R}^{n_{q_2}}$, $q = (q_1, q_2)$.

To estimate the parameters p from experimental data, we minimize the weighted sum of the squares of the residuals between measurement values η_i and model responses $h_i(t_i, x(t_i), p, q))$, $i = 1, \ldots, M$:

$$
\begin{aligned}
&\min_{p,x} \sum_{i=1}^{M} w_i \cdot \frac{(\eta_i - h_i(t_i, x(t_i), p, q))^2}{\sigma_i^2} \\
&\dot{y} = f(t, y, z, p, q) \\
&0 = g(t, y, z, p, q) \\
&0 = d(x(t_0), \ldots, x(t_f), p, q)
\end{aligned}
\tag{1}
$$

The quantities σ_i are the variances of the measurement errors. The weight $w_i \in [0; 1]$ specifies if measurement i is actually carried out or not. In parameter estimation we can assume that all the weights are 1, later in experimental design we will use the w_i as variables to choose the placement of the measurements. For the numerical solution of this kind of problems, we use the boundary value problem optimization approach suggested by Bock [2].

The parameter estimation problem (1) depends on the randomly distributed experimental data, hence the solution $\hat{p}$ are also random variables. Their uncertainty can be described by the variance-covariance matrix [2]

$$
C = (I \; 0) \begin{pmatrix} J_1^T J_1 & J_2^T \\ J_2 & 0 \end{pmatrix}^{-1} \begin{pmatrix} J_1^T J_1 & 0 \\ 0 & 0 \end{pmatrix} \begin{pmatrix} J_1^T J_1 & J_2^T \\ J_2 & 0 \end{pmatrix}^{-T} \begin{pmatrix} I \\ 0 \end{pmatrix},
$$

where J_1 resp. J_2 are the derivatives w.r.t. the parameters of the least squares terms resp. the constraints of problem (1) evaluated in the solution point.

Our aim is to compute an experimental design, i.e. controls q for layout, setup and processing and weights w for measurement selection, which yields — by parameter estimation from the experimental data—estimates with minimal statistical uncertainty. For this purpose we minimize functions on the variance-covariance matrix,

$$
\phi(C) = \operatorname{trace}(C) \text{ or } \det(C) \text{ or } \max\{\lambda : \lambda \text{ eigenvalue of } C\} \text{ or } \max C_{ii}
$$

Let $\xi := (q, w)$ denote the experimental design variables. Then we can formulate the experimental design optimization problem

$$\min_{q,w,x} \phi(C)$$

C is the variance-covariance matrix
in the solution point of the problem

$$\left[\begin{array}{l} \min_{p,x} \sum_{i=1}^{M} w_i \cdot \dfrac{(\eta_i - h_i(t_i, x(t_i), p, q))^2}{\sigma_i^2} \\[2mm] \dot{y} = f(t, y, z, p, q) \\ 0 = g(t, y, z, p, q) \\ 0 = d(x(t_0), \ldots, x(t_f), p, q) \end{array} \right] \tag{2}$$

Constraints:

$$\dot{y} = f(t, y, z, p, q)$$
$$0 = g(t, y, z, p, q)$$

$$lo \le \psi(t, x, p, q, w) \le up$$
$$0 = \chi(t, x, p, q, w)$$

$$w \in \{0, 1\}^M.$$

Remark 1 *The objective function of problem (2) is defined on the Jacobian $J = \begin{pmatrix} J_1 \\ J_2 \end{pmatrix}$ of the parameter estimation problem (1) which depends on derivatives of the solution of the dynamic system w.r.t. the parameters p, see [4].*

The experimental design problem (2) is a nonlinear inequality-constrained optimal control problem. The objective function is implicitly defined on derivatives of the solution of the dynamic model equations and not separable. For the numerical solution, we apply the direct approach of optimal control, which consists of parameterization of the control functions q_2, discretization of the state constraints, relaxation of the 0-1 variables, and a finite-dimensional parameterization of the solution of the dynamic system. For details we refer to [4]. We obtain a finite dimensional nonlinear constrained optimization problem: $\min \phi(\xi)$ s.t. $0 = \chi(\xi)$, $0 \le \psi(\xi)$

3 Derivative Based Optimization

To solve the experimental design optimization problem we choose the Newton-type method of Sequential Quadratic Programming (SQP). For details on this method we want to refer e.g. to the textbook [9]. We employ the SQP implementation SNOPT [3].

For the optimization, first derivatives of objective and constraints with respect to the optimization variables ξ are required. We consider directional derivatives for directions $\Delta\xi$ and apply the rule

$$\Delta\phi := \frac{d\phi}{d\xi}\Delta\xi := \lim_{h \to 0} \frac{\phi(\xi + h\Delta\xi) - \phi(\xi)}{h}$$

$$= \frac{d\phi}{dC}\Delta C, \quad \Delta C := \frac{dC}{dJ}\Delta J, \quad \Delta J := \frac{dJ}{d\xi}\Delta\xi. \tag{3}$$

The steps in (3) require intricate derivative computations. $\frac{d\phi}{dC}\Delta C$ and $\frac{dC}{dJ}\Delta J$ mean the differentiation of functions on matrices w.r.t matrices, see [4].

$\frac{dJ}{d\xi}\Delta\xi$ is the derivative w.r.t. ξ of the derivative w.r.t. p of the parameter estimation problem, e.g.

$$
\frac{dJ_1}{dq}\Delta q = -\mathrm{diag}\left(\frac{\sqrt{w_i}}{\sigma_i}\right) \cdot \left(\frac{\partial h_i}{\partial x}\frac{\partial^2 x_i}{\partial p \partial q}\Delta q + \frac{\partial^2 h_i}{\partial x \partial x}\frac{\partial x_i}{\partial p}\frac{\partial x_i}{\partial q}\Delta q + \frac{\partial^2 h_i}{\partial x \partial q}\frac{\partial x_i}{\partial p}\Delta q \right.
$$
$$
\left. + \frac{\partial^2 h_i}{\partial p \partial x}\frac{\partial x_i}{\partial q}\Delta q + \frac{\partial^2 h_i}{\partial p \partial q}\Delta q \right)_{i=1,\ldots,M} \tag{4}
$$

where $x_i := x(t_i, p, q)$. Note that for the computation of (4) we not only need the solution x of the DAE, but also first and mixed second derivatives:

$$
\frac{\partial x}{\partial p}(t_i, p, q), \quad \frac{\partial x}{\partial q}(t_i, p, q), \quad \frac{\partial^2 x}{\partial p \partial q}(t_i, p, q).
$$

We apply the backward differentiation formulae (BDF), a multistep integration method implemented in the code DAESOL [1], to solve the DAE systems. The derivatives of x are solutions of variational DAEs (VDAE). BDF schemes for these VDAE can also be considered as derivatives of the BDF scheme for the DAE if the same stepsize and order control is used. Hence we can compute the exact derivatives of the numerical approximation of x. Moreover, all these BDF schemes have the same structure, so the matrix decompositions for solution of the implicit problems can be applied simultaneously to all required first and second derivatives. For details on this approach, see [1] or [4].

The diverse VDAEs for the first and second derivatives contain first and second derivatives of the model functions f and g of the right hand side of the DAE. Further, first and second derivatives of the measurement model response functions h_i and of the nonlinear constraints are required. Usually, these functions are given as user defined subroutines. In our software VPLAN, we apply techniques of automatic differentiation based on the package ADIFOR [10] to compute the needed derivatives automatically.

4 Derivative Free Optimization

We use the derivative free multidirectional search method developed by Torczon [6], which is based on the iterative change of a simplex with $k+1$ points (where k is the number of the arguments in the function) so that the procedure "converges" to a minimal value point. Compared with the widely used simplex method [5, 7], this method searches k distinct directions in parallel, i.e. only the best point is kept in each iteration, which makes it "converge" faster and more reliable than the simplex method in case the function has many arguments. The multidirectional search method uses only function values of the objective function and is only able to treat simple-bounds-constraints.

5 Numerical Results

We compare the two optimization approaches for an experimental design optimization problem for the Diels-Alder reaction [8]. It is a chemical reaction with a catalytic and non-catalytic reaction channel. Aim is to determine the reaction velocities of both reaction channels. The model of this process can be formulated as an ordinary differential equation system. where the state variables model the molar numbers of the species. The model contains 5 parameters, the steric factors and activation energies of the reaction velocities and catalyst deactivation rate. Experimental design variables are the initial molar numbers, the concentration of the catalyst and the temperature and the weights for the placement of 10 measurements.

We want to plan two experiments for the most significant estimation of the model parameters. This leads to an optimization problem with simple-bounds-constraints on the experimental design variables as only constraints. Thus it is also possible to treat it with the derivative free optimization method. For each experiment we have 17 degrees of freedom, thus altogether 34 optimization variables for two experiments. We start the optimization procedures with an experimental layout with objective value 0.328328 for the A criterion $(\text{trace}(C))$.

The computation with VPLAN using the SNOPT SQP optimizer needs 167 SQP iterations which require 168 function calls and 384 derivative evaluations of the objective function to achieve convergence to the optimal solution with A criterion $= 0.0193858$. This computation runs 8.1 seconds user cpu time on a Pentium4 2.5 GHz under Linux.

The multidirectional search method terminates with an objective value of 0.019768 for the A criterion. To achieve this result, 30443 function evaluations are necessary in a user cpu time of 5 minutes on the same computer as above.

The objective value of the derivative based result is slightly better than the objective value of the derivative free result. The computational time of the derivative based method is 8 seconds compared to 5 minutes for the derivative free method.

Newton-type optimization methods such as the method of sequential quadratic programming require in particular derivatives of the objective function which is especially complicated for experimental design. Intricate computations are needed to provide the derivatives efficiently. Using this derivative based approach we have developed the software package VPLAN which can solve generally formulated problems of this class. It is inveigling to use easy to implement derivative free optimization methods instead. In this paper we have shown that, besides the drawback that such methods are restricted to problems with only simple-bounds-constraints, they require tremendously more computational time to achieve comparable results for nonlinear experimental design problems.

References

1. I. Bauer *et al.*, Numerical Methods for Initial Value Problems and Derivative Generation for DAE Models with Application to Optimum Experimental Design of Chemical Processes, *Scientific Computing in Chemical Engineering II*, F. Keil, *et al.*, eds., Springer-Verlag, Heidelberg, **2** (1999), 282–289.
2. H.G. Bock, Randwertproblemmethoden zur Parameteridentifizierung in Systemen nichtlinearer Differentialgleichungen, 1987, Bonner Mathematische Schriften 183.
3. P.E. Gill, W. Murray and M. A. Saunders, SNOPT: An SQP Algorithm for Large-Scale Constrained Optimization, *SIAM J. Opt.*, **12** (2002), 979–1006.
4. S. Körkel, Numerische Methoden für Optimale Versuchsplanungsprobleme bei nichtlinearen DAE-Modellen, Universität Heidelberg, 2002, http://www.ub.uni-heidelberg.de/archiv/2980.
5. J. A. Nelder and R. Mead, A Simplex Method for Function Minimization, *Comput. J.* **8** (1965), 308-313.
6. V.J. Torczon, Multi-Directional Search: A Direct Search Algorithm for Parallel Machines, Houston, TX, USA, vii+85, 1989.
7. W.H. Press, S.A. Teukolsky, W.T. Vetterling and B.P. Flannery, Numerical Recipes in C: The Art of Scientific Computing, Cambridge University Press, 1992.
8. R.T. Morrison and R.N. Boyd, Organic Chemistry, Allyn and Bacon, Inc., 4th ed., 1983.
9. J. Nocedal and S. J. Wright, Numerical Optimization, Springer-Verlag, Springer Series in Operations Research, New York, 1999.
10. C. Bischof, A. Carle, P. Khademi and A. Mauer, The ADIFOR 2.0 System for the Automatic Differentiation of Fortran 77 Programs, Center for Research on Parallel Computation, Rice University, Houston, TX, 1994, CRPC-TR94491.

Scalable SPMD Algorithm of Evolutionary Computation

Yongmei Lei[1] * and Jun Luo[1]

School of Computer Engineering and Science, Shanghai University, Shanghai 200072, China. ymlei@mail.shu.edu.cn, luojun9803@163.com

Abstract

This paper addresses two parallelization techniques used for evolutionary computation. We study the grid enabled evolutionary computation model,and the differences between the coevolution and the space decomposition based parallel evolutionary algorithm are described in detail. We propose master-slave mode and equality mode used for SPMD program implementation. In this paper, we also discuss the advantages and drawbacks of these two parallel computing model. Through comparing the solution precision attained between parallel evolutionary algorithms, we stress the excellence of space decomposition based evolutionary algorithms. Finally, successful experiment results are given to show the better optimization efficiency achieved through the parallel evolutionary algorithms.

1 Introduction

Grids are defined as infrastructure allowing flexible, secure, and coordinated resource sharing among dynamic collections of individuals, institutions and resources referred to as virtual organizations. The evolutionary computation model is satisfied with the condition of grid computing application. However, in the existing evolutionary algorithm, it is difficult to bring out the features to the maximum. In this paper we studied a new grid enabled high performance computing model. Then, in this research, the co-evolution and space decomposition parallel evolutionary algorithm is studied, implemented and compared. Here, we study a SPMD (Single Program Multiple Data) algorithm for doing transformation of evolutionary computation. We have implemented the algorithm on a cluster, and tested for real time performance of the algorithm.

The outline of the paper is as follows: Section 2 discusses the grid enabled computation from an algorithmic and theoretical point of view. presents Space

---

*Supported by "SEC E-Institute: Shanghai High Institutions Grid project".

Decomposition Based Parallel Evolutionary Algorithm Section 3 describes an implementation developed for this framework. Section 4 describes the results of our experiments on a real SMP cluster under different circumstances. Finally Section 5 concludes the paper.

2 Grid Enabled High Performance Computation Model

2.1 Motivation and Strategies for Evolutionary Computation on Grid Enabled Computing

Scalability is one of important features of grid enabled computing. Evolutionary computation is the best scalable computation model for the time being. Evolutionary algorithms (EAs) are stochastic search methods that have been applied successfully in many search, optimization, and machine learning problems. Main method and techniques used in the parallel evolutionary computation algorithm:

- Dividing the population into subgroups to achieve massively scalable parallelism;
- Using Space Decomposition Based Parallel evolutionary algorithm;
- Remembering best solution and sharing it among all processors to accelerate the optimization process;
- Selecting other bed solution and moving to new area to simulate global search.;
- Solving large problems via small algorithm by using the coevolution-like decomposition.

2.2 Coevolution-Type Parallelization

One of the merits of evolutionary algorithms is that it searches many nodes in the search space in parallel. This requires us to generate randomly an initial population of the search nodes. For comparison purpose, we first briefly review the parallelization of Multi-population coevolution-type parallel genetic algorithm(CoPGA) in master-slave mode.

Master and slave configuration has been chosen to carry out this implementation. There is one processor, namely master, which performs all the tasks that cannot be carried out in parallel. In addition, the master processor submits the jobs to the rest of processors, referred to as slaves, which carry out the tasks that can be done simultaneously.

2.3 Space Decomposition Based Parallelization

The space decomposition based parallel genetic algorithm (SDPGA) is based cellular partition methodology in combinatorial topology. The encoding is splicing and the decoding is decomposable, a long string can be deciphered independently through decoding its substrings in a composition way. parallel computing performance is limited by high communication latencies and by the fact that the computing resources may have different workloads and are possibly heterogeneous. But these problems can be overcome by SDPGA. It is robust and scalable for grid-enabled computation.

Individual space $\mathbf{H}_L$ can be discomposed of P subspaces: $\mathbf{H}_L = \mathbf{H}_{L1}$ $\vee$ $\mathbf{H}_{L2}$ $\vee \mathbf{H}_{L3}$ $\vee \dots \vee \mathbf{H}_{Lp}$. So the population space $\mathbf{X} = (X_1, X_2, \dots, X_N) \in \mathbf{H}_L^N$. We define the subpopulation: $\mathbf{X}_j = (X_1^{(j)}, X_2^{(j)}, \dots, X_N^{(j)})$, j=1, 2, ..., P; then $\mathbf{X} = \mathbf{X}^{(1)}$ $\vee \mathbf{X}^{(2)}$ $\vee \dots \vee \mathbf{X}^{(p)}$. Every subpopulation $\mathbf{X}^{(j)}$ can be calculated on grid nodes, and exchange information periodically.

3 Implementation Parallel SDPGA and CoPGA on SMPs Clusters

Since evolutionary algorithms are stochastic in nature, being able to collect this kind of statistic is very important. MPI has been used for the parallel implementations of the proposed Co-evolution and Space decomposition based parallel genetic algorithm.

3.1 Data Distribution and Load Balancing

Our parallel evolution computing program will be based on the SPMD model. So each process picks up a proper strip and starts executing EC simulations locally. Except for occasional message exchanges and global operations, the simulation proceeds asynchronously.

There are two ways of load balancing : a static balance and a dynamic balance. If the times required to do each job are similar, there is a great waste of time in the communication among processors. Therefore, a dynamic distribution of the tasks has been implemented.

3.2 SPMD Program Mode

Two different parallel implementations of the evolution algorithm have been developed and the final solution precision attained are compared in this paper. These implementations are described in this sections .

In the CoPGA algorithm ,it can be implemented by Master-slave program mode.Master evaluate population and select the best one rotate the population.Master communicates to all slaves.

SDPGA algorithm can be implemented by equality mode. By mapping every subspace to one processor.Every process deals with part of each individual space. Different processes achieve results with different precisions. The communication relationship among neighbor nodes. Practically useful simulations are feasible on a SMP cluster, because of scalability to a large number of processors.

4 Experiment and Discussion

4.1 Experimental Design

The experiments are carried out on a dedicated cluster. MPICH, a portable implementation of MPI standard, is used for providing communication functions in parallel computing environment.

The following algorithms using the proposed method and other methods are examined and compared: (1) Space decomposition based parallel genetic algorithm: This is the scalable, robust and adaptive method. (2) Coevolution diverse random algorithm: At the beginning, each subpopulation randomly creates its own parameter set. This algorithm is comparable to the SDPGA adaptive algorithm.

This group of experiments aims to exhibit the detailed evolution process of the SDPGA. Our experiments were organized as three groups with different purposes:

$$f_1(x) = 4x_1^2 - 2.1x_1^4 + \frac{1}{3}x_1^6 + x_1x_2 - 4x_2^2 + 4x_2^4, \; x_1, x_2 \in [-5, 5]$$

$$f_2(x) = \sum_{i=1}^{n} \left(\sum_{j=1}^{i} x_j^2 \right)^2, \; x_i \in [-100, 100]$$

$$f_3(x) = \sum_{i=1}^{n/4} \left\{ 100 \left(x_{4i-3}^2 - x_{4i-2} \right)^2 + (x_{4i-3} - 1)^2 + 90(x_{4i-1}^2 - x_{4i})^2 \right.$$
$$\left. + 10.1 \left[(x_{4i-2} - 1)^2 + (x_{4i} - 1)^2 \right] + 19.8(x_{4i-2} - 1)(x_{4i} - 1) \right\}, x \in [-50, 50]$$

4.2 SDPGA Algorithm Results at Different Computing Nodes

The very good scalability of this code enables us to investigate optimization systems that would be impossible using previously available computers. Solution precision attained to track different simulation size of evolution computing models on ZiQiang 2000 is shown Tn table 1.

Table 1. The evolution process of SDPGA applied to f3, $x \in [-50, 50]$

Iteration(k)	Computing Node	x1	x2	x3	x4	$\varepsilon = \lvert f^{(k)} - f^* \rvert$
100	node0	0.0003	-0.8195	4.0962	15.5649	1.8e+3
100	node1	0.0070	-1.2293	-4.0965	15.5716	1.8e+3
100	node2	0.0073	-1.2297	-4.0969	15.5720	1.8e+3
1000	node0	0.0003	0.0003	0.8195	0.8195	1.6e+1
1000	node1	0.4036	0.1988	1.0179	1.1076	5.8e+0
1000	node2	0.4099	0.2051	1.0243	1.1139	5.5e+0
10000	node0	0.0003	0.0003	-0.8195	0.8195	1.6e+1
10000	node1	0.7684	0.5893	-1.1971	1.4405	1.3e-1
10000	node2	0.7689	0.5896	-1.2004	1.4408	1.3e-1

As the encoding length L increases, the SDPGA algorithm can provide a successive refined, more and more accurate representation of problem variables.

4.3 Comparisons

To assess the effectiveness and scalability of the algorithms, The computational efficiency comparison results are shown in Table 2. Scalability and the

Table 2. The solution precision attained of the CoPGA and SDPGA when applied to the test suit, $\varepsilon = |f^{(k)} - f^*|$

Function	Iteration	ε of CoPGA	ε of SDPGA
f1(x)	100	1.9e-4	5.1e-7
f1(x)	1000	1.7e-4	5.0e-7
f1(x)	10000	1.9e-4	5.0e-7
f2(x)	100	2.6e+0	3.6e-7
f2(x)	1000	3.0e-1	3.6e-7
f2(x)	10000	3.6e-2	3.6e-7
f3(x)	100	5.4e+3	1.8e+3
f3(x)	1000	2.9e+3	5.5e+1
f3(x)	10000	2.9+3e	1.3e-1
f3(x)	50000	1.6e+1	1.3e-1

solution precision attained of CoPGA and SDPGA are compared . Clearly, from these tables that the SDPGA can always attains higher solution precision.

5 Conclusions

This paper addresses the computation model problem of grid enabled high performance computation. Its objective is to study the scalability and robust of evolutionary algorithm to implement computation on stateful resources. Evolutionary algorithm provides a modeling framework more scalable and robust than previous approaches.Although our project focuses on evolutionary computation ,the environment supports any application that is massively parallelizable and is robust (the success of the application does not depend on the success of any given subprocess). This paper contains a modern vision of the parallelization techniques used for evolutionary algorithms . The proposed co-evolutionary parallel algorithm and space decomposition based parallel evolutionary algorithm have been successfully applied to realistic case studies.The Scalability and the solution precision attained of CoPGA and SDPGA are compared .The experiment results also show that SDPGA can always attains higher solution precision.

References

[Dan04]   Daniel Sabban : Bringing Grid Web Services Together, Globus World, San Francisco ,CA, (2004)

[EM02]   Enrique Alba ,Marco Tomassini: Parallelism and Evolutionary Algorithms.IEEE Transactions On Evolutionary Computation, **6**, 443–462 (2002)

[GTI98]   G. Allen, T. Dramlitsch, Ian. Foster et al: Efficient Execution in Heterogeneous Distributed Computing Environments with Catus and Globus,Proc. SC01 (SC2001), Denver.(2001)

[HE02]   Hu X, Eberhart R: Adaptive particle swarm optimization: detection and reponse to dynamic system. IEEE Congress on Evolutionary Computation, Honolulu, Hawaii (2002)

[Ian98]   Ian Forster: Grid-Enabled MPI: Message Passing in Heterogeneous Distributed Computing Systems, Supercomputing 98 (1998)

[MRD01] Mark Baker, Rajkumar Buyya , Domenico Laforenza: The Grid :A Survey on Global Efforts in Grid Computing , School of Computer Science , Univ. of Portsmouth (2001)

[ZKY03] Zong-Ben Xu, Kwong-Sak Leung,Yong Liang:Efficiency speed-up strategies for evolutionary computation:fundamentals and fast-GAs. Applied Mathematics and Computation, **142**, 341–388 (2003)

Checkpointing RSIP Applications at Application-level in ChinaGrid *

Chunjiang Li[1], Xuejun Yang[1], and Nong Xiao[1]

School of Computer Science, National University of Defense Technology, Changso, China lcj@hnxinmao.com

In this paper, we checkpointing remote-sensing image processing (RSIP) applications in computational grid using a novel application-level uncoordinated checkpoint protocol based on Job Progress Description (JPD). The JPD of a job is composed by a Job Progress Record Object (JPRO) and a group of Job Progress State Objects (JPSO), these two kinds of objects act as checkpoint data for the job and the methods of them can be used as checkpoint APIs. Both the applications and the computing resources participate in the ChinaGrid project. Experiments show that using such checkpoint protocol, the RSIP Applications can be more robust when resource failure occurs.

Key words: Computational grid, RSIP applications, Checkpointing

1 Introduction

The computational Grid [1, 2] has a small mean time to failure. The most common failure modes include machine faults in which hosts go down and network faults where links go down. In order to reduce the recovery time of the jobs, checkpointing and recovery service is absolutely necessary, which can save partial results and job states, avoid restarting the job from the very beginning. Checkpoint and recovery (CPR) techniques have been studied for a few decades. But in computational grid, how to checkpoint jobs is still an open issue.

In this paper, we present an application-level checkpointing protocol for computational grid based on **Job Progress Description (JPD)**. Which requires each job to be divided into independent job progresses and the job states of each progress are defined by the user. At the end of each progress, checkpoint APIs are called to save the partial results and the job states to

*This work is supported by the Educational and Research Grid Project (China-Grid) of China under Grant No.CG2003-GA00103

stable storage. The partial results and the job states facilitate the recovery of the failed jobs. Then we use it for checkpointing a kind of application grid, remote-sensing image processing applications, which running on the grid platform of the ChinaGrid project. The ChinaGrid project is a grid computing project sponsored by the Ministry of Education of China, aiming at integrating computing resources, data and applications in multiple universities in China for research and educational purpose.

2 Related Work

A key requirement of Grid Checkpoint Recovery (GridCPR) service is recoverability of jobs among heterogeneous grid resources. In several meetings of GGF held recently, the Grid Checkpoint Recovery (GridCPR) Working Group presented some memo [3] and drafts [4, 5] which provided information to the grid community regarding a pro-posed architecture for grid checkpoint recovery services and a set of associated Application Programmer Interface (API). The consensus is that application level checkpointing protocol is the only one suitable for the grid environment. But their work was also very elementary, and there is still much work has to be done. In the work of the European DataGrid [6] project, they proposed an application level checkpoint protocol for the data processing applications [7], which divides the job into several job steps, and records user-defined job states at the end of each job step as the checkpoint data for job recovery.

3 Basic Idea of JPD-Based Checkpointing

In this section, we describe the **JPD**-based application level checkpointing mechanism for computational grid. Firstly we give following definitions:

Job Progress (JP) A group of continuous operations in a job. For example, a group of continuous statements in the source code of the job make up a Job Progress. So, the single threaded job can be divided into a series of continuous Job Progresses.

Independent Job Progress (IJP) If the running of one job progress does not depend on the running environment of the other job progresses, then this job progress is called independent job progress.

Job Progress States (JPS) The user defined states of the job at the end of each independent job progress. It can be expressed by a series of $<$ $var, value >$ pairs.

3.1 Checkpoint Data Set

We use two kinds of abstract data objects as the checkpoint data of a job the definition of these two kinds of objects is given below:

Job Progress States Object (JPSO) The object which records the job states for one **IJP**.

Job Progress Record Object (JPRO) The object which records the series of **IJP**s of a job and the latest **IJP** finished by the job.

So, during the running of a job, one **JPRO** and a series of **JPSO**s can record the job states, we call them as **Job Progress Description (JPD)**, which acts as the checkpoint data of the job. The structure of **JPD** is illustrated in Fig.3.

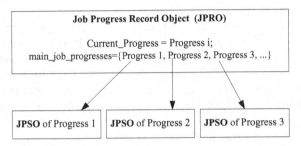

Fig. 1. Data Elements in **JPD**

This kind of organization of checkpoint data can benefit to the job control in computational grid. For in computational grid, an application consists multiple tasks, by checking the **JPRO** of each task, the job manager can grasp the information about the states of each task and can determine the progress of the whole application.

4 RSIPA:Remote-Sensing Image Processing Applications

The great advancement in remote-sensing technologies has brought new challenge to remote-sensing image processing(RSIP) [8]. First, because of the great amount of computation and the complex operation, remote-sensing image processing needs the capability of more large-scale computing. Second, fast remote-sensing image processing technology is of urgent need in many fields, and parallel processing is one of the effective methods to solve that problem. But traditional computing system can't meet the high-performance demands of remote-sensing image processing any more.

Grid is an ideal platform for remote-sensing image processing for it can integrate rich computation and storage resources. And computational grid can support traditional parallel applications. For example, using MPICH-G2[9], a standard MPI program can run in computational Grid, while the system running the programs could be Mainframe, Cluster, or heterogeneous computers that locate faraway, but all these are transparent to users.

With the requirements for the application of remote-sensing image processing in ChinaGrid which will provide with common services of "211 Project" of China ministry of education in the period of the Tenth Five-year Plan, we design and implement parallel remote-sensing image processing applications based on computational Grid.

5 Checkpointing RSIP Applications

As the discussion of the JPD-based checkpointing protocol in 3, in order to checkpointing RSIP applications with this protocol, the programs should be partitioned into IJPs at application level. For RSIP applications, it is easy to partition the core algorithms into IJPs. An Case of job partitioning is given in Fig. 1, it is a worker task of the PIWA-LOC [10] algorithm, and it partitioned the job into two IJPs. During the processing, when each task fulfills the calculation on half of the subimage, it invoke the checkpoint APIs to store the partial result into storage.

All the modules in the RSIP applications are programmed with C and MPICH. So,during the processing, if there are messages sending and receiving between the worker tasks, in order to avoid domino-effect, the process of each task would flush its message queue to avoid in flight message getting lost, and then all the processes would all synchronously checkpoint.

6 Experiments

In order to evaluate the effects of checkpointing RSIP application with JPD-based checkpoint mechanism, we conduct elementary experiments with PIWA-LOC Algorithm. We run this program on a cluster of 8 computing nodes, and record the run time with and without checkpointing. For evaluating the checkpoint mechanism, we add failure events to the computing nodes manually. There are totally 2 nodes failure occurs during the experiment.

In the experiment, the input image is 6000×6000. The run time with and without checkpointing as well as the recover time with and without checkpointing is given in Table 1.

Table 1. Experimental Results

	Without Checkpointing (s)	With Checkpointing (s)
Run time	11.97	14.2
Recover time	7.94	4.14

```
GridCPR_init();
current_progress = Jobprogress.load_next_progress ();
 //  get the identifier of next IJP;

switch(current_progress) {

case progress1:
      new JPSO current_JPSO;
      for(v=0;v<H/2;j++)
        for(u=0;u<W;u++)
          {
            ReverseSamplePoint(u,v,&x,&y);
            array=GetSampleMatrix(srcImage,x,y);
            dstImage[u][v]=Resample(array);
          }
      current_JPSO.save_value(v_low,0);
      current_JPSO.save_value(v_high,H/2);
      current_JPSO.save_value(dstImage,dstImage_low);
      Jobprogress.save_JPSO(current_JPSO);  // save current JPSO
      JPRO.current_progress=progress1;
      Jobprogress.save_JPRO();  //save current JPRO

case progress2:
      new JPSO current_JPSO;
      for(v=H/2;v<H;j++)
        for(u=0;u<W;u++)
          {
            ReverseSamplePoint(u,v,&x,&y);
            array=GetSampleMatrix(srcImage,x,y);
            dstImage[u][v]=Resample(array);
          }
      current_JPSO.save_value(v_low,H/2);
      current_JPSO.save_value(v_high,H);
      current_JPSO.save_value(dstImage,dstImage_high);
      Jobprogress.save_JPSO(current_JPSO);  // save current JPSO
      JPRO.current_progress=progress2;
      Jobprogress.save_JPRO();  //save current JPRO

}
      MPI_Send(dstImage,master);
```

Fig. 2. A Sample of Checkpointing RSIP Applications

7 Conclusion and Future Works

In this paper, we proposed the Job Progress Description based checkpoint mechanism for checkpointing at application level in computational grid, and applied this checkpoint mechanism to the Remote-Sencing Image Processing applications which act as a kind of public services in the ChinaGrid project. There are much work need to be done for consummating this checkpointing mechanism, such as how to combining message logging mechanism for checkpointing asynchronously, how to manage the checkpoint data generated in checkpointing, how to optimize the performance of checkpointing and recovery.

References

1. I. Foster and C. Kesselman, The Grid: Blueprint for a New Computing Infrastructure, Morgan Kaufmann Publishers, 1999.
2. I. Foster, The grid: A new infrastructure for 21st century science, *Physics Today* **54** (2002).
3. GridCPR Working Group, An Architecture for Grid Checkpoint Recovery Services and a GridCPR API, www.gridforum.org/Meetings/ggf7/drafts/GridCPR001.doc, 2003.

4. GridCPR Working Group, GWD-I: An Architecture for Grid Checkpoint Recovery Services and a GridCPR API, gridcpr.psc.edu/GGF/docs/draft-ggf-gridcpr-Architecture-1.0.pdf, 2004.
5. GridCPR Working Group, GWD-I: Use Cases for Grid Checkpoint and Recovery, gridcpr.psc. edu/GGF/docs/draft-ggf-gridcpr-UseCases-1.0.pdf, 2004.
6. DataGrid, European DataGrid Project, www.eu-datagrid.org, 2004.
7. A. Gianelle and R. Peluso and M. Sgaravatto, Job Partitioning and Checkpointing, *Technical Report DataGrid-01-TED-0119-0-3*, 2002.
8. S.L. Zhu and Z.K. Zhang, Gain and Analysis of Remote Sensing Image, Science Press, eijing, China, 2000.
9. N.T. Karonis, B. Toonen and I. Foster, MPICH-G2: A Grid-Enabled Implementation of the Massage Passing Interface, *J. Parall. Distrib. Computing* **63** (2003), 551-563.
10. Y.H. Jiang, X.J. Yang, H.D. Dai, and H.Z Yi, A Distributed Parallel Resampling Algorithm for Large Images, LNCS 2834/APPT2003, Xiamen, China, 2003, 608-518.

Fast Fourier Transform on Hexagons [*]

Huiyuan Li[1] and Jiachang Sun[1]

Laboratory of Parallel Computing, Institute of Software, Chinese Academy of Sciences, Beijing 100080, P. R. China. hynli@mail.rdcps.ac.cn, sun@mail.rdcps.ac.cn

Summary. We propose fast algorithms for computing the discrete Fourier transforms on hexagon. These algorithms are easy to implement, they reduce the computation complexity from $\mathcal{O}(M^2)$ to $\mathcal{O}(M \log M)$, where M is the total number of sampling points.

Key words: Fast Fourier transforms, algorithms, hexagon, non-tensor-product

1 Introduction

Encouraged by the success of fast Fourier transform (FFT) in information and computing sciences [1, 6] and the increasing application demand for complex geometry, we wish to extend FFT to efficiently solve problems on certain irregular domains [2, 7, 8, 10]. We find further motivation for the study of sampling schemes for multidimensional isotropic functions. In general, sampling 2-dimensional isotropic functions on hexagonal lattice is significantly more efficient than sampling such functions on square lattice [3]. Motivation also arises from the study of mesh generation for solving partial differential equations. As a compromise between rectangular grid and unstructured mesh, hexagonal tessellation is widely used in practical applications. It possesses simple data structure and more flexibility, it is easy to use, while it will bring out less effect of grid orientation.

The classic approaches for multidimensional discrete Fourier transform (DFT) can not be taken with functions periodically sampled on hexagon. To overcome this difficult, we formulate the hexagonal DFT in terms of the bilinear form of a periodicity matrix. Resorting to the periodicity matrix factorization, we derive the decimation-in-time (DIT) and the decimation-in-frequency (DIF) hexagonal FFT algorithms. From the technical point of view, the DIT and DIF hexagonal FFT algorithms are both composed of 3

---

[*]This project is supported by National Natural Science Foundation of China (No. 60173021) and Basic Research Foundation of ISCAS (No. CXK35281).

FFTs sampled on a square, a series of shifted 3-point FFTs and a certain enciphering/deciphering procedure. As a result, our algorithms can be conveniently and efficiently implemented by using existing FFT packages such as FFTPACK [9] and FFTW.

We give some numerical results in the last section, which demonstrate the high performance of our hexagonal FFT algorithms.

2 Discrete Fourier transform on hexagon

Let Ω be a centrally symmetric hexagon. We establish the 3-directional partition on Ω, and specify each knot by an integer pair (k_1, k_2), with the center knot indexed by $(0, 0)$. Given a positive integer N, we define

$$\Lambda_N = \big\{ (k_1, k_2)' \in \mathbb{Z}^2 : \ -N \le k_1, k_2, k_1 + k_2 < N \big\}, \tag{1}$$

which stands for the set of dotted knots indicated in Fig. 1. For a complex number sequence $u_{\mathbf{j}}$ defined on Λ_N, we define the hexagonal DFT [8],

$$\hat{u}_{\mathbf{k}} = \frac{1}{|\det(\mathbf{N})|} \sum_{\mathbf{j} \in \Lambda_N} u_{\mathbf{j}} \exp\left(-2\pi i\, \mathbf{k}' \mathbf{N}^{-1} \mathbf{j}\right), \quad \mathbf{k} \in \Lambda_N. \tag{2}$$

where the integer matrix $\mathbf{N} = \begin{pmatrix} 2N & -N \\ -N & 2N \end{pmatrix}$, and $|\det(\mathbf{N})| = \operatorname{card}(I_N) = 3N^2$.

Generally, $\mathbf{N}$ is referred to as the periodicity matrix. Due to the orthogonality, we find the inverse formula for (2),

$$u_{\mathbf{j}} = \sum_{\mathbf{k} \in \Lambda_N} \hat{u}_{\mathbf{k}} \exp\left(2\pi i\, \mathbf{k}' \mathbf{N}^{-1} \mathbf{j}\right), \quad \mathbf{j} \in \Lambda_N. \tag{3}$$

The periodicity matrix $\mathbf{N}$ induces an equivalent relation on $\mathbb{Z}^2$. We shall say that two integer vectors $\mathbf{m} \equiv \mathbf{n}$ (mod $\mathbf{N}$) if $\mathbf{m} = \mathbf{n} + \mathbf{N}\mathbf{r}$ for some integer vector $\mathbf{r}$. It is true that $\mathbf{j}, \mathbf{k} \in \Lambda_N$ with $\mathbf{m} \equiv \mathbf{n}$ (mod $\mathbf{N}$) implies $\mathbf{j} = \mathbf{k}$. In this case, we say Λ_N a period associated with $\mathbf{N}$. Specifically, we shall use the notation $(\mathbf{j})_{\mathbf{N}}$ to denote the vector which is both congruent to $\mathbf{j}$ and contained in Λ_N.

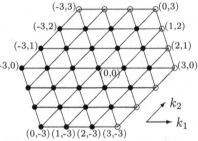

Fig. 1. The 3-directional partition and the hexagonal indices.

3 Fast Fourier transform on hexagon

In this section, we develop some algorithms to evaluate the hexagonal DFTs (2) and (3) conveniently and efficiently. Without loss of generality, we only consider the following "canonical" form,

$$X_{\mathbf{k}} = \sum_{\mathbf{j} \in \Lambda_N} x_{\mathbf{j}} e^{-2i\pi \mathbf{k}' \mathbf{N}^{-1} \mathbf{j}}, \quad \mathbf{k} \in \Lambda_N. \tag{4}$$

Let $\mathbf{v} = (1, -1)'$. The periodicity matrix $\mathbf{N}$ can be factorized as followings,

$$\mathbf{N} = \mathbf{PQ}, \qquad \mathbf{P} = N\mathbf{I}, \quad \mathbf{Q} = \mathbf{I} + \mathbf{vv}'.$$

Now $\mathbf{P}$ is a periodicity matrix for an ordinary square DFT, and its associated period $I_{\mathbf{P}}$ can be chosen in a general way,

$$I_{\mathbf{P}} = \{\mathbf{k} = (k_1, k_2)' : 0 \le k_1, k_2 < N\}.$$

Define $\mathbf{e} = (1, 0)'$ and $\Gamma = \{-1, 0, 1\}$. Then for any $\mathbf{j}, \mathbf{k} \in \Lambda_N$, there exist unique vectors $\mathbf{p}, \mathbf{m} \in I_{\mathbf{P}}$ and unique integers $q, n \in \Gamma$ such that [4, 5]

$$\mathbf{k} = (\mathbf{m} + n\mathbf{v})_{\mathbf{N}}, \qquad \mathbf{j} = (\mathbf{p} + N q\mathbf{e})_{\mathbf{N}}.$$

Noting that $\mathbf{j}' \mathbf{N}^{-1} \mathbf{k} \equiv N^{-1} \mathbf{p}' \mathbf{m} + (3N)^{-1} n(Nq + p_1 - p_2) \pmod{1}$, we rewrite the hexagonal DFT sum in (4) as

$$X_{(\mathbf{Qm}+n\mathbf{v})_{\mathbf{N}}} = \sum_{\mathbf{p} \in I_{\mathbf{P}}} e^{-\frac{2\pi i}{N} \mathbf{p}' \mathbf{m}} \Big(\sum_{q \in \Gamma} x_{(Nq\mathbf{e}+\mathbf{p})_{\mathbf{N}}} e^{-\frac{2\pi i}{3N} n(Nq + p_1 - p_2)} \Big).$$

Let $w = e^{-\frac{2\pi i}{3N}}$. Define $y_{q,\mathbf{p}} = x_{(Nq\mathbf{e_1}+\mathbf{p})_{\mathbf{N}}}$ and

$$Y_{n,\mathbf{p}} = \sum_{q \in \Gamma} y_{q,\mathbf{p}} w^{n(Nq + p_1 - p_2)}. \tag{5}$$

Then for any $n \in \Gamma$,

$$X_{(\mathbf{Qm}+n\mathbf{v})_{\mathbf{N}}} = \sum_{\mathbf{p} \in I_{\mathbf{P}}} Y_{n,\mathbf{p}} e^{-\frac{2\pi i}{N} \mathbf{p}' \mathbf{m}}, \quad \mathbf{m} \in I_{\mathbf{P}}. \tag{6}$$

This means the computation of (4) can be fulfilled by solving (5) and (6) successively. It is well known that (6) is the standard square DFT, which can efficiently computed by any existing FFT packages such as FFTW and FFTPACK. While (5) defines N^2 shifted 3-point DFTs, the computation of which are depicted by the butterfly symbol in Fig. 2.

In practice, the hexagonal FFT computations are normally performed *in place* in a 2-dimensional array. Let the input data x be stored in the $3N \times N$ array A in the natural order, i.e., $A(p_1 + Nq, p_2) = x_{(\mathbf{p}+Nq\mathbf{e})_{\mathbf{N}}}$, $\mathbf{p} \in I_{\mathbf{P}}$, $q \in \Gamma$. The in-place computation implies $y_{q,\mathbf{p}} = A(p_1 + Nq, p_2)$, $\mathbf{p} \in I_{\mathbf{P}}$, $q \in \Gamma$, this may divide the array A into three $N \times N$ subarrays indicated by Fig. 4. Denote $p = p_1 - p_2$ and define two arrays of real trigonometric function values,

$$wr(p) = \cos(2\pi p/3N), \ wi(p) = \sin(2\pi p/3N), \quad -N \le p \le 2N - 1.$$

We propose an in-place algorithm in Fig. 3 corresponding to the butterfly symbol in Fig. 2. This algorithm is expressed in terms of complex numbers, but all the multiplications are by real number or by i.

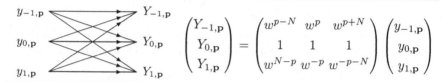

Fig. 2. The butterfly for the computation of the shifted 3-point DFT.

$$u = A(p_1 - N, p_2) + A(p_1 + N, p_2); \qquad A(p_1 - N, p_2) = A(p_1 - N, p_2) - A(p_1 + N, p_2)$$
$$A(p_1 + N, p_2) = A(p_1 + N, p_2) - A(p_1, p_2); \qquad A(p_1, p_2) = A(p_1, p_2) + u$$
$$u = wi(N - p) * A(p_1 - N, p_2) + wi(p) * A(p_1 + N, p_2)$$
$$A(p_1 + N, p_2) = wr(N - p) * A(p_1 - N, p_2) - wr(p) * A(p_1 + N, p_2)$$
$$A(p_1 - N, p_2) = A(p_1 + N, p_2) - i * u; \qquad A(p_1 + N, p_2) = A(p_1 + N, p_2) + i * u$$

Fig. 3. An in-place algorithm corresponding to the butterfly in Fig. 2.

The in-place computation of hexagonal FFT will be achieved after the in-place FFTs having been successfully performed on the subarrays of A. However, the final output X is scrambled (see Fig. 4). To get the naturally ordered output, a deciphering procedure is needed: the content of the entry $A(p_1 + Nq, p_2)$ should be moved to the location $A(m_1 + Nn, m_2)$, where

$$\mathbf{p} + Nq\mathbf{e} \equiv (\mathbf{Qm} + n\mathbf{v})_N, \quad \mathbf{m}, \mathbf{p} \in I_\mathbf{P}, \quad n, q \in \Gamma. \tag{7}$$

It is obvious that

$$(\mathbf{Qm} + n\mathbf{v})_N - (\mathbf{m} + Nn\mathbf{e}) \equiv (m_1 - m_2 + v)\mathbf{e} \pmod{N}.$$

The above equation means that for any given $\mathbf{p} \in I_\mathbf{p}$, the source location line $\mathbf{p} + \lambda\mathbf{e}$ periodically coincides with the destination location line $\mathbf{m} + \lambda\mathbf{e}$. Thus the deciphering procedure can also be performed *in place*.

$x_{-3,0}$	$x_{-3,1}$	$x_{-3,2}$
$x_{-2,0}$	$x_{-2,1}$	$x_{-2,2}$
$x_{-1,0}$	$x_{-1,1}$	$x_{-1,2}$
$x_{0,0}$	$x_{0,1}$	$x_{0,2}$
$x_{1,0}$	$x_{1,1}$	$x_{-2,-1}$
$x_{2,0}$	$x_{-1,-2}$	$x_{-1,-1}$
$x_{0,-3}$	$x_{0,-2}$	$x_{0,-1}$
$x_{1,-3}$	$x_{1,-2}$	$x_{1,-1}$
$x_{2,-3}$	$x_{2,-2}$	$x_{2,-1}$

input x

$(-3,0)$	$(-3,1)$	$(-3,2)$
$(-2,0)$	$(-2,1)$	$(-2,2)$
$(-1,0)$	$(-1,1)$	$(-1,2)$
$(0,0)$	$(0,1)$	$(0,2)$
$(1,0)$	$(1,1)$	$(1,2)$
$(2,0)$	$(2,1)$	$(2,2)$
$(3,0)$	$(3,1)$	$(3,2)$
$(4,0)$	$(4,1)$	$(4,2)$
$(5,0)$	$(5,1)$	$(5,2)$

indices of array A

$X_{-1,1}$	$X_{1,-3}$	$X_{0,-1}$
$X_{1,0}$	$X_{0,2}$	$X_{2,-2}$
$X_{-3,2}$	$X_{-1,-2}$	$X_{-2,0}$
$X_{0,0}$	$X_{-1,2}$	$X_{1,-2}$
$X_{2,-1}$	$X_{1,1}$	$X_{-3,0}$
$X_{-2,1}$	$X_{0,-3}$	$X_{-1,-1}$
$X_{1,-1}$	$X_{0,1}$	$X_{2,-3}$
$X_{-3,1}$	$X_{2,0}$	$X_{-2,-1}$
$X_{-1,0}$	$X_{-2,2}$	$X_{0,-2}$

output X

Fig. 4. The input x in array A is overwritten by the scrambled output X.

The above algorithm is obtained by *decimating* the output frequency sequence, and we call it decimation-in-frequency (DIF) hexagonal FFT algorithm. Similarly, letting $\mathbf{j} = (\mathbf{m} + n\mathbf{v})_N$, $\mathbf{k} = (\mathbf{p} + Nq\mathbf{e})_N$ yields

$$X_{(Nqe+p)_N} = \sum_{n\in\Gamma} \left(\sum_{m\in I_\mathbf{P}} x_{(\mathbf{Qm}+n\mathbf{v})_N} e^{-\frac{2\pi i}{N}\mathbf{p}'\mathbf{m}} \right) e^{-\frac{2\pi i}{3N}n(Nq+p_1-p_2)}.$$

Thus by almost reversing the above DIF algorithm, we can easily derive the decimation-in-time (DIT) hexagonal FFT algorithm. Due to page limit, we omit the details.

Now we analyze the arithmetic cost in the hexagonal FFT algorithms. Let L be the flop count in computing an $N \times N$ square FFT. Then the flop count in computing the hexagonal FFT is $3L + 24N^2$. If FFTPACK [9] is used for computing the square FFTs and if $N = 2^{p_2} 3^{p_3} 4^{p_4} 5^{p_5} 6^{p_6}$, then this flop count amounts only to $N^2(30p_2 + 54p_3 + 51p_4 + 81.6p_5 + 80p_6 - 12) + 36N$.

4 Numerical results

In this section, we present some numerical results concerning the hexagonal FFT algorithms in §3. Our experiments have been made under the Linux/g77 environment on PIII 1.5GHZ/1GB computers. We use the FFTPACK subroutines to implement square FFTs.

N	3	9	27	81	243	729
DFT	2.824E-5	5.619E-4	3.962E-2	3.535	525.13	54650.
DIF	1.937E-5	4.475E-5	3.001E-4	3.819E-3	4.188E-2	0.515
DIT	1.629E-5	4.074E-5	3.123E-4	3.621E-3	4.283E-2	0.527

Table 1. Time expenses (in seconds) of hexagonal FFT and hexagonal DFT.

N	4	8	16	32	64	128	256	512
DFT	4.611E-5	3.682E-4	5.061E-3	8.028E-2	1.427	26.521	734.73	12057.
DIF	1.918E-5	4.468E-5	1.040E-4	3.876E-4	1.669E-3	8.365E-3	4.318E-2	0.211
DIT	1.591E-5	3.720E-5	1.043E-4	4.017E-4	1.807E-3	8.764E-3	4.381E-2	0.221

Table 2. Time expenses (in seconds) of hexagonal FFT and hexagonal DFT.

Table 1 and Fig. 5 (left) indicate that as N trebles, the elapsed time for evaluating the hexagonal DFT by straightforward summation increases about 81 times, while the elapsed time for hexagonal FFTs only increases about 9 times. Table 2 and Fig. 5 (center) imply that as N doubles, the elapsed time for evaluating the hexagonal DFT by straightforward summation increases about 16 times, while the elapsed time for hexagonal FFTs only increases about 4 times. We also plot the elapsed time of the hexagonal FFTs on the right side of Fig. 5 as N increases successively from 100 to 6000. This plot shows us nearly linear curves. All the numerical results demonstrate that our hexagonal FFT algorithms is very efficient, they reduce the complexity of the hexagonal DFT from $\mathcal{O}(N^4)$ to $\mathcal{O}(N^2 \log N)$.

By performing round-trip FFTs (first FFT and then inverse FFT), we find that the errors of the hexagonal FFTs grow very slowly as N increases (see Fig. 6). Thus we conclude that our hexagonal FFT algorithms are efficient, accurate and stable.

362   Huiyuan Li and Jiachang Sun

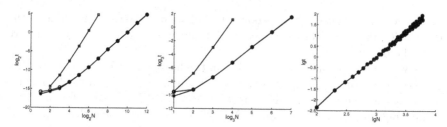

Fig. 5. Logarithm plots of the elapsed time for hexagonal DFTs and FFTs. Left: $N = 2^n$; Center: $N = 3^n$; Right: $N = 100n$. $\square$: DFT; $*$: DIF FFT; $\circ$: DIT FFT.

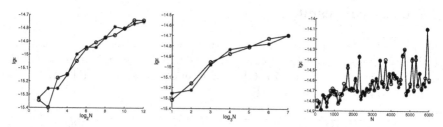

Fig. 6. The errors of the round-trip hexagonal FFTs. Left: $N = 2^n$; Center: $N = 3^n$; Right: $N = 100n$. $*$: DIF FFT; $\circ$: DIT FFT.

References

1. J. W. Cooley and J. W. Tukey. An algorithm for the machine calculation of complex Fourier series. *Math. Comp.*, 19:297–301, 1965.
2. D. E. Dudgeon and R. M. Mersereau. *Mutlidimensional Digital Signal Processing*. Prentice-Hall:Englewood Cliffs, NJ, 1984.
3. R. M. Mersereau. The processing of hexagonally sampled two-dimensional signals. *Proc. IEEE*, 67:930–949, 1979.
4. R. M. Mersereau and T. C. Speake. Unifed treatment of Cooley-Tukey algorithms for the evaluation of multidimensional DFT. *IEEE Transactions on Acoustic, Speech and Signal Processing*, 29:1011–1018, 1981.
5. R. M. Mersereau and T. C. Speake. The processing of periodically sampled multidimensional signals. *IEEE Transactions on Acoustic, Speech and Signal Processing*, 31:188–194, 1983.
6. D. N. Rockmore. The FFT: An algorithm the whole family can use. *Computing in Science & Engineering*, 2(1):60–64, 2000.
7. J. Sun. Generalized Fourier transformation in an arbitrary triangular domain. *Advances in Computational Mathematics*, to appear.
8. J. Sun. Multivariate Fourier series over a class of non tensor-product partition domains. *J. Comput. Math.*, 21:53–62, 2003.
9. P.N. Swarztrauber. Vectorizing the ffts. In G. Rodrigue, editor, *Parallel Computations*, pages 51–83. Academic Press, 1982.
10. J. L. Zapata and G. X. Ritter. Fast fourier transform for hexagonal aggregates. *Journal of Mathematical Imaging and Vision*, 12:183–197, 2000.

LBGK Simulations of Spiral Waves in CIMA Model

Qing Li and Anping Song

School of Computer Eng. & Sci., Shanghai University, Shanghai 200072, P.R.
China The Shanghai University Center for Nonlinear Sciences, Shanghai 200072,
P.R. China {qli,apsong}@mail.shu.edu.cn

Abstract. Simulations of emergence of spiral waves in the chloride-iodide-
malonic acid (CIMA) reaction-diffusion knietic model with a lattice Bhatnagar-
Gross-Krook (LBGK) method are discussed. We get a set of parameters, which
can obtain some typical spiral waves in CIMA model.

Keywords lattice Boltzmann method, reaction-diffusion system, pattern
formation, spiral waves

1 Introduction

The legendary mathematician Alan Turing set up formalism capable of describing formation of stationary concentration patterns by symmetry breaking
in reaction-diffusion systems [MK93]. Experiments on the CIMA reaction in
open gel reactors have revealed the existence of stationary space periodic concentration patterns, so-called Turing patterns [ISO95, GB00]. Lengyel and Epstein have developed a simple two-variable kinetic mechanism-CIMA model:

$$\begin{cases} \frac{\partial u}{\partial t} = \frac{1}{\sigma}\left(a - u - 4\frac{uv}{1+u^2} + \nabla^2 u\right) \\ \frac{\partial v}{\partial t} = b\left(u - \frac{uv}{1+u^2}\right) + d\nabla^2 v \end{cases} \tag{1}$$

In this kinetic model, the iodide and chlorite play respectively the role of
the activator and of the inhibitor [DA93]. Such patterns have been calculated
using a variety of systems of partial differential equations (PDE) [SJ02, QP01,
SW99, LK01, WHY01].

Spiral waves, other kinds of spatiotemporal organization, are simulated in
homogeneous chemical systems by using various cellular automata [JOS02,
GST90]. Instead of the qualitative cellular automata, we will demonstrate
how spiral waves can be emerged from a homogeneous system by using an
LBGK method.

The LBGK method has the advantage of reducing computational costs as compared to the integration of PDEs, because of a coarser space, time discretization and simple relaxation operation. Furthermore, the LBGK method has the advantage of parallelization, because of the local evolution scheme in each iterative step.

2 The LBGK method for reaction-diffusion equations

The reaction-diffusion equations can be generally written as follows:

$$\frac{\partial \rho_s}{\partial t} = D_s \nabla^2 \rho_s + R_s \qquad 1 \le s \le M \qquad (2)$$

where t is time and ∇^2 is the Laplacian operator with respect to the spatial coordinate x, M is the number of species and $\rho_s(x,t)$ is the mess density of the species s at time t and position x. D_s is the diffusion coefficient (in this paper, we assume that D_s is isotropic and independent of x). We will concentrate on the application of reaction-diffusion equations to models of spatially extended system in which several chemical species coexist, diffusing in space while chemically reacting. In such a case, R_s is the reaction term, which depends on ρ_s and the densities of the other species that react with s. There are several LBGK models for reaction-diffusion equations (see refs. [SW99, LK01]). In this paper, we use D2Q5 LBGK model [WHY01] to simulate the formation of spiral waves in CIMA kinetic model.

Consider $M = 2$, use u and v to stand for ρ_1 and ρ_2 respectively. Let us define $f_i(x,t)$ and $g_i(x,t)$ be the distribution functions of the two species with velocity e_i at some dimensionless time, t, and dimensionless position, x. The coordinate x only takes on a discrete set of values: the nodes of the chosen lattice. All the simulations performed in this paper are in two spatial dimensions. The nearest neighbor vectors defined as

$$e_0 = (0,0), \quad e_i = \left(\cos \frac{(i-1)\pi}{2}, \sin \frac{(i-1)\pi}{2} \right) \quad i = 1,2,3,4 \qquad (3)$$

The lattice Boltzmann BGK evolutionary equations for $f_i(x,t)$ and $g_i(x,t)$ can be written as

$$f_i(x + c\varepsilon e_i, t + \varepsilon) = (1 - \omega_1)f_i(x,t) + \omega_1 f_i^{eq}(x,t) + \varepsilon R_1(u,v)/5 \qquad (4)$$

$$g_i(x + c\varepsilon e_i, t + \varepsilon) = (1 - \omega_2)g_i(x,t) + \omega_2 g_i^{eq}(x,t) + \varepsilon R_2(u,v)/5 \qquad (5)$$

where $\varepsilon = \Delta t, c = \Delta x/\Delta t$. The equilibrium distribution functions are

$$f_i^{eq}(x,t) = \frac{u(x,t)}{5}, \qquad g_i^{eq}(x,t) = \frac{v(x,t)}{5} \qquad (6)$$

$$u(x,t) = \sum_{i=0}^{4} f_i(x,t), \qquad v(x,t) = \sum_{i=0}^{4} g_i(x,t). \qquad (7)$$

The relation between the diffusion coefficients, D_s in Eq.(2) and the relaxation factors, ω_s, in the LBGK evolutionary equation Eqs.(4,5) are

$$D_s = c^2 \varepsilon \left(\frac{1}{\omega_s} - \frac{1}{2} \right) \frac{2}{5} \qquad 0 < \omega_s < 2, \qquad s = 1, 2. \qquad (8)$$

3 Spiral Waves in CIMA model

In CIMA kinetic model Eq.(1), the reaction terms are

$$R_1(u, v) = \frac{1}{\sigma} \left(a - u - 4\frac{uv}{1 + u^2} \right), \qquad R_2(u, v) = b \left(u - \frac{uv}{1 + u^2} \right) \qquad (9)$$

Linear stability analysis shows that for realistic values of the parameters in CIMA model Eq.(1), the unique homogeneous steady state, $u = a/5$ and $v = 1 + a^2/25$, may undergo a diffusion-driven instability, even when the diffusion coefficients of the activator and inhibitor are of the same order.

For the simulation of spiral waves, we use periodic boundary conditions in both x and y directions. Because with periodic boundary conditions, spiral waves collide with themselves. Random initial concentration distribution of both species are used and there are about half of grid points that each point is assigned the following values (see Fig.1 at $t = 0$):

$$u(\boldsymbol{x}, 0) = \frac{a}{5}, \qquad v(\boldsymbol{x}, 0) = 1 + \frac{a^2}{25} \qquad (10)$$

such that satisfying $R_1(u, v) = R_2(u, v) = 0$.

We can obtain spiral waves if $d = 1.07$, $20 \leq a \leq 50$, $1.5 \leq b \leq 2.9$ and $8 \leq \sigma \leq 11$. Take $a = 30, b = 2.5, \sigma = 10$ and $\Delta x = 10, \Delta t = 1$, then compute ω_1 and ω_2 with Eq.(8). The results are shown in Fig.1. In Fig.1, the snapshots are taken after 0, 50, 100, 150, $\cdots$, 1700 timesteps.

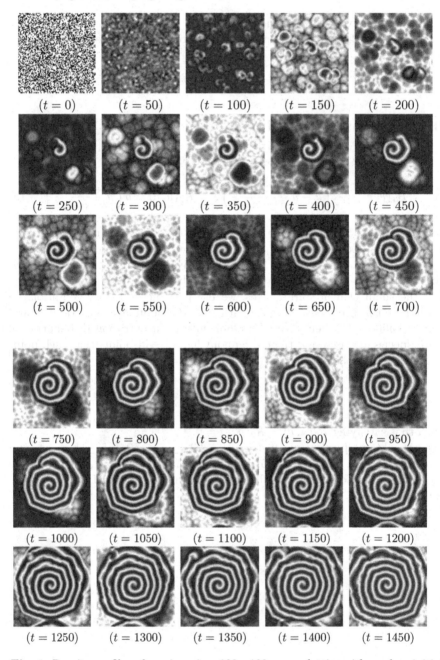

Fig. 1. Density profiles of species u in a 100×100 square lattice with random initial condition

We perform another simulation with the same conditions but with the determinate initial condition (see Fig.2 at $t = 0$).

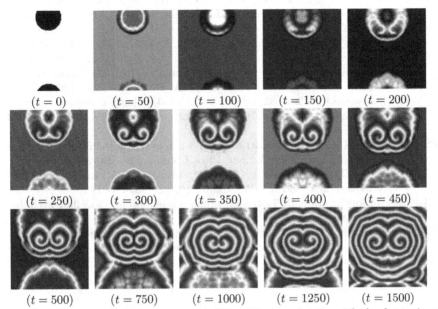

Fig. 2. Density profiles of species u in a 100×100 square lattice with the determinate initial condition

We have shown the existence of spiral waves in CIMA reaction-diffusion system with D2Q5 LBGK method. It should be pointed out that the outcomes, spiral waves, depend strongly on the initial conditions.

Acknowledgments

We are grateful to Prof. Qian Y.H. for his helpful discussions. This work is supported by National Science Foundation of China (Grant 60373071) and Fazhan Foundation of Shanghai(Grant 02AK11).

References

[Tur52] Turing A.M.: The chemical basis of morphogenesis. Phil. Trans. R. Soc. London, Ser. B **237**, 37–72 (1952)

[KCD91] Kepper P.D., Castets V., Dulos E., et al.: Turing-type chemical patterns in the chlorite-iodide-malonic acid reaction. Phisica D, **49**, 161–169 (1991)

[OS91]   Ouyang Q., Swinney H.L.: Transition from a uniform state to hexagonal and striped Turing patterns. Nature, **352**, 610–612 (1991)

[LE91]   Lengyel I., Epstein I.R.: Modeling of Turing structures in the chlorite-iodide-malonic acid-starch reaction system. Science, **251**, 650–652 (1991)

[JPD90]   Jensen O., Pannbacker V.O., Dewel G., et al.: (1990). Subcritical transitions to Turing structures. Physics Lett. A, **179**, 91–96 (1990)

[RBD99]   Rudovics B., Barillot E., Davies P.W., et al.: Experimental studies and quantitative modeling of Turing patterns in the (chlorine dioxide, iodine, malonic acid) reaction. J.Chem. Phys. A, **103**, 1790–1800 (1999)

[DCD93]   Dawson S.P., Chen S., Doolen G.D.: (1993). Lattice Boltzmann computations for reaction-diffusion equations. J.Chem. Phis., **98**, 1514–1523 (1993)

[BS00]   Blaak R., Sloot P.M.: Lattice dependence of reaction-diffusion in lattice Boltzmann modeling. Comput. Phys. Comm., **129**, 256-266 (2000)

[LZW01]   Li Q., Zheng C.G., Wang N.C., et al.: LBGK simulations of Turing patterns in CIMA model. J. Sci. comp., **16**, 121–134 (2001)

[MH90]   Markus M., Hess B.: Isotropic cellular automaton for modelling excitable media. Nature, **347**, 56–58 (1990)

[GST90]   Gerhardt M., Schuster H., Tyson J.: A Cellular automaton model of excitable media including curvature and dispersion. Science, **247**, 1563–1566 (1990)

Parallel Iterative CT Image Reconstruction on a Linux Cluster of Legacy Computers

Xiang Li, Jun Ni, Tao He, Ge Wang, Shaowen Wang, and Body Knosp

Center for Statistical Genetics Research, University of Iowa, Iowa City, IA 52242, USA
{xiang-li, jun-ni, tao-he, ge-wang, shaowen-wang, boyd-knosp}@uiowa.edu

Summary. The expectation maximization (EM) algorithm is one of the iterative reconstruction (IR) algorithms that enable to reconstruct superior CT images, compared with the conventional filtered back-projection (FBP) method. The EM-IR algorithm can also be used when the data is incomplete. The major disadvantage of the EM-IR is its high demand on computation and slow reconstruction. To improve the performance, we developed a parallel EM on a Linux cluster composed of legacy (recycled) and heterogeneous PCs. The system, speed-up and efficiency from our parallel computations are presented. The study provides basic insight into how to conduct medical image reconstruction using junk PCs to simulate a heterogeneous parallel system.

Key words: medical image processing, image reconstruction, parallel computing, LINUX cluster

1 Introduction

X-ray Computed Tomography (CT) is one of the medical imaging techniques that reconstruct images of an objects internal structure. It reconstructs the cross-sectional image by computing the distribution of the objects inherent quality of absorbing X-ray photons using a set of projection data, which record the number of photons measured after the X-ray passing through it. The iterative reconstruction (IR) technique is a branch of methods other than the conventional filtered back-projection (FBP) for CT image reconstruction. While the FBP method reconstructs the image by filtering the projection data and performs a back-projection once to the filtered data, the IR algorithm characteristically computes the final image through many loops of forward-projection and back-projection.

There are many IR algorithms available. Representative among them is the maximum likelihood (ML) expectation maximization (EM) [MK93, ISO95, GB00]. With ML-EM the image is obtained iteratively as an optimal estimate that maximizes the likelihood of the detection of the actual measured photons of a statistical modeling of the imaging system. The EM method can also be deterministically interpreted as the process of minimizing the I-divergence error between the estimated and measured projection data in nonnegative space [DA93]. The EM, together with other IR algorithms, is superior to the FBP method in terms of better image quality (contrast and resolution) under noisy projections [SJ02]. It is especially prominent in the positron emission tomography (PET) and single photon emission computed tomography (SPECT) where the noise level in the projection is high. The IR can also be used with incomplete projection data and for metal artifacts deduction where FBP fails [QP01]. Modern commercial PET and SPECT scanners have begun to include IR as a standard algorithm in the software package [SW99].

However, the major disadvantage of the IR is its high-demand on computation. For a large set of data from 3D volume scan it might take hundreds hours for IR to run on a single computer, preventing it from real time clinical applications. The reason is that besides updating the image a single iteration of the IR needs one forward-projection and one back-projection whereas the FBP only needs one back-projection. As an example for EM often at least 30 iterations are needed resulting in approximately 60 times longer than the FBP to reconstruct an image with similar quality. A practical approach to acceleration of the IR is to parallelize the computation using PC clusters. As the computing technology dramatically advances, many recent parallelisms were based on the second type [LK01, JOS02]. Clusters have been built on either WinNT or Unix/Linux platform and could be a hybrid of SIMD computers and MIMD system. For example, [JOS02] used a cluster of computers each with four processors. With the development of the distributed computing technology, client-server and peer-to-peer network models have been used to decentralize the image reconstruction system [LK01, JOS02, 10].

The paper presents a parallel EM-IR algorithm on a Linux-powered PC cluster. composed of recycled PCs with heterogeneous performance. The projection data were collected from a peripheral quantitative CT (pQCT) scanner STRATEC Xct2000, which is installed in the University of Iowa Hospitals and Clinics. Section2 briefly introduces the parallelization scheme. Section 3 reports the performance of our parallel implementation, followed by a discussion section.

2 Parallel EM-IR Algorithm

The cross-sectional image is sampled with pixels of equal size (Fig. 1(a)). Each pixel has a constant attenuation factor $x_j, j = 1, \ldots, N$ where is the total number of voxels. Estimated projection $\hat{b}_i, i = 1, \ldots, N$ is the line integral along the i-th ray path and $b_i, i = 1, \ldots, M$ is the measured projection, where M is the total number of rays. Let x and b represent the corresponding N-dimensional and M-dimensional column vector respectively; $A = (a_{ij})$ be the matrix mapping x to b, where a_{ij} measures the contribution of x_j to b_i. We have the following linear system: $Ax = b$. The CT image reconstruction problem is to find x. The EM algorithm is formulated as follows: $x_j^{(k+1)} = x_j^{(k)} \left[\sum Mi = 1 a_{ij}(b_i/\hat{b}_i) / \sum Mi = 1 a_{ij} \right], j = 1, \ldots, N$, where $x^{(k)}j$ is the attenuation factor of image pixel j at iteration step k; b_i is the projection data acquired by detector i; a_{ij} is the contribution of x_j to b_i; $b_i = < a_i, x^{(k)} >$ is the estimated projection data at step k, a_i denotes the i-th row of the matrix $A = (a_{ij})$ and $<,>$ the inner product. The reconstruction process is illustrated in Fig. 1(b).

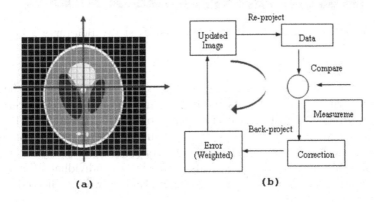

(a)   (b)

Fig. 1. (a) Cross-sectional image is sampled by pixels of equal size and (b) The iteration process

The most time-consuming part of the IR is the forward- and back-projections. The computation is in a piecewise manner for all projection data. As after the projection data is calculated after ray-tracing along a specific X-ray path, the data can be compared with the corresponding measured projection and then be back-projected. With the object-oriented view of the abstract definition of a "ray", each ray takes care of its three actions (forward-projection, get-comparison, and back-projection) and three pieces of data (estimated projection, measured projection, and the comparison between the two). The ray is implemented as a micro-forward-back-projector and its implementation facilitates the parallelization of the IR algorithm. The projection data can be decomposed into several sub-domains and be sent to different processors. The working node performs the computation of the micro-forward-back-projection. The master node gathers temporary images after back-projection and assembles them to generate the updated "global" image and then send the image back to all working nodes for next process of reconstruction. The communications between master and working nodes are achieved using MPI message passing library.

3 Performance Test Results

Fig. 2 shows the reconstructed standard arm phantom during iteration in the parallel computation.

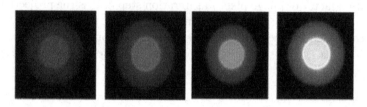

Fig. 2. From left to right: the reconstructed image at iteration times 5,6,7 and 8, respectively.

Table 1 gives the data of computing time, speedup and efficiency of the parallel reconstruction. Fig. 3 plots the benchmarks of performance in terms of speedup and efficiency. The projection data comprises 180 views of 310 samples each. There are 55800 data points. Speedup goes down because the number of processes exceeds the actual number of processors. The performance for small data set are strongly depends on the legacy individual PC's system behavior. As dataset increases, the total performance will be effected by the memory of each processor.

Table 1. The computational time, speedup and efficiency

NP	Time(seconds)	Speed-up	Efficiency
Sequential	185.0		
2	139.0	1.331	0.666
3	126.0	1.468	0.489
4	115.0	1.609	0.402
5	101.0	1.832	0.366
6	88.0	2.102	0.350
9	72.0	2.569	0.285
10	67.0	2.761	0.276
12	66.0	2.803	0.234
15	102.0	1.814	0.121

4 Conclusion

A parallel EM-IR algorithm is developed on a legacy Linux PC cluster composed of heterogeneous PCs. The architecture of such system is dedicated for

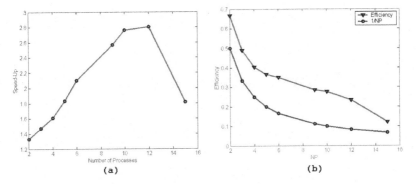

Fig. 3. (a) the Speedup vs. processor number and (b) the efficiency

small research groups or clinics. Our exploration shows a cluster of 12 legacy computers with an average CPU speed of 350MHz and a combined 576MB memory can even serve as introductory HPC system which speedup almost 2/3 of the reconstruction time with the comparison of sequential computing. The present work demonstrates the feasibility of utilizing "a cluster of recycled computers" to solve image reconstruction.

References

1. Rockmore J, Macovski A (1977) A Maximum Likelihood Approach to Image Reconstruction. Proc Joint Automatic Control Conf. 782-786
2. Shepp LA, Valdi Y (1982) Maximum Likelihood Reconstruction for Emission Tomography. IEEE Trans. Med. Imag. MI-1:113-122
3. Lange K, Carson R (1984) EM Reconstruction Algorithms for Emission and Transmission Tomography. J. Comput. Assist. Tomog. 8(2):302-316
4. Synder DL et al (1992) Deblurring Subject to Nonnegativity Constraints. IEEE Trans. Signal Processing 40:1143-1150S
5. Hutton BF et al (1997) A Clinical Perspective of Accelerated Statistical Reconstruction. European Journal of Nuclear Medicine 24 (7)
6. Wang G et al (1996) Iterative Deblurring for CT Metal Artifact Reduction. IEEE Trans. Med. Imag. 15:657-664
7. Leahy R, Byrne C (2000) Recent Developments in Iterative Image Reconstruction for PET and SPECT. IEEE Trans. Med. Imag. 19 (4)
8. Shattuck D et al (2002) Internet2-based 3D PET Image Reconstruction Using a PC Cluster. Phys. Med. Biol. 47:2785-2795
9. Vollmar S et al (2002) Heinzel Cluster: Accelerated Reconstruction for FORE and OSEM3D, Phys. Med. Biol. 47:2651-2658
10. Li X, He T, Wang S, Wang G, and Ni J (2004) P2P-enhanced Distributed Computing in Medical Image EM Reconstruction. the 2004 International Conference on Parallel and Distributed Processing Techniques and Applications (PDPTA)

Grid Communication Environment Based on Multi-RP Multicast Technology*

Xiangqun Li[1,2], Xiongfei Li[1,2], Tao Sun[1], and Xin Zhou[1]

[1] College of Computer Science and Technology, Jilin University, Changchun
    130025, China
[2] State Key Laboratory for Novel Software Technology, Nanjing 210093, China
    lxf@jlu.edu.cn

Summary. Based on the PIM-SM multicast protocol, the paper realized the grid
job submission by building a "shared-based tree". In the multicast group, the source
submits job and members provide resources. The paper also proposed two kinds of
"RP switch" methods: timing switch and gradually switch, in order to resolve bottleneck problem and capability decline of the shared-based tree. Simulation results
show that multicast makes grid job submission effective and RP switch methods can
improve the reliability of job submission, guarantee the QoS of grid communication.

Key words: grid, PIM-SM protocol, RP point, RP switch

1 Introduction

Multicast is the process of sending single packet to multiple destinations. It
can optimize the network performance, meet the transmission requirements
and put the new applications into practice, which unicast and broadcast can't
accomplish well.

These characteristics fit the needs of grid computing perfectly. For example,
submissions of jobs to computing farms, program and data distribution between computing resources. Using the multicast technology in the grid can
lighten the sender's burden and improve the QoS of grid [MK93].

Based on PIM-SM (Protocol Independent Multicast-Sparse Mode) protocol,
the paper built a multicast tree to implement grid job submission. The source
of group submits job and the members provide resources. The dynamic members may result in the tree's performance decrease, so we presented two kinds
of switch methods: Timing Switch and Gradually Switch to improve grid job
submission efficiency.

---

*Supported by the National Natural Foundation of China under Grant
No. 60373097; the Natural Science Foundation of Jilin Province under Grant
No.20020606; the Natural Science Foundation of Jilin University.

The rest of the paper is organized as follows. Section 2 implements the grid job submission based on a multicast tree. Section 3 presents Timing Switch and Gradually Switch. Section 4 is the simulation. Section 5 is the conclusion.

2 Grid Job Submission Based on Multicast

In the session of grid job submission, there are a lot of data communications. Unicast will increase the load on requester and broadcast will increase the network traffic, while multicast can solve the problem perfectly.

PIM-SM protocol [ISO95] is one of the most mature multicast routing protocols. After analyzing the protocol, we implemented PIM-SM protocol's main functions [GB00] [DA93] using raw Socket. The implementation includes forwarding multicast packets, managing control messages, building and maintaining multicast routing table, maintaining all timers and so on. Figure 1 is the framework of the PIM-SM protocol (*note:* RIB is the unicast routing table; MRIB is the multicast routing table).

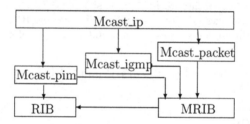

Fig.1. framework of PIM-SM protocol

Mcast_ip module accepts IP packets, analyzes type code and destination address, and triggers corresponding processes.

1. If the packet is a PIM-SM packet, Mcast_ip module triggers Mcast_pim module, which manages PIM-SM protocol's control massages, builds and maintains multicast routing table;
2. If the packet is an IGMP packet, Mcast_ip module triggers Mcast_igmp module, which manages IGMP protocol's control massages, gathers and maintains member information;
3. If the packet is a multicast packet, Mcast_ip module triggers Mcast_packet module. It forwards multicast data packets when successfully matching entry in MRIB; otherwise, it triggers register mechanism or discards the packet.

RP (Rendezvous Point) is the shared-based tree's root. SPT is a shortest path tree between source and RP, gained by unicast protocol. We described particular implementations in other papers.

3 Multi-RP Switch

3.1 introduce multi-RP mechanism

In grid environment, the providers can join or leave a group at any moment. The dynamic state makes the share-based tree's performance worse, so researchers proposed "multi-RP" mechanism, which need switch between the new RP and the old RP. Reference [SJ02] presented several switch methods. However,these methods require that the new RP has the knowledge of network topology and group membership, or needs more control messages making the switch process a bit complex.

In this paper, we proposed two switch methods: timing switch and gradually switch, supposing the new RP has been chosen by some means as in [QP01] [SW99] [LK01], and the old RP is valid.

3.2 switch mechanism

Timing switch mechanism: The old RP starts the timer and forwards new RP's address to the members and the source. Members send Join/Prune messages towards new RP to establish a new share-based tree; source sends Register message to establish new SPT. When the timer expired, new RP replaces the old one to forward multicast data packets. The timer is the key of timing switch.

Gradually switch mechanism: The old RP forwards new RP's address to the members and the source. If the new RP is not on the share-based tree (not a middle router), it joins the old RP. Members and source join the new RP and when a member received reduplicated multicast data packets, it prunes from the old one. The gradually switch process will not finish until all members join the new RP and prune from the old RP.

3.3 new control message

PIM-SM protocol assigns 4 bits for type code so that we can denote 16 kinds of different messages. Besides Register message, Register-Stop message and Join/Prune message [WHY01] which have been defined, we use New_RP message to inform group of new RP address, assigning 9 as its message type. Figure 2 shows the form of New_RP message.

0 3	4 7	8 15	16 31
PIM ver	Type	Reserved	Checksum
Group multicast address			
New RP address (IPv4)			

Fig.2. New_RP message

4 Simulation and Analysis

4.1 compare multicast with unicast

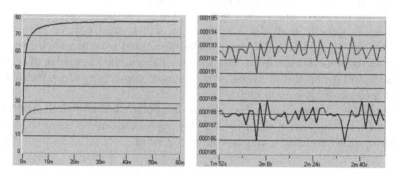

Fig.3. link utilization compare (X: time (minute); Y: utilization)
Fig.4. delay compare (X: test time; Y:delay)

The paper built a campus network using the simulate software that offers net equipments and protocol environments. Figure 3 is the average link utilization of unicast and multicast when submit jobs. The result shows unicast's link utilization is twice more than that of multicast's.

Figure 4 compares the data delay of different share-based trees, one tree roots on the old RP, and the other one roots on the new RP. The red line is old multicast tree's delay and the blue line is the new multicast tree's delay which is decreased by 2.59%.

4.2 RP switch simulation and test

Figure 5 shows members receive data by timing switch and gradually switch. The blue line expresses timing switch, and there is data loss. This is due to some members had not joined the new RP when the old RP stopped working. We will discuss the timer setting later in our paper. The red line expresses gradually switch, and there is reduplicated packets, which consists with our estimation.

Figure 6 is members' join delay statistics. The statistics shows join delay of majority of members is between 2 and 3 seconds in the campus network. When setting a longer timer the data loss decreased.

As a new generation network, grid communication attracts more and more researchers' attention and multicast technology is integrated into grid technology widely because it meets grid environment perfectly. The paper realized grid job submission based on multicast technology. Simulation shows that using multicast technology to submit job can decrease data delay, offer better

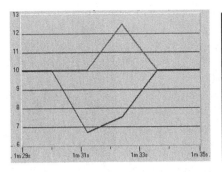 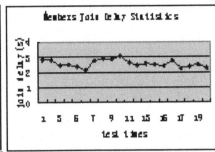

Fig.5. timing switch and gradually switch compare
(X: time; Y: member received data (packets/second))
Fig.6. member join delay statistics (X: test times; Y:delay(s))

quality of service in grid environment. The paper also presented two kinds of RP switch mechanisms: timing switch and gradually switch. Test results of two kinds switch show their advantages and disadvantages, we can use them according to circumstances.

References

[AM01]   Anirban Chakrabarti, G.Manimaran: A Case for Scalable Multicast Tree Migration, 2001IEEE, pp2026-2030

[B0307]   Berkeley, XORP PIM-SM Routing Daemon Version 0.3, www.xorp.org/releases/0.3/docs/pim/pim_arch.pdf; Jul(2003)

[B0311]   Berkeley, XORP Multicast Forwarding Engine Abstraction Version 0.5, www.xorp.org/releases/current/docs/mfea/mfea_arch.pdf, Nov(2003)

[Bea2000] [American] Beau Williamson write, jinxin Gu, qinliang Nan translate: Developing IP Multicast Networks volume 1, publishing house of electronics industry; Jun(2000)

[Mik02]   Mikael Prytz: RP Replacement of Shared Multicast Distribution Trees in a Shortest path routing network; Optimization and System Theory (2002)

[RFC2362] RFC2362, ftp://ftp.rfc-editor.org/in-notes/rfc2362.txt, (1998)

[Seb01]   Seborah, Estrin: RP Relocation Extension to PIM-SM Multicast Routing PIM Working Group,dragt-ydlin-pim-sm-rp-01, (2001)

[UCC]   University of Cambridge Computer Laboratory, multicast transport for grid computing; www.escience.cam.ac.uk/projects/multicast/

[WWH02]   Ting-Yuan Wang, Lih-Chyau Wuu, Shing-Tsaan Huang: A Callable Core Migration Protocol for Dynamic Multicast Tree, Journal of Information Science and Engineering 19 2002, pp479-501

Building CFD Grid Application Platform on CGSP

Xinhua Lin[1], Yang Qi[2], Jing Zhao[1], Xinda Lu[1], Hong Liu[2], Qianni Deng[1], and Minglu Li[1]

[1] Department of Computer Science & Engineering, Shanghai JiaoTong University, Shanghai 20030, P.R. China {lin-xh,zhaojing}@sjtu.edu.cn, {lu-xd,deng-qn,li-ml}@cs.sjtu.edu.cn
[2] Department of Engineering Mechanics, Shanghai JiaoTong University, Shanghai 20030, P.R. China {luvanaki,hongliu}@sjtu.edu.cn

Summary. The Computational Fluid Dynamics (CFD) Grid Application Platform (GAP) is one of five major application platforms in ChinaGrid. It provides a Grid environment for different CFD applications which need high performance computing. In this paper, first we introduce the project background of ChinaGrid General Support Platform (CGSP), and then compare CGSP with Service Domain (SD), which adopted as hosting environment for CFD GAP originally. At last we present how to build the CFD GAP on CGSP and extend its functions for specified CFD application requirements.

Keyword. CFD, GAP, CGSP, Service Domain

1 Introduction

Computational Fluid Dynamics (CFD) integrates computational mathematics, computer science, hydromechanics, computer visualization, and etc. Its main research areas include computational hydrodynamics, computational aerodynamics, computational thermo-fluids, numerical weather forecast, and etc [Anderson95]. The CFD Grid Application Platform (GAP) in ChinaGrid is been built to provide a Grid environment for different CFD applications which need high performance computing. Another meaning of word "GAP" is we are trying to "bridge" the "GAP" between the CFD engineers and computer scientists. Currently we are building the CFD GAP on CGSP.

1.1 CGSP Project Background

ChinaGrid General Support Platform (CGSP) is the fundamental of China-Grid. There are five application platforms running based on CGSP, such as image processing platform, CFD platform, bioinformatics platform, course on-line platform and large scale information processing platform. So CGSP is often called as "the platform of platforms".

CGSP is developed to meet the huge requirements from different research area, most of them need computational power, in China. CGSP is not only a software, like "Globus Toolkit", the de facto standard of Grid. We can describe the relationship between CGSP and Globus as platform and toolkit. The CGSP include parts of modules from Globus, but more than that, such as job management, enhanced information service and packaging existing command-line style programs to Grid Service automatically.

The members of CGSP development team are from the top 5 universities in China, and the demo version of CGSP will appear in October and the first version will be released at the end of this year.

2 Compare with Service Domain

Service Domain (SD)[Tan03] is developed by IBM, shipped with Emerge Technology Toolkit (ETTK) which available in IBM alphaWorks, and now it is combined into WebSphere platform. Basically, SD is a web service registration center and it can invoke the web service registered. We have developed a prototype system of CFD GAP, which adopted SD as the hosting environment to integrate different CFD Web Services and invoke them from SD.

2.1 Service Domain Background

SD is built from standard Web Services and Grid Services infrastructures, and it provides a higher level services for implementing Web Services and Grid solutions.

SD uses three interfaces, which include service entry interface, service attachment interface and service policy interface, to describes a service sharing and aggregation model: The service entry interface is provided as a web service with an externally known Uniform Resource Identifiers service address. All the Web Service can be registered, executed and administered in SD; The service attachment interface is generated from the information registered for the provided services to support different dispatching; The service policy interface is used to set the intelligence rules for the operations of a SD.

2.2 Compare CGSP with Service Domain

As we mentioned before, when we begin to building CFD GAP, we adopted SD as a hosting environment, and it seems work well. However, when we digging more deep into CFD Grid, we find something in SD is inconsistently with our requirements. At that time, we have a chance to participate the CGSP project, and the requirements of CFD applications are highly considered by the architects of CGSP. Based on the technical preview version of CGSP, CFD GAP works much better. Let's take a deep look at what is different between CGSP and SD.

Their goals are different. SD is developed for e-business, and it high considers the requirements from enterprise. For example in SD, users in same group can only have the same operations. It is a flaw in computational science area, where users want to have their own operations.

Their functions are different. SD provides a service management, which in "service triangle", SD plays the role of "service registry". It is for service registration, not responsible for service providing or service deployment. While, CGSP is a platform, playing the roles both "service registry" and "service provider". It packages the existing command-line style programs to web service or Grid service automatically.

The Grid standard They've adopted are different. SD supports the Open Grid Service Infrastructure (OGSI) [Tuecke03], which will be no longer supported in the next generation Grid standard. Globus Toolkit 4 will adopt a new standard called Web Services Resource Framework (WSRF) [Foster04]. While begin to design the CGSP, architects agree to support the WSRF, and so most of CGSP functional modules are followed with WSRF standard.

SD only support simple security infrastructure, users from different role groups simply enter the user name and password , then login into SD, invoke the web service registered in SD. This seem enough for enterprise usage. However, if put SD in wide area network, it will be cracked down soon by some evil programmers very easily. In CGSP, user need not only to enter the user name an password, but also the a certificate authority (CA) file get from the CA center. And every operation provided by CGSP needs to check the user ID.

Users login to SD and they find some web services are listed in web pages. These services are binding by administrators. If the services are limited, users can find the service they want in the services list, however, if the service quantity is huge, users will feel tired to find right service. CGSP provides more effective way to find the service. It use the classification code to search the service in information service modules.

In SD, after some sub hubs registered into main hub, main hub can dispatch jobs to these sub hubs, however, the way for dispatching is random. Though workload balancing is considered in SD, the supporting is very weak. CGSP provides the dynamic job scheduling and considers highly about the workload balancing.

The final flaw about SD is it is a commercial product, and not open source, while CGSP is developed by universities in China, and it is open source and free of charge.

3 CFD specified in GAP

In some sense, CGSP is a platform and CFD GAP is an application based on CGSP. CGSP provides the infrastructure for CFD GAP, and CFD adds some specified functions modules. We propose a concept triangle, which means easy to use for end users, easy to develop applications for application developers and experts, easy to manage for platform administrators.

3.1 Easy-to-use

Most of CFD applications are command-line style, and it is difficult to use for CFD beginners. The GAP adds the visualization layer to CFD applications, in another words, a friendly User interface (UI). by UI, end users have no need to remember what the parameters of applications mean and how to use them.

The other contribution of GAP is CFD work flow template, which can be used in more than 80% CFD applications. It includes three steps: preprocessing, calculation and postprocessing.

Preprocessing is the first step in building and analyzing a flow model. It includes building the model (or importing from a CAD package), creating a mesh. Some commercial preprocessing tools can be found, such as GAMBIT, G/Turbo and TGrid. However, in most cases, CFD researchers develop their own specific preprocessing programs.

After preprocessing, the Domain Decomposition divides the input mesh to some small meshes and the CFD solver does the calculations and produces the results. The most widely used commercial products are FLUENT and FloWizard. FLUENT is used in most industries. FloWizard is the first general-purpose CFD product for designers built by Fluent. Also, large numbers of specific programs are written by researchers.

Postprocessing is the final step in CFD analysis, and involves organization and interpretation of the data and images. most of commercial software includes full postprocessing capabilities. And some free and open source data visualizing tools are also widely used.

3.2 Easy-to-develop

The CFD GAP is not only for end users, but also for some CFD experts. what are the requirements from these experts? In fact they want to use GAP without extra learning cost, which means they want to use the Grid without knowing what is Grid and what is Globus. So GAP provides a very convenient way to package existing CFD applications into web service. After these experts login GAP, they upload the the CFD programs to GAP, binary code is enough, and then select the target machine, which will run the CFD program and publish the web service. That is all about the deploy a CFD web service in GAP. CFD experts do not have to know WSDL, XML, web service and etc.

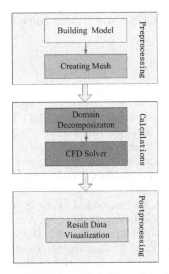

Fig. 1. General CFD Work Flow Template In Three Steps

Currently, GAP can support packaging sequence program running both on Windows and Linux. Parallel programs only running on Linux using MPICH can be packaged automatically, however, the other feathers will be implemented in the future.

CFD applications always have large data files, most of them are hundreds megabytes and some are even more than gigabytes. GAP provides two ways to deal with these large files. First way is moving the data file by ftp. Users can use a ftp applet in web browser move the big file. However, if the file is too big, GAP provide another way, which by using the data management in CGSP returning the URL of a data file. Users can input the Unique Reference Link (URL) as a parameter to CFD service, then the service will find the data file by URL automatically.

3.3 Easy-to-manage

All modules both from CGSP and GAP have the GUI for manage the the way very easily. Platform administrators can use these GUI to manage the CGSP and GAP very conveniently.

4 Conclusion

In this paper, we have introduced the CFD GAP in ChinaGrid from following two perspectives:

(1) CGSP is well-developed general Grid platform comparing with SD in some different aspects.

(2) CFD GAP is built on CGSP and it adds some CFD application specified functions.

References

[Anderson95] J. D. Anderson. Computational Fluid Dynamics: The Basics with Applications. McGraw Hill, US, (1995)

[Tan03] Y.-S. Tan, B. Topol, V. Vellanki, and J. Xing. Business service grid, part 1: Introduction, manage web services and grid services with service domain technology. Technical report, Grid computing in IBM develperworks (2003)

[Tuecke03] S. Tuecke, K. Czajkowski, J. Frey, S. Graham, C. Kesselman, T. Maquire, T. Sandholm, D. Snelling, Vanderbilt. Open Grid Services Infrastructure (OGSI) Version 1.0. Global Grid Forum (2003)

[Foster04] I. Foster, J. Frey, S. Graham, S. Tuecke, et al, Modeling Stateful Resources with Web Services,Technical report, Globus alliance (2004)

A New Grid Computing Platform Based on Web Services and NetSolve

Guoyong Mao, Wu Zhang, and Bing He

School of Computer Engineering and Science, Shanghai University, Shanghai
200072 PRC gymao@mail.shu.edu.cn zhang@staff.shu.edu.cn
he.bing@huawei.com

Abstract. The development in the technology of grid computing and Web
Services has made it possible to integrate various computational resources to
handle challenging scientific and engineering computations[NW03]. Combined
with the technology of Web Services and the existing NetSolve[Don02][CD97],
we present a new grid system for scientific computing. We analyze the system
architecture, illustrated the advantages, and discussed some methods to solve
problems encountered in the process of implementing this system.
Key words: Web Services, NetSolve, Grid Computing, SOAP

1 Introduction

It has become a new trend to integrate various computational resources to
handle challenging scientific and engineering computations. But until recently,
there are still some difficulties in using high performance computer and grid
technology on a large scale as there are few efficient developing tools and
programming environments.

Web Services is a technology that allows heterogeneous systems to be integrated rapidly, easily and at a lower cost than was possible ever before.
Moreover, this integration is based on messages derived from service semantics rather than network protocols. This enables a loose and flexible coupling
of business functions across the Internet within different parts of the same
organization and between distinct organization boundaries.

NetSolve is a grid computing middleware used in scientific computation, it is
developed by University of Tennessee and Oak Ridge National Lab. Based on
the Client-Server-Agent, it can facilitate users to solve complicated scientific
computation using computing resources that are distributed among Internet.
NetSolve can search computing resources available in the net, select the best
resource to handle problems, and return the results to users[HJ98].

In this article, we will discuss a new grid computing platform based on Web Services and NetSolve, which can combine the advantages of both Web Services and NetSolve[DGL89].

2 NetSolve System

The NetSolve system consists of Agent, Server and Client. Agent is the interface connecting NetSolve system. It keeps searching for servers that can satisfy requests from clients. Agent can be used to handle load balance, manipulate requests from clients, monitor the performance of the server and offer error acceptance mechanism.

Server is the computing center of the system. Server daemon waits for the requests from clients. Server can be run on single workstation, cluster, SMP or MPP. PDF(problem description file) is the key component of the server, which can be used to add existing software instances into system services.

Clients can access NetSolve system using simple APIs, requests are forwarded in this way. Agent will deal with these requests, it can select the best server to perform numerical computing. Once the server is selected, connection between clients and server are established, data are input from clients to server, and computation is started. Results will return from server to clients directly.

3 Hiberarchy of grid computing system

Many resources, like computing software and scientific packages, can be integrated into server of NetSolve system. Therefore, it has good scalability, and research works surrounding NetSolve are on the rise.

However, NetSolve system doesn't have a universal interface, it's interface protocol can only be applied in the clients of NetSolve system. Therefore, other systems can't communicate with NetSolve directly. Besides, it can't provide a friendly and all-purpose interface for people who want to use the high performance computing resources. Another limitation is: unless clients, agent, servers are located in the same local network, real IP address is needed for them, as they communicate with each other via IP address. But for security and limitation of IP resources concern, many clusters, like ZQ2000 high performance cluster computer of Shanghai University, doesn't have real IP address for each node of the cluster. Therefore, NetSolve clients from outside the local network can't access any server on the node of ZQ2000. With the help of Web Services, these two disadvantages of NetSolve can be settled easily, because Web Services may use HTTP protocol, which can offer a friendly interface to facilitate calling services and make real IP address no longer a necessity. Furthermore, many developing tools, like java and .net, support Web Services well, they can be used to develop software on scientific computing after the services provided by parallel computing environment is described using WSDL. Based on this idea, we bring forward a grid computing system, which is divided into 5 levels.

- **Software and hardware on Parallel cluster.** Computer, network, operation system and other necessary software and hardware are included in this level.
- **Parallel numerical computing environment.** Parallel programming software and parallel numerical computing software library like MPI, PETSC and Scalapack are included in this level. They serve as parallel programming and executing environment. The parallel numerical software library can improve the efficiency and robustness of parallel programs. The above two levels constitute the NetSolve server.
- **NetSolve agent.** NetSolve agent is included in this level.
- **Web Services server.** The Web Services server consists of service and service description. A service is a software module deployed on network accessible platforms provided by the service provider. It exists to be invoked by or to interact with a service requestor. While the service description contains the details of the interface and implementation of the service.
- **Application Development Environment.** Software systems supporting Web Services, like .net and java Web Services developing tool kits, are adopted in the application development environment.

4 Work flow and implementation of system

The work flow of system can be summarized into three steps:

1. **Query service.** Developing tools that supporting Web Services are used to develop software on scientific computing. These tools can be used to query which service exist in Web Services server. Service and it's interface described in WSDL format are returned to users from the server. WSDL documents are used by clients to generate stub needed in services calling. After that, clients can call the service, just like calling a common local process.
2. **Execution of service.** Clients send SOAP (simple object access protocol) [AN96]requests to Web Services server. On receiving the requests, the Web Services server converts it into a service execution request to NetSolve system. NetSolve system can then select the best server to execute the corresponding service. The Web Services server can also call other servers to execute services that are not provided by a given server.
3. **Return of service results.** NetSolve system returns the results of a service to Web Services server after it is completed. Clients can then get SOAP responses described in XML format from the Web Services server.

In our system, Web Services server provides interfaces for users, and these interfaces are provided in the form of standardized services. Application software of clients can be developed using developing tools that support Web

Services. The main task of Web Services interface is to provide call interface for external users, deserialize the SOAP requests that are in the form of XML, and turn SOAP requests into corresponding service call.

Web Services server turns services provided by NetSolve system into his service, which is needed by Web Services call interface. The Web Services server and NetSolve agent are installed on the same computer. All services provided by NetSolve servers are registered in NetSolve agent. We use java language to encapsulate services provided by NetSolve in the form of API into services of Web Services, and deploy these services in Web Services server. Parameters needed in calling these APIs are sent in the form of SOAP messages from clients to Web Services server. Query interface of Web Services provides call interface for users to query existing services. It offers URL hyperlink of WSDL file of Web Services to users in the form of web pages. WSDL file gives a detailed description about the service interface of system in the form of standardized XML file. Client program can be generated using WSDL file.

5 Testing environment and performance analysis

Our testing environment is based on ZQ2000 cluster computer of Shanghai University. A program designed for parallel solving linear equations has been added to the NetSolve library using PDF. Since such program is very common in scientific and engineering computation, we use this program to test the performance of system. The equation is Ax=b, where A is n*n sparse matrix. We did three groups of tests, the first is in a local network, the second in the campus net and the third in the Internet. In these tests, real IP address is only available in the computer where Web Services server and NetSolve agent are installed . This computer has two Ethernet cards, one with real IP address, the other with local IP address, which is located in the same local network with ZQ2000. The test result is shown in Fig 1, where the lowest line represents time needed in direct API calling; the middle line SOAP calling in the campus net; and the top line SOAP calling in the Internet. Obviously, calling service via SOAP is greatly determined by network status.

From the test we can see that the goals (easy to develop, no IP limitation, firewall free and higher security level) we brought forward is achieved in our grid computing platform.

With the combination of Web Services and NetSolve, We provide a development environment to apply grid technology in numerical computing, which has been successfully applied in the ZQ2000 high performance cluster computer of Shanghai University. However, prototypical as it is, there is still much to do about this development platform before we can apply it on a large scale. At the same time, the technology of Web Services is developing very fast, so we can look forward to the emergence of new and better technology. Hence, we should constantly improve our system to catch up with this fast development so as to provide a better platform in numerical computing.

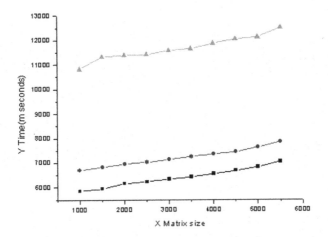

Fig. 1. Time needed in different environments as matrix size increased

Acknowledgements. This paper was supported by "SEC E-Institute: Shanghai High Institutions Grid" project, the fourth key subject construction of Shanghai, and the Key Project of Shanghai Educational Committee (No.03AZ03).

References

[CL03]   Cai,X.L., Liang,Y.Q. (2003) Web Services technology, architecture and applications. Publishing house of electronics industry, BeiJing

[Don02]   Dongarra,J. (2002) Logistical Quality of Service in Net-Solve.http://www.cs.utk.edu/netsolve/

[DSS02]   Dorian,A., Sudesh,A., Susan,B. (2002) User's Guide to NetSolve v1.4.1. http://www.cs.utk.edu/netsolve/

[MMJ01] Martin,G., Marc,H., Jean-Jacques,M., Henrik,F. N. (2001) SOAP Version 1.2,W3C Working Draft 9. http://www.w3.org/TR/SOAP12/.

[HJ98]   Casanova,H. and Dongarra,J. (1998) Applying NetSolve's network enabled server, IEEE Computer. Sci. & Eng., 5:3, 57-66

[ICJ02]   Ian,F., Carl,K., Jeffery,N., Steven,T. (2002) Grid services for distributed system integration. IEEE computer,35(6):37-46

Uncertainty Analysis for Parallel Car-crash Simulation Results [*]

Liquan Mei[1] and C.A. Thole[2]

[1] School of Science, Xi'an Jiaotong University, Xi'an 710049, P. R. China
   lqmei@mail.xjtu.edu.cn
[2] Fraunhofer Institute for Algorithms and Scientific Computing, Schloss
   Birlinghoven, 53754, St. Augustin, Germany

Abstract. Small changes in parameters, load cases or model specifications for crash simulation may result in huge changes in the results, characterizing the crash behavior of an automotive design. For a BMW test case, differences between the position of a node in two simulation runs of up to 10 cm were observed, just as a result of round-off differences in the case of parallel computing. The paper shows that numerical properties of the simulation codes as well as bifurcations in the crash behavior in the certain parts of the design are reasons for scatter of simulation results. The tool DIFF-CRASH^{TM} was developed to compare simulation results and cluster those nodes of the car model, which show similar scatter among the simulation runs and then trace back to a certain part to remove the uncertain behavior. DIFF-CRASH^{TM} is the only activity using data mining technology for crash simulation uncertainty analysis.

Keywords: Uncertainty analysis, Crash Simulation, Data Mining, Clustering

1 Introduction

Nowadays the car manufacturing industry relies heavily on simulation results. By simulation the number of real prototypes is reduced, the insight into the features of the actual design is increased and the turn-around time between model changes is much shorter than in the case of real tests. Numerical crash simulation is the most computer-time consuming simulation task in car design. Therefore it is obvious that crash simulation codes were among the first industrial simulation codes, which were ported onto parallel distributed memory architectures [TS1][CG1]. Using the mpp-versions of industrial crash simulation codes the engineers made a surprising discovery for certain models: The

---

[*]This work was funded by the German Ministry for Education and Research (BMB+F), NSF of China 10471109, SRF for ROCS SEM of China .

result of numerical simulation changed from one parallel execution to the next by more than 10 cm for the node positions, although the input decks and the simulation parameters were identical. For the test case provided by BMW consisting of about 60.000 shell elements , the maximal differences between several simulation runs on state of 80ms are larger than 10cm. Actually this observation has stopped car manufacturing companies from using mpp-system for crash simulation for more than 5 years[Mal].

Geometric scatter analysis is performed in weather forecast, and stability analysis is usually performed by stochastic variation of some design parameters and correlation analysis for some key measures like the intrusion. As part of the AUTOBENCH Project (1998-01) and the AUTO-OPT Project(2002-05)[KT1], the reasons for the scatter of the results were investigated in detail. It turned out that numerical properties of the simulation codes as well as certain features of the car design may be responsible for the "butterfly effects". Typical sources of instabilities are buckling and contact of different parts under an angle of 90. Due to the nature of crash simulation many parts might show uncertainty behavior. Usually only a small subset has a real impact on those values which measure the crash behavior (like intrusion). Measuring the scatter of simulation results for these characteristic values is a first step. In order to improve the design it is necessary to trace this scatter back to its origin in space and time.

During the investigation an analysis tool named DIFF-CRASH^{TM} was developed which allow to measure the scatter and trace back to its origin. it enables the engineer to understand the mechanisms of propagation and amplification of scatter during the crash itself as a basis for the improvement in stability of the car design.

2 Uncertainty analysis functions of DIFF-CRASH^{TM}

DIFF-CRASH^{TM} is a tool for the detailed analysis of the scatter of simulation results. Currently it supports PAM-CRASH and LS-DYNA , two of the four leading commercial simulation codes. For a detailed investigation of the uncertainty of simulation results, DIFF-CRASH^{TM} performs data mining operations on the binary output files of PAM-CRASH and LS-DYNA and generates additional values per point and time step. These results are added as scalar and vector functions to the binary output files and may be visualized using the code specific tools PAMVIEW and LS-POST as well as the GNS-Animator visualization tool.

DIFFCRASH^{TM} supports series of uncertainty analysis functions. Here we introduce several of them. One challenge for this work is the problem size. A typical crash simulation code consists of 500,000 nodes and for a dependence analysis up to 150 time steps may be used. The data bases of objects to be compared therefore contain 70 million objects. Fortunately, the geometric position of the nodes provides some structure, which can be exploited.

PD3MX

Function PD3MX indicates the maximal 3D distance between two of the positions of the same node at the same state but of different runs

$$PD3MX(n,t) = \max_{i,j}\{\|P_i(n,t) - P_j(n,t)\|\} \tag{1}$$

where $P_i(n,t)$ is the position of node n in time t for the ith run.
Parallel execution implies that each processor is responsible for a certain part of the model. At interfaces between several processors the sequence for the summation of the simulation results is not fixed. Different sequence may lead to small differences for example for force calculation at interface nodes. Here PD3MX function is used to measure the scatter of parallel crash simulation results, an then try to find some interesting part which have strong uncertainty behavior.

PD3IJ

Function PD3IJ shows the sequence numbers I and J of the two runs, which have the maximal distance

$$PD3IJ(n,t) = i(n,t) + 0.01 \times j(n,t) \tag{2}$$

where $i(n,t)$ and $j(n,t)$ satisfied

$$PD3MX(n,t) = \|P_{i(n,t)}(n,t) - P_{j(n,t)}(n,t)\| \tag{3}$$

Function PD3IJ allows to identify the most extreme two runs in some area so that we can do further research about this two most extreme runs to understand what it's going on.

Similarity Function

Let $X(p,t,s)$ be the position of node p at time t during the crash simulation process of run s out of a total number of S simulation runs. $X(p,t) = \{X(p,t,s)|s \in (1 \cdots S)\}$ is the ordered set of node positions and

$$M_{ij}(p,t) = \frac{\|X(p,t,i) - X(p,t,j)\|}{\max\limits_{k,l}(\|X(p,t,l) - X(p,t,k)\|)} \tag{4}$$

is the related scaled distance matrix. The similarity function $sim(X1,X2)$ then defines a relation between two sets of node positions. A computed value of 1 indicates that the scatter of the nodes is the same and a value over a certain threshold b_{sim} indicates a significant relation between the scatter of these two node position sets. If the scatter of the node positions of two nodes at the two time steps $(p1, t1)$ and $(p2, t2)$ is similar, the scatter at these two points is itself likely to have the same origin.

For a given reference (p_{ref}, t_{ref}) at any node and time step (p, t), DIFF-CRASH implements a function, which returns $sim(X(p, t), X(p_{ref}, t_{ref}))$.

The Similarity function was based on singular vector idea, but it is more convenient to show the results in visualization tools because it use a scalar instead of a vector. it can be used to identify the source of the part with indeterministic behavior for parallel simulation runs.

Clustering

Clustering is to propose to develop algorithms that automate the process of finding the number of group(clusters), and find high-quality solutions to group memberships. Data Mining for scientific applications in particular cluster analysis is used in many different application areas, especially for simulation data here.

One of the most difficulties known from the literature [JD1][Ha1] is the fact that the assignment to clusters is not a deterministic process. One node might be related to two others, which are not related with each other. Therefore this node might be assigned to any of the two reference clusters. In order to reduce these effects, the clustering process for each time state is performed in the following steps .

· Perform clustering $Cluster(N)$ for this reference time step

· Find the center node for each cluster

· Perform the second $Cluster(N)$ to check for all related nodes to these center nodes

· Sort center nodes by the size of clusters to get a new node sequence.

For the details of each of these steps see [MT1] .

Measure of distance is needed for clustering process, here for the simulation data the measure is based on our similarity function.

3 Uncertainty Analysis for Test Cases

In this paper, we only consider the BMW test case which contains about 60,000 elements crashed at 30 mph against a fixed wall. 17 runs were performed using mpp PAM-CRASH on an IBM SP-2 for the same input deck without any variation. Figure 1 (left) shows a PD3MX result for the parts of the fire wall close to the driver's feet at 81 ms, here we discuss the uncertainty of this part for example. The color indicates the scatter of the results at each point. Red areas are those with the largest values.

Figure 1(right) shows the result of the cluster algorithm as colors on the test case. Each color represents a new cluster. In total the algorithm identifies 13 different clusters for the whole car. One cluster dominates the fire wall in the area of the driver's legs for those nodes with the largest scatter.

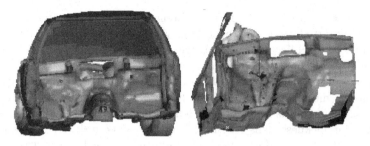

Fig. 1. PD3MX result (left) and the Clustering result(right) for 17 runs on the fire wall parts

According to Figure 2, the "green" cluster shows up between 40 and 50 ms for the first time on the fire wall and gets larger. This cluster starts when the shock absorber pipe hits the rest of the motor carrier near 28 ms. At this point in time the shock absorber pipe puts pressure on the motor carrier parallel to the direction of this part. This causes a buckling effect on the part of motor carrier. The clustering function indicates, that this critical buckling causes the substantial scatter at the fire wall.

Fig. 2. evolution of the clusters with time(70ms, 60ms, 50ms, 40ms, 38ms, 28ms)

Original

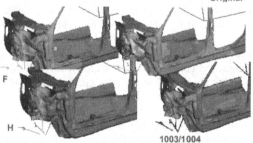

1003/1004

Fig. 3. Scatter of several variants of the original model showing a substantial improved behavior

PD3IJ results also show a slightly different position in the area where the engine block hits the axle part near 24ms which causes the axle to slip up the engine block in a different way. This has substantial impact on the way how the axle impacts the drivers foot area.

Our work aims at stable car model. As a result of this analysis, some attempt was done to modify the parts in the areas of the origin of the instability. Here small change was done on the element of the part of motor carrier by changing the thickness, density or node position respectively. As a consequence the scatter of the result due to parallel computing on the fire wall was reduced substantially, which was shown in figure 3.

Conclusions and perspectives

DIFF-CRASH is the only activity using data mining technology for crash simulation uncertainty analysis. The paper has shown that DIFFCRASH can be useful in measuring the scatter of parallel car crash simulation results and identifying the origin of the scatter of simulation results in crash simulation. As a result of the analysis using DIFF-CRASH the motor carrier was modified in the areas of the origin of the instability. As a consequence the scatter of the result due to parallel computing on the fire wall was reduced substantially.

This application is only a first step using data mining technology in the context of industrial crash simulation. Car manufacturing companies store the complete outputs of their crash simulation runs. This provides an excellent basis for data mining applications, like design of experiments, optimization, parameter fitting for coarser models used in concept studies.

References

[CG1]   Clinckenmaillie, J.,Galbas, H.G.: High scalability of parallel PAM-CRASH with a new contact search algorithm. Lect. Notes in Comp. Sci.,**1823**:439-444(2000)

[Ha1]   Hamerly, G.: Learning structure and concepts in data through data clustering, doctoral thesis of Uni. of California, San Diego (2002)

[JD1]   Jain, A.K., Dubes, R.C.: Algorithms for Clustering Data, Prentice Hall (1988)

[KT1]   Kuhlmann,A., Thole, C.A., Trottenberg, U.: AUTOBENC/AUTO-OPT: Towards an integrated construction enviroment for virtual prototyping in the automotive industry. Lect. Notes in Comp. Sci. **2840**:686-690(2003)

[Ma1]   Marczyk, J.: Principles of simulation-based computer-aided engineering. FIM Publications, Barcelona (1999)

[MT1]   Mei, L., Thole, C.A.: Clustering algorithms for parallel car-crash simulation analysis, Proceedings of High Performance Computing Conference 2003, Hanoi

[TS1]   Thole, C.A., Stuben, K.: Industrial simulation on parallel computers. Parallel Computing **25**:2015-2037(1999)

Distributed Partial Evaluation on Byte Code Specialization

Zhemin Ni, Hongyan Mao, Linpeng Huang, and Yongqiang Sun

Department of Computer Science and Engineering,
Shanghai Jiaotong University, Shanghai 200030, P.R. China
{zhmni, mhy, lphuang, sun-yg}@sjtu.edu.cn

1 Introduction

Web applications based on Java have platform-independent and portable features, however, its interpreted features limit the execution speed. Just-in-Time and Hotspot compiling techniques are employed to enhance the running performance, the speed up obtained is not observable and at the expense of memory. The partial evaluation technique as an automatic program transform technique is used to specialize generic programs into specific implementation for given parameters. Jspec specializes Java programs using Tempo and Harissa, a Java-to- C compiler [MY99], but the specialized programs are no longer applied to other application programs. Masuhara and Yonezawa [RHE02] give the Byte Code Specialization. There is a little study about partial evaluation building on object-oriented languages and heterogeneous environments. This paper presents the byte code specialization to partially evaluate Java instruction sequences, which availably deals with virtual function dispatch, control-shift instruction, etc. Constant propagation, function unfolding and structure transform are used to simplify the source program and yield the effective byte code program. The distributed partial evaluation framework is given and analyzed that client utilizes the specialized class file of the server, so locally speeds up the running process. The ray tracing application is performed to investigate the improvement of performance, that the speedup is achieved as expected that shows partial evaluation is beneficial to the performance of program.

2 Partial Evaluation

2.1 Partial evaluation

Partial evaluation is an automatic program transformation technique, which improves efficiency of a program by performing as much computations as possible before all the inputs become available [CD93]. Using partial evaluation technique constructs a partial evaluator that produces the specialized version of a given program for the known parameters. A partial evaluator propagates constant values, folds constant expressions or unfolds function calls, besides that, it eliminates all verifications concerning the validity of the arguments with the available context.

Given a power function computing x^n, if the exponent n is known (for $n = 5$), we can get a specialized version $power\_5(x)$ with respect to the parameter x, which eliminates the condition statements and operation statements on constant, and is much faster than the general program $power(5, x)$.

2.2 The partial evaluation based on byte code specialization

The platform neutrality of Java language is convenient for running on heterogeneous hosts, and that programs are simple and portable. However Java programs also give rise to performance problems due to the interpreted feature. Accordingly, we combine partial evaluation into Java Virtual Machine (JVM) to optimize Java programs. The Java-oriented partial evaluator, called Jmix, is proposed and implemented which uses the Byte Code Specialization (BCS) [MY99] by parsing and specializing Java Machine Virtual Instructions of class files. The source byte code program is transformed into the specialized byte code program with respect to a given usage context.

Online specialization approach is adopted that records the state of any variable, instruction and stack, and the byte code program is treated as a specialized object. By parsing the structure of Java class files, Jmix gets the constant pools and keeps in the Vector object. In addition, the specialized instruction sequence is obtained. Comparing with other partial evaluators [MY99, RHE02], Jmix has some observable features. (1) Jmix can evaluate virtual function dispatch, stack access, inclined method, and array bound. (2) The control-shift instructions are solved through the static semantic of the byte code in the undetermined conditions. (3) Jmix also handles the specialization of multi-objects besides the specialization of the methods of a single object. (4) In terms of byte code, it's easy to prove the correctness of the evaluation.

3 Partial evaluation working in distributed environments

3.1 The framework of distributed partial evaluation

The distributed partial evaluation is partial evaluation technique extended to geographically dispersed environments in order to enhance the efficiency of distributed application. According to Jmix and web technologies, we construct a distributed partial evaluator namely DJmix. Jmix only performs the specialization in a single host. DJmix not only possesses the capability of Jmix but also manages the communication and collaboration of heterogeneous environments. By means of the partial evaluation and the parallel execution, the performing efficiency increases as compared to running in the single host. The distributed partial evaluation framework is composed of a name server, a client and many of servers. DJmix built on Jmix is platform-independent, hence DJmix can be hosted on any operating system like Windows, Unix and Linux with JVM. If the service in client includes methods and classes to be specialized, then the specialization process occurs locally. If the service is in server a, which will partially evaluate and return the specialized service to client, otherwise server b performs the specialization.

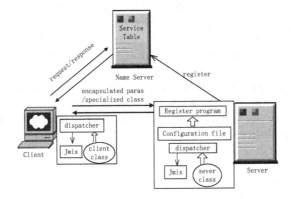

Fig. 1. The distributed partial evaluation framework

- The name server keeps a table, which includes all the configuration information of each server. The client and server contain a dispatcher and a Jmix, in addition each server has a register program and a configuration file.
- The dispatcher is responsible for communicating with the name server and other servers. The dispatcher of the client constructs a request message, including the specialized service needed and known parameters, and sends to the name server, and in turn the name server returns the queried results.
- The configuration file describes the information relevant with class files, and is registered in the service table of the name server by the register program. The configuration file contains the following information encapsulated with XML.

3.2 Experiment results

In distributed computing environments, lots of applications are involved with considerable computations and memory demands. Here, we introduce a case study about the ray tracing to investigate partial evaluation. Ray tracing is a well-known object space rendering technique for image synthesis [And96]. The ray tracing is more suitable for optimizing with partial evaluation.

The local computer as the client contains DJmix and the ray tracing program, and the ray tracing program runs on top of DJmix. The client loops over all the pixels of an image and dispatches tasks to the servers. The name server and the server are responsible for registry and complicated computations, and returns results to the client. Thus, every subtask of the ray racing is specialized by partial evaluator. Every program separately runs in the hosts without DJmix and with it. The speedup achieved is from 2 to 5 [MHS03], and it's observed that the more complex images, the higher speedup is achieved.

There is little work on partial evaluation with respect to object-oriented languages and distributed environments. Jspec specializes Java programs using Tempo and Harissa, a Java-to-C compiler, but the specialized programs are no longer applied to other application programs [CHM98, MY99]. DJmix is the first distributed partial evaluator building on Java byte code specialization adapted to distributed environments.

The partial evaluation based on byte code specialization is analyzed and the distributed partial evaluation framework is described. The experiment of the integration of the ray tracing with DJmix is discussed that it's interesting to optimize the performance of more web applications and produce more economic value.

References

[CD93] Consel, C. and Danvy, D.: Tutorial notes on partial evaluation. In ACM Symposium on Principles of Programming Languages, Pages 493-501 (1993)

[CHM98] Consel, C., Hornof, L., Marlet R., Muller G. and Thibault S.: Tempo: Specializing Systems Applications and Beyond. ACM Computing Survey, **30**, Issue 3es (1998)

[MY99] Masuhara, H., and Yonezawa, A.: Generating optimized residual code in run-time specialization. In Proceedings of International Colloquium on Partial Evaluation and Program Transformation (1999)

[And96] Andersen, P.H.: Partial Evaluation Applied to Ray Tracing. Software Engineering in Scientific Computing, pp. 78-85, DIKU report D-289 (1996)

[RHE02] Reynld, A., Hidehiko, M., Eijiro, S. and Akinori, Y.: Supporting Objects in Run-Time Bytecode Specialization. In ASIA-PEPM'02 (2002)

[MHS03] Mao, Y., Huang, P., and Sun, Q.: Partially Evaluating Grid Services by DJmix. GCC, China (2003)

Scientific Visualization of Multidimensional Data: Genetic Likelihood Visualization *

Juw Won Park, Mark Logue, Jun Ni, James Cremer, Alberto Segre, and Veronica Vieland

Center for Statistical Genetics Research, University of Iowa, Iowa City, IA 52242, USA
{juw-park, mark-logue, jun-ni, james-cremer, alberto-segre, veronica-vieland}@uiowa.edu

Summary. Although many computer graphic technologies have been developed for visualizing multidimensional multivariate data, the scientific visualization used by research scientists to interpret genetics data is very promising technique. In this paper, we present our research in a scientific visualization on linkage analysis data to enhance the performance or the efficiency of genetic likelihood research.

Key words: scientific visualization, multidimensional data, linkage analysis, computer graphics, genetic data, biostatistic

1 Introduction

Linkage analysis methods are generally used to scan the human genome for disease gene loci, special locations on a chromosome where a gene is located. Since the parameters that fit a complicated probability model are difficult to estimate, genetic researchers are left guessing what parameter values to use before the study is performed and left wondering what effect that choice had at the end of the study. Recently, distributed computing has been implemented to allow a broader view of this complex parameter space: visualization can help researchers interpret and navigate these results.

The statistic method we used for linkage analysis requires the computation of the probability, or likelihood, of the pattern of inheritance observed in the marker data. The method calculates a scaled version of probability, known as HET-LOD or HLOD. The HLOD function is a function of 6 parameters over the pedigree data. The first step of HLOD computation is MLIP (Multiprocessor LIPed: Multiprocessor version of Jurg and Ott's LIPED program).

---

*This project is supported by ITS (Information Technology Services) at The University of Iowa and Center for Statistical Genetics Research at The University of Iowa.

MLIP results in 6-dimensional data representing LOD values based on the full variations of the ranges of 5 parameters. Using these LOD scores and the variations of remaining parameter values, the HLOD values can be calculated. The result data set is 7-dimensional data set representing HLOD and other 6 model parameters.

It currently takes significant time and efforts to examine the data set because the data is not only enormous in size but also multidimensional. Visualization can facilitate researchers discovery process by displaying the 3-D surfaces in interactive or guided manner. It helps researchers understand/reveal relations among parameters that would not have been easy to see without visualization of the data. As researcher doesnt make the assumptions at the outset of the study, the effect of each parameter can be determined more reasonably and precisely.

2 Visualization of Multidimensional Multivariable Data

Through out the study in multidimensional multivariable data visualization history, some techniques attempt to show all dimensions as one display, whereas others focus on interactive techniques, in which the user interactively selects subsets for display by using an input device such as mouse or keyboard. During the many years, a lot of new multidimensional multivariable data visualization techniques have been invented and some of them started using both techniques at once. They display all dimensions at once and they interactively change views too. The following lists some techniques that have been invented over last three decades.

Multidimensional Display
Stick Figure Icon, Autoglyph, Color Icon, Hierarchical Axis(Histogram plot), Dimension Stacking, Parallel Coordinates, XmdvTool[2]
Interactive Techniques
Brushing, Hyperslice, Worlds within Worlds[3]

We can see that our visualization tool falls in the category of interactive technique as we display the 3D subset of 7D data interactively.

3 Development of Prototype Visualization Tool

Based on the model parameters, the HLOD values are calculated and stored in a file along with six parameter values. As the standard 3D graphics can only

---

[2]Matthew Ward integrated four popular multidimentional data visualization techniques and they are scatterplot, dimension stacking, star glyph, and parallel coordinates.

[3]A.K.A. AutoVisual. This technique was developed by Clifford Beshers and Steven Feiner from Columbia University.

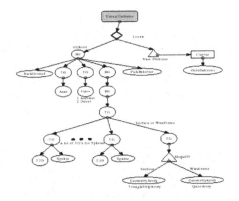

Fig. 1. Scene Graph

show a surface defined by HLOD values and two other parameters at a time, we developed a tool that visualizes the surface and allows users to change the view of the surface as a function of all six parameters simultaneously.

3.1 Input Data Format

The input file is 7-dimensional data representing HLOD values on the full variations of the six model parameters. A grid over all parameters with varying "density" is used for the parameter variation. Disease Frequency, for example, has six different values in the range from 0.001 to 0.8 and our data contains 0.001, 0.01, 0.1, 0.3, 0.5, and 0.8. The restrictions imposed on the relations among three penetration values are 1)$T_0 < T_1 \leq T_2$, or 2)$T_0 \leq T_1 < T_2$. In other words, T_1 is greater than or equal to T_0 and T_2 is greater than or equal to T_1 but T_0, T_1, and T_2 cannot be the same values at the same time.
When we visualize the data, we assume that the given input file contains a grid of 7-dimensional points without repeated parameter values. The input is a 75MB text file for one marker and the input will ultimately grow up to 20GB binary file for multiple markers in the future. The system we used for this visualization has P4 2.8GHz CPU, 1GB of RAM, and 64MB ASUS NVIDIA GeForce3. Our current data set contains the data from one marker and the data is generated from UI College of Public Health.

3.2 Development of Tool using Java3D

As it is reading data from a file, the max HLOD value and all different values of each parameter are determined and stored to the memory. Once the reading process is done, the visualization tool creates Scene Graph. Fig. 1 shows the Scene Graph Tree. Right hand side sub-tree of Locale comes from Java3D default class.

This tool can display selected data set in 3 forms Points, WireFrame, and Surface. Of course, user can set the display form anytime and Points can be displayed with either WireFrame or Surface at the same time. However, it does not display WireFrame and Surface at once. Fig. 2 shows all five possible displays. Even the WireFrame or Surface shows more perceivable image, Points display is important because neither wireframe nor surface is pickable.

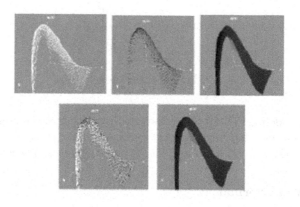

Fig. 2. Five Different Display Options

User interface has six dropdown boxes and four slider bars. Two dropdown boxes to select parameters whose values will be mapped on x-axis and y-axis, each of other four dropdown boxes is used to select 3rd, 4th, 5th, and 6th parameter name respectively. At the bottom of the user interface, there is a table to display global max, local max, and picked values. Global max shows the maximum HLOD value among all HLOD values and six parameter values associated with max HLOD value. The local max shows the largest HLOD value among the HLOD values that are currently displayed on the screen and six parameters that yield local max HLOD. Picked values are the values associated with the point user picked on the screen. The slider bars are used to change the values for the other 4 parameter variables. The user can manipulate the values of the slider bar values to see their effect on the surface. To get the optimal view of the function in any given circumstance, user can zoom and rotate the image with mouse. Fig. 3 shows the user interface with surface display view.

Animation feature is associated with last slider bar. When user clicks on animation button, it increases and decreases slider bar automatically and the display panel shows the images accordingly with adjustable animation interval.

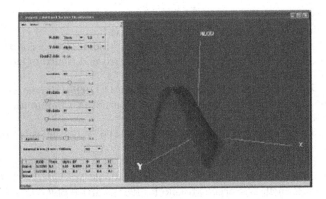

Fig. 3. User Interface with Surface Display

4 Discussion and Future Work

Our current tool visualizes the surface and allows users to change the view of the surface as a function of all six parameters simultaneously. Through this feature, we hope to be able to show the interdependencies that we did not expect to see without visualization.

This interactive navigation/exploration over multidimensional space allows researchers to see effects of all of the parameters without resorting to complicated statistical methods. Providing the surface for the data set points gives more 3-D realistic view to user and helps user understand the surface represented by the data set.

Further work could result in a general tool for the display other functions of genetic parameters such as population risk, and sibling risk. The viewer could also be expanded to display multiple graphical views at once. This would help researchers determine the parameter effects on multiple functions simultaneously. In addition, visualizing the surface in fine detail, or in denser parameter grid, may lead to the statistics that have more power to detect disease genes. We also would like to expand the dimension of the visualization beyond three.

References

1. Chen S, Doolen GD (1998) Lattice Boltzmann Method for Fluid Flows. Annual Review of Fluid Mechanics 30:329-364
2. Frisch U, Hasslacher B, Pomeau Y (1986) Lattice-gas Automata for the Navier-Stokes Equations. Physics Review Letter 56:1505-1508
3. McNamara GR, Zanetti G (1988) Use of the Boltzmann Equations to Simulate Lattice-gas Automata, Physics Review Letter 61:2332-2335
4. Chen H, Chen S, Matthaeus WH (1992) Recovery of the Navier-Stokes Equation Using a Lattice-gas Boltzmann Method. Physics Review A 45:5539-5542

5. Qian YH, D'Humieres D, Lallemand P (1992) Lattice BGK Models for Navier-Stokes Equations. Europhysics Letter 17:479-484
6. Bhatnagar PL, Gross EP, Krook M (1994) A Model for Collision Processes in Gases. I. Small Amplitude Processes in Charged and Neutral One-component Systems. Physics Review 94:511-525
7. Hou S, Zou Q, Chen S, Doolen G, Cogley AC (1995) Simulation of Cavity Flow by the Lattice Boltzmann Method. Journal of Computational Physics 118:329-347
8. Lin CL, Lai YG (2000) Lattice Boltzmann Method on Composite Grids. Physical Review E 62:2219-2225
9. Satofuka N, Nishioka T (1999) Parallelization of Lattice Boltzmann Method for Incompressible Flow Computations. Computational Mechanics, 23:164-171
10. Betello G, Richelli G, Succi S, Ruello F (1992) Lattice Boltzmann Method on a Cluster of IBM RISC System/6000 Workstations. First International Symposium on High Performance Distributed Computing 242-247
11. Skordos PA (1995) Parallel Simulation of Subsonic Fluid Dynamics on a Cluster of Workstations. 4th High Performance Distributed Computing 6-16
12. Kandhai D, Koponen A, Hoekstra AG, Kataja M, Timonen J, Sloot PMA (1998) Lattice-Boltzmann Hydrodynamics on Parallel Systems. Computer Physics Communications 111:14-26
13. MultiProcessing Environment (MPE) Tools, Argonne National Laboratory, MPI home page (http://www-unix.mcs.anl.gov/mpi/index.html)
14. Ni J, Lin C, Zhang Y, He T, Wang S, Knosp BM (2004) Performance Evaluation of a Parallel Lattice Boltzmann Method for Cavity Flows Using Cluster Computing. the 2004 International Conference on Parallel and Distributed Processing Techniques and Applications (PDPTA'04), Las Vegas, Nevada, Hamid R. Arabnia (Ed.), Vol. I, CSREA Press 10-16

Building Intrusion Tolerant Software System for High Performance Grid Computing *

Wenling Peng[1,2], Huanguo Zhang[1], Lina Wang[1], and Wei Chen[1]

[1] School of Computer, Wuhan University, Wuhan 430079, Hubei, China
[2] Network Center,Gannan Teachers College,Ganzhou 341000,Jiangxi,China
peng_wenling@yahoo.com.cn

Abstract The intrusion tolerant software system is a novel concept to high performance grid computing, and it can provide an intended server capability and deal with the impacts caused by the intruder exploiting the inherent security vulnerabilities. In the paper, we describe and analyze the hypothesis about intrusion tolerance software system to high performance grid computing. Then an intrusion tolerance software technology and architectures is presented by detecting the return address or node of each module and adding "HoneyCode" nonfunctional code that is similar to "Honeypot" to confuse intruder, so that the grid security and performance are enhanced.

Key words: intrusion tolerant, software system, grid computing, honey code

1 Introduction

In recent years the concept of grid computing is gaining more and more popularity with the rapid development of the Internet and the increased number of powerful computers. In a grid environment various kinds of resources can be congregated as a single unified system to solve large-scale resource-intensive problems, despite of their geographical locations. Since resources are geographically distributed, heterogeneous in nature, owned by different organizations, Grid resource management faces the security problems in high performance grid computing. Therefore, As more attentions are paid to the network security, software architecture is emerging as an important discipline for software engineers. As the size and complexity of software system grows, the overall design of system structure becomes more important than the selection of algorithms and data structure. Especially, the complexity of the

---

*This research was supported by the National Natural Science Foundation of China(60473023, 60373087, 90104005) and the National High Technology Research and Development Program of China (863 Program) (2002AA141051).

present software systems virtually cannot totally avoid the existence of security vulnerabilities. External and internal intruders could attempt to disrupt the application by compromising a software block. Although intrusion prevention techniques such as firewalls and anti-virus software can partly prevent attackers from exploiting these vulnerabilities, vulnerabilities is inevitable. Intrusion tolerance is a novel concept for software architecture and a new design paradigm which potentially provide capability for dealing with residual security vulnerabilities.

To support software module/component applications, we believe that the intrusion tolerant software system is necessary to provide robust communication and module management services that can tolerate misbehaving software module members and members whose machine has been compromised.

1.1 Intrusion tolerant Software System and Grid Computing

Currently, software security plays the vital role to grid computing among all common computer security problems. In Ref. [1], these concepts are essentially centralized, in order to keep their implementation simple and verifiable. The centralized approach is not suitable for distributed, local autonomic and concurrent system.. Consequently, building of a Network or trusted computing base (TCB) which is network computing management, composed of a set of cooperating TCB. Within each site of the distributed system, a local TCB is responsible for the authentication of local users, and for the access control to local objects.

We believe, an intrusion tolerant software system is one that can continue to function correctly and provide the intended services to the user in a timely manner even in the face of an attack. In other words[1], an intrusion-tolerant software system is capable of self-diagnosis, self-repair, and reconstitution (recur to by network computing management, if possibly), while continuing to provide service to legitimate clients (with possible degradation) in the presence of intrusions.

This means that the intrusion tolerant software system is designed to eliminate the impact caused by the intrusion and the damage can be automatically repaired. Unlike intrusion prevention and detection, the intrusion tolerant software system can potentially counteract unknown intrusions exploiting vulnerabilities.

1.2 Software Architectures

The architecture of a software system is defined in terms of computational components and relationship and interactions among those components.There are numerous other architectural styles and patterns. Some are general, and others are specific to particular domains. We briefly make use of these important categories:distributed processes and main program/subroutine organizations. With the development of the network, most of the currently grid software system are based on distributed architecture.

2 Basic Intrusion Method of Module Software

Software is an inherently complex system. Software architects and coders can hardly prevent residual and unintended faults and/or security vulnerabilities in a software system even with the best efforts. Security vulnerabilities in software on networked systems provide attackers an avenue to penetrate those systems. The source of these security weaknesses is usually traced to poor software development, non-secure links between computing systems and applications. The attacker can utilize these to launch intrusions. e.g. virus ,Trojan horse, buffer overflow attacks etc. An attack can render the attacked software module/component unavailable and further control the action of compromised component. The key method of intrusion software is enumerated below:injection,modification,extraction, malicious code, system disruption,etc.

3 The Fundamentals of Intrusion Tolerant Software

3.1 Intrusion Tolerance versus Fault Tolerance

To discuss intrusion tolerance, we need to compare it with fault tolerance in some aspects, because intrusion tolerance is similar to fault tolerance. Similarity exists in that both disciplines aim to have the system continue providing acceptable service in the presence of anomalies. Some of the key fault tolerance concepts and approaches can be used as a basis for developing intrusion tolerance solutions. Redundancy can be achieved by replication, diversity, and reconfigurations one such notion.

However, despite some similarities, there are also a few differences:

- The software security intrusion is caused by deliberate user action but failure occurs accidentally.
- Attacker has to actively identify vulnerability in the system while system is always vulnerable to failure. They have to spend time and effort to cause a security intrusion.

Hence, fault tolerant software techniques and methods absolutely or directly applied to obtain intrusion tolerant software and new solutions are needed.

3.2 Mechanisms

Despite adherence to perfect software engineering principle, it is hardly possible to avoid bug or vulnerability. The grid software systems may contain numerous vulnerabilities. Researches believe that designing an intrusion software system with maximal security assurance requires that planned activities or servers can be involved the following strategies [2]:vulnerability avoidance,vulnerability removal,vulnerability blocking,and intrusion tolerance.Intrusion tolerance is the ultimate line-of-defense.

4 Strategy of Intrusion Tolerant Software System

Based on the above discussion, an intrusion software tolerant system has to be capable of reorganizing itself, preferably autonomous, in order to mitigate the effects of an intrusion[3]. Because software security risks come in two main factors: architectural problems and implementation errors. We give some properties to build an intrusion tolerant software system; the goal is to enable the creation of tolerant intrusion, distributed, dynamically adaptive software system. The following guidelines can help building intrusion tolerant software system.

4.1 Immune Code Method

Intrusion containment, that is, the ability to restrict damages caused by an intruder. We build an intrusion tolerant software system similar to human immune system (HIS).

In [4], When lymphocytes detect any foreign material that they can recognize, the immune reaction happens. The main characteristic of the immune system lies in the processing procedure, including generate detectors, detect and eliminate pathogens, and memorize the infections. This explains why the immune system is always robust even if one part of itself is disabled or the immune system itself is being attacked.

Here, we construct an antigen software source code so as to restrict the influence of the attack. In the software system, inject attack code or change the return address is an common-used intrusion method, so the buffer overflows seem to be the most common problems to software system. Actually, in Ref. [5], buffer overflow attacks exploit a lack of bounds checking on the size of input being stored in a buffer array. In addition, most file-infector viruses insert a single jump instruction that transfers control to the virus code (stored at the end of the program)

Hence, if the return address cannot be changed, then the attacker has no way of invoking the injected attack code, and the attack method is thwarted. To be effective, detecting that the return address has been altered must happen before a function returns, we may place an "antigen" word next to the return address on the stack. When the function returns, the control module or TCB first checks to see whether the antigen word is intact before jumping to the address pointed to by the return address word. This approach assumes that the return address is unaltered *iff* the antigen word is unaltered.

Once the antigen word is altered, the "immune reaction" is called, i.e. TCB replace it with other proper modules that can provide with same function, or changes the return address to the point of security code.

4.2 "Honeycode"-Adding Honey Nonfunctional Code

Depending on the complex software architecture and respecting the constraints of complied software, we can insert proper no-ops or other nonfunctional sequences of instructions at random locations in compiled code. This could perhaps potentially affect timing relations at execution-time and would slightly change the physical location of instruction, so that it will cost more time and energy to perform intrusion.

We may construct codes that appear vulnerable as bait to attack, but actually offer no access to valuable data and administrative controls. Because adding nonfunctional code, "Honeycode" can lure attackers away from valuable program segments or modules, and properly transform the attack flow and object to the attacker, "Honeycode" can help to enhance prevention against some attacks and protecting software code organization, so it will decrease the probability of successful intrusion. It can also help to aware of if the software program has been attacked through detection of these "Honeycode" code.

5 Conclusion

No matter how carefully the high performance grid computing software system is designed and intensively tested, it can not totally avoid unanticipated vulnerabilities. Intrusion tolerance is an ambitious goal to solve this problem, the above discussion can not tolerate all intrusion aiming at the existing vulnerabilities, we must make further effort to discuss and design the particular challenging problems as well as future research directions.

References

1. Deswarte Y, Blain L, Fabre J C. Intrusion Tolerance in Distributed Computing Systems. Proceedings of the International Symposium on Security and Privacy, New York: IEEE Press,May 1991, 110-121.
2. Stavridou V, Dutertre B, Riemenschneider R A, et al. Intrusion Tolerant Software Architectures. Proceedings of DARPA Information Survivability Conference and Exposition (DISCEX II'01)Volume II, Anaheim, California, June 12-14, 2001,1230-1241.
3. Wang F, Uppalli R, Killian C. Analysis of Techniques For Building Intrusion Tolerant Server Systems. Proceedings of IEEE Military Communications Conference (MILCOM 2003), Boston, MA, Oct 13-16, 2003, 729-734.
4. Liang Yi-wen, Li Huan, Kang Li-shan. A Multi-Agent Immunology Model for Security Computer. Wuhan University Journal of Nature Sciences. 2001, 6(1-2): 486-490.
5. Cowan C, Pu C, Maier D, et al. StackGuard: Automatic Adaptive Detection and Prevention of Buffer-Overflow Attacks, Proceedings of the 7th USENIX Security Symposium, San Antonio, TX, January 1998 , 63-78.

Model Driven Information Resource Integration

Pengfei Qian and Shensheng Zhang

CIT Lab, Computer Science & Tech Department, Shanghai Jiaotong University,
Shanghai 200030, China

Abstract: In this paper, a kind of technology research is introduced, which adopts model driven method to realize distributed and heterogeneous information resource integration. To implementing distributed and heterogeneous resource integration, the model driven resource integration (MDRI) method which based on model driven architecture(MDA) is put forward, and the definitions of the SSRM,SIRM,SIURM are discussed, then the unified mapping and transformation process(SSRM→SIRM→SIURM)is also presented, all of which have been the basis of interoperation among the different application domains.
Key word: Meta-data,XML,MDA

1 Introduction

The interoperation of information among different application domains is an imminent requirement of cooperative commerce development. Accordingly, a unified information resource model must be constructed to strongly support the semantic interoperation and the integration of distributed and heterogeneous information resource. In this paper, a method named MDRI(Model Driven Resource Integration)which based on MDA(Model Driven Architecture)is introduced to realize distributed and heterogeneous information resource integration.

MDA as a kind of software development method is brought forward by OMG(Object Management Group). Unified software model is established by the MDA to finish the separation of the function and the interrelated realization technology, and then improves the software development efficiency. At the same time, the architecture of the MDA also provides an effective realization method for the information resource integration of distributed and heterogeneous application domains[BHWB03].

2 Model Driven Resource Integration Method: MDRI

The MDRI method is a supplement of MDA technique. The core of MDRI is using resource models to describe all kinds of information resource. A unified resource model is built to describe the whole domain information resource, and some specific resource models are used to describe the different heterogeneous sub-domain resource. The integration of the distributed and heterogeneous information resource is the integration of these specific resource models into the unified resource model. MDRI realizes the unified description of distributed and heterogeneous information resource, and establishes the rule of model exchange and transformation, and simultaneously makes it more convenient to the search and release of distributed and heterogeneous information resource. Generally speaking, the whole realization process of MDRI includes 4 steps:

(1) Locating the information resource that is to be integrated and confirming the original storage form and access form of information resource, for example: XML, Excel, Text File, Relational Database, E-mail etc.

(2) According to the original storage form and access form of the heterogeneous information resource, extracting the special model description from all kinds of information resource and then set up different resource models respectively. All of these models named SSRM (Storage Specific Resource Model) are dependent of the concrete information resource storage and access forms.

(3) According to the model transformation rule, transforming the SSRM models of all kinds of information resource to the local unified models. These models named SIRM (Storage Independent Resource Model)are independent of the concrete information resource storage and access forms. The metadata of the SIRM model is defined through the description of the SSRM model, and then the original information resource is transformed to the XML document on the basis of the metadata of the SIRM model. This step mainly solves the problem of heterogeneous information resource integration.

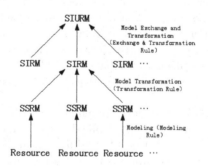

Fig. 1. Unified Transformation Process of Distributed and Heterogeneous Information Resource

(4) According to the model exchange protocol and the model transformation rule, transforming all kinds of SIRM models to the unified platform independent model named SIURM (Storage Independent Unified Resource Model) which contains semantic information. These SIRM models are distributed over different application domains, and conform to the different domain schemes. This step mainly solves the problem of distributed information resource integration, and simultaneously the process of model transformation is the process of the matching and transformation of domain schema, the combination of information resource, the establishment of resource indexes, the definition of content released work-flow and the realization of inter-operation [Mik02][JRSFS03].

According to the four steps of the MDRI method, figure1 shows the unified transformation process of distributed and heterogeneous information resource. In the MDRI approach, UML (Uniform Modeling Language) is regarded as model description language; XML, XML Schema is the storage form of model & model structure respectively; XMI (XML Metadata Interchange) is used to realizing resource exchanging & model transformation between SIRM models and SIURM model.

3 Conclusion

Combining with the MDA, this paper introduces a method named MDRI to implement the process of the united mapping and transformation of distributed and heterogeneous information resource. Some definitions are listed, which include SSRM, SIRM and SIURM. Further, four steps are given to implement MDRI, these steps are: (1) locating resource, (2) building SSRM, (3) Transforming SSRM to SIRM, (4) Transforming SIRM to SIURM.

References

[BHWB03] Bing Qiao, Hongji Yang, William C. Chu, Baowen Xu. Bridging Legacy Systems to Model Driven Architecture. Proceedings of the 27th Annual International Computer Software and Applications Conference (COMPSAC), 2003. p136-152.

[Mik02] Mikaël Peltier etc. MTrans, a DSL for Model Transformation. Proceedings of the Sixth International Enterprise Distributed Object Computing Conference(EDOC),2002. p56-70.

[JRSFS03] Jana Koehler, Rainer Hauser, Shubir Kapoor, Fred Y. Wu, Santhosh Kumaran. A Model-Driven Transformation Method. Proceedings of the Seventh IEEE International Enterprise Distributed Object Computing Conference(EDOC),2003. p161-180.

DNA-based Parallel Computation of Addition*

Huiqin Qu[1] and Hong Zhu[1]

Intelligent Information Processing Laboratory, Fudan University, China
huiqin.qu@fudan.edu.cn, hzhu@fudan.edu.cn

Summary. This paper gives a method to implement additions with one common addend in parallel by DNA computing. The main idea is to divide every addition into bit addings, and perform the bit addings of different additions in parallel. The key bio-operation used in this paper is "back-to-back" cut, which can be accomplished by restriction endonucleases.

keywords: DNA computing, bit adding, addition addings

1 Introduction

The concept of molecule computer has been popular since Adleman solved the directed Hamiltonian path problem[Adl94] by controlling DNA. After that, many researches on performing addition with DNA molecules have been done, e.g. [GFB96][GFB96][HR01] [GPZ97][QL98]. This paper showed another method to implement additons with the advantages that the result of the former addition can be the addend of the next addition, and it can accomplish additions in parallel, on condition that these additions have one addend in common. The key bio-operation used in this paper is "back-to-back" cut which stems from the paper that proves the ability of DNA computing is equal to that of Universal Turing Machine(UTM)[Rot96].

The rest of the paper is organized as follows: we first review the structure of DNA and the key operations needed in session 2. In session 3, we present the scheme for designing DNA sequence for the addends in addition. The process of the addition and the potential difficulties of the process are narrated in session 4. Finally, we discuss the advantages and shortcomings of the method in session 5.

---

*Supported by a grant from the Ministry of Science and Technology (grant No. 2001CCA03000), National Natural Science Fund (grant No. 60496321 and No. 60273045) and Shanghai Science and Technology Development Fund (grant No. 03JC14014)

2 Structure of DNA and operations used in this paper

2.1 Structure of DNA

A DNA strand is a sequence of four types of nucleotides distinguished by one of four bases they contain, and the bases are denoted A, C, G, T. The two ends of the strand are distinct and are conventionally denoted as 3' end and 5' end. Two strands of DNA can form a Double-Stranded(DS) DNA, if the respective bases are Watson-Crick complementary to each other $-A$ matches T and C matches G, also 3' end matches 5' end. For example, two single stranded DNA molecules 5'-$ACCTGC$-3' and 3'-$TGGACG$-5' can form a DS DNA molecule $\frac{5' - ACCTGC - 3'}{3' - TGGACG - 5'}$. If a DNA strand contains small amount of nucleotides, we call it oligo. We refer to the single stranded overhang at the end of a DS DNA as "sticky end".

2.2 Cut by restriction endonucleases,ligation and remove

Restriction endonucleases can recognize some special sequence of nucleotides in DS DNA and cut it at some place (maybe produce sticky ends). Some endonucleases cut DNA at the place some nucleotides away from the recognition site, for example, the class IIS restriction endonuclease [Rot96][BPA01][GGA98]. We call two adjacent special sequences recognized by two endonucleases in a strand "back-to-back area". See Fig. 1.

Fig. 1. the operations of cut and and ligation

The left part of Fig. 1 shows the function of class IIS restriction endonuclease. There is an example of "back-to-back" area in this part too. The two adjacent light grey regions are two recognition sites, and their cut sites are pointed to by the two arrowheads respectively.

Two DNA strands that have complementary sticky ends can combine together to form one DNA strand. E.g. $\frac{5' - ATTGCA - 3'}{3' - TAA - 5'}$ and $\frac{5' - CTA - 3'}{3' - CGTGAT - 5'}$ can compose a DNA strand $\frac{5' - ATTGCACTA - 3'}{3' - TAACGTGAT - 5'}$. The right part of Fig. 1 shows the operation of ligation. We refer to the operation "remove" as removing away DNA of specifically length from other DNA molecules, using gel electrophoresis.

3 Sequences for addends

3.1 Sequence for left addends

There are two addends "a" and "b" in an operation of addition, and "a" is called left addend and "b" right addend. In this paper, we can complete additions with the same right addend in parallel, such as $a + b$, $a' + b$, $a'' + b$, $\ldots a^{(l)} + b$. We take the addition "$a + b$" for example. $a_n a_{n-1} \cdots a_0$ and $b_m b_{m-1} \cdots b_0$ are the binary representations of a and b respectively. In brief, the method for implementing the operation $a + b$ is as follows: put a into the tube, and divide b into bits, then perform the addition bit by bit, i.e. first calculate $a_0 + b_0 + c_0$, and then $a_1 + b_1 + c_1$. The process goes on until $a_k + b_k + c_k$, $k = \max\{m + 1, n + 1\} + 1$ is performed. We call the step $a_i + b_i + c_i$ bit-i adding for short in this paper, and c_i is the carry in bit-i adding. The DNA sequence for a is also used as the sequence to store the result of addition. For parallel addition, $k = \max\{$bit number of all addends$\} + 1$, i.e. we calculate $a + b$, $a' + b$, $a'' + b$, $\ldots a^{(l)} + b$ together, and in this situation, k is the maximal of all the bit numbers of a, a', a'', $\ldots$, $a^{(l)}$, b plus 1. For $a_n a_{n-1} \cdots a_0$, we prepare such circular DNA molecules showed in Fig. 2, and add them into the tube.

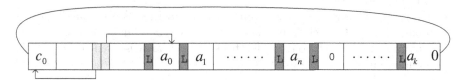

Fig. 2. the DNA that encodes "a"

As showed in Fig. 2, in addition to a_0, a_1, $\cdots a_k$, there is a sub-sequence to encode the carry $c_0, c_0 = 0$. The two light grey regions are the "back-to-back" area. The first cut site is c_0 , and the second is a_0. We regard "a" as a k-bit string, and the bit larger than n is zero. The sequence for the last bit(the last "0") is adjacent to the sequence for the carry directly, i.e. the DNA strand is a circuit.

3.2 Oligos for right addend

We divide $b_m b_{m-1} \cdots b_0$ into bits $b_0, b_1, \cdots, b_m$ and $b_{m+1}, \cdots b_k$, $b_i = 0, m < i < k+1$. We prepare four kinds of oligos for each $b_i, 0 \leq i \leq k$, corresponding to $c_i = 0 \& a_i = 0$, $c_i = 0 \& a_i = 1$, $c_i = 1 \& a_i = 0$ and $c_i = 1 \& a_i = 1$ respectively(for bit 0, c_0 only have the value 0). a_i' and c_{i+1} are the results of $a_i + b_i + c_i$. See Fig. 3.
The oligos for b_i are composed of five parts as follows:

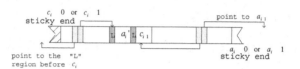

Fig. 3. the oligos for b_i

- A sticky end complementary with the incomplete c_i (the left end of the oligo).
- A sticky end complementary with the incomplete a_i (the right end of the oligo).
  The two sticky ends can combine the oligo for b_i with the proper DNA molecule. For example, the first oligo in Fig. 3 will combine with the DNA molecule that stands for $c_i = 0, a_i = 0$.
- The subsequence encoding the result of $a_i + b_i + c_i$, including the new a_i', $a_i' = (a_i + b_i + c_i) \bmod 2$, and the carry c_{i+1}, $c_{i+1} = \lfloor (a_i + b_i + c_i)/2 \rfloor$.
- The first back-to-back area(on the left in Fig. 3). One of the arrow points to the region marked "L" before c_i(notice that the cut site "L" is not in the oligo itself), the other points to the region marked "L" before a_i' in the oligo. The function of this back-to-back area is to cut away c_i, which will be useless later.
- The second back-to-back area(on the right in Fig. 3). One of the arrow points to a_{i+1}, which is in the circular DNA molecule, the other arrow points to c_{i+1} in the oligo. This back-to-back area is used in bit-$(i + 1)$ adding, i.e. $a_{i+1} + b_{i+1} + c_{i+1}$.

3.3 Bit-i adding

We take bit-1 as example. See Fig. 4. The circular DNA molecule presented in step 1 is the result of bit-0 adding. In step 2, we add the endonucleases, and the old a_0 is cut away, leaving two sticky ends. Remove away the residues. In step 3, put the four kinds of oligos for b_1 into the tube. In step 4, put ligases into the tube, and the oligos will combine with proper DNA sequences and form circular DNA molecules again. In Fig.4, we give an example that the oligo with $c_1 = 1$ and $a_1 = 1$ sticky ends combines with DNA sequence with $c_1 = 1$ sticky end and $a_1 = 1$ sticky end. The oligo brings two back-to-back areas to the circular DNA molecule. The previous one is used to cut away the useless c_1 , as showed in step 5, and the second is used in the next bit adding. In step 6, we get the circular DNA molecules that are ready for the next bit adding.

3.4 Bit-k adding

For bit-k adding, it must make preparation for the next operation of addition. After bit-k adding, the circular DNA molecule in the solution is the result of

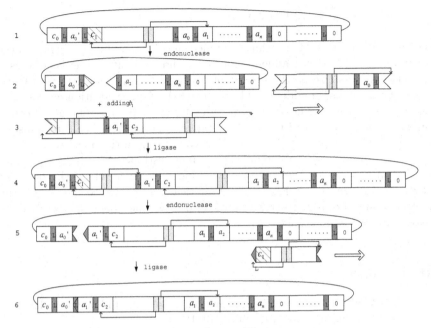

Fig. 4. bit-1 adding

$a + b$, and it is required that the DNA molecules are suitable for another operation of addition, that is to say, we can calculate $(a + b) + c$ after we get the result of $a + b$. bit-k adding also have 6 steps with most of which are similar with that in bit-i adding. Fig. 5 shows step 3 and 6 in bit-k adding to explain why it can meet with requests above.

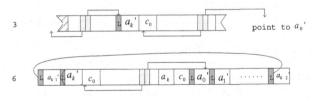

Fig. 5. bit-k adding

The second back-to-back area in oligo for bit-k is be between c_0 and a_0' (5, step 3), so that the next operation of addition can be performed from its bit-0 adding. The oligo for b_k brings in another $c_0 = 0$. The circular DNA molecule in step 6 is the result of bit-k adding, thus the final result of the whole add operation. It is easy to see that we can start another add operation based on this circuit. So this method can accomplish continuous operations of addition.

4 Analysis and further work

The whole operation of addition is divided into k bit addings, where k is the bit number of the larger addends plus one, and each bit adding is composed of constant bio-steps. So the step number of addition by this method is linear to the bit numbers of addends. The result of the former addition can be used as one addend in the next addition directly, and we only generate "correct answer" and do not need to select results. Furthermore, we can make use of the massively computing ability of DNA computing in this method.

Before the process of the computing, we must estimate how large the result will be(in this paper. We should try to find method in which the molecule that keeps the result can lengthen automatically according to the result. And another point we need to pay attention to is to find the proper codings to avoid unexpected ligations.

References

[Adl94]   Adleman, L.: Molecular Computation of Solutions to Combinatorial Prob-
          lems. Science, **266** 1021–1024 (1994)
[GFB96]   Guarnieri, F., Fliss, M., Bancroft, C.: Making DNA add. Science,
          273(5272), 220–223 (1996)
[GPZ97]   Gupta, V., Parthasarathy, S., Zaki, M.J.: Arithmetic and Logic Opera-
          tions with DNA. Proceedings of the 3rd DIMACS Workshop on DNA
          Based Computers, 212–220 (1997)
[QL98]   Qiu, Z.F., Lu, M.: Arithmetic and logic operations for DNA computers.
          Second IASTED International Conference on Parallel and Distributed
          Computing and Networks, 481–486 (1998)
[HR01]   Hug, H., Schuler, R.: DNA-based parallel computation of simple arith-
          metic . 7th international Workshop on DNA-Based Computers. Tampa,
          U.S.A., 159–166 (2001)
[Rot96]   Rothemund, P.: A DNA and restriction enzyme implementation of Turing
          Machine. In : DNA based computer . American Mathematical Society,
          ISBN 0-821800518-5 (1996)
[Kar97]   Kari, L.: DNA computing: the arrival of biological mathematics. The
          mathematical Intelligencer, **19.2** (1997)
[BPA01]   Benenson, Y., Paz-Elizur, T., Adar, R., Keinan,E., Livneh, Z., Shapiro,
          E.: Programmable and autonomous computing machine made of biomole-
          cules. Nature , **414**, 430–434 (2001)
[GGA98]   Gheorghe, P., Grzegorz, R., Arto, S.: DNA Computing. New Computing
          Paradigms, Springer-Verlag (1998)

Measurements and Understanding of the KaZaA P2P Network

Jun Shi[1], Jian Liang[2], and Jinyuan You[1]

[1] Department of Computer Science and Engineering, Shanghai Jiao Tong University, China {shijun, you-jy}@cs.sjtu.edu.cn
[2] Department of Computer and Information Science, Polytechnic Univ., USA jliang@photon.poly.edu

1 Introduction

KaZaA[1] network is a file sharing system based on a peer to peer distributed technology called "FastTrack". KaZaA has become one of the most important application in Internet today. With over 3 million satisfied users, KaZaA is significantly more popular than Napster and Gnutella ever was. An understanding of the KaZaA's protocol, architecture and signaling traffic is of critical importance for the P2P research community. But to date, little has been known about the specifics of the KaZaA design. We have built a measurement platform that has enabled us to gain significant insights into KaZaA. The paper is complementary to [2] and [3], which focus in KaZaA file sharing traffic.

KaZaA has two classes of peers, Ordinary Nodes (ONs) and Super Nodes (SNs). The SNs peers are more powerful and have great responsibilities. As shown in Fig.1, each ON is assigned to a SN. When an ON launches the KaZaA application, the ON establishes a TCP connection with a SN. The ON then uploads to its SN metadata for the files it is sharing. This allows the SN to maintain a database which includes the identifiers of all the files its children are sharing, metadata about the files, and the corresponding IP address of the ONs holding the files. An SN is not a dedicated server, instead, it is typically a peer belonging to an individual user. For each file that it is sharing, the metadata that an ON uploads to its SN includes: the file name, the file size, the ContentHash and the file descriptors [4]. The ContentHash plays an important role in the KaZaA architecture. KaZaA hashed every file to a signature, which becomes the ContentHash of the file. If a download from a specific peer fails, the ContenHash enables the KaZaA client to search for specific file automatically without issuing a new keyword query. When a user wants to find files, the user's ON sends a query with keywords over the TCP connection to its SN. For each match in its database, the SN returns the IP address and metadata corresponding to the match. Each SN also maintains long-lived

TCP connections with other SNs, creating an overlay network among the SNs. When a SN receives a query, it may forward the query to one or more of the SNs to which it is connected.

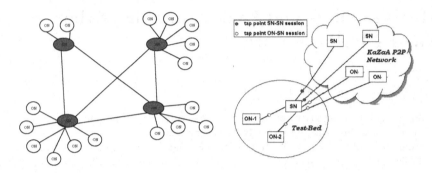

Fig.1. Supernode and Ordinary nodes in KaZaA network   **Fig.2.** Test-bed configuration

2 Measurements

As shown in Fig. 2, we built a test-bed consisting of three workstations, each with a KMD version 2.0 client software installed. We patiently waited until KaZaA promoted one the three nodes to a SN. At startup all three workstations functioned as ONs in the KaZaA network. When one of the workstations was promoted to a SN, we manipulated the Windows Registries in the other two ONs so that each of the two registries listed only the promoted SN. In this manner, we forced both ONs to become children of the SN. The test-bed also connects to the larger KaZaA P2P network through connections from the test-bed SN to other SNs and other ONs in the KaZaA network. We then deployed software traffic monitors to capture all of the traffic inbound to and outbound from the SN for each of its ON and SN connections. We then did an offline traffic analysis based on our understanding of the KaZaA signaling protocol. We have also developed our own version of a KaZaA client, which emulates the behavior of the official KMD client for signaling traffic. We first explored the degree of connectivity of a SN. Specifically, from our SN node, we studied the number of simultaneous connections to ONs and to other SNs. Fig.3 presents measurement result taken on two different days for different durations. Each graph in Fig.3 shows the evolution of the number of simultaneous TCP connections from ONs to our test SN and from SNs to our test SN. For both the ON and SN connections, the number of connections begins at one and climbs to a threshold, around which it subsequently vacillates. For the number of simultaneous SN-ON connections, depending on the day, this threshold is in the 100-160 connection range. Since on a typical day there are

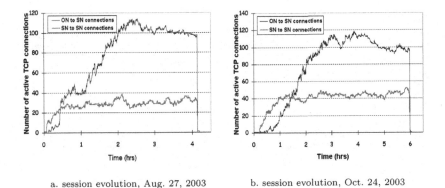

a. session evolution, Aug. 27, 2003   b. session evolution, Oct. 24, 2003

Fig.3. Evolution of SN-SN and ON-SN connections with time

roughly 3 million peers, we therefore speculate that there are on the order of 30,000 supernodes in the KaZaA network at any given time. We also observe that for the number of simultaneous SN-SN connections, depending on the day, the threshold is in the 30-50connection range. Thus, at any given moment, each supernode is roughly connected to 0.1% of the total number of supernodes. Our measurement study has determined the KaZaA overlay is highly dynamic. Although, as observed in Fig.3, the number of simultaneous connections vacillates around a threshold, the individual connections change frequently. We performed measurements on the duration of ON-SN TCP connections and SN-SN TCP connections. We plot the distribution of connection lifetime for these two types of TCP connections in Fig. 4.

We attribute the large number of short lifetime ON-SN connections to two factors. First, we have observed that initially at startup, an ON probes candidate SNs listed in its Supernode List Cache with UDP packets for possible connections. The ON then initiates simultaneous TCP connections with the available SNs in its SN list. Out of these successful connections, the ON selects one SN as the final choice and ir disconnects from other SNs. Hence, some ON-SN connections are short-lived. A second reason for short-lived ON-SN connections is that many ONs are KaZaA-Lite clients. KaZaA-Lite clients hop supernodes during the query process. Each such hop generates a short-lived connection. We conjecture the short lifetime of SN-SN connections is due to (1) SNs searching for other SNs with currently small workloads, and (2) long-term connection shuffling, to allow users to query a large set of SNs over long time scales and (3) at times, SNs in the overlay connect to each other just for the purpose of exchanging SN list caches. One crucial characteristic of the topology is that nodes (SNs and ONs) employ to select neighbors. We know that a newly connecting ON receives a list of 200 SNs from its parent SN. As already discussed, this list is a subset of all the SNs in the parent SN cache. The contents of this list sent from the SN to the ON influence the ON's future

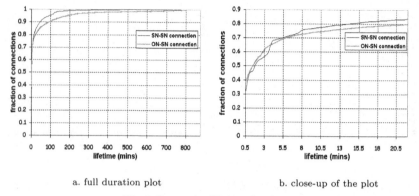

a. full duration plot   b. close-up of the plot

Fig.4. Connection lifetime distribution

decisions about which SNs to connect to; this in turn affects the overlay topology. Based on our measurements, we hypothesize that KaZaA peers mainly use two criteria for ON-to-SN and SN-to-SN neighbor selection. One of these criteria is supernode workload. Each KaZaA ON chooses a parent SN from the local Supernode List Cache in the Windows Registry. One of the information fields in this list is the average workload of the supernode [4]. It is unclear how the value in this field is calculated. Nevertheless, in our experiments, the KaZaA client displayed a marked preference for SNs with low value for the workload field. Fig. 5 illustrates this preference. The second criteria is

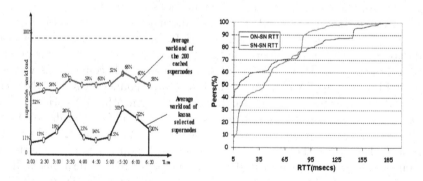

Fig.5. Preference for Supernodes with less Workload   **Fig.6.** Round-Trip Time measurement

based on locality, that is, nodes (both ONs and SNs) appear to choose overlay neighbors that are in some sense close. We have performed experiments to investigate locality. The experiment uses Ping to measure the round-trip time (RTT) from the SN in our testbed to its ON and SN neighbors. Fig.6 shows the distribution of the RTT with respect to the percentage of neighboring SN peers. We observe that about 60% of the connections between neighboring SNs have RTT less than 50 msec. It is instructive to compare these values with some typical RTT values for IP datagrams on the Internet. Also it can be observed that almost 40% of the ON-SN connections have a RTT less than 5 msec, with the other 60% having RTTs more or less uniformly distributed over hundreds of milliseconds. Thus it appears that KaZaA takes locality into account when dynamically creating the overlay network.

3 Conclusion

The supernodes form the backbone of the KaZaA network. The SNs frequently exchange possibly subsets of these lists with each other. Thus, the KaZaA backbone is self-organizing and is managed with a distributed, but proprietary, gossip algorithm. SNs establish both short-lived and long-lived TCP connections with each other. Each SN maintains a database, storing the metadata of the files its children are sharing. SNs do not exchange metadata with each other. The KaZaA client connects with one or more the SNs in this list and obtains new lists. It appears that the entries in these lists are biased with locality; the provided SNs are close to the ON with respect to various locality metrics. The ON-to-SN connections are formed in decentralized, distributed manner and appear to take locality in account.

One important design lesson learned from the KaZaA experiment is that the large-scale P2P should exploit the heterogeneity of the peers. Peers differ in up times, bandwidth connectivity, and CPU power. To exploit the heterogeneity, the peers should be organized in a hierarchy of two or more peers, which the peers in the higher tiers being more powerful. Many design decisions taken by the creators of KaZaA seem to be have been done without careful consideration. We conjecture that there is significant room for improving the search performance in two-tier unstructured P2P file sharing systems.

References

1. KaZaA Homepage http://www.kazaa.com/
2. Gummadi KP, Dunn RJ, Saroiu S, Gribble SD, Levy HM, and Zahorjan J (2003) Measurement, Modeling, and Analysis of a Peer-to-Peer File-shraing Workload. In Proc. the 19th ACM Symp. on Operating Systems Principles.
3. http://iptats.globalcrossing.net/
4. Leibowitz N, Ripeanu M, and Wierzbicki A (2003) Deconstructing the KaZaA Network. In the 3rd IEEE Workshop on Internet Applications, Santa Clara, CA.
5. KaZaA P2P FastTrack File Formats http://home.hetnet.nl/~frejon55/

An Application of Genetic Algorithm in Engineering Optimization

Lianshuan Shi [*1], Lin Da[2], and Heng Fu[3]

[1] Computer Department, Tianjin University of Technology and Education, Tianjin 300222, P.R. China shilianshuan@263.net
[2] Beijin Jiaotong University, Beijing, 010021, China
[3] Information College, Tianjin University, Tianjin 300222, China

Summary. Genetic Algorithm is used to solve the sizing optimization of structure. First of all, the mathematical model of the problem is developed by Statically Determinate Supposition. Then, the genetic algorithm is used to solve the mathematical model. Because only one time structure analysis is needed, the total times of structure analysis is decreased. An efficient method is given for using genetic algorithm to solve structural optimization.

Keywords: Genetic algorithm; Structural optimum design; Discrete variable

1 Introduction

According to the kinds of the design variables, structural optimum design is divided into two kinds, the discrete variable and continuous variable. The discrete variable optimum design means all or a part of the design variables of the mathematical model of structure optimization takes values in some discrete sets. The discrete optimum design problem belongs to the combinatorial optimization problem. The combinatorial optimization is a NP-difficult problem, and the computational times to find the optimum solution grows exponentially with the increasing of the number of design variables. Chai Shan and Sun Huanchun [1, 2] given the delimitative combinatorial algorithms for $(0, 1)$ programming problem with the same discrete sets for all design variables. Shi Lianshuan etc. [5] popularized the algorithm to discrete optimum design problem, in which, the number of the design variables is arbitrary, and a sequential delimitative and combinatorial algorithm is given. At the sequential delimitative and combinatorial algorithm, the primal problem is converted to several smaller sequential sub-problems with lower dimensions and similar structures by recurrent method. In the process of optimizing, the Multi-level Generating

---

*The author is a visiting professor at Department of Computer Science, University of Otago, Dunedin, New Zealand

method is used to generate the new combinations, and the multilevel delimitative and combinatorial algorithm is used to search optimum solution. In the process of delimiting at every level, the objective function and constraint functions are converted into new functions by the linear transformations. These new functions are used to delimit by the delimitative algorithm [1]. Both the constraint functions and the unified constraint function are used to delimit so that the computational efficiency is higher than only using the unified constraint function to delimit. A few examples show that the algorithm has higher accuracy. The algorithm was used to the optimum design of structure with discrete variables. But when the dimension of problem is larger, the computational times are more. The succeed application field of the genetic algorithm is combination optimization. When the genetic algorithm is applied to structural optimization, a lot of structure analyses are needed. For example, suppose there are 10 individuals in the population when genetic algorithm is used to structure optimization problem, then when doing 30 times circles, 300 times structure analyses are need. If the numbers of population increase, the number of structural analyses is more. The efficiency of the algorithm is lower. In order to overcome the difficult that a lot of structural re-analyses are need when genetic algorithm is used to solve structure optimization problem, a new algorithm is given. Firstly the mathematical model of the problem is given according to the Statically Determinate Supposition. Then, the genetic algorithm is used to solve the mathematical model. Because only one time structure analysis is needed, the total times of structure analysis is decreased. An efficient method is given for using genetic algorithm to solve structure optimization. Several examples show the algorithm is feasible and efficient.

2 The mathematical model of truss structure with discrete variables

$$P \ find \quad A = (A_1, A_2, ..., A_n)^T$$
$$min \quad W = \sum_{i=1}^{n} \rho_i A_i l_i$$
$$st \quad \underline{\sigma_i} \leq \sigma_i \leq \overline{\sigma_i}, \quad i = 1, ..., n$$
$$\delta_j = \sum_{i=1}^{n} F_{ij}^V F_{il} l_i / \overline{\delta_j}$$
$$l = 1, 2, ..., n_L, \quad j = 1, 2, ..., n_D$$
$$A_i \in S(p) = \{s_{i,1}, s_{i,2}, ..., s_{i,n_i}\}, i = 1, ..., n$$

where, F_{ij}^V is the visual internal force of the ith element according to the j displacement constraint, F_{il} is the force of the ith element under the lth load case, n_L is the number of load cases, n_D is the total number of displacement constraints, n is the element number. Let $\tau_i = F_{ij}^V F_{il} l_i / E$, Under the condition of the statically determinate supposition, $\sum_{i=1}^{n} F_{ij}^V F_{il} l_i / E \overline{\delta}_j$ is const. At a round of optimizing, the model P is changed into a non-linear programming problem with reciprocal design variables. Because the design variables take discrete values, it is a combinatorial optimization problem, the genetic algorithm can be used to solve this model.

3 Algorithm

At the algorithm, the gene of the chromosome is expressed directly by the cross-sectional area values that can decrease the error that caused by data transformation. The classical genetic algorithm is used to solve the solution. The examples show the algorithm is feasible and efficient, and the time of performing is shorter.The algorithm steps are as follow.
(1) Initialize the cross-sectional area variable $A0$;
(2) Structure analysis and give the mathematical model of problem;
(3) Use GA to solve the mathematical model and obtain the optimum solution A of the current round;
(4) If $A = A0$, then goto (5); Else goto 2;
(5) Stop Optimizing, the optimal value is $A^* = A$.

4 Example

In this paper, two classical structure optimization problem are given, one is 10-bar plane truss, another is 25-bar space truss.

Example 1. A plane truss with 10 bars, 6 nodes and 10 design variables. Its figure is the same as shown in [6]. $\rho = 27150.68 N/m$, $E = 68.97 GN/m^2$. The allowable stress are $\overline{\sigma} = 17243.5 N/cm^2$ and $\underline{\sigma} = 17243.5 N/cm^2$ for all elements. There are 2 static load cases. Load case 1: $P_{2y} = -4.45 \times 10^5$, load case 2: $P_{4y} = -4.45 \times 10^5$. The allowable displacement in y direction of each moveable node is $50.8 mm$. The allowable discrete set is $\{0.645, 1.936, 6.416, 19.355, 32.258, 51.613, 96.774, 96.774, 109.677, 141.935, 154.838, 187.096, 200.000\}$. The optimum results under the cases of different initial values are given at $Tab.1$, the comparation between the current algorithm and other algorithms is given in the $Tab.2$. The optimizing result is satisfied.

Example 2. A 25-bar space truss is subjected to two independent load conditions. Its figure is the same as shown [6]. The material properties are $E = 68.97 GN/m^2$, $\rho = 0.027688 N/cm^2$. In order to maintain the symmetry of the structure, $x_2, y_2, z_2, x_4, y_4, z_4$ are taken as independent shape variables. The

Table 1. The searching times and optimizing result at every circle of 10-bar truss

Initial Value(cm^2)	First	Second	Third	Four	Five	Six	Seven	Eight
200	187.10	187.10	200.00	187.10	200.00	187.10	200.00	200.00
200	51.613	32.258	32.258	19.355	19.355	32.258	6.4516	6.4516
200	200.00	200.00	200.00	200.00	200.00	154.84	154.84	154.84
200	96.774	96.774	96.774	96.774	96.774	96.774	96.774	96.774
200	19.355	6.4516	19.355	6.4516	6.4516	6.4516	6.4516	6.4516
200	51.613	32.258	32.258	19.355	19.355	19.355	6.4516	6.4516
200	109.67	109.67	96.774	96.774	96.774	51.613	51.613	51.613
200	109.67	96.774	109.67	109.67	109.67	154.84	96.774	141.94
200	96.774	96.774	96.774	109.67	109.67	1419.4	109.67	141.94
200	51.613	51.613	32.258	32.258	32.258	19.355	19.355	6.4516
Optim.Value(N)	27969	26233	25762	25223	25411	25100	20389	22714

Table 2. The comparison between three algorithms

	No	Initial Value	RDA	SDCA	GA
Cross	1	141.9	200.0	187.1	200.0
sec.	2	141.9	0.645	0.645	0.645
area	3	141.9	141.9	154.8	141.9
(cm^2)	4	141.9	96.77	96.77	96.77
	5	141.9	0.645	0.645	0.645
	6	141.9	0.645	0.645	0.645
	7	141.9	51.61	51.61	51.61
	8	141.9	141.9	141.9	141.9
	9	141.9	141.9	141.9	141.9
	10	141.9	0.645	0.645	0.645
Weight(N)			22737	22737	22741

RDA:The Relative Difference algorithm;
SDCA:The Sequential Delimitative and Combinatorial algorithm.

allowable discrete set of area variables is $S = \{0.51613, 0.64516, 1.9355, 4.5161,$ $6.4516, 12.903, 19.355, 25.806\}cm^2$. The discrete interval of shape variables is $1cm$. The displacements in x, y directions at nodes 1 and 2 are not larger than $0.889cm$. The load cases,allowable stresses and the linking conditions of cross-sectional area variables and the coordinate variables are the same as [6].

The optimum results under the cases of different initial values and the comparation between the algorithm and other algorithms are given at Tab.3. The optimizing result is satisfied.

5 Conclusions

Genetic Algorithm is used to solve the sizing optimization of structure. Firstly, the mathematical model of the problem is given by the statically determinate

Table 3. The optimal results under the cases of different initial values and the comparison between the current algorithm and other algorithms

	No.	Initial values						HTRA	SDCA	CA
	1	25.806	19.355	25.806	0.5161	0.5161	19.355	0.5161	0.5161	0.5161
	2	19.355	12.903	19.355	25.806	12.903	19.355	12.903	12.903	12.903
Cross	3	4.5161	19.355	4.5161	12.903	19.355	19.355	19.355	19.355	19.355
Sec.	4	12.903	4.5161	4.5161	0.6451	19.355	19.355	0.5161	0.5161	0.5161
Area	5	12.903	4.5161	12.903	19.355	19.355	19.355	0.5161	0.5161	0.5161
(cm^2)	6	19.355	4.5161	19.355	0.6451	19.355	19.355	4.5161	4.5161	4.5161
	7	19.355	12.903	19.355	19.355	19.355	19.355	12.903	12.903	12.903
	8	6.4516	19.355	6.4516	19.355	19.355	19.355	19.355	19.355	19.355
Weight(N)								2679.7	2679.9	2679.9
Iterative times	2	2	2	2	2	2		1	2	

HTRA:The Hypo-times Radial Algorithm;
SDCA:The Sequence Delimitative and Combinatorial Algorithm;
CA:Current algorithm.

supposition. Then, the genetic algorithm is used to solve the mathematic model. Because only one time structure analysis is need, the total times of structure analysis is deduced. An efficient method is given for using genetic algorithm to solve structure optimization.

Acknowledgments

This work was supported by the TianJin Natural Science Foundation of China under the grant No.02360081 and the Education Committee Foundation of Tianjin under the grant No.20022104

References

1. Chai, S., Sun H.: An application of delimitative and combinatorial algorithm to optimum design of structures with discrete variables. Struc. Optim. **11** (1996) 151-158
2. Chai, S., Sun H.: A two-level delimitative and combinatorial algorithm for the discrete optimization of structures. Strut. Optim. **13** (1997) 250-257
3. Cheng, G., et al.: Genetic algorithm and its application [M]. The People Post Press, Bei Jing(in chinese) (2001)
4. Kalyanmoy D.: Multi-objective optimization using evolutionary algorithms[M]. UK:Wiley(2003)
5. Shi, L. et al.: An application of a sequential delimitative and combinatorial algorithm to the discrete optimum design of structures. Journal of Dalian University of Technology (in Chinese), **39** (1999) 592-596
6. Sun, H., et al: Discrete Optimum design of structure. The Press of Dalian University of Technology, DaLian(in Chinese) (2002)

A High Performance Algorithm on Uncertainty Computing

Suixiang Shi[1], Qing Li[2], Lingyu Xu[2], Dengwei Xia[1], Xiufeng Xia[1], and Ge Yu[1]

[1] School of Information Science & Engineering, Northeastern University, Shenyang 110004, China ssx@mail.nmdis.gov.cn
[2] School of Computer Science & Engineering, Shanghai University, Shanghai 200072, China

Abstract. In this paper, we develop a high performance algorithm which is adapted to uncertainty computing and give a new combination rules coming from the D-S and supply a gap that Dempster ignoranced. The evidence sources are adapted in different cases. The credibility of the evidence changes along with the different focus element. So, we give various credibility for every focus element to increase precision. The new method improves the precision and gets rid of disconvergent answer.

Keywords. uncertainty computing, information fusion, Dempster-Shafer, evidence theory, credibility

1 Preface

Uncertainty evidence computing is a important branch of high performance computing. The effective method of multi-dimensional uncertainty information fusion, its main principal is so combine conclusions from independent sources to produce the consistent result, that implement inter-information complementary. Its disadvantage is that the results of combination occasionally differ from thinking of human beings, and one of the reasons is that the evidences conflict and the credibility of evidence sources differ each other. Researchers have paid attention to and continuously improved that, Yager R. and D. Dubois take attention to the problem, give an aggregate additive operator of consistency factors and analysis about that, present a kind of direction on priory level when exist different evidence priority combination rules[MK93, ISO95]; after that researchers introduced priority parameters to D-S rules according to situation of one's regions[GB00, DA93, SJ02, QP01], so guarantee to increase the authority of reliable evidences in the case of conflicting evidences; references[SW99, LK01, WHY01, JOS02] add an assistant rule, so correct the wrap of conclusion.

We also considered the problem mentioned above, but the difference with references is in that the sight of the paper emphasize to simultaneously improve both combination rules and credit degree, cooperatively use in decision-making, as a result supply a gap that references ignored in a certain extents. we present a kind of improve rule which is capable of simultaneous description of multi-evidence sources that are special in different cases and an evidence source which gives different authorities to different cases, introduce specialization factor μ to scale the authority that the evidence decide the case. The experiment result gives a significant improvement.

2 Analysis of D-S rule

2.1 Definitions

Let's Θ is a frame of discernment:

Definition 1: Suppose function m satisfies $2^{\Theta} \to [0,1]$, and satisfies

$$m(\emptyset) = 0, \quad \sum m(A_i) = 1$$

Then m is defined as basic probabilistic function, and if $A_i \neq \emptyset$ in any $A_i \in \Theta$, then A_i is defined as focal element.

2.2 Combination rules

On the evidences m_1, m_2 satisfy

$$m_1 \oplus m_2(A) = c^{-1} \sum_{A = A_i \cap A_j} m_1(A_i) m_2(A_j) \tag{1}$$

where

$$c = \sum_{A_i \cap A_j \neq \emptyset} m_1(A_i) m_2(A_j) = 1 - \sum_{A_i \cap A_j \neq \emptyset} m_1(A_i) m_2(A_j).$$

3 Improved evaluation of evidence and its combination method

3.1 Definitions

Let's Θ is a discrimination frame:

Definition 2: Suppose m is the *mass* function of any evidence E, conversion $\hat{E}$ of E is defined as follows:

$$\hat{m}(A_i) = m(A_i)\mu_i, \qquad A_i \neq \Theta \tag{2}$$

$$\hat{m}(\Theta) = m(\Theta) + \left(1 - \sum \mu_i\right) \tag{3}$$

where $\mu_i \in [0,1]$ is called as the specialization factor of focal element, from (2) and (3) satisfy $\sum \hat{m}(A_i) = 1$, and $\hat{m}$ is considered as basic probability distribution function. E is called original evidence, $\hat{E}$ is called specialization evaluation evidence, when at any i have $\mu_i = 1$, then E and $\hat{E}$ is consistent, $m = \hat{m}$; when at any i have $\mu_i = 0$, could regard as ineffective evidence, $\hat{m}(\Theta) = 1$, conclusion correspondents to the ignorance.

3.2 Improved combination rule

On the m_1, m_2, correspond $\hat{m}_1, \hat{m}_2$ the estimated evidence combination:

$$
\begin{aligned}
\hat{m}_1 \oplus \hat{m}_2(A) &= c^{-1} \sum_{A = A_i \cap A_j} \hat{m}_1(A_i)\hat{m}_2(A_j) \\
&= c^{-1} \sum_{\substack{A = A_i \cap A_j \\ A_i, A_j \neq \Theta}} m_1(A_i)\mu_{1i}m_2(A_j)\mu_{2j} \\
&+ c^{-1} \sum_{\substack{A = A_i \neq \Theta \\ A_j = \Theta}} m_1(A_i)\mu_{1i}\left[m_1(\Theta) + \left(1 - \sum \mu_{2j}\right)\right] \\
&+ c^{-1} \sum_{\substack{A_i = \Theta \\ A = A_j \neq \Theta}} \left[m_1(\Theta) + \left(1 - \sum \mu_{1i}\right)\right]m_2(A_j)\mu_{2j}, \quad A \neq \Theta
\end{aligned}
$$

$$\hat{m}_1 \oplus \hat{m}_2(\Theta) = c^{-1}\left[m_1(\Theta) + \left(1 - \sum \mu_{1i}\right)\right]\left[m_2(\Theta) + \left(1 - \sum \mu_{2j}\right)\right] \tag{4}$$

where $c = 1 - \sum_{A_i \cap A_j = \Phi} \hat{m}_1(A_i)\hat{m}_2(A_j) = \sum_{A_i \cap A_j \neq \Phi} \hat{m}_1(A_i)\hat{m}_2(A_j)$, μ_{1i} is specialization factor of m_1 on A_i, μ_{2j} is specialization factor of m_2 on A_j, and if exist at least a $\mu \neq 1$, then $c \neq 0$, $\hat{m}_1, \hat{m}_2$ don't conflict completely.

4 Experiments

In other to validate the effectiveness of the method, we have applied it to the grant deterministic tide forecast of eastern sea, at present the tide forecast data source mainly come from weather satellite and tide observation station on the land, for the general data processing data equalization according to the traditional equation of tidewater mechanics can calculate rather exact values; for the high uncertainty problems independently estimate the data each from satellite and tide observation station, present probabilistic results, after that combine the local conclusions in which include uncertainty, finally evaluate them, as a result increase the reliability of disaster forecast. We have experiments on the possible situations: A = global non general tide, B = local non general tide, C = general tide and the frame of discernment $\Theta = \{A, B, C\}$, correspondent possible collection is

$$\{\Phi\}, \{A\}, \{B\}, \{C\}, \{A, B\}, \{A, C\}, \{B, C\}, \{A, B, C\}.$$

According to experience of experts and verification data, specialization factor is like as table 1.

After two information source each independently present combine them, at the same time according to the principal of disaster prevent, need to identify when the greatest combination belief is the one of $\{A\}, \{B\}, \{C\}$, it is effective, otherwise consider it is ineffective, and analysis the original data again, we pick 71 examples(show examples proportional to practical samples), test effective fusion results, and the comparison with traditional D-S theory is like table 2.

Table 1. The Valuable Focus Elements Specialization Quantification

Focus credibility quantification	$\{\Phi\}$	$\{A\}$	$\{B\}$	$\{C\}$	$\{A, B\}$	$\{A, C\}$	$\{B, C\}$	$\{A, B, C\}$
Information source specialization parameters	μ_{i0}	μ_{i1}	μ_{i2}	μ_{i3}	μ_{i4}	μ_{i5}	μ_{i6}	μ_{i7}
Weather satellite($i = 1$)	—	0.9	0.7	0.7	0.75	0.7	0.7	—
Ground tide observation station($j = 2$)	—	0.6	0.95	0.6	0.6	0.6	0.72	—

Table 2. Results Comparison from Two Methods

Results	Conclusion focus element	Statistics	Our method	Traditional method
	$\{A\}$ 33	Right decision	7	12
		False decision	26	21
		Exact rate	78.8%	63.7%
Examples 71	$\{B\}$ 27	Right decision	22	16
		False decision	5	11
		Exact rate	81.8%	59.2%
	$\{C\}$ 11	Right decision	7	6
		False decision	4	5
		Exact rate	63.6%	54.5%

Using the method of multi-dimensional fusion of evidences, is able to effectively collect multi kind of uncertainty evidences; final exactness of decision making lies on evidence source level and complexity of the problems. Therefore a majority of evidence sources are different on the authority of the different problems, conclusions in table display that exactness in the great degree depend on evaluation of specialization of evidence sources, through experiment discovered that the more different the specializations between information sources, the more complex the problem, in that case the combination rule of the paper is able to concentrate reliable information, hence advantage of the results relatively appear more obvious; because in the experiment have no absolute reliable evidences, hence don't absolutely conflict between evidences. The method is especially significant for the property of uncertainty of ocean weather forecast that goes beyond the ability of human experts.

Acknowledgements. This research is partially supported by National key Project (Grant 2004BA608B-03-03-03, 2001-608B-08-08).

References

[YAG87] Yager R.: Using Approximate Reasoning to Represent Default Knowledge. Artificial Intelligence, **31**(1), 99–112 (1987)

[DP88] Dubois D., Prade H.: Default Reasoning and Possibility Theory. Artificial Intelligence, **35**(2), 243–257 (1988)

[CR98] Chen L., Rao S.S.: Modified Dempster-Shafer theory for multicriteria optimization. Engineering Optimization, **30**(3/4), 177–201 (1998)

[Mur00] Murphy C.K.: Combining belief functions when evidence conflicts. Decision Support Systems, **29**(1), 1–9 (2000)

[MR98] Murphy, Robin R.: Dempster-Shafer theory for sensor fusion in autonomous mobile robots. IEEE Transactions on Robotics and Automation, **14**(2), 197–206 (1998)

[LC02] Lefevre E., Colot O.: Vannoorenberghe P. Belief function combination and conflict management. Information Fusion, **3**(2), 149–162 (2002)

[MR96] Murphy, Robin R.: Adaptive rule of combination for observations over time. in Proc of IEEE International Conference on Multi-sensor, Fusion and Integration for Intelligent Systems 1996 IEEE, Piscataway, NJ, USA, 125–131 (1996)

[PS00] Pigeon Luc,Solaiman Bassel,Toutin Thierry etc.Dempster-Shafer theory for multi-satellites remotely-sensed observations. in Proceedings of SPIE-The International Society for Optical Engineering, SPIE, Bellingham, WA, USA. 4051, 228–236 (2000)

[WHY01] Wang Z., Hu W.D., Yu W.X.: A combination rule of conflict evidence based on proportional belief assignment, Acta Electronica Sinica, **29**(z1):1852–1855 (2001)

[JOS02] Josang A.: The consensus operator for combining beliefs, Artificial Intelligence, **141**(1-2), 157–170 (2002)

Parallel Unsteady 3D MG Incompressible Flow

C. H. Tai[1,2], K. M. Liew[1,2], and Y. Zhao[2]

[1] Nanyang Centre for Supercomputing and Visualization, Nanyang Technological
University, Nanyang Avenue Singapore 639798, Republic of Singapore
`jontai73@pmail.ntu.edu.sg`
[2] School of Mechanical and Production Engineering, Nanyang Technological
University, Nanyang Avenue Singapore 639798, Republic of Singapore
`mkmliew@ntu.edu.sg`

Abstract. The development and validation of a parallel unstructured tetrahedral non-nested multigrid method for simulation of unsteady 3D incompressible viscous flow are described in this paper. The Navier-Stokes solver
is based on the artificial compressibility method (ACM) and a higher-order
characteristics-based finite-volume scheme on unstructured multigrid. Unsteady flow is calculated with a matrix-free implicit dual time stepping scheme.
The parallelization of the multigrid solver is achieved by multigrid domain
decomposition approach (MG-DD). There are two parallelization strategies
proposed in this work, the first strategy is a one-level parallelization strategy
using geometric domain decomposition technique alone, the second strategy
is a two-level parallelization strategy that consists of a hybrid of both geometric domain decomposition and data decomposition techniques. The Message-
Passing Interface (MPI) Library is used for communication of data and loop
arrays are decomposed using the OpenMP standard. The parallel codes using
single grid and multigrid are used to simulate steady and unsteady incompressible viscous flows for a 3D lid-driven cavity flow and flow over a 3D circular
cylinder for validation and performance evaluation purposes. The speedups
and efficiencies obtained by both the parallel solvers are reasonably good for
all test cases.
Key words: Parallel multigrid computing, domain decomposition, unsteady
incompressible flow, unstructured tetrahedral grid.

1 INTRODUCTION

The demands of CFD to tackle more and more complex problems found in
engineering applications at system level, particularly in the aerospace and
automotive industries, coupled with the emergence of new, more powerful
generations of parallel computers, have led naturally to work on parallel computing. Algorithms and strategies that successfully map structured grid codes

onto parallel machines have been developed over the previous decade and have become quite established. Extension of the capabilities of these structured grid codes to include unstructured grid codes requires new algorithms and strategies to be developed. Multigrid techniques have been demonstrated as an efficient means for obtaining steady-state numerical solutions to both the compressible Euler and Navier-Stokes equations on unstructured meshes in two and three dimensions [LSD92][M92][PMP93].

This work attempts to parallelize the serial unstructured multigrid code by multigrid domain decomposition approach (MG-DD), using the Single Program Multiple Data (SPMD) and Multiple Instruction Multiple Data (MIMD) programming paradigm, to efficiently simulate unsteady incompressible viscous flows. The aim of this study is to develop a highly parallel multigrid solver so as to avoid very large single processor memory requirement in sequential processing and to reduce the simulation time for simulating unsteady flows with complex geometries.

2 MATHEMATICAL FORMULATION

The 3D Navier-Stokes (NS) equations for incompressible unsteady flows, modified by the ACM [C67], can be re-written in vector form with dimensionless parameters:

$$C\frac{\partial W}{\partial \tau} + K\frac{\partial W}{\partial t} + \nabla \cdot \overrightarrow{F_c} = \nabla \cdot \overrightarrow{F_v} \tag{1}$$

where

$$W = \begin{bmatrix} p \\ u \\ v \\ w \end{bmatrix} \overrightarrow{F_c} = \begin{bmatrix} U \\ uU + p\delta_{ij} \\ vU + p\delta_{ij} \\ wU + p\delta_{ij} \end{bmatrix} \overrightarrow{F_v} = \begin{bmatrix} 0 \\ \frac{1}{Re} \cdot \nabla \cdot u \\ \frac{1}{Re} \cdot \nabla \cdot v \\ \frac{1}{Re} \cdot \nabla \cdot w \end{bmatrix}$$

$$K = \begin{bmatrix} 0\,0\,0\,0 \\ 0\,1\,0\,0 \\ 0\,0\,1\,0 \\ 0\,0\,0\,1 \end{bmatrix} C = \begin{bmatrix} 1/\beta\,0\,0\,0 \\ 0\ \ 1\,0\,0 \\ 0\ \ 0\,1\,0 \\ 0\ \ 0\,0\,1 \end{bmatrix} \tag{2}$$

where W is the flow field variable vector, $U = u \cdot \overrightarrow{i} + v \cdot \overrightarrow{j} + w \cdot \overrightarrow{k}$, $\overrightarrow{F_c}$ and $\overrightarrow{F_v}$ is the convective and viscous flux vectors, respectively, being a constant called artificial compressibility, K is the unit matrix, except that the first element is zero, and C is a preconditioning matrix.

3 NUMERICAL METHODS

The method of artificial compressibility is adopted for solution of the incompressible NS equations [C67]. Eq (1) is discretized on an unstructured tetrahedral grid. A cell-vertex scheme for finite volume spatial discretization is adopted here. The convective term is discretized for every edge in the mesh using a high-order upwind characteristics-based scheme using an edge-based data structure [TZ03]. And the viscous term is calculated based on a cell-based method. A matrix-free implicit dual time-stepping method is used to advance the solution in time.

4 PARALLEL-MULTIGRID IMPLEMENTATION

This work develops two parallelization strategies for the multigrid solver using the one-level and two-level parallelization strategies by geometric domain decomposition technique alone and a hybrid of both geometric domain decomposition and data decomposition techniques. The flowcharts presented in Fig. 1 and Fig. 2 depict both the parallelization strategies The multigrid domain decomposition (MG-DD) approach is adopted for the multigrid parallelization.

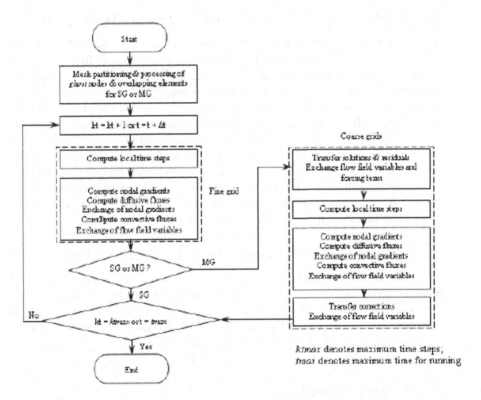

Fig. 1. Flowchart depicting geometric domain decomposition parallelization strategy alone using MPI only.

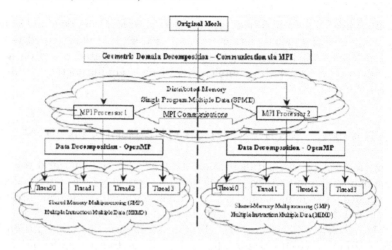

Fig. 2. Two-level parallelization strategies using MPI with OpenMP.

4.1 First-Level Parallelization Strategy - Domain Decomposition

METIS [KKM98] is employed to decompose the flow domain into a set of S sub-domains that may be allocated to a set of P processors. In the MG-DD approach, non-nested multigrids are independently generated first. Then domain decomposition of the finest grid is performed, follows by decomposition of the coarse grids guided by the fine-grid partitions and load balancing in the coarse mesh is reasonably well ensured. An algorithm has been developed to identify the ghost nodes, overlapping elements and to write the individual grid files with local numbering for each partition. The main concept of this algorithm is that those elements along the inter-processor boundaries and with nodes having different partition numbers are considered as overlapping elements which are cut through by partition lines. And those nodes that formed these elements are a mixture of core and ghost nodes.

4.2 Second-Level Parallelization Strategy - Data Decomposition

Data decomposition technique can be achieved using OpenMP standard [OM99] and it decomposes loop iteration arrays into a set of data that are distributed to the respective threads within an MPI process. In this work, the data are distributed using a programming model known as "program level decomposition", each thread uses its rank and the number of threads to determine which part of the data it will process.

4.3 Multigrid Parallelization

The MG-DD approach is adopted in this study. This means that the non-nested multigrids are independently generated first [TZ03]. Then domain decomposition of the finest grid is performed, which is followed by decomposition of the various coarse levels of grids guided by the finer grid partitions. This is achieved by using the fine grid partitions to infer the coarse level partitions (i.e. the coarse grid is to inherit its partition from that of the corresponding fine grid) and load balancing in the coarse mesh is reasonably well ensured. In the multigrid process, the flow field variables and residuals of the coarse grid nodes are obtained directly from their corresponding fine grid nodes while ghost nodes are updated by inter-processor communication.

5 COMPUTATIONAL RESULTS

In this section, the test cases employed to examine the convergence behavior and speed-up characteristics of the unstructured parallel-multigrid solver are steady and unsteady incompressible viscous flow for a 3D lid-driven cavity flow and flow over a 3D circular cylinder. Both serial and parallel computations are performed on an SGI Origin 3400 machine with 32 processors.

5.1 Three-dimensional Lid-Driven Cavity Steady Flow

In this computation, the Reynolds number specified is 400. A three-level multigrid is used. The parallel solutions obtained agree well with the numerical results of Jiang et al. [JLP94]. In the one-level parallelization strategy, the performance results for both parallel SG and MG are shown in Fig. 3, in term of speedup characteristics, parallelization efficiency and percentage computation and communication time. It is clear that the speedup is not that significant for the MG as compared to the SG. This is mainly due to the additional overheads of transferring variables to the coarser grids, which increases the communication time. In this study, the purpose of using the two-level parallelization strategy is to study the effectiveness and viability of this new strategy; therefore only parallel-SG computation is used to make a comparison with the speedup of one-level parallelization strategy. The performance in term of relative speedup of different combinations of OpenMP threads with fixed MPI processors are shown in Fig. 4. The two-level hybrid MPI-OpenMP implementation is seen to give slightly poorer scalability, although the actual differences are never more than 10

5.2 Viscous Unsteady Flow Past a Three-dimensional Circular Cylinder

The test case considered for high Reynolds number unsteady flow is the viscous flow past a three-dimensional circular cylinder for Reynolds number of 200. A three-level MG is used to compute the flow. Fig. 5 shows the lift and drag coefficients on the circular cylinder versus non-dimensional time for both parallel SG and MG. Although both of the flows took the same non-dimensional time to become fully periodic, the number of sub-iterations for

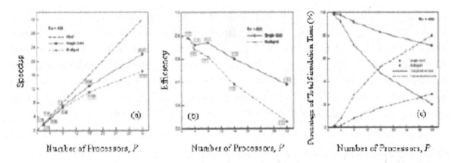

Fig. 3. (a) Speedups; (b) Efficiency and (c) Comparison between computation and communication time for both parallel SG and MG (Re = 400).

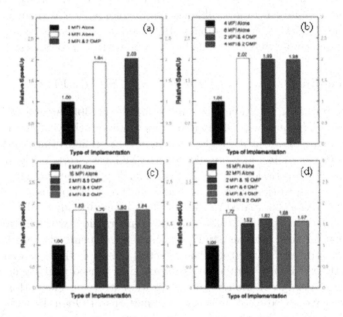

Fig. 4. Relative speedup from (a) 2 to 4 processors; (b) 4 to 8 processors; (c) 8 to 16 processors and (d) 16 to 32 processors for MPI alone and hybrid MPI-OpenMP implementations.

MG is much less than the SG. The performance results for unsteady flow of Re = 200 using both parallel SG and MG is based on the speedup characteristics, efficiency of parallelization and comparison between percentage computation and communication time, as shown in Fig. 6.

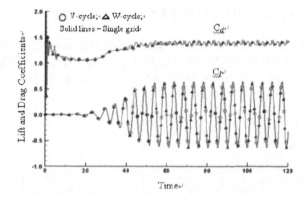

Fig. 5. Lift and drag coefficients versus time for flow over a circular cylinder.

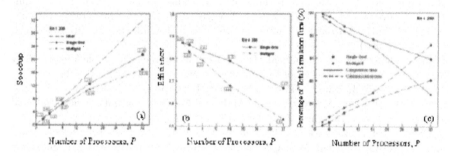

Fig. 6. Performance for unsteady flow past a circular cylinder of Re = 200; (a) Speedup, (b) Efficiency and (c) Comparison between percentage computation and communication time.

6 CONCLUSIONS

A 3D parallel unstructured MG Navier-Stokes code for unsteady incompressible flows has been successfully developed and validated. The method of parallelization by both the method of one-level and two-level parallelization strategies by geometric domain decomposition technique alone and hybrid of both geometric domain decomposition and data decomposition techniques using both the Single Program Multiple Data (SPMD) and Multiple Instruction Multiple Data (MIMD) programming paradigm were successfully implemented into both the single grid and multigrid codes. The convergence rates are found to be independent of number of partitions. The MG method has lower speedups and poorer scalability than the SG method, due to rapid rise in the time spent on communication as a result of denser coarse grids used.

7 ACKNOWLEDGEMENTS

This research work is supported by a research scholarship provided by Nanyang Technological University (NTU) and Sun Microsystems Pte. Ltd. The provision of computing facilities by Nanyang Centre for Supercomputing and Visualization (NCSV), NTU is acknowledged.

References

[LSD92] Lallemand, M., Steve, H. and Dervieux, A. "Unstructured Multigridding by Volume Agglomeration". Computers and Fluids 21, pp397-433. (1992)

[M92]   Mavriplis, D. "Three-dimensional Multigrid for the Euler Equations". AIAA Journal 30, pp1753-1761. (1992)

[PMP93] Peraire, J., Morgan, K. and Peiro, J. "Multigrid Solution of the 3D Compressible Euler Equations on Unstructured Tetrahedral Grids". International Journal for Numerical Methods in Engineering 36, pp 1029-1044. (1993)

[C67]   Chorin A., A Numerical Method for Solving Incompressible Viscous Flow Problems, Journal of Computational Physics, 2 (1), 12 -26. (1967)

[TZ03]   Tai C.H., Zhao Y., Parallel Unsteady Incompressible Viscous Flow Simulation Using An Unstructured Multigrid Method, Journal of Computational Physics, 192, pp.277-311.(2003)

[GLS94] Gropp W., Lusk E., Skjellum A., Using MPI: Portable Parallel Programming with the Message-Passing Interface, The MIT Press, Cambridge, Massachusetts. (1994)

[OM99]   OpenMP home page, OpenMP: Simple, Portable, Scalable SMP Programming. http://www.openmp.org. (1999)

[KKM98] Karypis G., and Kumar V., Metis: A Software Package for Partitioning Unstructured Graphs, Partitioning Meshes, and Computing Fill-Reducing Orderings of Sparse Matrices, Version 4.0, University of Minnesota, Department of Computer Science. (1998)

[JLP94] Jiang B., Lin T. L., Povinelli L. A., Large-scale Computation of Incompressible Viscous Flow by Least-Squares Finite Element Method, Computer Methods in Applied Mechanics and Engineering, 114, 213-231. (1994)

Research of Traffic Flow Forecasting Based on Grids

Guozhen Tan *, Hao Liu, Wenjiang Yuan, and Chengxu Li

Department of Computer Science and Engineering,
Dalian University of Technology, Dalian 116024, P.R. China
gztan@dlut.edu.cn

Summary. The complexity of traffic road network and the huge amount of traffic flow data results in the complexity of traffic flow forecasting. Gird computing technology integrates the grid resources and provides the traffic flow forecasting problem with resource sharing and coordination abilities. In this paper, according to the correlation theory, a traffic flow forecasting algorithm based on back-propagation(BP) neural network for single road section has been put forward, followed with a grid computing model to meet the high-performance requirement of the forecasting process. Making full use of the coordination ability between the multi-nodes of the grid computing, this method solves the precious problems in traffic flow forecasting, such as low efficiency, low real-time, etc.

Key words: neural network, traffic flow forecasting, Gird, coordination

1 Introduction

Urban Traffic Control and Route Guidance System (UTCRG) is considered as one of the essential parts of Intelligent Transportation Systems (ITSs). Based on the accurate real-time traffic flow data of main intersection, UTCRG is used for coordinating the transportation capabilities, such as dynamic traffic guidance, dynamic traffic control and dynamic choose of the optimal path. The performance of the traffic control and guidance is highly decided by the traffic flow forecasting. Artificial neural network has many advantages, such as good adaptability, high precision and needlessness of accurate mathematical expression, so has been widely used in many fields [AT00][YMJ99].
ITSs is an ever-changing, non-liner complicated system. The road traffic information collected by the sensors in the whole road network is enormous and continuously increases. The complexity of traffic network and ITS' demands

---

*This work was supported in part by Grand 2002CB312003 of High Tech. Research and Development (973) Program, China.

of forecasting precision is a challenge to the real-time and high efficiency of
the forecasting technology. With the demand of high precise forecasting result, the forecasting of the neutral and the complexity of study extremely
increased. In the previous traffic forecasting system, the system's response
time will increase much when the computer training the neutral network as
to the road section forecasting requests. The complex computation and the
processing ability of the computers will be the bottleneck of the whole ITS.
With the objective of the coordination and sharing of computation resource,
storage resource, data resource, information resource and knowledge resource,
the advent of grid computation provide a new method to address the bottleneck, which develops rapidly accompanying with the internet, specially for
complex computation [FK99][5]. The multiple computers' coordination based
on grid computation technology can provides efficiency and reliable computation ability and implement the real-time forecasting by sharing of multiple grid
nodes computation ability. Because of the characteristics of the grid [FK99],
the whole computation environment can continuously expand its computation
ability according to requirement of the forecasting system.

2 Traffic Flow Forecasting Model Based on BP Neural Network

In practical traffic flow forecasting, traffic flow has certain relations with the
past ones of the current road section. Also, as a part of the road net, traffic flow
is certainly influenced by the ones of upstream and downstream road sections.
So future road section's traffic flow can be forecasted by the past ones of the
following road sections: the current road section, the upstream road section
and the downstream road section. Owing to the difference of different road
section's traffic conditions, choosing the traffic flow of different road sections in
different time segments, which has strong correlation with the being forecasted
ones as the neural network's inputs is good for improving forecasting precision
[4]. The different time segment's traffic flow is stochastic, so it can be treated
as stochastic variable. According to the correlation coefficient between two
stochastic variables, the correlation between different road sections' traffic
flow can be calculated. Assume $u_i(t)$ is the traffic flow of road section i in
future time segment t, $u_j(t - n)$ is the traffic flow of road section j in time
segment $t - n$. $u_i(t)$ is the neural network's output, and chooses $u_j(t - n)$
which has strong correlation with $u_i(t)$ as the neural network's inputs. The
correlation coefficient ρ between $u_i(t)$ and $u_j(t - n)$ is:

$$\rho = \frac{E((u_i(t) - \overline{u_i(t)})(u_j(t - n) - \overline{u_j(t - n)}))}{\sqrt{E(u_i(t) - \overline{u_i(t)})^2 E(u_j(t - n) - \overline{u_j(t - n)})^2}} \tag{1}$$

In equation (1), $E(X)$ is the variable X's mathematical expectation, $\overline{u_i(t)}$
and $\overline{u_j(t - n)}$ are the mathematical expectations of $u_i(t)$ and $u_j(t - n)$. The

range of ρ is $[-1, 1]$. The bigger the value of $|\rho|$ is, the stronger correlation between $u_i(t)$ and $u_j(t - n)$ is. $|\rho| = 1$ denotes that $u_i(t)$ and $u_j(t - n)$ are linear correlative, $|\rho| = 0$ denotes that $u_i(t)$ and $u_j(t - n)$ are not correlative. By this way, this paper chooses $u_j(t - n)$ which has strong correlation with $u_i(t)$ as the BP neural network's inputs, and founds a traffic flow forecasting model based on BP neural network. The BP neural network used in this paper is a 3-layer structure: one input layer, one middle layer and one output layer. The inputs is the traffic flow which has strong correlation with the being forecasting ones, the outputs is forecasting traffic flow.

Basing on the upper researches, and using the Globus Toolkit 3 development environment, the functions of neural network training and traffic flow forecasting are implemented and deployed according to the OGSA criterion, and as grid services to realize the urban traffic flow forecasting.

3 Grid Architecture for Traffic-flow Forecasting

The Globus Toolkit which based on the OGSA architecture is used for the computing environment of traffic flow forecasting [FKJS02][5]. Due to OGSA, the new service oriented grid architecture combined with web service technology, the traffic-flow forecasting is provided to the user as grid service.

The traffic-flow forecasting system based on grid is composed of four layers: user layer, proxy layer, grid service layer, Data middle layer. The proxy layer and grid service layer is built in the GT3 Grid Platform, on which they provide grid basic service. Show as Fig. 1.

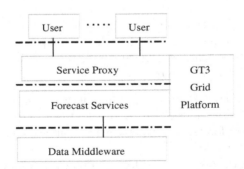

Fig. 1. Grid architecture for traffic flow forecasting

The user layer is the user interface to the forecasting. It provides the input and output of the forecasting via the browser for the user. The proxy layer is a connection between user layer and service layer. The proxy nodes accept input request and publish the output results to the user layer via web service. Meanwhile the proxy node is responsible for selecting the best service node from the grid service register server and implement the optimal performance of the forecasting system. The grid service layer is composed of many forecasting service nodes which consists of multiple heterogeneous computer supporting GT3. The service node, on which forecasting program based on neural network are deployed answer for the computation of the forecasting service. There is a special grid node, grid service register node, which just publish the information of forecasting service nodes in the grid application service layer. The Data Middleware layer provides the data of traffic-flow the forecasting service needed.

4 Traffic Flow Forecasting Method Based on Grid

In order to coordinate all the forecasting service nodes, it is important to dynamically manage these nodes. The grid service register node is the core of service nodes' dynamic management. The service register node deploys the grid index service which all the grid service of whole the platform. The forecasting and study services of all the nodes are registered to grid service register node and periodically communicate with it to guarantee the proxy node can always get the available forecasting and study service. The service nodes communicate with the index service in the register node by the service data, which contains the GSH of forecasting service and provides the state information of forecasting and study service nodes such as CPU type, frequency, the usage of memory, the usage of CPU. The amount of nodes deploy the forecasting service and study service can change according to actual running condition. The proxy node selects the best service node via accessing the service data in the register service node.

Because the flow forecasting and the study of neutral network can always run on the relatively optimal nodes by the selection of proxy nodes, the service nodes' balance can be achieved and guarantee the real-time forecasting. Fig. 2 shows the procedure of the forecasting service.

a. The data collection and processing middleware gather and handle the traffic-flow data the sensors send and store them to the database.
b. The user send the forecasting request, whose contents are the roads needing forecasting to the proxy service.
c. The proxy gets the current state information of all available service nodes by subscribing the service data concerning the service node information in register service.
d. The proxy nodes arrange the grid nodes by the predefined selection policy of service node such as optimal CPU performance, optimal memory, optimal bandwidth. For forecasting of single road, the request will be sent to the optimal node; for massive roads, the request will be sent to different nodes by some ratio according to the different performance of the gird nodes. While waiting for the result, if the service data that the proxy

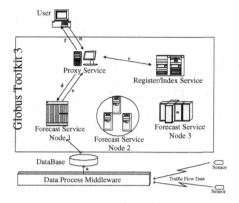

Fig. 2. The procedure of the forecasting service

node subscribe show the forecasting node invalidate, the proxy node will select the service node and submit the request over again.

e. The forecasting service get the parameter list of the response and the historic flow data from the database, compute and return the result according to the road information requested.

f. After receiving the forecasting result, the proxy service inverts it into the format user wanted and returns it as web service.

5 Experiment Results and Conclusions

By the SCOOT system, this paper gets real traffic data of DaLian, and chooses three neighbor road sections' traffic data of Gaoerji Road from 7 : 30 to 8 : 30 in one month (30 days per month). The traffic data is partitioned based on time segment of 5 minutes, so 360 records of traffic data can be got. After correlation analysis, the following conclusion can be drawn: the traffic flow to be forecasted has strong correlation with the ones in the past first and second time segments of the current road section, has strong correlation with the ones in the past second and third time segments of the upstream road section, and has strong correlation with the ones in the past fourth and fifth time segments of the downstream road section. Assume to forecast certain road section's traffic flow in time segment $t \sim t+5$, the neural network's inputs are the current road section's traffic flow in time segments $t-5 \sim t$ and $t-10 \sim t-5$, the upstream road section's traffic flow in time segments $t-10 \sim t-5$ and $t-15 \sim t-10$, and the downstream road section's traffic flow in time segments $t-20 \sim t-15$ and $t-25 \sim t-20$. Using the upper forecasting model and grid system structure, we do the traffic flow forecasting in an experiment environment, the results are shown in Fig. 3:

According to the test, the traffic-flow forecasting system acts good real-time and reliable performance and can expand computation ability without

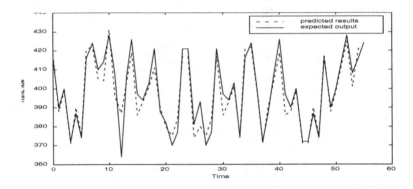

Fig. 3. Forecasting results

affecting the running of the system when the training of neutral network and a lot of forecasting request happen meanwhile. It is an available method to speed up the development of ITS through grid technology. The computation schema studied in this paper can also apply to other component such as route layout, the shortest path selection in time-dependent network.

References

[YN98]   Yun, S.Y., Namkoong, S.: A Performance Evaluation of Neural Network Models in Traffic Volume Forecasting. Mathematical and Computer Modeling, **27(6)**, 293–310 (1998)

[KSA02]   Kalaitzakis, K., Stavrakakis, G.S., Anagnostakis, E.M.: Short-Term Load Forecasting Based on Artificial Neural Networks Parallel Implementation. Electric Power Systems Research, **63**, 185–196 (2002)

[TG02]   Tan, G.Z., Gao, W.: Shortest Path Algorithm in Time-Dependent Networks. Chinese J. Computers, **25(2)**, 165–172 (2002)

[FK99]   Foster, I., Kesselman, C.: The Grid: Blueprint for a New Computing Infrastructure. Morgan Kaufmann Publishers, Inc., San Francisco, CA (1999)

[DCL02]   Du, Z.H., Chen, Y., Liu, P.: Grid Computing. Tsinghua University Press, Beijing (2002)

[FKJS02]   Foster, I., Kesselman, C., Jeffrey, M.N., Steven, T.: The Physiology of the Grid-An Open Grid Services Architecture for Distributed Systems Integration (2002)

[F02]   Foster, I.: The Grid: A New Infrastructure for 21st Century Science. Physics Today, **55(2)**, 42–47 (2002)

[FK98]   Foster, I., Kesselman, C.: The Globus Project: A Status Report. Proc. IPPS/SPDP'98 Heterogeneous Computing Workshop, 4–18 (1998)

A Mobile Access Control Architecture for Multiple Security Domains Environment

Ye Tang[1], Shensheng Zhang[1], and Lei Li[1]

CIT Lab, Computer Science Department, Shanghai Jiaotong University, Shanghai 200030, China, {tangye,sszhang,lilei}@cs.sjtu.edu.cn

Summary. The characteristics of multiple security domains environment include multifactor, dynamic, heterogeneity, and openness, which make its resources at high security risks. Thus, it is very critical for mobile access to realize the efficient resources access control in multiple security domains environment. This paper aims at the problems of dynamic privilege assignment for mobile users in cross-domain access, proposes a mobile access control architecture for multiple security domains environment (MACAMSDE), and introduces flexible access control mechanisms. In addtion it discusses the enabling key technologies.

Key words: mobile access, access control architecture, multiple security domains

1 Introduction

With the development of mobile technologies, the time is going into the age of m-commerce, which allows people to access resources by moving across organizational boundaries and to collaborate with others within/between organizations and communities. The characteristics of multiple security domains environment include multifactor, multi-domain, heterogeneity, dynamic, and openness, which makes its resources at high security risks. Thus, it is very critical to realize the efficient mobile resources access control in multiple security domains environment.

The existing secure distributed architectures mainly focus on single security domain, and they are not applicable for large scale distributed mobile access control in multiple security domains environment. Some researchers propose a two phase policy integration mechanism [SJEA03]. [VSSK03] describes the use of trust negotiation to enable two parties with no pre-existing relationship to perform sensitive transactions. Al-Muhtadi etc. propose an authentication framework to addresses poor security problem in ubiquitous computing environment [ARC02]. Some requirements are presented for an access control system that simultaneously supports mobility, collaboration, and peer-to-peer, illustrate a solution [FDK02]. Some researchers propose a solution based on trust management for pervasive computing environments [KFJ01]. All of them only provide single security mechanism, or only concentrates on single security service.

This paper aims at the problems of dynamic privilege assignment for mobile users, proposes a mobile access control architecture for multiple security domains environment (MACAMSDE), and introduces flexible access control policies and mechanisms.

The paper is organized as follows. Section 2 describes MACAMSDE and its mobile access mechanisms. Section 3 briefly discusses MACAMSDE and gives the formal definition of MACAMSDE. The key supporting technologies of MACAMSDE is presented in section 4. Section 5 concludes the paper.

2 Mobile Access Control Architecture for Multiple Security Domains Environment

2.1 MACAMSDE Framework

MACAMSDE provides two resource access control mechanism. One is direct access control between mobile users and security domains. The other is access control amongst different security domains when there is the resource sharing or collaboration among them. These access control relationships are shown in Figure 1.

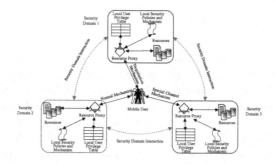

Fig. 1. MACAMSDE Framework

Access from mobile users to destination security domain and collaboration interaction among security domains both are mediated by resource proxies. The resource proxies encapsulate local resources of security domains and provide access control interfaces, which makes individual security domain keeping autonomous. They are charge of security policies collaboration for inter-domain accesses, and provide multiple access mechanisms to mobile users.

2.2 Mobile Access Mechanisms of MACAMSDE

MACAMSDE provides mobile users three kinds of access mechanisms: normal mechanism, special channel mechanism and negotiation mechanism.

- Normal Mechanism: Mobile user has pre-assigned account of destination security domain, and accesses the resources according to the account when he moves to the destination security domain.
- Special Channel Mechanism: When mobile user does not posses the pre-assigned account in a new security domain, special access channel with limit privilege is provided to access public information resources, such as default account. Mobile user can obtain some common information browsing services or resources without providing his real identity information.
- Negotiation Mechanism: Mobile user may have not the required privilege with his accounts when he moves to a new security domain. He sends access privilege request to the administrator of destination security domain, and negotiates with the administrator.

3 Formal Definitions of MACAMSDE

In order to describe MACAMSDE clearly, we give its formal definitions as follows:

Definition 1 (mobile user), mobile user is the set of users who can accomplish resource access activities in multiple security domains, written as U_m. Mobile user is the subject and participant of mobile access control activities. Mobile user set is the intersection of user sets of security domains in which mobile user can access their resources, which can be denoted as $U_m = \bigcap\limits_{i=1}^{n} U_i$, where U_i is user set of the ith security domain, and n is the number of security domains which mobile user can access.

Definition 2 (resource set, resource subset), resource set is the set of all resources in all security domains which support mobile access, written as R. It is the object of mobile access control activities. If $r \subseteq R$, r is resource subset of R. Resource set is union of resource subsets, denoted as $R = \bigcup\limits_{i=1}^{n} r_i$, where r_i is a resource subset, and n is the number of resource subsets.

Definition 3 (credential set, credential subset), credential set is the information set that is used to prove the identity of mobile users in security domains, written as C. If $c \subseteq C$, c is credential subset of C. Credential set can be denoted as $C = \bigcup\limits_{i=1}^{m} c_i$, where c_i is credential subset, and m is the number of all credential subsets. Credential can be password or certificate etc.

Definition 4 (operation set, operation subset), operation set is a finite set, written as OP. Its every element is a kind of operation on resource set. We define $OP = \{create, erase, use, modify, read, write\}$. If $o \subseteq OP$, o is operation subset of OP. Operation set can be denoted as $OP = \overset{m}{\underset{i=1}{\cup}} o_i$, where o_i is operation subset, and m is the number of all operation subsets. Mobile user has different operation subset on different resource subsets in different security domains.

Definition 5 (privilege set, privilege), privilege set is the subset of $U \times R \times 2^{OP}$, written as P. Every element of P is a privilege.

Definition 6 (security policy), security policy is a set of rule which defines the relationship between U and R. It is written as of security domain, which can be mapped on P.

Definition 7 (security domain, destination security domain), security domain is a logical management structure, which employs single and consistent SP, written as SD. SD is comprised of U, R and SP. Destination security domain is the security domain which provides resources to mobile user, written as DSD.

Definition 8 (authentication), authentication is a procedure that mobile user u_m proves his identify by providing his credential c, written as AO. AO can be described as $AO \equiv \Gamma(U_m, C), \Gamma : U_m \times C \rightarrow RE$, where RE is the result of authentication.

Definition 9 (local resource authorization), local resource authorization is a procedure that local security domain decides what kind of privileges local user can obtain to access local resource r according to local security policy $SPl, SP_l \subseteq SP$.

Definition 10 (authorization across security domain), authorization across security domain is a procedure that local security domain decides which privileges remote user can obtain to access local resource r according to local security policy SP_m, $SP_m \subseteq SP$.

Definition 11 (mobile authorization), mobile authorization is a procedure that security domain decides whether or not mobile user u_m can access resource r according to security policy SP. Mobile authorization combines local resource authorization and authorization across security domain.

4 Key Supporting Technologies of MACAMSD

The implementation of MACAMSDE needs many supporting technologies which not only make multiple security domains environment secure scalable and flexible, but also provide mobile users convenience in complex environment. Here we just discuss several key supporting technologies:

(1) Smart card, it is a portable security device with a microprocessor. In MACAMSDE, smart card can be used to store various sensitive data such as passwords, certificates and private keys.

(2) Agent, it is an autonomous, interactive and reflective object. It can be used to implement the resource proxy in security domain to encapsulate all kinds of resources access interfaces, which can keep autonomy of security domain.

(3)Single Sign-On, it supports user to authenticate himself only once and transparently access relevant resources. The resource proxy of destination security domain authenticates mobile user once in multiple security domains environment. It realizes the transparent distributed authorization.

(4)GSS-API, it provides distributed security service universally. The resource proxy can be implemented with GSS-API to enhance the scalability and flexibility of MACAMSDE and integrate heterogeneous security domains.

(5)XML, it is universal standard style to describe data and information. A universal information exchange style can be constructed to describe all kinds of security data according to XML syntax.

5 Conclusion

This paper proposes a mobile access control architecture for multiple security domains environment (MACAMSDE), gives the formal definition of MACAMSDE, and discusses the key supporting technologies of MACAMSDE. It employs different security control mechanisms according to different security attributes of resources in security domains, and satisfies the requirements of flexibility and scalability in dynamic environment.

References

[ARC02] Al-Muhtadi, J.; Ranganathan, A.; Campbell, R.etc. ;A flexible, privacy-preserving authentication framework for ubiquitous computing environments, 22nd International Conference on Distributed Computing Systems Workshops, July 2002, pp:771 - 776

[FDK02] Fenkam, P.; Dustdar, S.; Kirda, E.; Reif, G.; Gall, H.;Towards an access control system for mobile peer-to-peer collaborative environments, Eleventh IEEE International Workshops on Enabling Technologies: Infrastructure for Collaborative Enterprises, June 2002, pp:95 - 100

[KFJ01] Kagal, L.; Finin,T.; Joshi, A.;Trust-based security in pervasive computing environments, Computer , Vol. 34 , Issue: 12 , Dec. 2001 pp:154 - 157

[VSSK03] Vawdrey, D.K.; Sundelin, T.L.; Seamons, K.E.; Knutson, C.D.;Trust negotiation for authentication and authorization in healthcare information systems, the IEEE Proceedings of the 25th Annual International Conference on Engineering in Medicine and Biology Society, Vol. 2, Sept. 2003, pp:1406 - 1409

[SJEA03] Shafiq B.; James B. D. Joshi1; Elisa Bertino etc.. Optimal Secure Interoperation in a Multi-Domain Environment Employing RBAC Policies.https://www.cerias.purdue.edu/tools_and_resources/bibtex_archive/archive/2003-24.pdf (2003)

Inversion of a Cross Symmetric Positive Definite Matrix and its Parallelization

K.A.Venkatesh

Post Graduate Department of Computer Applications, Alliance Business Academy, Bangalore 560 076, India Email:venki@rediffmail.com

Summary. The aim of this paper is to provide a new algorithm to find the inverse of a cross symmetric matrix and its parallel implementation.

Key words: Cross symmetric matrix, Positive definite matrix, CREW PRAM.

1 Introduction

In most of the situations, inverting matrix is essential in order to get the solution vector. Examples arise in statistics [5, 6, 7], in numerical integrations in superconductivity computations [8], and in stable sub space computation in control theory. The recent years have seen increasing interest in computing inversion of a matrix in parallel. Inversion algorithms for general non singular matrices are mostly based on the availability of complete LU factorization. This paper presents a new approach for computing inverse of a Cross Symmetric matrix and its parallelization in CREW PRAM.

2 Cross Symmetric Matrix

A square matrix of order n is said to be cross symmetric matrix if

(i) $a[i, j] = a[\, n - i + 1, n - j + 1\,]$ for $1 = i, j = n.$ and
(ii) $a[i, j] = a[j, i].$

Centro symmetric was introduced by A.C.Aitken[1] and coincides with the definition of cross symmetric matrix defined by Graybill[2]. J.R.Weaver[3] defined the cross symmetric matrix A as $A = RAR$, where R is a reversal matrix. as A.Cantoni and P.Butler[4] imposed an additional condition to the definition of Aitken and Graybill that the matrix is symmetric also. In this paper we deploy the definition of Cantoni and Butler. The transition matrix of finite Markov-Process is one of the best examples of centro symmetric positive definite matrix. Normally, one can find centro symmetric matrices in the estimation of BLUE(Best Linear Unbiased Estimator). The cross symmetric matrix can be viewed as

$$\begin{pmatrix} \lambda & c[i] & \mu \\ c[j] & M[j] & c[j] \\ \mu & c[j] & \lambda \end{pmatrix}$$

where λ, μ are scalars, $c[j]$ is a n-dimensional vector over the field of real or complex numbers, $M[j]$ is a square matrix of order 2n-1.

3 Known Results

A centro symmetric matrix of order 2n and 2n+1 shown as below

$$\begin{pmatrix} A & BJ \\ JB & JAJ \end{pmatrix} \begin{pmatrix} A & X & BJ \\ Y^{\prime} & a & Y^{\prime}J \\ JB & JX & JAJ \end{pmatrix}.$$

Where A and B are matrices of order n, X,Y are vectors and a is any scalar, J is a matrix of order n having 1 in the (i, n - i + 1)th place for every i and zero else where. Pre-multiplication of a matrix by J interchanges ith row and (n-i+1)th row for every i where as post multiplication interchanges the corresponding columns.

Let A, B, C, D, P, Q, L1, L2, U1, U2 be matrices of order n, Li being lower triangular and Ui upper triangular, X, Y, u, V, l1, l2, u1, u2 be vectors and a, b, c scalars. Write,

$$R = \begin{pmatrix} A & X & B \\ Y^{\prime} & a & V^{\prime} \\ C & U & D \end{pmatrix} L = \begin{pmatrix} L & X & B \\ Y^{\prime} & a & V^{\prime} \\ C & U & D \end{pmatrix} U = \begin{pmatrix} A & X & BJ \\ Y^{\prime} & a & Y^{\prime}J \\ JB & JX & JAJ \end{pmatrix}.$$

and

M =(D - C A-1 B) - (a - Y' A-1 X) -1 (U - C A-1 X) (V' - Y' A-1 B)

RESULT 1:

With notation as above and assumption that A is non singular and (a - Y' A-1 X) = 0 , R has LU decomposition iff A and M have LU decomposition

RESULT 2:

If A is non singular and (a - Y' A-1 X) = 0 then R has LU decomposition if A and (D - C A-1 B) have LU decomposition and one of (U - C A-1 X) and (V' - Y' A-1 B) is zero vector

4 Algorithm

In this section, I will be providing a sequential algorithm to compute the inverse of cross symmetric matrix and the functions utilized in this algorithm.

Algorithm CSInverse (Input; output)
Input : cs matrix of real entries of order N
Output : cs inverse matrix of order N // N, IM, L, T, K, Z, I, J : Integers
/* RM, CJT, XJ, XNJT, ANS1, ANS2, BJ, DJ1, BJ1, CNJT,
CNJ, XJT, XNJ, ANS3, BNJT, EJ1, BNJ, M1, M2 - Matrices */

```
Begin                                        KNJ ← [-(P2/S)]
    If N is even then                        Write KJ ← KNJ
Begin                                        SCALAR(KJ,XJ;ANS1)
    IM ← N/2                                 SCALAR(KNJ,XJ;ANS2)
    For i ← 1 to IM do                       ADD(ANS1,ANS2;ANS3)
    For j ← 1 to IM do                       SCALAR(-1,ANS3;BJ)
    MI[i,j] ← A[IM + i -1 ,IM + j-1 ]        Write BJ
    Det ← MI[2,2] * MI[1,1,] - MI[1,2]* MI   TRAN(BJ;BJT)
    Det ← ( 1/ Det)                          Rev(BJ;BNJ)
    TI ← MI[1,1]                             Write BNJ
    RM[1,1] ← MI[2,2]                        REV(BJT,BNJ)
    RM[2,2] ← TI                             MULT1(XJ,BJTANS1)
    RM[1,2] ←(-1)* MI[1,2]                   MULT1(XNJ,ANS2;DJ1)
    RM[2,1] ←(-1) * MI[2,1]

    SCALAR(Det,RM;RM)                        Write DJ1
    Z ← 2; Write RM                          SUB(RM,DJ1;EJ1)
END                                          Write EJ1
ELSE BEGIN                                   END WHILE
    IM ← ( N + 1) /2                         Z ← Z + 2; M ← M -1 ;
    M2 ← A[IM,IM]                            L ← L -1 ; T ← T + 1 ;
    RM[1,1] ← ( 1/ M2)                       RM[1,1] ← KJ
    Z ← 1                                    RM[1,Z] ← KNJ
END                                          RM[Z,1] ← KNJ
    L ← IM; M ← L-1; T ← 1 ;                 RM[Z,Z] ← KJ
    WHILE(Z¡N) DO                            FOR i ← 1 to (Z-2) do
        FOR i ← 1 TO Z do                        RM[1,i+1] ← BJT[1,i]

        CJT[1,1] ← A[M,L-1+i]                    RM[i+1,1] ← BJ[1,1]
        Write CJT[1,1]                           RM[Z,i+1] ← BNJT[1,i]
        REV(CJT;CJ)                          FOR j ← 1 to ( Z-2) do
        TRAN(CJT;CJ)                             RM[i+1,j+1] ← EJ1[i,j]
        MULT 1(RM,CJ;XJ)                     Write RM
        Write XJ                         END
```

TRAN(XJ;XJT)
REV(XJT;XNJT)
TRAN(XNJT;XNJ)
MULT2(CJT,XJ;G1)
P1 ← A[IM-T,IM-T]-G1
MULT2(CNJT,XJ;G2)
IF (N is even) then
    P2 ← A[IM-T,IM+T-1]-G2
ELSE
    P2 ← A[IM-T,IM+T]-G2
S ← (P1 * P1)-(P2* P2)
KJ ← (P1 / S)

The following functions are deployed in the above algorithm

1. **Algorithm REV**(A; C)
   Input: column (row) vector A
   Output: C, Reverse of the given vector A,
   BEGIN
       For i ← 1 to M do
       For j ← 1 to N do
       C[i,j] ← A[M+1-i, N+1-j]
   END

2. **Algorithm MULT1** (A, B; C)
   Input: A and B are conformal
   matrices
   Output : C is the output
   BEGIN
       For i =1 ← M do
       For j = 1 ← P do
       C[i,j] ← 0
       for k = 1 to N do
       C[i,k] ← C[i,j] + A[i,k] * B[k,j]
   END

3. **Algorithm MULT2** (A, B; DD)
   Input: A and B are column (row)
   row (column) conformal matrices
   Output: DD, real value
   BEGIN
       D ← 0
       For i ← 1 to m do
       For j ← 1 to p do
       For k ← 1 to N do
       D ← D + A [i, k]* B[k, j]
       DD ← D
   END

In addition to above functions addition, subtraction, transpose, scalar multiplication of matrices have been used.

5 Parallelization

In this section, the parallel implementation of the above algorithm in CREW PRAM is presented.

```
MULT1(A,B:array:M,N,P: integers: VAR    MULT2(A,B:array:M,N,P: integers: VAR
C:array)                                DD:float )
  BEGIN                                   BEGIN
    FOR i ← 1 TO M DO in parallel           D ← 0
    FOR i ← 1 TO P DO in parallel           FOR i ← 1 TO M DO in parallel
    BEGIN                                   FOR i ← 1 TO P DO in parallel
      C[i,j] = 0;                           For K ← 1 TO N DO
      For K ← 1 TO N DO                         D ← D + A[i,k]*B[k,j]
          C[i,j] ← C[i,j] + A[i,k]*B[k,j]   DD ← D;
    END;                                    END;
  END;
```

Similarly we can employ CREW PRAM with n2 processors to add and subtract two matrices of same order A and B.

6 Conclusion and Future Work

In this paper I have presented the new approach to find the inverse of a cross symmetric matrix. In order to find the inverse of any square matrix of order n, the number of elements required is n(n+1) / 2 elements, where as in this method one has to use much less than n(n+1) / 2 elements. The advantage of this method is that only the distinct entries of the matrix is used. Therefore this method works faster than any other existing methods when the order of the matrix is much higher. The error analysis and implementation of this algorithm on various parallel architectures will be carried out.

Acknowledgement

I thank Prof. P.Nagabhushan Department of studies in Computer Science, Mysore University, Prof.A.K.Chellappa of Vellore Institute of Technology and extending my sincere gratitude to the management of Alliance Business Academy Bangalore, who supported in my all endeavors.

References

1. A.C.Aitken. *Determinants and Matrices*. **Oliver and Boyd, 1939**
2. F.Graybill. *Introduction to Matrices with Application in Statistics*. **Wadsworth, Belmount, 1969**
3. J.R.Weaver. *Centro Symmetric Matrices, their basic properties, Eigen values and Eigen Vectors*. **Amer. Math. Monthly, 92, (1985) 711-717**
4. A.Cantoni and P.Butler. *Eigen values and Eigen vectors of a Centro symmetric matrices*. **Linear Algebra and Applications, 13,(1976), 275-278.**

5. *F.L.Bauer and C.Reinsch. Inversion of positive definite matrices by Gauss - Jordan method. In J.H.Wilkinson and C.Reinsch, editors,* **Linear Algebra, Vol II of Handbook for Automatic Computation, pages 45 - 49. Springer-Verlag.**

6. *J.H. Maindonald. Statistical computation.* **Wiley, New York,1984.**

7. *P.McCullagh and J.A.Nelder. Generalized Linear Models.* **Chapman and Hall, 1989**

8. *M.T. Heath, G.A. Geist and J.B.Drake. Early experience with Intel iPSC/860 at ORNL.* **Technical Report ORNL/TM-11655, Oak Ridge National Laboratory,TN,USA, September 1990. 26 pp.**

Intelligent File Transfer Protocol for Grid Environment

Jiazeng Wang and Linpeng Huang

Department of Computer Science, Shanghai Jiao Tong University, Shanghai 200030, P.R. China wangjz@sjtu.edu.cn huang-lp@cs.sjtu.edu.cn

Abstract. Grid has becoming a new computing paradigm, making intensive data services available. However, the lacking of suitable data transfer mechanism specific to grid environment and the difficulty of manual management of large scale data sets decrease the usability of data grid. In this paper we present a new file transfer method that can both enhance data transfer efficiency and simplify data grid management. We use extended sliding window algorithm to increase the network utility and multiple data stream transfer to realize parallel data transfer, fractional data transfer, and faster retransmission. To simplify and automate data grid management, we elevate FTP to a semantic web service level. Simulated results proved that our method improves data transfer in grid environment.
Key words: Data Transfer, Grid FTP, Semantic Web Service, DAML-S

1 Introduction

Grid[ICS01] is defining a new paradigm in the computing world. Different from conventional distributed systems which focus only on resource sharing and message passing, grid puts more emphasize on large scale data sharing, high speed data trans-portation, and in some cases ironclad system security[Dee04][Tuch04]. With grid technology, data-intensive services are becoming feasible so that scientific data are moving quickly into this realm. Examples include National Virtual Observatory pro-ject (NVO)[Ann02], Laser Interferometer Gravitational Wave Observatory (LIGO)[Bar02], Earth System Grid (ESG)[Ber01]. Grid greatly simplifies the configuration and management of large amount datasets. In addition, grid has integrated technologies in other areas such as Web Service, middle-ware technology, public encryption management in a so smooth fashion that access remote data is a household task. Finally, grid provides handful tools such as the Globus Toolkit[Grid03] to make the environment programmable.

Although grid makes large scale data sharing available, it does not make it ideal. There are two problems encountered in todays data grid applications.

First, the file transfer protocol (FTP)[RFC959], a popular data sharing mechanism, was not designed for the grid. The second problem is that data sharing is often a human initiated action. In a grid environment where the data size is often terabytes or petabytes, managing data manually is intolerable.

In this paper, we describe a mechanism called intelligent file transfer protocol (IFTP) that can enhance data sharing in grid. In order to overcome the shortcomings of traditional FTP, we adopted a more flexible sliding window algorithm[AST96] and multiple data stream transmission. To our experience, by expanding the size of sliding window we greatly improved the data transmission efficiency between two parties, and by using parallel transfer we obtained a more flexible FTP configuration specific to our grid environment. To minimize human intervention in a data sharing session, we elevated the FTP functionalities from normal commands to web services. In our environment, data providers register available data to a central registry machine and data consumers lookup data source information in this registry. Registry then helps to establish the direct connection between the server and client. To fully automate the whole process and thus minimize the corresponding human labor, we applied semantic web service technology[MKT03] that describes the services in DAML-S[DAML][DAML-S], enabling the discovery and interaction between web services.

The paper proceeds as follows. In section 2, we discuss the new requirements of data transfer in our grid environment, and in section 3 we discuss how to enhance performance by using adaptive sliding window and multiple data streams. Section 4 covers how we provide FTP services as a web service and the register-lookup-transfer process. Section 5 includes the system architecture and our demonstration results. Section 6 reports the related work, and finally in section 7 we conclude.

2 System Requirements

In a typical data grid, data is generated at a high speed and data sources usually lo-cate at geographically distributed sites. System configuration is dynamic, because new data sources would be added into the grid at some unpredicted time, because data providers maybe collapse so that the shared data vanish at once, and because some machines can both be data providers and data consumers and change their roles in a single session frequently. Thus, in an ideal grid data transfer should be both fast and simple. Other requirements include flexible data transfer policy, parallel data transfer, etc. In summary, the system requirements are:

- Transfer large amounts of data (TB or above) at a high speed
- Access to large amounts of data located at different locations
- Allow a process to use multiple data streams to obtain data
- Data consumers are allowed to get a portion of data

- Retransmission mechanism
- Access data from multiple data sources simultaneously
- Dynamic and automatic data management

3 Performance Enhancement

3.1 Configurable Window Size

Sliding window protocol[AST96] is used in TCP sessions to enable a burst of data frames to be transmitted as a single group.

Fig. 1. A sliding window with size 4.

The size of sliding window has a significant impact on the network efficacy. Suppose every data frame is k bits, the sender has a sending window of size s, the underlying network transmission rate is b bits/s, and the round trip time between a pair of sender and receiver is R seconds, then the time used to transfer the data is:

$$k * s/b \qquad (1)$$

And the total time used in a round trip is (we assume the time used to transmit the acknowledgement data frames can be omitted):

$$k * s/b + R \qquad (2)$$

As a result, the efficacy of the network can be calculated by:

$$f(s) = (k * s)/(k * s + bR) \qquad (3)$$

The window size is currently represented as a 16-bit TCP field in the TCP header and henceforth smaller than 64K bytes. In our implementation, we followed the defi-nitions in RFC 1323, expanding the window size to 32 bits and using a scale factor to carry the size value in a 16-bit option field. Every time a new connection is about to establish, the two parties extract this field and calculate the sliding window size ac-cording to the factor. Once they start to communicate, however, the window size can not be changed.

Theoretically, the data transfer efficiency can be increased with respect to window size because $f(s)$ is monotonic increase. In practice, we use adaptive window size. Large window consumes much memory and the underlying vir-tual memory system will come into play which leads to excessive disk I/O and henceforth decreases the systems usability. In addition, since the grid en-vironment has a heterogeneous structure, hard coding the window size and applying it to the entire grid is not a wise choice.

3.2 Multiple Data Stream Transfer

Multiple data streams can further utilize the bandwidth provided by the grid environment. Classic FTP use single stream In a multiple data stream transfer process, a data receiver first negotiates with its data provider about how to split the file, and then the two participants start a multithreaded data transfer session.

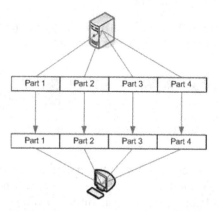

Fig. 2. A file transfer session using 4 parallel data streams.

Using multiple data streams has several advantages over single stream mechanism. First, transmission efficiency is increased. Second, by partitioning a large file into several parts, we get a more effective retransmission mechanism. Third, data consumer can ask for only a portion of data. Finally, by changing the above structure slightly, we can implement stripped data transfer as described in system requirements.

The number of streams can be set by users so that users can get maximal control over the transfer procedure. However, in our implementation, we do not allow manual setting of the number of streams. Instead, we use the following formula to calculate the number of streams:

$$number of streams = bandwidth * delay/window \; size \qquad (4)$$

We made the decision according to the following two criteria. First, if a user initi-ates too many data streams, system overhead (such as cache, context switching) is huge. Second, our purpose is to provide satisfactory data transfer mechanism that requires less human intervention.

4 Deliver FTP as Web Service

Though networked computing environment greatly increases interoperability, it is often assumed that there is a programmer sitting behind the backdrop

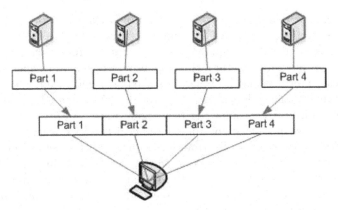

Fig. 3. A data consumer set 4 data streams, with each stream attached with a different data provider.

to make things happen. For example, resources are accessed by their URLs, and port numbers are precompiled in the software and can only be changed by updating its configuration files. We tried to change such condition in our grid by promoting FTP from normal commands to semantic web service. The establishment of a connection is consists of three relating procedures. First, providers submit their data information into service profile and inform the registry. Then, consumers compiler their request and send their requests to registries. The registry matches requests with services and returns a list of service candidate to the con-sumers. Finally, consumers connect with providers, and start the data transfer process. The whole process is depicted in figure 4.

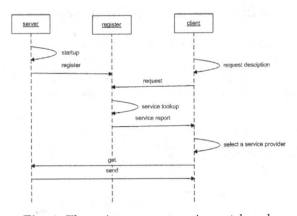

Fig. 4. The register-request-service match cycle

The role of service profile is to describe the web service capability. We define web service capability as functions that perform some transformation, and we represent such functions in DAML-S. Since DAML-S is a high level representation of web services and FTP functionality is straightforward, it is easy for the registry to match requests with services. To fully take advantage of the power of this infrastructure, the data providers should frequently update their service information. This is accomplished by using a deamon process that monitors the changes take place in the providers. When the data information has been changed, the system can reconfigure itself. We divided our system into three parts: providers, consumers and registries, as usually done in the community. Conventional FTP transaction involves providers and consumers only, but now these two parties communicate with the help of registries. By semantics, we mean that we defined some ontology about file information in DAML-S, which bridges the gap between an infrastructure of web services based on WSDL and the semantic web.

```
<!-- This is the FileTransferService -->
<service:Service rdf:ID="FileTransferService">
    <!-- Reference to the Profile -->
    <service:presents rdf:resource="#Profile_FileTransfer_Service"/>
    <service:describedBy rdf:resource="&FileTransfer_process;#FileTransferProcessModel"/>
    <service:supports rdf:resource="&FileTransfer_grounding;#FileTransferGrounding"/>
</service:Service>
```

Fig. 5. File transfer service described in DAML-S

5 System Architecture and Performance

The above discussion forms a firm basis upon which a new FTP infrastructure can be built. Sliding window algorithm is embedded into a local ftp server. And multiple data stream transfer is implemented as a multi-thread program. Since the number of streams is automatically calculated, no more negotiation

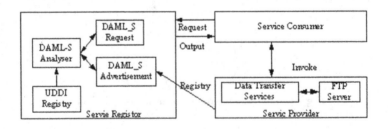

Fig. 6. System architecture of IFTP

message between provider and consumer is needed. To simply the implementation of the web service architec-ture, we use one central registry. DAML-S is a high level description language, so DAML-S virtual machine is needed. A DAML-S virtual machine is composed of a request compiler, which reads and analyses DAML-S service profiles, and UDDI registry program. We put this virtual machine in the registry machine to simplify the system design. In a larger system, optional design exists. For example, the virtual machines can be installed on service providers, or we can set aside some idle machines to act as DAML-S brokers. To test the effect of window size on transmission time,

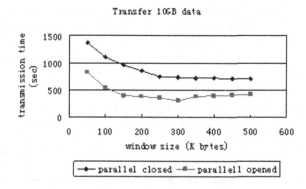

Fig. 7. Transfer a 10GB data set with different window size and multiple data stream option

we use a 10GB dataset with the multiple data stream option closed. The time used to transfer the dataset decreased when the window size increased. From the result we learned that by slightly expand-ing the window size we could get a dramatic performance enhance. The original window size limit is 64K bytes, and by expanding it to 150K bytes, we get a 29% saving in transmission time. With parallel transmission opened, the time further decreased. But because the number of stream increases when the window size decreases, the performance enhancement is unnoticeable. This is because a multi-threaded program leads to more system overhead, and our test data is small relative to the capability of parallel transfer.

6 Related Work

GridFTP[Grid03] is a new standard suitable for file transfer in grid environment. Its features are: GSI and Kerberos support, third-part control of data transfer, parallel data transfer, striped data transfer, partial file transfer and reliable data transfer. However, it does not address system management problems. RootFTP is based on multiple data stream transmission. The number of parallel streams is chosen by users. BBFTP is another program using multiple data streams. Unlike RootFTP, BBFTP implements other features such as data compression, big window, automatic retry, etc. Our approach is similar with that of BBFTP, but we allow adaptive window size. Semantic web service[Ber01][MKT03][DAML][DAML-S] is originally designed for intelligent e-business agents. It extends web service by allowing services to com-municate with each other via DAML-S scripts.

This paper presents our research on efficient data transfer in grid environment. Our contributions include expanding current sliding window size, exploiting multiple data streams, and using semantic web service to automate data grid management.

Acknowledgements. This paper is supported by SEC E-Institute: Shanghai High Institutions Grid project 200308 and Shanghai Information Grid project.

References

[Ann02] Annis, J., Zhao, Y., Voeckler, J., Wilde, M., Kent, S. and Foster, I. Applying Chimera Virtual Data Concepts to Cluster Finding in the Sloan Sky Survey. Super-computing, 2002.

[Bar02] Barish, G. and C.A. Knoblock. An Expressive and Efficient Language for Information Gathering on the Web. Proceedings of the Sixth International Conference on AI Planning and Scheduling (AIPS-2002) Workshop, 2002.

[Ber01] Berners-Lee, T., James Hendler and Ora Lassila. The Semantic Web. Scientific American, May 2001.

[RFC959] RFC 959. File Transfer Protocol.

[ICS01] I. Foster, C. Kesselman, S. Tuecke. The Anatomy of the Grid: Enabling Scalable Vir-tual Organizations. International J. Supercomputer Applications, 15(3), 2001.

[Dee04] E. Deelman, G. Singh, M. P. Atkinson, A. Chervenak, Chue Hong, Kesselman, S. Patil, L. Pearlman, and M. Su. Grid-Based Metadata Services. 16th Interna-tional Conference on Scientific and Statistical Database Management (SSDBM04), June 2004.

[Tuch04] R. Tuchinda, S. Thakkar, Y. Gil, E. Deelman. Artemis: Integrating Scientific Data on the Grid. Proceedings of the Sixteenth Innovative Applications of Artificial Intelligence, July 2004.

[Grid03] The Grid2003 Project, http://www.ivdgl.org/grid2003/.

[AST96] Andrew S. Tanenbaum. Computer Networks. Prentice Hall. 1996.

[MKT03] Massimo Paolucci, Katia Sycara, Takahiro Kawamura. Delivering Semantic Web Services. CMU-RI-TR-02-32, Robotics Institute, Carnegie Mellon University, May 2003.

[DAML] DAML Joint Committee. DAML-OIL Language. http://www.daml.org.

[DAML-S] The DAML Services Coalition. DAML-S: Web Service Description for the Seman-tic Web. In ISWC2002, 2002.

A Parallel Algorithm for QoS Multicast Routing in IP/DWDM Optical Internet*

Xingwei Wang[1], Jia Li[1], Hui Cheng[1], and Min Huang[2]

[1] Computing Center, Northeastern University, Shenyang, 110004, P.R.China
  wangxw@mail.neu.edu.cn
[2] School of Information Science and Engineering, Northeastern University,
  Shenyang, 110004, P.R.China

Abstract. A parallel algorithm for QoS multicast routing in IP/DWDM optical Internet is proposed. Given a multicast request and a required end-to-end delay interval, the algorithm could find a cost suboptimal QoS multicast routing tree by integrating multicast routing and wavelength assignment. It has been proved that finding such a tree is NP-hard. Hence, a coarse-grain parallel genetic simulated annealing algorithm is proposed to build the tree. The proposed algorithm takes load balancing into account as well. Simulation results have shown that the proposed algorithm is both feasible and effective.

Key words: IP/DWDM optical Internet, QoS, multicast, coarse-grain parallel genetic simulated annealing algorithm, load balancing

1 Introduction

IP/DWDM optical Internet is considered to be a promising candidate for Next Generation Internet (NGI) backbone [PM01]. Supporting multicast and QoS is very critical in NGI.

In consideration of the uncertainties of both the user QoS requirement and the characteristics of network status, flexible QoS should be accommodated. Thus, delay interval [AT00] is introduced to describe the end-to-end delay requirement instead of bounded delay.

Given a multicast request and a required end-to-end delay interval, the algorithm could find a cost suboptimal QoS multicast routing tree by integrating

---

*This work is supported by the National Natural Science Foundation of China under Grant No.60473089, No. 60003006 and No. 70101006; the Natural Science Foundation of Liaoning Province in China under Grant No. 20032018 and No. 20032019; the Modern Distance Education Engineering Project of China MoE.

multicast routing and wavelength assignment. It has been proved that finding such a tree is NP-hard [YMJ99]. Hence, a coarse-grain parallel genetic simulated annealing algorithm is proposed to build the tree.

2 Basic Assumption

IP/DWDM optical Internet can be modeled as a graph, consisting of the set of nodes representing optical nodes and the set of edges representing optical fibers that connect the nodes. Assume each node exhibits multicast capability, equipped with an optical splitter at which an optical signal can be split into an arbitrary number of optical signals. In consideration of the still high cost of wavelength converter, only some nodes are equipped with full-range wavelength converters. Assume the conversion between any two different wavelengths has the same delay at any optical node with converter. For each edge, there are two oppositely directed fibers for data transmission along its two directions. Each directed fiber is called a link. The number of wavelengths that a link can support is finite, and it may be different from that of others.

3 Model Description

Given a graph $G(V, E)$, where V is the set of nodes and E is the set of edges. If wavelength conversion happens at node $v_i \in V$, the wavelength conversion delay at v_i is $t(v_i) = t$, otherwise, $t(v_i) = 0$. The set of available wavelengths, delay and cost of edge $e_{ij} = (v_i, v_j) \in E$ are denoted by $w(e_{ij}) \subseteq w_{ij} = \{\lambda_1, \lambda_2, \ldots, \lambda_{n_{ij}}\}$, $\delta(e_{ij})$ and $c(e_{ij})$ respectively, where w_{ij} is the set of supported wavelengths by e_{ij} and $n_{ij} = |w_{ij}|$.

A multicast request is represented as $R(s, D, \Delta)$, where $s \in V$ is the source node, $D = \{d_1, d_2, \ldots, d_m\} \subset V$ is the set of destination nodes, and Δ is the required end-to-end delay interval of users. Suppose $U = \{s\} \cup D$.

The proposed algorithm is to construct a multicast routing tree from the source to all the destinations, i.e. $T(X, F)$, $X \subseteq V$, $F \subseteq E$.

The total cost of T is defined as follows:

$$Cost(T) = \sum_{e_{ij} \in F} c(e_{ij}). \tag{1}$$

To balance the network load, edges with more available wavelengths should be selected firstly. The edge cost function is defined as follows:

$$c(e_{ij}) = n - |w(e_{ij})|. \tag{2}$$

$$n = \max_{e_{ij} \in E} \{n_{ij}\}. \tag{3}$$

Let $P(s, d_i)$ denote the path from s to d_i in T. The communication delay between s and d_i along T, denoted by PD_{sd_i}, can be represented as follows:

$$PD_{sd_i} = \sum_{v_i \in P(s,d_i)} t(v_i) + \sum_{e_{ij} \in P(s,d_i)} \delta(e_{ij}). \tag{4}$$

The delay of T is defined as follows:

$$Delay(T) = \max\{PD_{sd_i} | \forall d_i \in D\}. \tag{5}$$

Let $\Delta = [\Delta_{low}, \Delta_{high}]$, user QoS satisfaction degree is defined as follows:

$$Degree(QoS) = \begin{cases} 100\% & Delay(T) \leq \Delta_{low} \\ \frac{\Delta_{high} - Delay(T)}{\Delta_{high} - \Delta_{low}} & \Delta_{low} < Delay(T) < \Delta_{high} \\ 0\% & Delay(T) \geq \Delta_{high} \end{cases} \tag{6}$$

Different QoS requirements can be satisfied by adjusting Δ.

4 Algorithm Design

The coarse-grain parallel genetic simulated annealing algorithm combines coarse-grain Genetic Algorithms (cgGA) with Simulated Annealing Algorithms (SAA) [4], i.e., the Boltzmann accepting criterion of SAA is introduced into the substitution strategy of cgGA. In cgGA, a whole population is divided into several subpopulations. Each subpopulation evolves independently, exchanges useful individual information with others regularly to boost the evolution, and thus exhibits inherent parallelism.

4.1 Chromosome Expression

A chromosome is denoted by binary coding. Each bit of the code corresponds to one node in $G(V, E)$. The graph corresponding to solution S is $G'(V', E')$. Let the function $bit(S, i)$ denote the ith bit of S, $bit(S, k) = 1$ iff $v_k \in V'$. The code length equals $|V|$. Construct the minimum cost spanning tree $T'_i(X'_i, F'_i)$ of G'. T'_i spans the given nodes in U. However, G' may be unconnected, thus S corresponds to a minimum cost spanning forest, also denoted by $T'_i(X'_i, F'_i)$. It's necessary to prune the leaf nodes not in U and their related edges in T'_i. The result is denoted by $T_i(X_i, F_i)$, and assign wavelengths to T_i.

4.2 Wavelength Assignment Algorithm

The objective is to minimize the delay of the tree by minimizing the number of wavelength conversions, making $Degree(QoS)$ high. It is based on the wavelength graph (WG) ideas [5]. If T_i is a tree, assign wavelengths to it.

Construct WG for T_i. In WG, $N * w$ nodes are created, where $N = |X_i|$, $w = |\cup_{e_{ij} \in F_i} w(e_{ij})|$. All the nodes are arranged into a matrix with w rows and N columns. Add edges in WG, where a vertical edge represents wavelength conversion, assigning wavelength conversion delay as its weight, and a horizontal edge represents an actual edge in T_i, assigning propagation delay as its weight. The WG construction method is shown in [6].

Treat WG as an ordinary topology graph, find the shortest paths from the source column to each destination column using Dijkstra algorithm, then construct the multicast routing tree. In addition, the problem of more than one wavelengths being selected on one edge should be solved to utilize the resource efficiently. Map the paths in WG back to the paths and wavelengths in T_i, thus the wavelength assignment is completed.

4.3 Fitness Function

Fitness function $f(S)$ is determined by $Cost(T_i)$ and $Degree(QoS)$:

$$f(S) = (Cost(T_i) + [count(T_i) - 1] * \rho)/(Degree(QoS)). \tag{7}$$

$count(T_i)$ is the number of trees in T_i, ρ is a positive constant. If T_i has more than one tree, add a penalty value to the cost of T_i and take a smaller value for $Degree(QoS)$.

4.4 Initial Population Generation

The initial population is composed of the chromosomes generated randomly.

4.5 Initial Annealing Temperature Selection

Set the initial annealing temperature $t_0 = K\delta$, K is a sufficiently large number, and $\delta = C_g - C_u$, C_g is the total cost of the current graph, and C_u is the cost of the subgraph composed of all the nodes in U.

5 Simulation Research

The simulation research has been done on some actual network topologies, including NSFNET and CERNET.

5.1 Tree Cost Evaluation

Contrast solutions obtained by the proposed algorithm with the optimal cost solution obtained by exhaustive search. The simulation result is shown in Table 1. The obtained solutions by the proposed algorithm are rather satisfied.

Table 1. Simulation result on the tree cost

Session	Ratio of session	Obtained tree cost vs. optimal cost		
number	nodes in the network	$\leq 1\%$	$\leq 5\%$	$\leq 10\%$
1	42.9%	80.5%		13.2%
2	57.1%	96.0%	0.2%	
3	71.4%	85.7%	0.5%	1.1%

5.2 QoS Evaluation

The delay of the tree obtained by the proposed algorithm and its counterpart obtained without considering $Degree(QoS)$ are compared. Fig. 1 shows that the QoS of the multicast routing tree is improved effectively and efficiently.

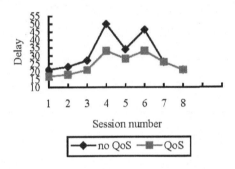

Fig. 1. QoS evaluation

A parallel algorithm for QoS multicast routing in IP/DWDM optical Internet is presented. Given a multicast request and a required end-to-end delay interval, the proposed algorithm could find a cost suboptimal QoS multicast routing tree. It takes load balance into account as well. Simulation results have shown that it is effective and efficient.

References

[1] Green P.: Progress in optical networking. IEEE Communication Magazine, **39**, 54–61 (2001)
[2] Dean H. L., Ariel O.: QoS routing in networks with uncertain parameters. IEEE/ACM Transactions on Networking, **6**, 768–778 (1998)
[3] Ramaswami R., Sivarajan K. N.: Routing and wavelength assignment in all-optical networks. IEEE/ACM Transactions on Networking, **3**, 489–500 (1995)
[4] Wu H. Y., Chang B. G., Zhu C. C., et al: A multigroup parallel genetic algorithm based on simulated annealing method. Jounal of Software, **11**, 416–420 (2000)

[5] Chlamtac I., Farago A., Zhang T.: Lightpath (wavelength) routing in large WDM networks. IEEE Journal on Selected Areas in Communications, **14**, 909–913 (1996)

[6] Wang X. W., Cheng H., Li J., *et al*: A multicast routing algorithm in IP/DWDM optical Internet. Journal of Northeastern University (Natural Science), **24**, 1165–1168 (2003)

A Parallel and Fair QoS Multicast Routing Mechanism in IP/DWDM Optical Internet*

Xingwei Wang[1], Cong Liu[1], Jianye Cui[1], and Min Huang[2]

[1] Computing Center, Northeastern University, Shenyang 110004, P.R. China
  wangxw@mail.neu.edu.cn
[2] School of Information Science and Engineering, Northeastern University,
  Shenyang 110004, P.R. China

Abstract. Quality of Service (QoS) requirement is denoted by the range to support the flexible and heterogeneous QoS. According to the microeconomics, a Kelly/PSP-model-based pricing strategy is presented to support the inter-group fairness. The Equal Link Split Downstream (ELSD) method is adopted to support the intra-group fairness. Based on the parallelized Firing Coupled Neural Network (FCNN), the multi-constrained QoS multicast routing algorithm is designed. Simulation results have shown that the proposed mechanism is both effective and efficient, and its runtime efficiency is higher than the corresponding serialized one.

Key words: IP/DWDM optical Internet, QoS multicast, routing, fairness, neural network, microeconomics

1 Introduction

IP/DWDM optical Internet is one of the critical networking techniques of the Next Generation Internet (NGI), and Quality of Service (QoS) and multicast are essential capabilities of NGI [1]. With the gradual commercialization of Internet, paying for network usage is becoming inevitable. For group applications, both inter-group and intra-group fairness should be guaranteed [2, 3]. According to the principle of "who gets the profit, who pays", "pricing by quality" and "pricing with quality" should be performed. In addition, due to the inaccuracy and dynamics of the network status information and the difficulty in accurately describing the user QoS requirement, the flexible QoS should be supported. Further, different members of the group often have different QoS requirements, thus heterogeneous QoS should be considered.

*This work is supported by the National Natural Science Foundation of China under Grant No.60473089, No.60003006 and No.70101006; the Natural Science Foundation of Liaoning Province in China under Grant No.20032018 and No.20032019; the Modern Distance Education Engineering Project of China MoE.

In this paper, the discussion is focused on a parallel and fair QoS multicast routing mechanism in IP/DWDM optical Internet. QoS requirement is denoted by the range to support the flexible and heterogeneous QoS. Combining price-based Kelly model and game-theory-based PSP model, a microeconomics-based pricing strategy is proposed, not only supporting inter-group fairness, but also helping the wavelength resource allocation approach Pareto optimum under Nash equilibrium [2, 3]. Using Equal Link Split Downstream (ELSD) [4] method to apportion the cost among group members, the intra-group fairness is supported. Based on the parallelized Firing Coupled Neural Network (FCNN) [5], the multi-constrained QoS multicast routing problem is solved, not only exploiting the inherent parallelism in FCNN to improve the runtime efficiency, but also adapting to the scalability of the network size and the problem complexity, at the same time meeting with QoS requirement. Combining the above, a parallel and fair QoS multicast routing mechanism is established.

2 Parallel and Fair QoS Multicast Routing Mechanism

2.1 Model Description

IP/DWDM optical Internet can be modeled as a graph $G(V, E)$, where V is the set of nodes representing wavelength routers or Optical Cross-Connect (OXC) and E is the set of edges representing optical fibers. Assume each node $v_i \in V$ possesses the unlimited multicast capability, and is associated with the following three parameters: $d_{v_i}, j_{v_i}, l_{v_i}$, representing its delay, delay jitter and error rate respectively, $i = 1, 2, 3, \cdots, |V|$. Each edge $e_{ij} \in E$ is associated with the following four parameters: $c_{e_{ij}}, w_{e_{ij}}, d_{e_{ij}}, l_{e_{ij}}$, representing its cost, number of supported wavelengths, delay and error rate, $j = 1, 2, 3, \cdots, |V|$. Let $\Lambda(e_{ij})$ represent the set of available wavelengths of e_{ij}.

A QoS multicast request is denoted by $R = (s, D, Dlb, Dub, Elb, Eub, W_n)$, where $s \in V$ is the multicast source node, $D \subseteq \{V - \{s\}\}$ is the set of multicast destination nodes, Dlb and Dub denote the lower and upper bound of delay respectively, Elb and Eub denote the lower and upper bound of error rate respectively, W_n is the number of the required wavelengths. In the established multicast tree, $T(s, D)$, for $\forall d \in D$, the following relations exist:

$$Delay(P_T(s, d)) = \sum_{v_i \in P_T(s,d)} d_{v_i} + \sum_{e_{ij} \in P_T(s,d)} d_{e_{ij}}. \tag{1}$$

$$Jitter(P_T(s, d)) = \sum_{v_i \in P_T(s,d)} j_{v_i}. \tag{2}$$

$$Error(P_T(s, d)) = 1 - \prod_{v_i \in P_T(s,d)} (1 - l_{v_i}) \prod_{e_{ij} \in P_T(s,d)} (1 - l_{e_{ij}}). \tag{3}$$

$$\Lambda(P_T(s,d)) = \bigcap_{e_{ij} \in P_T(s,d)} \Lambda(e_{ij}). \tag{4}$$

$$\Lambda(T(s,D)) = \bigcap_{d \in D} \Lambda(P_T(s,d)). \tag{5}$$

$$Cost(T(s,D)) = \sum_{e_{ij} \in T(s,D)} c_{e_{ij}}. \tag{6}$$

where $P_T(s,D)$ represents the path from s to d in $T(s,D)$.
The following constraints should be met with:
1) Delay constraint: $Dlb \leq Delay(P_T(s,d)) \leq Dub$.
2) Jitter constraint: $0 \leq Jitter(P_T(s,d)) \leq Dub - Dlb$.
3) Error rate constraint: $Elb \leq Error(P_T(s,d)) \leq Eub$.
4) Wavelength constraint: $W_n \leq |\Lambda(T(s,D))|$.
The objective is to minimize the cost of the established tree with the QoS requirement guaranteed, that is

$$\texttt{Minimize}(Cost(T(s,D))) \ s.t. \ \texttt{constraints 1), 2), 3) and 4).} \tag{7}$$

Note that the formulas (4) and (5) should follow the wavelength continuity constraint. For the nodes with wavelength conversion capabilities, define a wavelength conversion function WCF. Thus, not only the wavelength continuity constraint is broken, but also these two formulas can still be used.

$$\Lambda(e_{ij}) = WCF(\Lambda(e_{ij})). \tag{8}$$

2.2 Kelly/PSP-Model-Based Pricing Strategy

A Kelly/PSP-model-based [2, 3, 6] pricing strategy in IP/DWDM optical Internet is proposed, adopting the following wavelength price formula:

$$q_{e_{ij}} = k_1 \frac{1}{w_{e_{ij}}} \frac{a_{e_{ij}}}{(1 - Y_{e_{ij}})^{b_{e_{ij}}+1}} + k_2 X_{e_{ij}}. \tag{9}$$

where $q_{e_{ij}}$ is wavelength price on the edge e_{ij}, $X_{e_{ij}}$ is the number of the currently available wavelengths on e_{ij}, $Y_{e_{ij}}$ is wavelength occupation ratio on e_{ij} computed as the formula (10), k_1 and k_2 are two weight coefficients, $a_{e_{ij}}$ and $b_{e_{ij}}$ are two experience-based wavelength pricing parameters on e_{ij}, determined by the formulas (14) and (15).

$$Y_{e_{ij}} = w_{e_{ij}} - X_{e_{ij}}/w_{e_{ij}}. \tag{10}$$

Use the formula (11) to compute $c_{e_{ij}}$.

$$c_{e_{ij}} = W_n \times q_{e_{ij}}. \tag{11}$$

The wavelength price can belong to three different regions, i.e., low, sound, and high [6]. Due to the wavelength occupation ratio is low in the low price region, denoted by η_0 , define the low-inclined price as π_0. Due to the wavelength occupation ratio is high in the high price region, denoted by η_1, define the high-inclined price as π_1 . Thus,

$$\pi_0 = k_1 \frac{1}{w_{e_{ij}}} \frac{a_{e_{ij}}}{(1 - \eta_0)^{b_{e_{ij}}+1}} + k_2 w_{e_{ij}}(1 - \eta_0). \tag{12}$$

$$\pi_1 = k_1 \frac{1}{w_{e_{ij}}} \frac{a_{e_{ij}}}{(1 - \eta_1)^{b_{e_{ij}}+1}} + k_2 w_{e_{ij}}(1 - \eta_1). \tag{13}$$

Solving the equations (12) and (13), get the formulas (14) and (15).

$$a_{e_{ij}} = e^{\ln w_{e_{ij}} + \ln \frac{\pi_0 - k_2 w_{e_{ij}}(1-\eta_0)}{k_1} + \frac{\ln[\pi_0 - k_2 w_{e_{ij}}(1-\eta_0)] - \ln[\pi_1 - k_2 w_{e_{ij}}(1-\eta_1)]}{\ln(1-\eta_1) - \ln(1-\eta_0)}}. \tag{14}$$

$$b_{e_{ij}} = \frac{\ln[\pi_0 - k_2 w_{e_{ij}}(1 - \eta_0)] - \ln[\pi_1 - k_2 w_{e_{ij}}(1 - \eta_1)]}{\ln(1 - \eta_1) - \ln(1 - \eta_0)} - 1. \tag{15}$$

When the wavelength usage on the edge is in the low price region, use π_0 as price, reducing the computing overhead. When the wavelength usage on the edge is in the sound price region, get its real-time price by the formula (9). When the wavelength usage on the edge is in the high price region, auction wavelength and its base price is π_1 , the one who offered the highest price won, if no one participated, the auction aborted.

2.3 Fair Multicast Cost Apportion

To provide the intra-group fairness, it is necessary to apportion multicast cost among group members fairly. According to the principle of "who gets the profit, who pays", ELSD is adopted.

$S(v_i)$ is the set of v_i and all its downstream multicast destination nodes. $IC(v_i)$ is the apportioned cost of node v_i from its upstream. $OC(e_{ij})$ is the cost apportioned to the downstream output edge e_{ij} . For each edge, has

$$OC(e_{ij}) = IC(v_i) \times |S(v_j)|/|S(v_i)|. \tag{16}$$

$$IC(v_j) = c_{e_{ij}} + OC(e_{ij}). \tag{17}$$

The cost apportioned by node v_i is $IC(v_i)/|S(v_i)|$.

2.4 Algorithm Design and Simulation Study

In this paper, the proposed algorithm is based on the Single Instruction Multiple Data (SIMD) shared memory model. Based on the parallelized FCNN, the multi-constrained QoS multicast routing algorithm is discussed in [7]. After the wavelength pre-assigned tree is established, if it is unnecessary to auction wavelength or auction succeeded, perform wavelength assignment to the multicast tree actually, and then apportion that tree cost among group members by ELSD.

Limited to the experiment testbed, simulation researches have been done over some actual virtual topologies [7], and the performances of the proposed algorithm and its corresponding serialized version are compared (see Fig.1).

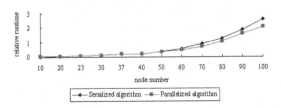

Fig. 1. Runtime comparison between parallelized algorithm and serialized algorithm

In conclusion, the proposed mechanism is both effective and efficient. It can not only meet with the needs of Internet commercialized operations with fairness support, but also provide the flexible and heterogeneous QoS multicast routing to applications.

References

[1] Chakraborty D., Chakraborty G., Shiratori N.: A Dynamic Multicast Routing Satisfying Multiple QoS Constraints. International Journal of Network Management, **13**, 321–335 (2003)

[2] Kelly F.P., Maulloo A.K., Tan D.K.H.: Rate Control for Communication Networks: Shadow Prices, Proportional Fairness and Stability. Operational Research Society, **49**, 237–252 (1998)

[3] Nemo S., Raymond R.-F.L., Andrew T., Aurel A.L.: Pricing, Provisioning and Peering: Dynamic Markets for Differentiated Internet Services and Implications for Network Interconnections. IEEE Journal on Selected Areas in Communications, **18**, 2499–2513 (2000)

[4] Shai H., Scott S., Deborah E.: Sharing the "Cost" of Multicast Trees: an Axiomatic Analysis. IEEE/ACM Transactions on Networking, **5**, 847–860 (1997)

[5] Zhang J.Y., Wang D.F., Shi M.H.: A Multiple Constrained QoS Routing Based on Firing Coupled Neural Networks. Journal of China Institute of Communications, **23**, 40–46 (2002)

[6] Fu X.M., Zhang Y.X., Ma H.J., *et al*: A Market-Based Approach to Allocate Bandwidth for Computer Networks. Acta Electronica Sinica, **27**, 127–132 (1999)

[7] Cui J.Y.: Research and Simulated Implementation of Neural-Network-and-Microeconomics-Based Fair QoS Multicast Routing Algorithms in IP/DWDM Optical Internet. MA Thesis, Northeastern University, Shenyang (2004)

A Study of Multi-agent System for Network Management

Zhenglu Wang and Huaglory Tianfield

School of Computing and Mathematical Sciences Glasgow Caledonian University,
70 Cowcaddens Road Glasgow, G4 0BA, United Kingdom zhengluw@hotmail.com,
h.tianfield@gcal.ac.uk

Abstract. Network systems are getting more and more complex and relatively inefficient. Therefore, networks will need to be controlled and managed, namely network management (NM) technologies are necessary. As the network systems require scaleable, flexible and economic solutions, the conventional NM methods are becoming more and more inadequate to solve these problems. The emerging intelligent agent paradigm seems to be a solution. In this paper, we discuss and investigate the feasibilities of constructing an agent-based paradigm for network management.

1 Introduction

Networks are becoming larger and more and more complex. It is a challenging task to manage the fast growing multi-protocol networks. Today's network management is dominated by a platform-centered paradigm based on the Client/Server (C/S) technologies. For example, SNMP (simple network management protocol) follows the C/S paradigm, typically associated with massive transfers of management data, which cause considerable strain on network throughput and processing bottlenecks at the manager host [1]. This centralized approach has drawbacks in scalability, reliability, efficiency, and flexibility [2], [3]. It is unsuitable for large and heterogeneous networks. So an open management architecture is required.
Agents and agent systems have been studied for over a decade, because it is regarded as an important tool for building a wide range of systems [4], [5].

2 Agent and Multi-agent Systems

As defined by Tianfield, an agent is an autonomous entity packaged of a set of capable computational entities, three of which, i.e. for internal scheduling, problem solving and social communication routing, respectively, are normative and others are optional [6]. An agent is illustrated in Fig. 1.

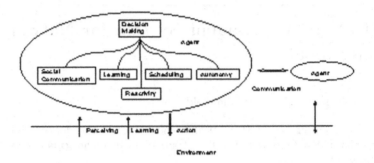

Fig. 1. Agent

A mature system is normally made of several different agents, involving their distributed data, knowledge, or control. A multi-agent system is composed of a collection of, possibly heterogeneous, computational entities which have their own problem-solving capability and are able to interact in order to reach an overall goal. Multi-agent systems emphasize both the autonomy of individual agent and cooperation between agents. Coordination is an important characteristic of multi-agent systems. Coordination is that of using particular mechanism to manage the inter-dependencies between the collaborative activities of agents [7].

3 Network Management and Current Issues

From the perspective of network end users, they expect fast, secure and reliable connections; from the perspective of network managers they would like to easily configure and control network access and resources; and as corporate managers they expect a low usage cost. [8] proposed a summary of users' network management requirements and reported in [9]: namely, controlling corporate strategic assets, controlling complexity, improving services, balancing various needs, reducing down time, and controlling Costs.

Because of its more or less natural heterogeneity, networks are difficult and complex to manage. Management systems must handle a wide and increasing number of devices, resources, protocols over larger and larger areas.

Nowadays, the wide adopted Network Management paradigm according to which the management application periodically accesses the data collected by a set of software modules placed on the network devices, by using an appropriate protocol. The Simple Network Management Protocol (SNMP) has become the standard protocol of management for IP networks [10]. The basic components of such systems of network management are management stations, network nodes (they may be computing and storage equipments as well as interconnection devices), A module called agent runs on each node to monitor its status and collect data to be processed, and a management protocol, which is used for transferring the management information among the SNMP-agents and the management stations.

However, SNMP is typical centralized paradigm which is generally appropriate for the applications with a limited need for distributed control, do not require a frequent polling of MIB variables, and need only a limited amount of information.

4 An Architecture of Multi-agent system for Network Management

According to the agents' specific advantages which have been mentioned in early section, we propose a framework for network management based on multi-agent systems. This framework system is divided into two parts. First part, several agents are placed on network devices to manage the local evens and reduce the workload for the management station. This local control subsystem is an OAA-like (Open Agent Architecture) structure [11]. A facilitator agent is defined as a "server agent", which is responsible for controlling the communication, coordinating the interactivities among agents. A user interface agent is responsible for accepting input, sending requesting to the facilitator for delegation to appropriate agents, and displaying the results of the distributed computation.

The other two categories of agents are application agents and meta-agents. The former are responsible for providing a collection of services of a particular sort. These services could be domain-independent, user-specific, or domain-specific. While facilitator processes domain-independent coordination strategies, mate-agents are used to assist the facilitator agent in coordinating the activities of other agent by using domain- and application- specific knowledge or reasoning. Interagent Communication Language (ICL) is to set up the communication between agents and facilitator. Fig. 2 illustrates the local control sub-system structure.

The second part, for the whole network management system, a different architecture is constructed to achieve task-driven self-organizing cooperation. Each local control sub-system can be viewed as a low-level actuator or middle manager which encompasses particular basic application agents. It also has a facilitator, which contains a mechanisms to deal with which controlling the communication, coordinating the interactivities among local systems. It also can facilitate the negotiation process between facilitator agents. The architecture is briefly shown in Fig. 3.

In the local control sub-systems, the whole agent systems can be written in different language and run in different platform. Each agent or agent system

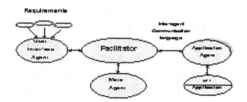

Fig. 2. Agent-based architecture for local control

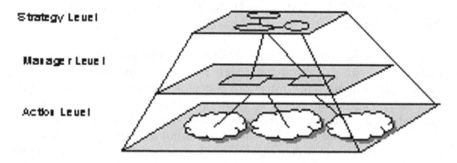

Fig. 3. Multi-level of network management system

just need to register their capabilities to the facilitators of same local sub-system or higher level. Facilitators maintain a knowledge base that records the capabilities of a collection of agents, and uses that knowledge to assist requesters and providers of service in making contact. Part of this registration is the natural language vocabulary that can be used to talk about the tasks that agent can perform.

The high-level coordination system includes several deliberative facilitators which can accomplish the following tasks: to provide global strategy, conflict resolving mechanism and society organization mechanism.

In sum, this network management paradigm has following attributes: Openness: the components of the system can be written in multiple programming languages and run on multiple platforms. The components that form an application can also be run on multiple platforms and operation systems; Scalability: Both of local sub-system and global system have full functional capabilities to deal with the local and global tasks. Compare to the centralized NM system, it is great reduction in the workload for central NM station; Flexible Hierarchy: The hierarchical relationship between sub-systems is dynamic. It means any subsystem can initiate the coordination in self-organization way.

This paper presents a brief review of constructing multi-agent based network management system and addresses some issues arising in current NM system and main benefits deriving from adoption multi-agent technology. The research in NM is driven towards more intelligent and autonomous. The multi-level structure is a promising solution which can contain both of the flexibilities of small-scale agent systems and powerful complexity solving of large scale multi-agent systems.

Much work remains to be done in exploring the potential of this framework. For example, how to establish an effective communication mechanism, so that agents could rapidly exchange information and take reaction, and maximally configure the local environment to provide a source of information.

References

1. Gavalas D, Greenwoodc D, Ghanbarib M, Mahony M (2002) Computer Networks 38: 693–711
2. Zhang D, Zorn W (1998) Computer Networks and ISDN Systems 30: 1551–1557
3. Yemini Y (1993) IEEE Communication Magazine 34: 20–29
4. lonso E, Colton S, Kudenko D, Moreau L, Schroeder M and Stathis K (2001) The Interdisciplinary Journal of Artificial Intelligence and the Simulation of Behaviour 1: 1–4
5. Maybury M (2001) International Conference on Intelligent User Interfaces Proceedings IUI: 3-4
6. Tianfield H (2000) International ICSC Symposium on Multi-Agents and Mobile Agents in Virtual Organizations and E-Commerce (MAMA'00):1574–374
7. Wooldridge J (2002) An Introduction to Multi-agent Systems. John Wiley and Sons Ltd.
8. Stallings W (2002) Communication Networks Management. Prentice Hall.
9. Terplan K (1996) SNMP SNMPv2 and RMON Practical Network Management. Addison-Wesley.
10. Prem Kumar G, Venkataram P (1997) Computer Communications 20: 1313-1322
11. Cohen R, Cheyer A, Wang M, Baeg C (1994) Proceedings of the AAAI Spring Symposium on Software Agents 1-8

A Practical Partition-based Approach for Ontology Version

Zongjiang Wang, Shensheng Zhang, Yinglin Wang, and Tao Du

Department of Computer Science and Engineering, Shanghai Jiao Tong University,
China {microw,sszhang}@sjtu.edu.cn

Abstract. Ontologies–as a formal, shared and common understanding of a domain–
are developed for knowledge sharing and reuse. However, due to the distributed and
dynamic nature of the web, there are many versions and different variations of
ontologes. The evolution of the ontology may cause the incompatibilities problem.
To solve the incompatibilities issue, an ontology version system is needed. In this
paper, we present a flexible approach to build such a system. Our method consists
of two tasks. First, we create a mechanism to support multiple ontology versions.
Every ontology version contains a version of ontology module. Secondly, we explain
how to partition the whole ontology into small modules according to the ontology
dependency.

Key words: ontology version, ontology partition, monoversion ontology multiversion ontology

1 Introduction

Ontologies are often seen as basic building blocks for the Semantic Web, as
they provide a reusable piece of knowledge about a specific domain. However,
those pieces of knowledge are often not static, but evolve over time [JJ00].
The evolution of ontologies causes operability problems, which will hamper
the effective reuse.

To solve the incompatibilities, we present a flexible approach to build an
ontology version system. Our method includes two tasks. Firstly, we create a
mechanism to support multiple ontology versions. Every ontology version is
composed of a version of ontology module. Finally, we explain how to partition
the whole ontology into small modules according to the ontology structure.

2 Ontology Version approach

2.1 Causes and Consequences of ontology change

A versioning methodology for ontologies copes with changes in *ontologies*. To examine the causes of changes, we will have to look at the nature of ontologies. According to Gruber (1993), an ontology is a *specification of a shared conceptualization of a domain*. Hence, changes in ontology are caused by either: changes in the domain, changes in the shared conceptualization, and changes in the specification.

Versioning support is necessary because changes to ontologies may cause incompatibilities problem, which means that the changed ontology can not simply be used instead of the unchanged version [MD01]. We see that a versioning methodology is necessary to take care of the following relations:between succeeding revisions of one ontology;between the ontology and instance data,related ontologies,and related applications.

2.2 Ontology version framework

In our approach a representation of a shared conceptualization of a domain is called ontology version. A multiversion ontology is defined as a set of logically independent and identified ontology versions . Formally, an ontology version is defined as a pair composed of the ontology version identifier and the set of versions of all the objects contained in the multiversion ontology, one version per object [MD01].

After examining the causes of ontology change, we classify the ontology change operations into two kinds: non-versioning operation and versioning operation. The former does not create a new ontology version. We can use a change log to deal with it. A versioning operation creates a new ontology version. This is the key problem that we need to solve.

In the simplest case a transaction concerns one ontology version. A non-versioning transaction queries or updates an ontology version, causing its evolution independently of the evolution of the other ontology versions. It corresponds exactly to the notion of transaction in monoversion ontology. A versioning transaction creates a new ontology version. It is addressed to an ontology version, the parent ontology version, and it creates a child ontology version, which is a logical copy of the parent. Thus, the set of ontology versions is organized as a tree, called derivation tree. Once created, the new ontology version will evolve autonomously, according to the non-versioning transactions addressed to it[MD01].

To summarize, there are two levels of operation on a multiversion ontology. At the upper level the user create or deletes a specified ontology version ah the lower level he reads, writes, creates or deletes a specified object in a specified ontology version.

2.3 Ontology object version identification

Since a child ontology version usually differs only partially from its parent, versions of the same module contained in different ontology versions may have identical values. To avoid redundancy, this value has to be physically shared by several ontology versions. This may be done by associating, for a module, several identifiers of ontology versions with one value that they share. However, the following problem arises: when a new ontology version is created, its identifier must be associated with one value of every module stored in the multiversion ontology. In a large ontology, the associating process would be inadmissibly long. To solve this problem, in the ontology version approach, ontology version identifiers are constructed in a special way. They are called ontology version stamps or simply stamps [JWH87].

As the multiversion ontology is organized as a tree of ontology versions, the stamp of an ontology version is constructed in such a way that it makes it possible to identify all the ontology version's ancestors. If an ontology version is the n-th child of its parent, whose stamp is p, then the child stamp is p.n. the root ontology version is stamped 0.

The following example shows how stamps are used to identify object versions. Consider a multiversion ontology, composed of four ontology versions. Its derivation tree is shown in Figure 1. An object A is stored in the multiversion ontology. From the logical point of view, one version of A appears in each ontology version. Thus each object version of a may be seen as a row of the relation Object_version_of_A presented in Table 1.

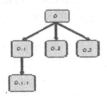

Value	Ontology Version Stamp
A_0	0
A_1	0.1
A_2	0.2
A_2	0.3

Table 1 Relation Object_version_of_A

Fig. 1. Ontology version derivation trees

However to avoid the replication of values of versions of A which are identical in several ontology version, like A_2 in ontology versions 0.2 and 0.3, this relation is implemented as shown in Table 2. Each row of this table, named Oid_A, may represent several object versions of A. For instance the last row of Table 2 implements object versions (0.2, A_2) and (0.3, A_2) of A.

3 Structure-based Ontology partitioning

This method consists of two tasks that are executed in four independent steps [Bat03]: Step 1. Create Ontology Graph: In the first step a dependency graph

Value	Ontology Version Stamp
A_0	0
A_1	0.1
A_2	0.2,0.3

Table 2 The multiversion object A

is extracted from an ontology source file [VA03]. Step 2. Determine Strength of Relations: In the second steps the strength of the dependencies between the concepts has to be determined. Following the basic assumption of our approach, we use the structure of the dependency graph to determine the weights of dependencies. Step 3. Determine Concept Islands: The proportional strength network provides us with a foundation for detecting sets of strongly related concepts. For this purpose, we make use of the 'island' algorithm [Bat03]. Step 4. Assign Isolated Concepts: Depending on the nature of the dependency graph it may happen that some nodes cannot be assigned to an island. As the definition of a partitioning does not allow unassigned classes, we have to assign these concepts to a module as well.

Discussion: Figure 3 shows how an ontology be partitioned. From this figure, we can see the whole ontology is partitioned small pieces: A_0, A_1, and A_2. Each piece of ontology can hold some even smaller pieces (i.e. in A_0 module, it has B, E, F, H).

The different steps described above lead us to a partitioning of the input ontology into modules that satisfy the formal conditions. One of the main problems with the approach as described above is the fact that we have to determine the size of modules that we want to be generated. The reason is that the optimal size of modules heavily depends on the size and the nature of the ontology.

In our experiments, we use a method to dynamically adjust the partition granularity to reduce the storage redundancy. (Figure 3) For example, in the module A_0, we can partition it into even smaller pieces according to the ontology change. Using this method, we can not only easily manage the ontology change, but also store the whole ontology efficiently.

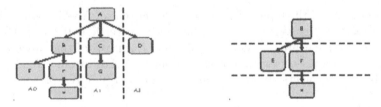

Fig. 2. The result of ontology partition   **Fig. 3.** The granularity of one ontology module

This paper proposed a practical method to manage multiple ontology versions, discussed the partition of a large ontology. The described system is not yet finished and should be developed further. We believe that it will significantly simplify the management of ontology changes and will store the ontologies in an efficient way.

References

[Bat03]   V. Batagelj. Analysis of large networks - islands Presented at Dagstuhl seminar 03361 August/ September 2003.

[HM03]   H. Stuckenschmidt and M. Klein. Integrity and change in modular ontologies. In Proceedings of the International Joint Conference on Artificial Intelligence - IJCAI' 03, pages 900–905, Acapulco, Mexico, 2003. Morgan Kaufmann.

[JJ00]   Jeff, Heflin and J. Hendler.: Dynamic ontologies on the web. Seventeenth National Conference on Artificial intelligence (AAAI-2000), Austin, TX, 2000.

[JWH87] J.Banerjee, W. Kim, H.J. Kim, and H. F. Korth: Semantics and Implementation of Schema Evolution in Object-Oriented Databases. SIGMOD Record (Proc. Conf. on Management of Data), 16(3):311–322, May 1987.

[MAD02] Michel Klein, Atanas Kiryakov, Damyan Ognyanov, and Dieter Fensel.Ontology versioning and change detection on the web. In 13th International Conference on Knowledge Engineering and Knowledge Management (EKAW02), Sig"uenza, Spain, 2002.

[MD01]   M. Klein and D. Fensel.Ontology versioning for the Semantic Web. In Proceedings of the International Semantic Web Working Symposium (SWWS), pages 75 – 91, Stanford University, California, USA, July 30 – Aug. 1, 2001.

[RAD02] R. Volz, A. Maedche, and D. Oberle.Towards a modularized semantic web. In Proceedings of the ECAI'02 Workshop on Ontologies and Semantic Interoperability, 2002.

[VA03]   V. Batagelj and A. Mrvar. Pajek - analysis and visualization of large networks. In M. Jnger and P. Mutzel, editors, Graph Drawing Software, pages 77–103. Springer, 2003.

Grid-based Robot Control

S.D. Wu[1] and Y.F. Li[1]

Dept. of Manufacturing Eng. and Eng. Management, City University of Hong
Kong, Kowloon, Hong Kong s.d.wu@student.cityu.edu.hk,
meyfli@cityu.edu.hk

bstract. Grid, as the new generation Internet information infrastructure,
presents a new solution for robot control. In this paper, we present our research on grid-based robot system which aims at organizing and integrating
robot's control capabilities to construct a RCG(Robot-Control-Grid), which
can provide control ability for robot tasks. Multiple robots located in different
places can collaborate to perform distributed robot tasks, and the portals corresponding to various robot systems are designed to user for robot accessing
and controlling. Also the security solutions of the RCG are discussesed. The
experience shows that the RCG is a feasible and effective mode for robot's
work and management.

Keywords. Grid, robot, OGSA, grid service, collaboration, security

1 Introduction

The most important part of a robot system is its controller. Nowadays an obvious technical trend of robot system is to construct controller applying open PC
software and hardware infrastructure, which can take full advantages of the
relevant PC techniques to enhance robot controller's performance effectively.
Grid, as the new generation Internet information infrastructure, has been researched widely with much attention in a variety of fields[FKT01],[DCL02].
In terms of robot control system, grid also emerges as a new solution with
great potential.

Grid is an integrated computing and resource environment[FC98] and it is
good at exporting easy-available, standard and economical computing abilities
from various computing resources. DIS(Distributed Instrumentation System)
is a representative grid application field. DIS manages the expensive instruments located in different places, and presents the corresponding interfaces for
remote control and accessing, which greatly promotes the using-efficiency and
is convenient to users. There is a famous DIS project Xport[MBH00] which
is developed for the shared instrumentation laboratory for macromolecular

crystallography. Similarly, robot is also a kind of complex instrument, so that the idea of constructing a robot control grid is feasible and valuable.

2 Grid and Its Technical Foundation

The robot control has its own characteristics and flows. Hence, a basic precondition to construct the RCG is that grid should do be able to provide technical support and foundation according to robot system's specified demands.

On one hand, what user respected is to take robot to perform and complete certain works. On the other hand, for robot, user's all demands can be deemed as some different kind of tasks. So the robot controller, the core of robot system, should not only be able to provide a control interface to the user, but also contain related logical functions modules which are responsible for conducting the tasks submitted by user.

Under this circumstance, the key of the RCG is to construct robot controller based on grid infrastructure. An effective work mode of the RCG is service-based robot control architecture, that is to say, user submits tasks to service, and then service executes the tasks, which drives robots running. The OGSA(Open Grid Services Architecture), characterized by grid service, provides great support to build service-based robot control architecture for the RCG[FKN02].

OGSA is the latest grid system architecture in which anything is abstracted as grid service. In this way, a robot connected into grid is also a kind of grid service, which is very favorable to take grid service as a universal interface to manage and use robot. Hence, in the RCG, grid service plays a very important role, which acts as the interface image of grid capabilities, and all robots' control functions are implemented and deployed by grid services. Some characteristics of grid service possessed are of great help to satisfy the robot control's demands, for instance, grid service is stateful, which is useful to keep the medium information during robot running. Another important feature is that grid service is able to change dynamically such as expending or shrinking, this means that when executing some complex tasks, the required more advanced and abstract robot service can be formed by combining multiple simple and basic robot services, which is a great advantage for the collaborative control of multiple robots.

3 Model of Robot-Control-Grid

In the support of grid service, this paper proposed grid service-based architecture model for the RCG, illustrated as Fig.1. Since the core factor of grid service played, the control to robot mostly is incarnated as the managing, scheduling and utilizing of robot grid services.

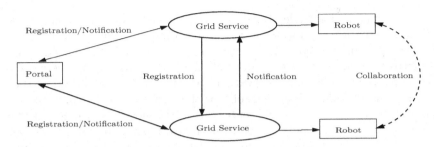

Fig. 1. Service model of the RCG

Generally, a robot control system is complex. If we simplify the robot control system moderately to care for only the robot motion control function, we could have this function mapped to a grid service to implement and the running of this grid service will drive robot motioning. In fact, a robot system consists of multiple grid services corresponding to multiple functions. For instance, besides the most important robot motion control service, the motion simulation grid service is also very important. These two grid services work relatively independently, however the data used in simulating comes from the motion control grid service, which requires enabling communicating and interacting among gird services. Here the collaboration of grid services in the RCG is involved naturally.

The collaboration of multiple robots is a key problem should be resolved since there must be multiple robots in the RCG. Fortunately, the characteristics of grid environment present great advantages to robots' collaborative control. Although the control relations for multiple robots looks complicated, the collaboration of multiple robots still can be translated naturally as the collaboration of multiple grid services in grid environment, which is similar in solution with the collaboration of multiple grid services for single robot control. The two kinds of collaboration solutions have their own means respectively. However, they are accordant in the technical way.

For complex robot tasks, there are two key issues for robot collaboration. One is how to decompose complex tasks and distribute the pieces of subtasks. The detailed guidelines depend on task's properties and robot's motion algorithms. For instance, for instruction-driven industrial robots, each robot's subtask can be separated relatively easily from the collaboration task. The other issue is the collaboration rules when multiple robots collaborate. Although there are no general rules, there exists a basic rule called "waiting-synchronizing": on the process of multiple gird services running, the starting of a service's instruction may require another service completing a specified instruction. Thus it has to wait for that grid service to synchronize. Usually, the user can custom the collaboration relations among grid services when submitting a task.

In the RCG, the grid services used for single or multiple robots control both require communication each other. For each grid service participated, according to its function and the role played in robot collaboration, it needs to register to the grid services from which it desires to get interested information. At the same time, it may accept other service's registration and send messages they required to them. Here two circumstances are taken to analyze the collaboration relations. One is for the grid services for single robot control and, the other is for multiple robots control. For the former, for the motion control grid service and simulation grid service, when the motion control grid service has completed sending pulse to robot control card in an interpolation cycle, it should send current robot joints position data to the simulation grid service to conduct simulation. Obviously, in this situation the communication content is large. For the latter, the collaboration of multiple robots, when a grid service for a robot completes an instruction or a piece of subtask, it may be required to send notification message to another gird service which is responsible for another robot. When this message arrived, the grid service will decide what next step to do according to the information contained in message. For instance, if it knows that the partner robot is running ok, it also will execute the next instruction orderly according to the collaboration rule configured in advance. This kind of communication load is low in some extent because the transmission content mainly consists of specified commands and signals.

After resolving the collaboration mechanism, the grid service's communication, the base of the collaboration, is another important problem should be settled subsequently. Globus presents registration and notification based communication techniques for grid services. Because the PUSH technique is used in sending the notification message to the registered services, the notification communication mechanism shows a higher efficiency.

The other issue of communication is the format of communication message, which can be resolved finely with service data techniques. Service data is an important feature of grid service, which can not only organize the basic information of grid service, but also define data format for communication content according to the demands of concrete application. The registration of grid service, actually, is to subscribe the desirable SDE(Service Data Element). So SDE plays an important role in the communication of grid services. Here the content of SDE in different circumstances in the robot control is listed. For simulating service, the SDE contains mainly the position data of robot joints. For the grid services for robot collaboration, the SDE contains the following items: signals specified for collaboration, which are defined in common and can be understood by all the participated grid services, currently executed instruction, potential next instruction, position data in specified moment, the states of partner robots and any other information related to collaboration. The SDE that the user portal cares contains the global states of robot grid, the current states of each robot, the interaction flows and traces of grid services, the progress of task or subtasks and various error or warning information, etc.

4 User Portal Design

Portal is the user entry to control robot in the RCG, and it provides a universal robot control interface. Basically, it should possess at least five function interfaces listed below.

- Querying and selecting of robot grid services. For a robot, maybe there are several suits of grid services by which different performance could be used. Thus the portal should give user an interface to query and select desirable grid services, and according to the characteristics of the selected grid service to prepare compatible robot task.
- Submitting robot task. In general, the robot task is in fact a piece of program, which is composed of robot instructions and relevant syntax sentences. Through a task submitting interface, user can submit task program to robots.
- Configuring of collaboration and work parameters. When the task is submitted, before starting control, some robot parameters such as running speed, times and simulating switch are required to setup. Furthermore, a more important step is to configure the collaboration relations among multiple robots and grid services. In some extent, this configuration may be complex. The robots will work under the collaboration rules configured by user.
- Management and monitoring of robot tasks. When robots are running, the user can manage task such as to start, stop, pause, or resume robot task, or turn off the robot controller and so on when interventions are needed.
- Information feedback to user. In the process of robot running or collaborating, the related detail information and key message should be feed back to portal client in time, which make user master the robots' states and provide support for determining whether to take an action.

5 Security Solutions

Robot is a kind of expensive and high-precision device. A robust robot grid has to present appropriate security solutions to guarantee robot working securely. Here are the solutions for the fundamental security demands of the RCG.

Privacy. In the interaction between user and grid services, what is transmitted is the robot task, a set of robot instructions and other codes. In general, there isn't special privacy demand for robot task, because the robot instruction is public. However, if user wishes the robot task keeps privacy, some encryption and decryption algorithms can be selected for use.

Integrity. The task user submitted represents user's specified control intention. If the task is intercepted and tampered by malicious user in the transmission, the means of the task will change, which is not allowed to happen. So the task has to ensure its integrity, which can be guaranteed by digital signature.

Authentication. The authentication is a compulsory security requirement for robot grid application. For the robot, grid must ensure the users submitted tasks are who they claim to be. In other words, a robot should be protected from malicious users who try to impersonate one of the parties in the secure robot control. On the other hand, for robot user, grid must assure that the robot is what user desires to control, then user will really perform controlling actions. So in the RCG, the authentication demand is mutual between robot and user, which can be implemented by digital certificates technique.

Authorization. Based on the authentication, authorization is a higher level security setting. In fact, for the robot control, these authenticated legal users may be authorized different level security rights according to user's role or robot task's requirement. For instance, some advanced grid services may only be allowed to be invoked by specified high level user. The authorization setting can be implemented by the ACL(Access Control List) mechanism.

In addition, the security of robot tasks is also should be considered. Generally the task is composed of program codes and robot instructions, so the checks of preventing viruses or malicious codes or illegal instructions are necessary practically.

6 A Simple Demo

Taking PT series industrial robots as control objects, this paper constructs a grid-based robot control instance under the basic architecture of the RCG. PT series robot includes 32 instructions, including motion instructions such as PTP, MOV and CIRCLE, and program control syntax instruction such as FOR, WHILE and RETURN, and welding and conveying used instructions. In our previous work, we have built the basic control framework for PT series robots which can run well locally[WCH01]. To construct grid-based robot control system, we take the robot instruction interpreter, which directly drives robot motioning, and motion simulating functions as samples, implement them applying grid service in Globus platform. By collaboration, these two grid services make up of a little-scale robot control system in grid environment. The interpreter is the kernel of PT series robot system, and the interpreter grid service is responsible for interpreting and executing the robot program submitted by user. Because the program is composed of instructions introduced above, the implementation of every instruction is the most important foundation for the interpreter services. The simulating grid service is to trace and visualize the robot's motion process via 3D animation with light, in which the simulating data, robot joints' coordinate position, comes from the sensors when instruction interpreter service is running. After completing and deploying these two grid services, user can submit compatible tasks to the interpreter grid service to control robots. Fig. 2 is the simulating illustration of PT500 robot with different view angles and the collaboration of two robots.

Fig. 2. Simulation views of PT500 and PT600 robots

This paper proposed the RCG and discusses the key issues. Considering the inherent characteristics of robot control, this paper studies how to apply grid techniques to resolve the problems related to robot control, and a simple robot control system demo is implemented in Globus platform. The work in this paper has proven that the idea, control robot in grid environment, is feasible and valuable, and presents appropriate approaches and solutions for building robot control system in grid environment.

Our works are just at the beginning. For the future work, we will pay more attention to the design of the physical and logical modes of grid robots, and practical and powerful collaboration mechanisms for multiple robots, etc.

References

[FC98]   I. Foster, C. Kesselman: The Grid: Blueprint for a Future Computing Infrastructure. Morgan Kaufmann, San Francisco, CA (1998)

[MBH00] Donald McMullen, Randall Bramley, John C. Huffman et al.: The Xport Collaboratory for High-Brilliance X-ray Crystallography, tech. report, http://www.cs.indiana.edu/ngi/sc2000 (2000)

[FKT01] I. Foster, C. Kesselman, S. Tuecke: The Anatomy of the Grid: Enabling Scalable Virtual Organizations. International J. Supercomputer Applications, 15(3) (2001)

[WCH01] Wu Shandong, Chen Yimin, He Yongyi: Development Technique of High Level Software System for Industrial Robot Controller Base on WIN 9X/NT, Journal of Shanghai University, Vol.7, No.6, 532-535 (2001)

[DCL02] Du zhihui, Chen Yu, Liu Peng: Grid Computing, Beijing: Tsinghua University Press (2002)

[FKN02] I. Foster, C. Kesselman, J. Nick et al.: Grid Services for Distributed System Integration. Computer. 35(6) (2002)

Group-based Peer-to-Peer Network Routing and Searching Rules

Weiguo Wu, Wenhao Hu, Yongxiang Huang, and Depei Qian

Department of Computer Science and Technology, Xi'an Jiaotong University, Xi'an 710049, China

Abstract. In the Peer-to-Peer file-sharing system, nodes at the edge of the network take part in the applications and share their resources. As the scale enlarged, search efficiency became the major problem. We researched on the organization and the search algorithm of the Peer-to-Peer file-sharing system to improve its efficiency. A group-based peer-to-peer network model is presented in this paper. In this model, peers with the same interest are organized into one group. All peers attempt to select the best neighbors by measuring other peers. Compared with Gnutella-like P2P systems, the group-based system can achieve high performance and reduce the loads of network as requests are transmitted in scope of the group instead of the whole system. Also we present the value-headed walk search algorithm and the measurement of downloading files among multiple answers. In value-headed walk, the peer forwards query to a subset of its neighbors based on the value of them. Results from theoretical analysis and simulations show that group-based P2P model and value-headed walk search algorithm could greatly reduce the amount of messages generated by searching files in P2P system.

Keywords Peer-to-Peer, Search algorithm, Group-based organization

1 Introduction

In P2P file-sharing system, it is extremely important to achieve high search efficiency. Napster uses a centralized search mechanism, and it is clear that such approaches have single point failure and it is difficult to keep the indices up-to-date [1]. Gnutella uses a flooding-based search mechanism. Its disadvantage is that each query generates heavy loads, which makes it poorly scaled [1]. Structured P2P systems use distributed hash table (DHT) to search an exactly described file, this search mechanism requires only O(log n) steps. But it can't support keyword searches.

In this paper, a group-based P2P file-sharing system is presented. Files are assumed to be divided into a number of themes. Peers interested in the same theme are organized into one group. The group-based P2P system could get these advantages:

(1)Reducing the search traffic and the load of peers.

(2)Reducing the time to search files.

(3)Supporting keyword searches in a group.

2 Group-based P2P files-sharing system model

In the group-based P2P model, peers are organized into groups by taking into account their interests. The collection of thematic categories can be defined by information retrieval or other fields. Any peer is just interest in several themes [2]. Peers learn the value of other peers and decide whom to connect or when to add or drop a connection in the group. The group can also match some special demands such as checking on the memberships and controlling the access.

2.1 Join Algorithm

When a peer wants to join the P2P system, it must find at least one active peer and then get its routing information to send out neighbor-queries. Research in Napster and Gnutella found that there is a significant amount of heterogeneity in these systems; bandwidth, latency, availability, and the degree of sharing vary between three and five orders of magnitude across the peers in the system [3]. If a peer receives the neighbor-query, it will measure the value of the requester to determine whether to accept.

```
if( neighborNum < maxNum ) {
    ACCEPT Y;
}
else if( value(Y) > value(X) ) {
    DROP X;
    ACCEPT Y;
}
else {
    REJECT Y;
}
```

Fig. 1. Join Algorithm

The measurements of the peers in the network and in groups are different. In the network the peer measures its neighbor according to the delay between the two peers and in a group the peer measures its neighbor both the delay and the number of files provided by its neighbor.

The measurement of a neighbor's value in network is:

$$\text{Value(netNeighbor)} = 1/\text{Delay} \qquad (1)$$

And the measurement in group is:

$$\text{Value(groupNeighbor)} = \text{FileNum}/\text{Delay} \qquad (2)$$

Using this join algorithm, those peers with abundant files has privilege to select high-evaluated neighbors, so their requests would be delivered among high-evaluated peers and get the right answer quickly.

2.2 Download Measurement

Bandwidth is one of the major factors for system performance. If request has detected several answers, the requester should measure which peer is the best for download.

The measurement of peer's efficiency for download is: λ is the size of the file.

$$\text{Efficiency(Download)} = 1/(\lambda/\text{Bandwidth} + \text{Delay}) \qquad (3)$$

Theoretically, the available bandwidth is most suitable to be used to measure the efficiency of transferring files. But there are several problems with measuring the available bandwidth in P2P system: The first, available bandwidth can significantly fluctuate over short period of time; the second, available bandwidth is determined by measuring the loss rate of an open TCP connection [3]. The bottleneck link bandwidth could be used as the approximation to the available bandwidth and these measurement techniques have been shown to be accurate and fast in simulation. Furthermore, in some parts of the Internet, available bandwidth is frequently equal to bottleneck link bandwidth [6].

3 P2P Search Algorithms

In this subsection, Flooding and Random Walk algorithms for the group-based system are analyzed and a value-headed walk routing mechanism is presented.

With Flooding, every incoming message is sent out on every outgoing line except the one it came from. It obviously generates vast numbers of duplicate requests. Flooding will work better in the group-based system than in Gnutella, because a group member is more likely to answer the query than an arbitrary peer in the system. With Random Walk, a query message is forwarded to a randomly chosen neighbor at each step until the object is found. The requesting peer sends out k query messages to an equal number of randomly chosen neighbors. Simulation results in [5] show that the messages are reduced by more than an order of magnitude compared to the standard flooding scheme. The disadvantage of random walk is that the performance is uncertain. With Value-headed walk, the source peer selects K highest evaluated paths to forward queries. The intermediate peers forward each query to the highest valued path at each step. Requests quickly converge at high-evaluated peers that can reduce the cost of time. The disadvantage of this algorithm is that the queries are frequently sent back to the high evaluated peers.

4 Simulation

In this section, two simulations are used to observe the efficiency of different search algorithms. In the flooding simulation, the number of messages generated once searching in P2P system is concerned. In the walk simulation, the effect of different files distributions is considered.

4.1 Flooding Simulation

The flooding simulation is used to evaluate the group-based organization compared with the Gnutella-like system. In this simulation, we use the 80-20 files distribution. Each file has four replicas. In group-based organization, each peer is interested in about 12.5 percent of themes. All peers interested in the same theme are classified into one group.

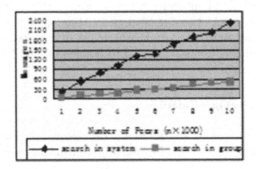

Fig. 2. Flooding simulation in Group and System

Both the two simulators use the same initial topology. Figure 4 presents the result of this simulation. As expected, the traffic generated by searching in a group is much slighter than searching in system. Although joining into a group generates additional messages, the cost is acceptable because searching files is more frequently. If files matching the request are sparsely located in network, the group-based model can achieve most advantage.

4.2 Walk Simulation

In this simulation, we consider the performance of the random walk algorithm and value-headed walk algorithm. In both random walk simulator and value-headed simulator, the requesting peer sends out eighteen query messages. We simulate their search efficiency at random files distribution and 80-20 distribution. On the left, Figure 5 shows the result of random distribution. The value-headed walk algorithm can get slight better result. But additional messages generated when exchanging the value information. On the right, Figure 5 shows the result of simulation with 80-20 files distribution. In this simulation, the amount of traffic generated by value-headed walk algorithm is greatly lower than the random walk.

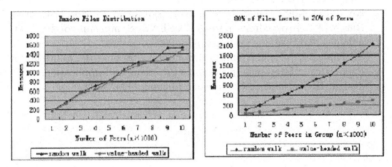

Fig. 3. Left: Walk Simulation with Random Files Distribution; Right: Walk Simulation with 80-20 Files Distribution

Measurements in P2P system indicate that client-like and server-like behavior can clearly be identified. In Gnutella, approximately 26 percent of the users shared no data. And 7 percent of the users together offered more files than all of the other users combined [3]. It is obvious that in this system condition, value-headed walk algorithm could achieve higher efficiency.

5 Conclusions

This paper presents an overview of group-based P2P files sharing system. The group-based P2P system could reduce the traffic generated by searching. For those P2P systems that have the characteristics of small-world structure and power-law structure [7], some sever-like peers shared more resources than other peers. So, the application of the value-headed walk routing mechanism could enhance their performance based on our simulations.

Acknowledgements. This work is supported by 863 Project (No. 2002AA104550) and CERGrid Project (No. CG2003-CG008).

References

[1]   Dejan S. Milojicic, Vana Kalogeraki, Rajan Lukose, Kiran Nagaraja, Jim Pruyne, Bruno Richard, Sami Rollins, Zhichen Xu. Peer-to-Peer Computing. HP Laboratories Palo Alto.

[2]   Murali Krishna Ramanathan1, Vana Kalogeraki, Jim Pruyne. Finding Good Peers in Peer-to-Peer Networks. HP Laboratories Palo Alto.

[3]   Stefan Saroiu, P. Krishna Gummadi, Steven D.Gribble. A Measurement Study of Peer-to-Peer File Sharing Systems.

[4]   Arturo Crespo, Hector Garcia-Molina. Routing Indices For Peer-to-Peer Systems. In ICDCS, 2002.

[5]   D.Tsoumakos and N. Roussopoulos. A Comparison of Peerto -Peer Search Methods. In WebDB, 2003.

[6]   Kevin Lai, Mary Baker. Nettimer: A Tool for Measuring Bottleneck Link Bandwidth. Proceedings of the 3rd USENIX Symposium on Internet Technologies and Systems, March 26-28, 2001.

[7]   M. Ripeanu, I. Foster, and A. Iamnitchi. Mapping the gnutella network: Properties of large-scale peer-to-peer systems and implications for system design. IEEE Internet Computing Journal, 6(1), 2002.

Research on the Parallel Finite Element Analysis Software for Large Structures

Weiwei Wu[1], Xianglong Jin, and Lijun Li[2]

[1] Institute of Automobile Engineering, Shanghai University of Engineering Science, Shanghai 200336, China wuzilin@263.net
[2] High Performance Computing Center, Shanghai Jiaotong University, Shanghai 200030, China jxlong@sjtu.edu.cn

Abstract. A parallelization strategy for finite element analysis software is proposed to overcome the difficulties imposed by the increasing scale of analysis. By means of parallel exploitation on commercial FEA software, a parallel FEA system for large structure is constructed at very low cost. Real tests prove the accuracy and scalability of the system.
Key words. Finite Element Analysis, High Performance Computing, Sparse Matrix, Message Passing Interface

1 Introduction

Being a computing technique, Finite Element Analysis has been broadly used in solving differential equations, structural analysis, electromagnetic analysis and so on. The computing scale of FEA becomes larger and larger, because of the complexity of modern structures and the pursuit for precision. It makes some utilities such as large structure optimization, long-term dynamical response analysis impossible on single PC or workstation. On the other hand, supercomputer has caught a rapid development in the last several decades. Because of its computing ability, fast I/O bandwidth and huge storage, supercomputer has taken large scale scientific computing as its main application area for a long time. Parallel FEA software has been concerned and developed in last years.

The project aims at a parallel linear FEA system. Attention has been devoted to the system efficiency in this project. Around the world, to solve sparse linear equations and eigenvalue problems in parallel efficiently is still an active research area. Different algorithms vary a lot in efficiency and accelerating capability. Unlike serial FEA, parallel programs must take account of data distribution and efficient parallel algorithms. The form of the sparse matrix can be changed by permutation, so to improve the computing efficiency.

2 System Architecture

A strategy of parallelization for authoritative commercial software is chosen to develop parallel FEA software rapidly in relatively low cost. Commercial software has advantages such as versatile and convenient pre or post processing, rich element library and detailed status records. Such serial commercial software adopts advanced programming techniques, which make them running stably and the analytical scale is only limited by external storage. These characteristics cannot be achieved in a short time. So the parallelization of only mathematical problem solving procedure is a low cost method to achieve a larger analysis scale and a shorter solving time. MSC.Nastran, the well-known FEA software for its performance, stability and integrated exploitation tools, is chosen for the parallelization. Till this time, four modules: static analysis, transient analysis, modes analysis and buckling analysis, has been recomposed to run on nation made supercomputers.

Theoretically, linear structural mechanics analysis can be concluded to the solving of large sparse algebra equations $K * p = u$, and modes analysis the equation $K * x = \omega^2 Mx$. In the practical tests, these two basic solving processes occupy a large part of time and computation of the whole analysis. Even for the simplest static analysis, the solving of linear equations still takes about 40 to 60 percent of the total time.

From above, the parallelization of solving these functions is the most rewarding and simple way to construct parallel FEA software. The system based on this strategy is shown on Fig1, the dashed line denotes the original analysis flow and solid line denotes the procedure after exploitation. The exploitation is divided into three parts: the design of parallel program, the reassembly of Nastran solving procedure and the design of graphical interface. The design of parallel program is the difficulty and pivot of the project. DMAP programs are assembled to change the analysis procedure to deploy the parallel program on supercomputer. PCL is the language chosen to develop graphical user interfaces.

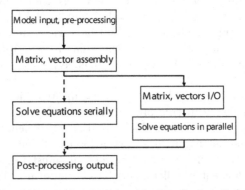

Fig. 1. Analysis flow before and after the parallelization

3 Parallel Algorithms for Large Sparse Matrix

The parallel programs solving linear sparse equations and eigenvalue problem are included in the project. The direct and iterative algorithms for linear sparse equations are developed respectively. The object computer is a distributed memory supercomputer, and all the programs are composed in C and FORTRAN, with MPI interface. BLAS subroutines are deployed in the programs.

All the direct method stems from the idea of Gaussian elimination. Because the stiffness matrix is symmetric and positive definite, the multifrontal algorithm[MK93, DA93], which occupies less memory, is used to realize the Cholesky factorization. Fig2 shows a sparse matrix and its elimination tree, where the 'x' denotes the nonzero element and the little circle denotes the element should be filled in the factorization procedure. The elimination starts from leaf nodes to the root, the computation in the same level can be carried out in parallel.

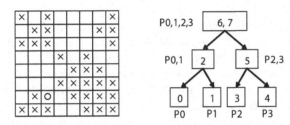

Fig. 2. A sparse matrix and its elimination tree

The data of matrix is stored or distributed over all CPUs. A float number array is used to store all the nonzero values in the matrix, and other two integer arrays store the index and row number. According to the algorithms described above, the data of the same level in the elimination tree should be distributed in different CPU as much as possible to facilitate the parallel computing. The number besides the node in Fig2 denotes the CPU in which the data is stored. The factorization starts from nodes 0, 1, 3, 4, whose data distribute on CPU 0, 1, 2, 3. After the factorization, the equations can be solved by the general backward substitution.

It is discovered in benchmark that large amount of memory is occupied by the direct method. The efficiency decreases when exterior memory is used, so iterative method[ISO95] is also programmed. Iterative programs take much less memory and exhibit a high accelerating capability. The iterative method has different data distributing with the direct method.

The partition and permutation[GB00] must be carried out both in direct and iterative methods. The partition is to distribute the data to all the CPUs. It should ensure the load balance and minimize the message passing between

CPUs during computation. The purpose of permutation is to change the form of matrix to that described in Fig2 by changing the index of rows and columns, and to preserve the sparsity in the factors.

Arnoldi/Lanczos is the most effective algorithm to extract a portion of eigenvalues of a large matrix. Implicitly Restart Lanczos algorithm[SJ02] modulates the original Lanczos method to improve the stability and speed. Implicitly Restart Lanczos algorithm is effective to extract the largest portion of eigenvalues, but the smallest frequencies and the corresponding modes are always demanded in dynamic analysis, So the original problem $Kx = \omega^2 Mx$ is transferred to $K^{-1}Mx = x/\omega^2$. If frequencies around a constant σ are needed, the problem can be transferred to $(K - \sigma^2 M)^{-1}Mx = \lambda x$, where $\lambda = 1/(\omega^2 - \sigma^2)$. This kind of transfer will import the problem of solving linear equations. When the direct method is used, the efficiency is relatively high because of the repetitious solving.

4 Results and System Performance

The accomplished static, dynamic, modes and buckling analysis modules are tested by structures in typical conditions. The outcome is compared with that of Nastran, and the relative error of displacement is below 10^{-10}. The corresponding outcome such as tension and velocity has a very low relative error too. Analyzer can set the precision of iterative method. Because it can be affected by many factors, experiences are required to give out correct parameters. It is discovered that iterative method need more time to get the same precision with the direct method, even with the optimized parameters.

Fig3 is part of the results got on a supercomputer in Shanghai. The model has about 40000 elements, and the matrix dimension is about 200000. Figure (a) and (b) show the time and accelerating ratio of a static analysis with direct and iterative method. Because the memory in each CPU is relatively small, the direct solver needs 8 CPUs to start working. The direct solver is always faster than the iterative solver, but the latter occupies less memory, which means it can solve the same problem by less CPUs. The direct and iterative method all shows a satisfying accelerating capability. The similar result can be got in transient response analysis and buckling analysis. Figure (c) shows the convergence of the two iterative methods, conjugate gradient and GMRES. Conjugate gradient algorithm has a better performance because the stiffness matrix is symmetric and positive definite. Figure (d) shows the time and accelerating capability of modes analysis. The accelerating ratio is not so satisfying because most time is consumed in backward substitution, which has a poor parallel performance.

The system is also assembled and tested on an IBM supercomputer, and shows similar results. Because of the enough memory on each computing node, it only needs 2 CPU to start the direct solver, and the total solving time is

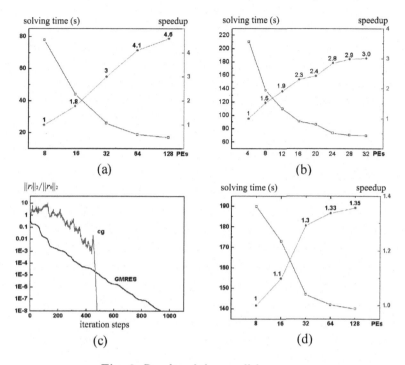

Fig. 3. Results of the parallel program

reduced by 1 magnitude. This reflects the portability and scalability of the system.

References

[JL92]   Joseph, W.H., Liu.: The Multifrontal Method for Sparse Matrix Solution: Theory and Practice. SIAM Review, 34, 82–109 (1992).

[Saa96]   Saad, Y.: Iterative Methods for Sparse Linear Systems. PWS publishing, New York (1996)

[Gup96]   Gupta, A.: Fast and Effective Algorithms for Graph Partitioning and Sparse Matrix Reordering. IBM Journal of Research and Development, 41, 171–183 (1996)

[Gup97]   Gupta, A., Karypis, G., Kumar, V.: A Highly Scalable Paralle Algorithm for Sparse Matrix Factorization. IEEE Transactions on Parallel and Distributed Systems, 8, 502–520, (1997)

[LS96]   Lehoucq, R. B., Sorensen, D. C.: Deflation Techniques for an Implicitly Restarted Arnoldi Iteration. SIAM J. Matrix Analysis and Applications, 17, 789–821 (1996)

An Amended Partitioning Algorithm for Multi-server DVE

Yanhua Wu and Xiaoming Xu

Dept. of Computer Science, Shanghai Jiaotong University, Shanghai 200030, China. {wyhross,xmx}@sjtu.edu.cn

Abstract. In this paper,We present an amended algorithm named AMD algorithm base on DLS partition algorithm. AMD algorithm keeps the advantage of the DLS algorithm and gets rid of its drawbacks. In our algorithm, All the servers take the highest possibility to handled the loads averagely, which keeps the system run stably .All the server get a fair load scare at any time. At the same time our amend algorithm can also reduce the overload times of the whole system. This will improve the system's security and stability.

Key words: Partition, distributed virtual environments, network load

1 Introduction

Distributed Virtual Environments (DVE) involves a large number of users (or avatars) enter into this virtual environment to interact with each other. A 3D virtual environment is a virtual world base on internet. It contains many high-resolution 3D graphics sceneries to simulate the real world. The most important application instances of DVE include Multiplayer online games and distributed simulations. With the developments of computer graphics fields, network technologies and CUP power, Internet-based massively multiplayer online games(MMOGs) have attracted significant attentions as a new stream of the entertainments industry. Many research works[1][2][3]propose the multi-servers architecture to share the work load of the huge number of avatars in the virtual environment. The virtual environment is divided into many areas named partitions. Each partition is assigned to a special server. All avatars belong the partition are managed by its associating server. However the avatars in the system can move freely. Some areas may affect more avatars then others, which may lead to the overload of some server. A DSL algorithm is presented in[2]. It is solve the avatar's free move and local server overload problem. But it has a drawback. The load (avatars) can't well-proportioned distribute to all the servers and once a server appear overload a time and it will appear repeatedly at the same time many other serves still keep empty. In this paper, we propose an amendatory algorithm named AMD algorithm based on the DSL algorithm. The AMD algorithm will be explained in detail in the follows sections. The rest of the paper is organized as follows: Section II describes the related works. In section III we discuss detail the partitioning algorithm proposed in [2] and gave out the describe of AMD algorithm. Simulation studies are described in Section IV and section V draw the conclusion.

2 Dynamic load sharing algorithm(DSL)

In [2], A dynamic load sharing algorithm (DSL) is proposed. The overloaded server will transfer some of its load to other non-overloaded servers. In addition, a client migration scheme is proposed to reduce the overhead time of this algorithm. The DLS algorithm is a decentralized sender-initiated load sharing algorithm. Each server in the system will periodically check its load to perform the corresponding load sharing action as shown in Fig 2. At the same time, all the server waits to receive load sharing requests from other servers, as shown in Fig2. In Fig.2, if a server S_i is overload, $L_i \dot{\iota} \delta$, it will attempt to share its load with the servers in its k-level domain $D(S_i, k)$ [2].

```
1.  begin
2.     OL_i = L_i - δ ;
3.     k = 1; /*level of domain*/
4.     while((OL_i > 0) or (k <= MAX_LEVEL_i)) do
          /*send load_sharing to neighbors only*/
5.        for(each S_j∈D(i,1)) do
6.           EL_j = send(load_sharing,OL_i,S_j,k);
7.           if(EL_j > 0)
8.              transfer(EL_j,S_j);
9.              OL_i = OL_i - EL_j;
10.             if(OL_i <= 0) break; /*exit for*/
11.          end if
12.       end for
13.       k = k + 1;
14.    end while
15. end
```

Fig. 1. DSL algorithm for S_i to share its load

3 Amend algorithm(AMD)

The algorithm discussed above can effectively share load among all the system servers. However it still has some drawbacks. The load share algorithm is executed only when overload appears. After the algorithm has executed the server still keep a saturate state. Once another avatar added, for example a avatars move into this partition or some new avatars created in this partition, the server will overload again and the algorithm has to executes repeatedly. This means once a server get to overload it will overload again and again in the follow time. At the same time there are many other servers have much redundant load ability. So the load is not distributed averagely in all the servers. We expect an ideal state that all the servers hold the same load. To make this ideal state be true we refined the DSL algorithm and give out a new algorithm. We name our algorithm *Amend algorithm*(AMD).

We define θ as the Temp_max_load and it is initialized with $1/m\delta$, which mean theδ is divided into m levels. In each level all servers get an average load number. When one server find that all the servers' loads get to θ (k = MAX_LEVEL) it will use the Notify(θ) to notify all the other servers that Temp_max_load enter into the next level.

The Amend algorithm load sharing pseudo-codes are:

Begin $OL_i = L_i - \theta$; k =1; /*level of domain */
While $((OL_i \text{ } \text{¿} 0)$ or (k ¡=MAX_LEVEL$_i$
)) do /*send load_sharing to neighbor*/
for *(each $S_j \in D(I,1)$) do*
$EL_j = send(load\_sharing, OL_i, S_j, k)$;
If*($EL_j > 0$)*
Transfer(EL_j, S_j);
$OL_i = OL_i - EL_j$;
End if
If *($OL_i <= 0$)*
break;/ exit for */* End if End for k=k+1;
End while
/* The temp_max_load turn to the next level */
If $k\text{¿}=MAX\_LEVEL_i$

$$\theta = \frac{1}{m}\delta + \theta$$

if($\theta \geq \delta$)
break /* exit */
end if
for *(each $S_j \in D(I,MAX\_LEVEL_i)$)* do *Notify(θ)*;
End for
End if
End

4 Experiments and discussion

We define the overload times and the Variance of servers' load:

$$\overline{L} = \frac{\sum\limits_{1}^{n} L_i}{n},$$

Variance $(L) = \sum\limits_{i=1}^{n} (\overline{L} - L_i)^2$, n: Server number

SD(L)=sqrt(Variance(L)).(SD:Standard Variance).

SD(L) and Variance(L) show the servers' load difference. In our simulation we set the server number 9 and δ as 30. Once $L(i)¿\delta$, overload number add 1. In simulations the variance of servers' load are logged to check the uniform of the server load. Avatars are added into the DVE environment averagely along the running time until the DVE reaches its saturation state. Once an avatar enters into the DVE it gets a random position. Because the position belongs to the area charged by a special server, the avatar is allocated to a special server. In our simulation the avatars can move freely from one point to another point. The destination point is created with an random algorithm. Three simulations were taken in our experiment. We recorded the overload times and the load variance of the DSL algorithm and AMD algorithm. Table 1. show the overload times of DSL and AMD algorithm.

Avatars	DSL	AMD
800	216	97
1000	307	164
1500	475	211

Table 1. The overload number for DSL and AMD

From table1 we can easily find that the AMD algorithm can effectively reduce the overload times of the DVE system compared to DSL algorithm. We think the load level skill lead to this result. Fig2. Gave out the Stand Variance of L(i) under different algorithm. From fig2 we can see that the AMD algorithm hold a low stand variance of load. This can make the whole DVE system run placidly and all the servers keep a more fair load scale.

5 Conclusions

We present an amend algorithm named AMD algorithm base on DLS partition algorithm [2]. AMD algorithm keeps the advantage of the DLS algorithm and

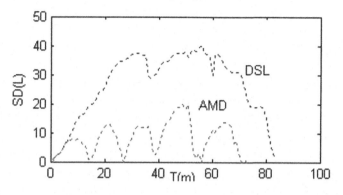

Fig. 2. Standard Variance of Load

get rid of its drawbacks. In our algorithm, All the servers take the highest possibility to handled the loads averagely, which keep the system run stably .All the server get a fair load scare at any time. At the same time our amend algorithm can also reduce the overload times of the whole system. This will improve the system's security and stability.

References

1. Zona Inc.: Set of massive multiplayer online games 2002, www.zona.net/news/2002mogreport.html.
2. T. Nguyen and B. Duong, A dynamic load sharing algorithm for massively multiplayer online games. Networks 2003, ICON2003, The 11th IEEE Intern. Conf., 131-136, 2003.
3. W. Cai, P. Xavier, S. Tumer, and B. Lee, A scalable architecture to support interactive games on the internet, Proc. of the 16th Workshop on Parallel and Distributed Simulation, 60–67, 2002.

High Performance Analytical Data Reconstructing Strategy in Data Warehouse

Xiufeng Xia[1,2], Qing Li[3], Lingyu Xu[3], Weidong Sun[2], Suixiang Shi[1], and Ge Yu[1]

[1] School of Information Science & Engineering, Northeastern University, Shenyang 110004, China xiaxiufeng@syiae.edu.cn
[2] School of Computer Science & Engineering, Shenyang Institute of Aeronautical Engineering, Shenyang 110034, China
[3] School of Computer Science & Engineering, Shanghai University, Shanghai 200072, China

Abstract. Many special requirements in on-line analytical processing and data mining need to access the historical available data of data warehouse. The computing performance is related to the analytical data reconstruction. We first analyze the differences between historical and current available data in data warehouse due to the time causation, and then advance the code equivalence and code coincidence. The high performance arithmetic by reconstructing the historical data to current analytical dada by using code reverting and abstracting methods are discussed.

Keywords. data warehouse, on-line analytical processing data mining, code equivalence, code coincidence, code reverting, code abstracting.

1 Introduction

The necessity and backup strategy of historical available data in data warehouse have been discussed in bibliography, the fundamental concepts were defined for available data, analytical data, and survival data at first, then the construction strategy and key technology of historical backup data—code replacement were introduced, finally the historical available data backup strategies based on rough time set was discussed. In this background, we will focus on the construction issues of historical available backup data in this paper.

Many special on-line analytical processing (OLAP) and data mining (DM) application need accessing the historical data in data warehouse (DW), the reconstruction issues of historical available data must be involved, and the key topic of data reconstruction is the correct mapping of code data. The accessing method of the detailed data, which has been transited to other storage media, has been discussed briefly in bibliography[MK93], at the same time, it also mentioned that the access cost is expensive and the access procedure is inconvenient and complex. In bibliography [ISO95, GB00, DA93] the backup and restore strategy are usually based on the functions provided by traditional OLTP DBMS. The reconstruction issue of available data is not mentioned in the published papers.

In addition, the reconstruction issue of available data is often arising during the adjustment of data valid period. As expatiated in bibliography, the adjustment of data valid period has great impacts on the backup strategy, if the prolongation of validity period is longer than one backup period, some survival data must be adjusted to analytical data, the decision must be made that if the historical data on external storage backup one (or more) backup period ago are needed to be rewrite into current system, *i.e.*, the reconstruction of analytical data.

2 Difference between survival data and analytical data

The backup file (survival data) refers to data set formed after the code replacement of current primary file (analytical data) in the last backup period, the differences of them can be represented in structural and non-structural aspects.

2.1 Non-structural difference

In most cases, the structures of backup file and primary file are same, the difference between them is only the variation of code attribute value in primary file in the near past, which makes the data in them inconsistent. For one code attribute, the non-structural difference exhibits in three ways: (1) New code value is inserted in primary file; (2) Old code value is deleted in primary file; (3) Code value is modified in primary file.

The three states have different influences on the data reconstruction strategy, the settlement of fist two states is simple, but it is not vice versa for the third state, since the data discrepancy has occurred between analytical data and survival data, the chief task for the implementation of reconstruction is to remove these discrepancies (refer the third part in this paper for detail).

2.2 Structural difference

Structural difference may have tremendous (sometimes disastrous) impact on the reconstruction of available data. Of course, for any DW system, the possibility for major structural adjustment is bare in one backup period (Since the reconstruction issue of DW system is involved, when the application changes dramatically, reconstruction will be of no help, that means the life cycle of the system is over and new system must be designed), that is to say, the performance adjustment and improvement of any DW system will be "slow". The structural difference exhibits in three ways:

1. New attribute is inserted in primary file.
2. Old attribute is deleted in primary file.
3. Some attributes are modified in primary file.

If only some attributes are inserted or deleted in primary file, the transformation from survival data to analytical data (import) will not be affected, but if some attributes in primary file are modified (such as modification of attribute name, type or constraints etc), then the reconstruction of available data may be accomplished by DWA manually (first DWA adjusts the survival data according to the specific situation, then follows the normal procedure). No details will be given due to the size of the paper.

3 Expansion of data dictionary

In order to do code reversion and code extraction automatically, application system must affirm if the two files are code equivalence, code compatibility or code incompatibility, the data dictionary must be expanded for the support of the detecting algorithm.

3.1 Global code dictionary

The global code dictionary contains all the code attributes, code owner (code file name) and code related operation. The dictionary is created during DW system design and maintained automatically by related utility programs in the DW running period (illustrated in table 1).

Table 1. Structure of global code dictionary

Attribute name	Type	Length	Meaning
Code name	Text	50	The name of code attribute
Type	Text	20	The type of code
Code file name	Text	50	The name of code file
Last update time	Date	8	The time of last update of the file
Update tag	Text	1	Category of update (0-insert,2-delete,3-modify)
Other attributes			Code constraint, code primary attribute etc.

As the metadata of DW, the dictionary is created during the system design, and all the update tags of codes are set to "0" (initial state). With the running of the system, code value and its meaning may have been changed, then the update tag must be adjusted at the same time. The attribute "code file name" is used to describe the code file containing the code and its meaning. In subject oriented DW design, code file may be as the same as the primary file, of course, it can also be independent one.

3.2 Global primary file directory

In order to define the range and determine implementation possibility of data reconstruction and prompt DWA preparing historical data, the building of global host file directory is necessary. Its format should be according to the organization below:

Table 2. Structure of global primary file dictionary

Attribute name	Type	Length	Meaning
Primary file name	Text	50	Primary file name in current system (analytical data file name)
Last backup time	Date	8	The date of the primary file backup
Backup file name	Text	50	File name before backup plus the time stamp
Reconstruction tag	Text	1	If the file is involved in current analytical data reconstruction
Reconstruction completion tag	Text	1	If the file has been dealt with in current analytical data reconstruction
Last reconstruction time	Date	8	Last analytical data reconstruction time

Before data reconstruction, which files should be reconstructed must be decided automatically or manually by DWA, the forming of available reconstruction directory can be done by setting the value of the reconstruction tag. In addition, reconstruction should be done orderly according the decided (valid) primary file directory, the reconstruction completion tag can be used to guarantee the correct and efficient execution of reconstruction operation.

4 Analytical data reconstruction

Analytical data reconstruction is not just simple the survival data write back operation, code reversion and code extraction issues are always involved under the situation that the primary files and historical backup files are code equivalence, code compatibility and code incompatibility, and then data write back operation should be executed. Usually, the process of analytical data reconstruction is complex and generally includes the operations:

Define reconstruction range: Select the survival data needing to be loaded from historical backup file into current system (historical available data reversion). Commonly two strategies can be used, first is that all the historical available data be complete restore, the second is part restore. Choice on the two strategies are often made with the participation of DWA or by special OLAP and DM programs automatically to build up available reconstruction directory.

Define validity: Define the life span of survival data. According to historical available data backup strategy mentioned in bibliography[1], the reconstruction range of analytical data should be multiples of backup period.

Determine code equivalence and code compatibility: if primary files and backup files are code equivalence or code compatibility, code reversion method can be used to revert the available content in backup file and insert them into current primary files. If primary files and backup files are code incompatible, code extraction can be used to do preparation for analytical data reconstruction and new analytical data utilization.

The companion papers depict the necessity and key technologies of historical available data backup and analytical data reconstruction in data warehouse systems. The methods introduced in the paper include: the construction and expansion of data dictionary, code replacement, backup strategy based on rough-time set, data transit and rewrite, in addition with the reconstruction technology of analytical data using code reversion, code extraction, code reset and code expansion methods in the case of code equivalence, code compatibility and code incompatibility, which have been applied in "Management information system of especial equipment supervised inspection" and "Engineering data management system of civil main airplane", it has been proved that they can solve the issues such as strong data archive effect, adjustment of valid period, analytical data reconstruction and the access of historical backup data for some special long time trend analysis etc. The implementation of these technologies meets the demands of enterprises and improves the usability of data warehouse, which gives effective support for decision-making.

Acknowledgements. This research is partially supported by National key Project (Grant 2004BA608B-03-03-03, 2001-608B-08-08).

References

[EM00] Efrem G., Mallach.: Decision Support and Data Warehouse System. McGraw-Hill Comp. Inc., 25–28 (2000)

[INM96] Inmon W.H.: Building the Data Warehouse. John Wiley & Sons, Inc., 136–139 (1996)

[IH94] Inmon W., Hackathorn R.D.: Using the Data Warehouse. John Wiley & Sons, Inc., 78–88 (1994)

[WAN98] Wang S.: Data warehousing & on line analytical processing. Beijing Science Press, 156–158 (1998)

High Performance Historical Available Data Backup Strategy in Data Warehouse

Xiufeng Xia[1,2], Lingyu Xu[3], Qing Li[3], Weidong Sun[2], Suixiang Shi[1], and Ge Yu[1]

[1] School of Information Science & Engineering, Northeastern University, Shenyang 110004, China xiaxiufeng@syiae.edu.cn
[2] School of Computer Science & Engineering, Shenyang Institute of Aeronautical Engineering, Shenyang 110034, China
[3] School of Computer Science & Engineering, Shanghai University, Shanghai 200072, China

Abstract. Considering the self-restriction of computer system, the usability of data warehouse, the executing efficiency of on-line analytical processing and data mining and the user's special demands in future, it is necessary to create a historical available data backup in data warehouse in which there is a mass of data has been saved. This is the first of the companion papers in which introduces the necessity and organizing strategies for constructing historical data backups after analyzing the usual and formal approaches of data backup in data warehouse. The process and important techniques of creating historical data backup are expatiated in detail, then the data backup strategies based on rough time set are advanced.

Keywords. data warehouse, historical data, on-line analytical processing, data mining, backup strategies, rough time set

1 Introduction

The data warehouse(DW) technology and derived technologies, such as on-line analytical processing(OLAP) and data mining(DM), have transited from the theoretical research period into practical application era, and more and more enterprises and organizations are benefiting from these greatly[MK93]. Since data warehouse stores huge volume data that can be classified as internal/external, historical/current or detailed/summarized information[ISO95], in addition to the continuous accumulation of historical data and introduction of new subjects to new demands. The data in warehouse is not only refreshed periodically, but also developed[GB00], which makes the system usability decrease and the expense increase.

There are close relations between data stored in data warehouse and that in traditional OLTP database. The larger the DW data volume and the finer the data granularity, the more accurate decision support information and effective application of OLAP and DM will be obtained[DA93]. But since DW always stores data spanning a long period and the capacity of online storage devices are limited, in addition with, the system security, reliability, cost and efficiency must be considered, A efficient backup system must be implemented for data warehouse system.

Bibliography[ISO95] talked of that the detailed low granularity data which has been exported to offline storage device are needed to be analyzed by some special requirements, and pointed out that the accessing cost is expensive and the procedure is complex and laboursome. Due to the article length, the paper will focus on the necessity of available historical data backup, especially the principle and method of construction.

2 Construction of historical backup data

Since there are obvious differences between the historical backup data and current data, data must be reconstructed, the construction procedure is generally divided into three steps including structure design, data preparation and backup execution.

2.1 Structure design

Structure design refers to how to construct a complete cover (attribute) set of backup data with the consideration of historical backup during the design phase of DW system. Since the historical available data on offline storage device is different from current data in logical structure, generally, the structure must be specified first, then the data is organized dynamically before the backup.

Two factors have impact of structure design, the first one is that the meaning of each attribute must be clear and exact in all backup data, *i.e.*, data should not exists as code format. The other one is that code attribute should be only replaced by available dependency set during code replacement, not all the primary attributes in the code file, especially the auxiliary attributes, in order to prevent the extra data redundancy. Structure design should follow the steps shown below:

- Defines the application demand of historical data, especially the demands for special OLTP and OLAP;
- Defines the backup data (primary file);
- For all the codes in the primary file, extracts available dependency set of each code attribute from corresponding code file;

- For each available dependency set, executing code replacing operation on structure, not on data content, to generate the new data structure, called historical backup file structure;
- Give each new structure a filename including the time stamp, *i.e.*, adding time stamp on filename.

2.2 Data preparation

Data preparation refers to the process extracting survival data from current DW system. If there are no special needs, historical data backup up can be done annually, *i.e.*, the outdated data backup is moved to removal offline storage device at the begin or end of a year, except the DW system growing rapidly. The preparation should follow the steps shown below:

- The adjustment and verification of data analysis period and survival period;
- The records exceeding the analysis period in primary file are inserted to the historical backup files;
- The outdated data is removed from all primary file;
- Execute code replacement operation against all historical backup files.

2.3 Backup execution

Backup execution refers to process that moving the prepared data to removable external storage device. The Execution should follow the steps shown below:

- Detect the storage capacity of the removable external storage device, and compare its size with the backup data volume;
- Execute the backup according to the designed strategy;
- Modify the historical backup data dictionary and appends part data contents;
- Mark the external storage media and entry the file.

Of course, the organization of backup file may has various forms, depending on the data scale and utilization of historical available data. As a rule, one independent removable storage device should store the independent historical backup data; it can be all the files in a backup period or one file of all the backup period. In some special cases, if the volume of historical backup data is too huge to be accommodated by an independent removable external device, the data file should be divided into multiple files automatically by backup program.

3 Organization of backup files and strategy

3.1 The organization of backup files

Although the large capacity removal external storage device suitable for data backup has been manufactured by current computer hardware technology, for historical available data, three kinds of backup file organization can be applied on the same storage media with the consideration of factors, such as management and utilization independence etc.

No matter what kind organization is adopted, the data scale must be estimated at the beginning of design, *i.e.*, the rows and required DASD (direct access storage device)[DA93] must be calculated for any known table.

Backup data is organized by backup period in the first scheme, that is to say, all historical data during one backup period are stored on one independent storage, of course it is under an supposed condition, but normally it can be practical in the era that huge capacity removal external storage device can be made. The advantage is that good device independence, convenient management and marking time stamp on backup media. Single file storage is used in the second scheme, *i.e*, the same backup file spanning several backup period is stored on the same storage media, the advantages include good data independence, convenient management and marking time stamp at file level. Backup files are organized according file types, *i.e.*, classified by business pattern or subject. This scheme falls between the first and second schemes.

In some unusual circumstance, no matter what kind organizations are adopted, the data volume may be too huge to be stored on one independent storage media, though backup program can detects the size of storage media automatically and prompts the user to change other storage media, one independent file must be stored on multiple storage media, so the management of them must be carried by groups and the data dictionary of historical backup file must be altered at the same time.

Since the volume of a single historical backup file has exceeded the size of an independent removal external storage media, it means that huge storage space has been occupied by many backup files, or we can say that large amount of "outdated" data has been stored in the system, shortening historical data backup period may improve the DW usability significantly. One historical data backup method based on rough time set is proposed in this paper, which can be used to deal with the problem.

3.2 A historical data backup strategy based on rough time set

The data increasing speed is varied between different DW. In any case, the backup operation for historical data should not be frequent. The increasing speed of DW, the application demands of available data and the size of backup file etc should be considered to select a storage media with appropriate size. The performance of backup media is all right and the capacity is large enough, the independence of backup data can be achieved by current technology. But backup strategy based on rough time set should be adopted with the consideration of DW usability and running efficiency of OLAP and DM.

Rough time set is only composed of three elements: season, half year and year, that is to say the operations of historical data backup are carried out every season, half year or year in DW system. Of course, different historical backup data dictionary should be adopted according to various backup strategy, reflected mainly on the "time stamp" attribute, the backup file name should also adjusted correspondingly.

Though historical available data is only used in special circumstance, it is an important part of DW system, it has been proved by practice ignoring the processing of historical available data will bring the user many inconveniences and may cause great damage. At present, DW, OLAP, DM and derived technologies have just come into practical application era (of course, there are still some area not mature and perfect in theory), but many enterprises and organizations are building or planning to building their own DW systems, the DW designer should recognize the importance of historical available data issue to be able to meet the special future demands.

Acknowledgements. This research is partially supported by National key Project (Grant 2004BA608B-03-03-03, 2001-608B-08-08).

References

[EM00]   Efrem G., Mallach.: Decision Support and Data Warehouse System. McGraw-Hill Comp. Inc., 75–78 (2000)

[INM96]   Inmon W.H.: Building the Data Warehouse. John Wiley & Sons, Inc., 66-67 (1996)

[IH94]   Inmon W., Hackathorn R.D.: Using the Data Warehouse. John Wiley & Sons, Inc., 151–157 (1994)

[WAN98]   Wang S.: Data Warehousing & on Line Analytical Processing. Beijing Science Press, 33–40 (1998)

[XWY01]   Xia X. F., Wang G.R and Yu G.: Study of Weak DSS Design Based on Time Rotation Strategy. Proceedings of The Northeastern Asian IT Symposium, 213–219 (2001)

[XWY02]   Xia X. F., Wang G. R. and Yu G.: Study of Operational Data Store Design Based on Time Rotation Strategy. Proceedings of 1st International Conference On All Optical Network and Embedded Internet & Information Fusion Technology, **2**:45–51 (2002)

A Modeling Method of Cloud Seeding for Rain Enhancement

Hui Xiao[1], Weijie Zhai[1,2], Zhengqi Chen[3], Yuxiang He[1], and Dezhen Jin[4]

[1] Laboratory of Cloud-Precipitation Physics and Severe Storms (LACS), Institute of Atmospheric Physics, Chinese Academy of Sciences, Beijing 100029, P.R. China hxiao@mail.iap.ac.cn
[2] Louhe Occupation Technical College, Luohe 450003, P.R. China
[3] Shaanxi Province Weather Modification Center, Xi'an 710015, P.R. China
[4] Jilin Province Weather Modification Center, Changchun 130062, P.R. China

Abstract. A modeling method for evaluating rain enhancement of cloud seeding with liquid carbon dioxide (hereinafter LC) coolant and silver iodide (AgI) ice nuclei has been developed. The method has been used to simulate a field experiment. Modeling results indicate that cloud seeding with LC and AgI in the appropriate part of cloud can induce notable change to cloud microphysical and dynamical processes, accelerating updraft velocity, speeding up formation of rain water, changing rainfall distribution, and finally increasing total rainfall. Different seeding agent like LC and AgI has different seeding effect. The mechanism of seeding LC to increase rainfall is analyzed.

Key Words: rain enhancement, cloud seeding, liquid CO_2, modeling method

1 Introduction

Cloud seeding has been widely carried out for rain enhancement in many countries all over the world (e.g. [LIS03]). The ability and efficiency to adjust some physical processes in appropriate clouds have been demonstrated and verified in laboratories, field observations and modeling simulations. The development in cloud physics, dynamics, statistics, computer modeling simulation, and high technology has greatly improved weather modification activities at present ([ORV96][BRU99]).

Two kinds of seeding agents are usually used for cold cloud seeding. One like silver iodide (AgI) provides ice-forming nuclei, and the other is a kind of coolant such as dry ice (solid CO_2). The most important advantage of coolant agent is that the number of ice crystal formed is nearly independent of the temperature, vapor, and supercooled water content in cloud. Since dry ice is usually made to little ballets before its seeding since dry ice ballet has a rapid falling speed and it must be dropped from cloud top or high altitude,

[FUK96] first put forward a new technique to directly seed liquid CO_2 (LC) horizontally at lower level of the supercooled area of cloud (hereinafter LC technique). The new technique of seeding LC by aircraft was experimented to enhance precipitation in Yan'an region of Shaanxi Province in the fall of 2002. The actual seeding rate of LC was 9 g s^{-1}. The purpose of this study is to compare the influence of the newly proposed LC seeding agent with current widely used AgI seeding agent on the evolution of cloud dynamics, cloud microphysics, and precipitation.

2 Numerical model and experiment scheme

The three-dimensional compressible non-hydrostatic convective cloud numerical model (IAP-CSM3D) with detailed seeding processes of LC and AgI agents was developed, based on our previous convective cloud model ([HON99][XWZ04]).

A field experiment of LC seeding by aircraft on Sept. 13, 2002, to enhance rainfall in Yan'an area was simulated in this study, based on the rawinsonde sounding data observed at Yan'an at 0800, September 13. Simulations were integrated to 120 min.

In the "natural cloud" modeling experiment (hereinafter Experiment NTL), no seeding agent was added to the modeling clouds. Three seeding experiments were simulated. First, Experiment LC1: the LC was sequentially released for 6 min in a limited area with a theoretical seeding rate of 0.6 g s^{-1} on a 30 km flight course from south to north with temperature level ranging from $-8°C$ to $-10°C$ at 15 min after initialization of the modeling cloud, corresponding to about 9 g s^{-1} of LC seeding rate and 2700 g of total dosage taken in field experiments according to [FUK96]. Second, Experiment LC2: the experiment scheme is the same as Experiment LC1 but for a theoretical seeding rate of 0.06 g s^{-1}, corresponding to 0.9 g s^{-1} of actual seeding rate and 270 g of total dosage of LC. Third, Experiment AgI2: the experiment scheme is the same as Experiment LC2 but for the seeding agent of AgI. The LC seeding rate in Experiment LC1 is the same as that used in real field and the AgI seeding rate in Experiment AgI2 is the same as that used usually.

3 Effects of cloud seeding on cloud microphysics

The influence of seeding LC or AgI on cloud and precipitation will be studied. The time evolution of maximum cloud water content in Fig.1a shows that cloud water content in seeded clouds decreases rapidly during 20 - 55 min (i.e. 5 - 40 min after LC and AgI seeding) due to cloud water glaciation induced by seeding agents, comparing with that in natural (NTL) clouds. The result indicates that artificial seeding ice particles into cloud can have supercooled liquid water in the cloud decrease.

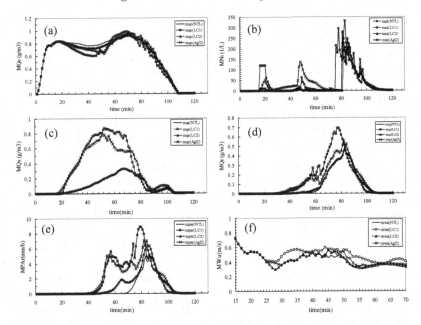

Fig. 1. Time evolutions of (a) maximum cloud water content MQc, (b) maximum ice crystal concentration MNi, (c) maximum snow content MQc, (d) maximum rain water content MQr, (e) maximum rain intensity MPAr, and (f) maximum updraft velocity MWu in the simulated clouds of Experiments Ntl, LC1, LC2, and AgI2, respectively

Comparing with that in natural clouds, ice crystal contents (Figure omitted) and number concentrations (Fig.1b) produced by LC or AgI increase sharply after seeding, and the seeding of LC is very effective, similar to that of AgI. As a comparison, ice crystals in natural clouds are appeared very late and reached a quite small peak only. Due to seeding with LC and AgI, a great number of ice crystals are produced in seeded clouds and grown up by vapor deposition processes. Snow in seeded clouds is formed automatically in a few minutes after formation of ice crystal and at about 20 min after the modeling initialization, while snow in natural clouds is formed later (Fig.1c). The snow continues rapidly growing by vapor deposition processes in clouds. Therefore, cloud seeding is able to speed up formation of ice crystal and also shorten the auto-conversion time from ice crystal to snow. Fig.1d illustrates that seeding LC or AgI is propitious to earlier appearance of rain water, increase of rain water peak value, and extending of rain water appearance period in seeded clouds. Similarly, the modeling results also indicate that seeding LC and AgI at appropriate part in clouds can initiate earlier rainfall and increase rainfall intensity (Fig.1e).

By analyzing microphysical processes that produce rain water in the clouds, it can be known that rain water comes into being mainly by cloud-rain-water auto-conversion processes in natural clouds and then grows by rain-and-cloud-water coalescence processes, but in seeded clouds, after seeding by LC and AgI, rain water begins to form via big ice particles (mainly graupel and snow) falling below the freezing level and then melting into rain water, and rain water further continues growing up by rain-and-cloud-water coalescence processes under the freezing level.

The time evolution of maximum updraft velocity in seeded and natural clouds is given in Fig.1f. Comparing with natural clouds, updraft velocity in seeded clouds after LC and AgI seeding obviously increases 0.1 - 0.2 m s^{-1} at 25 min after the modeling initiation (10 min after seeding). Especially, for the larger seeding dosage situation like Experiment LC1, the excess of updraft velocity goes on till at 65 min. The results indicate that seeding LC or AgI can cause a mildly dynamical effect and strengthen updraft in seeded clouds.

The radar observations took by Yan'an X-band radar before and after LC seeding on Sept. 13 show that the radar echo reflectivity above the actual seeding area strengthened 5 - 10 dBz and the top of the cloud rose 500 m after seeding of LC, higher than those before seeding. At the same time, the radar echo under the seeding area also became stronger. Meanwhile, by analyzing the field data collected by airborne PMS (Particle Measuring System), it can be seen that the average concentration of ice-snow crystals in diameter larger than 25 μm in the downwind of the seeding area was 28.6 L^{-1}, one order of magnitude larger than the average background concentration of 3.2 L^{-1} before seeding. All the facts have supported the above modeling results.

4 Effects of cloud seeding on rainfall distribution

Fig.2 presents surface distribution of 10-min accumulative rainfall for natural (NTL) clouds and two seeded clouds. By comparing 10-min accumulative rainfall distributions of LC1 and AgI2 seeded clouds (Fig.2b, c) with those of natural clouds (Fig.2a), it can be seen that seeding LC and AgI can produce

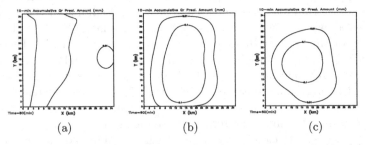

| | (a) | (b) | (c) |

Fig. 2. 10-min accumulative rainfall distributions in various clouds. (a) natural (NTL) cloud at 80 min, (b) LC1 seeding cloud at 60 min, and (c) AgI2 seeding cloud at 60 min

rainfall earlier, lengthen rainfall period, and especially change temporospatial distribution of rainfall.

5 Conclusion

Seeding of LC or AgI in appropriate part of the cloud can induce a notable change to cloud microphysical processes and cloud dynamical effects. The seeding of cloud can produce a large number of ice particles, speed up formation of rain water and enlarge rain water content peak value accompanying with accelerate updraft velocity. As a result, the seeding of cloud can also lengthen rainfall appearance time, change rainfall distribution, and increase total rainfall. Different seeding dosage of LC has different effect on rain water appearance time and rainfall amount. The seeding effect of LC seems better than that of AgI.

Rain formation mechanism of seeding cloud was analyzed. Rain water in natural clouds comes mainly from cloud-rain water auto-conversion processes and rapidly grows by rain and cloud water coalescence processes, but in seeded cloud, after seeding of LC, rain water first grows through the process of large ice phase particles (mainly graupel and snow) falling below the freezing level and melting into rain, then by cloud-rain water coalescence, rain water further grows to have rainfall increased and rainfall period prolonged.

Acknowledgements. This work was supported by Chinese National Natural Science Foundation under Program Grants 40333033 and 40175001, Chinese National Scientific Key Programs under Grants 2001BA610A-06-05 and 2001BA904B09, the Key Foundation of Chinese Academy of Sciences under Grant Y2003002, and the Key Foundation of IAP/CAS under Grant 8-4605.

References

[BRU99]   Bruintjes, R.T. (1999) A review of cloud seeding experiments to enhance precipitation and some new prospects. Bull. Amer. Meteor. Soc., 5, 805-820.

[FUK96]   Fukuta, N. (1996) Project mountain valley sunshine-progress in science and technology. J. Appl. Meteor., 35, 1483-1493.

[HON99]   Hong, Y.C. (1999) Study on mechanism of hail formation and hail suppression with seeding, ACTA Meteor. Sinica, 57, 30-44 (in Chinese).

[LIS03]   List, R. (2003) WMO weather modification activities, a fifty year history and outlook. Proceedings – 8th WMO Sci. Conf. on Wea. Mod., Casablanca, Morocco, 1-10.

[ORV96]   Orville, H.D. (1996) A review of cloud modeling in weather modification. Bull. Amer. Meteor. Soc., 77, 1535-1555.

[XWZ04]   Xiao, H., Wang, X.B., Zhou, F.F., Hong, Y.C., Huang, M.Y. (2004) A three- dimensional numerical simulation on microphysical processes of torrential rainstorms, Chinese J. Atmos. Sci., 28, 385-404 (in Chinese).

A High-level Design of Mobile Ad Hoc System for Service Computing Compositions

Jinkui Xie[1] * and Linpeng Huang[2]

[1] Dept. of Computer Science and Engineering, Shanghai Jiao Tong University,
Shanghai 200030, PR China, jkxie@cs.sjtu.edu.cn
[2] Dept. of Computer Science and Engineering, Shanghai Jiao Tong University,
Shanghai 200030, PR China, huang-lp@cs.sjtu.edu.cn

Abstract. This paper presents a high-level mobile ad hoc system for service computing compositions based on agents. The system is open and universal and tallies with Universal Plug and Play (UPnP) architecture standard. The high dynamics of such self-organizing system make high performance and high efficiency for service computing. The system which is based on Abstract State Machines theory is implemented in the specification language AsmL. In the system, any service component is viewed as an agent or a unit of several agents, so the composition of services can be viewed as a mobile ad hoc system of agents. The system has been used to depict a number of Web Service applications. The result shows that the system is flexible and practicable in modelling Web Service compositions.

Keyword: Service Computing Compositions, Mobile Ad Hoc System, Abstract State Machine, Agent

1 Introduction

This paper presents a high-level design model for service compositions. The model constructs the whole service compositions system as a collection of communicating agent subsystems. Each subsystem's communication structure is analogous in the essence. The model is open and has a Universal Plug and Play (UPnP) architecture [UPnP]. We set up the model with Abstract State Machine (ASM), and give some descriptions in AsmL [AsmL], namely the Abstract State Machine Language. The ASMs are able to simulate arbitrary algorithms in the step-for-step manner, and have been used to specify various architectures, protocols and numerous languages [GGV02]. The main contribution of the paper is the openness and universality of the high-level design model.

*This research is supported by China - 863 Program (No. 2001AA113160) and Shanghai Information Grid Project (No. 03dz15027).

The remainder of this paper is organized as follows. In section 2, we give a brief introduction to ASM. Section 3 is devoted to an agent-based service compositions framework. Sections 4 discusses an abstract communication model, which involves a collection of agents. Section 5 constructs a service compositions abstract machine based on agents, which is a high-level mobile ad hoc system. Finally, Section 6 provides some concluding remarks.

2 Abstract State Machines

Our method is based on the ASM theory. The following gives a brief description of ASM sufficient for our purposes in this paper [BS03]. More details can refer to [ASM].

Definition 1: A *signature* Σ is a finite collection of function names.

Definition 2: A *state* A for the *signature* Σ is a non-empty set X, the superuniverse of A, together with an interpretation f^A of each function name f of Σ.

Definition 3: An *abstract state machine* M consists of

1) a signature Σ,

2) a set of initial states for Σ,

3) a set of rule declarations,

4) a distinguished rule name of arity zero called the *main rule name* of the machine.

Definition 4: A machine M can make a *move* from state A to B (written $A \Longrightarrow_M B$), if the main rule of M yields a consistent update set U in state A and $B=A+U$. The updates in U are called *internal updates*, and B is called the *next internal state*.

Let M be an ASM with signature Σ. A *run* of M is a finite or infinite sequence A_0, A_1, ... of states for Σ such that

1) A_0 is an initial state of M,

2) for each n, either M can make a *move* from A_n into the next internal state A'_n and the environment produces a consistent set of external or shared updates U such that $A_{n+1}=A'_n+U$, or M cannot make a *move* in state A_n and A_n is the last state in the run.

3 Service Compositions Based on Agents

Research in the Web Service model is a hot area. In our design model, Web Service infrastructure can be divided into three layers: the Service Requestor Layer, the Middle Agent Layer, and the Service Layer [EK04]. Moreover, in this system, the Agent category has three subcategories: the Service Requestor Agent (SRA), the Service Provide Agent (SPA), and the Coordination Agent (COA). Among these subcategories, the agents are flexible and can transform their roles in the Web Service context, so the whole composition system is a Universal Plug and Play (UPnP) architecture.

A rational agent as the member of this system needs to balance its individual rationality and benevolence in facilitating to the growth of the group utility. Agents often use negotiation mechanisms, and an agent also needs to obey its commitments and the conventions which regulate the group behaviors within the system. Coordination means achieving coherence in the group activities and thus providing that the solution of a problem or the accomplishment of a task is obtained with less effort, less resources consumed, and better quality. Communication stands for the ability to exchange the pieces of information within an encounter in a uniform way and using shared terminology [EK04]. In the model, there are three basic operations:

1) $agent_1 \circ agent_2$: it represents a composite service that performs the service $agent_1$ followed by the service $agent_2$, i.e., $\circ$ is an operator of *sequence*.

2) $agent_1 + agent_2$: it represents a composite service that behaves as either service $agent_1$ or service $agent_2$. Once one of them executes its operation the other service is discarded, i.e., $+$ is an operator of *choice*.

3) $agent_1 \| agent_2$: it represents a composite service that performs the services $agent_1$ and $agent_2$ independently, i.e., $\|$ is an operator of *concurrence*.

So, the service compositions S can be defined as:

$S := \epsilon \mid agent$

$agent := a \mid agent_1 \circ agent_2 \mid agent_1 + agent_2 \mid agent_1 \| agent_2$

where: ϵ represents an empty service, i.e., a service which performs no operation; a represents a agent *constant*, used as an atomic or basic service in this context.

4 Abstract Communication Model

For the service compositions' purpose, we introduce A *distributed ASM* (DASM) involves a collection of *agents* [GGV02]. Agents are represented in global states as well. They are elements of a dynamically universe *Agent* that may grow and shrink. With each agent we associate a program defining its behavior. A static universe *Program* abstractly represents the set of all agent programs collectively forming the distributed program of A. Agents may perform their computation steps concurrently. A single computation step of an individual agent is called a *move* of this agent.

Formally, a run ρ of a distributed ASM A is given by a triple (M, λ, σ) satisfying the following four conditions: (1) M is a partially ordered set of moves where each move has only finitely many predecessors. (2) λ is a function on M associating agents with moves such that the moves of any single agent of A are linearly ordered. (3) σ assigns a state of A to each initial segment Y of M, where $\sigma(Y)$ is the result of performing all moves in Y; $\sigma(Y)$ is an initial state if Y is empty. (4) The *coherence condition*: If x is a maximal element in a finite initial segment X of M and $Y=X\text{-}x$ then $\lambda(x)$ is an agent in $\sigma(Y)$ and $\sigma(X)$ is obtained from $\sigma(Y)$ by firing $\lambda(x)$ at $\sigma(Y)$.

With AsmL, we can build a runtime environment for running distributed ASMs, which has the functionality for simulating concurrent agents' communications in this model [GGV02].

To keep track of the currently active agents, a variable *Agents* of type *set of Agent* is updated each time an agent is created or discarded.

class Agent
var Agents **as** Set **of** Agent = {}

Each agent's *state* evolves in sequential steps with each invocation of its *Program*. Each agent a has a *mailbox* of *messages* and a method *Insert Message* that is used by other agents to send messages to a.

type Message
class Agent
  Program()
  var mailbox **as** Set **of** Message = {}
  InsertMessage(m **as** Message)
    mailbox(m) := **true**

RunAgents()
  forall a **in** chooseSubset(Agents)
    Program(a)

Now we give more detailed description to some basic functionalities among SPA, SRA and COA. Based on these basic functionalities, their other individual functionalities can be extended. All behaviors of agent can be controlled by *COMMUNICATOR*.

class COMMUNICATOR **extends** Agent

type MESSAGE
class COMMUNICATOR
  ResolveMessage(m **as** MESSAGE) **as** Set **of** MESSAGE

type ADDRESS
class COMMUNICATOR
  var addressTable **as** Map **of** ADDRESS **to** Set **of** ADDRESS

destination(m **as** MESSAGE) **as** ADDRESS
class COMMUNICATOR
  Transform(m **as** MESSAGE, dest **as** ADDRESS) **as** MESSAGE
  ResolveMessage(m **as** MESSAGE) **as** Set **of** MESSAGE
    return {Transform(m,a) | a **in** addressTable(destination(m))}

class COMMUNICATOR
  var routingTable **as** Map **of** ADDRESS **to** Agent

class COMMUNICATOR
  Recipient(m **as** MESSAGE) **as** Agent?
    return routingTable(destination(m))

class COMMUNICATOR
  external ReadyToDeliver(m **as** MESSAGE) **as** Boolean

5 Service Compositions Abstract Machine

The Web Service is a dynamic large system, so its architecture must be open and flexible to any component in the web. The component may be a single agent or a unit of several agents, and it can joint or unjoint the Web Service in any time, so the whole service compositions system is a mobile ad hoc system. Additionally it provides a uniform view to the service compositions, which reflects the Universal Plug and Play (UPnP) property. The UPnP is an industrial standard for dynamic peer-to-peer networking defined by the UPnP Forum.

So it is rational to build a high-level executable Service Composition Abstract Machine with union several existing machine into a new one [Zam00].

type Service
type Label
type ServiceProgram
class COA **extends** Agent
  service **as** Service
  pgm **as** ServiceProgram
  var communicator **as** COMMUNICATOR
enum SPAMode
  exited
  raised
  halted
  running

class SPA **extends** Agent
  service **as** Service
  manager **as** COA
  var communicator **as** COMMUNICATOR
  var pc **as** Label
  var mode **as** SPAMode
  var subSPA **as** Set **of** SPA
class ServiceInstance **extends** SPA

Some instructions may be executed without incrementing the program counter; others cause the program counter to jump to a new position in the program.

type Instruction
Execute(intr **as** Instruction, p **as** SPA)
instr(pgm **as** ServiceProgram, lbl **as** Label) **as** Instruction
class SPA
  Program()
    if mode = running **then**

```
    let instr = instr(manager.pgm, pc)
    Execute(instr, me)
  else
    skip
```

As a coordinator, a COA has two independent jobs. One is to activate new service instances when *activating* messages are received. The other job is to handle message traffic.

```
class COA
  Program()
    ActivateServiceInstance()
    HandleMessageTraffic()
  HandleMessageTraffic()
  ReceiveIncomingMessages()
  ForwardOutgoingMessages()
```

In this paper, we start with the Abstract State Machine (ASM) theory. For the service compositions, we provide a Universal Plug and Play (UPnP) architecture based on agents, which is open and universal. Then we introduce a distributed ASM (DASM) involves a collection of agents, and build a runtime environment for running the distributed agents by AsmL. Finally, based on the agents, we construct a mobile ad hoc service compositions model. The future work of the model will focus on some complicated interactions in the service compositions.

References

[ASM]   Abstract State Machines (ASM), the academic Web site,
        http://www.eecs.umich.edu/gasm.

[AsmL] AsmL, the ASM Language, the website,
        http://research.microsoft.com/fse/asml

[Zam00] A.V. Zamulin. Specifications in-the-Large by Typed ASMs, International
        Workshop on Abstract State Machines, Monte Verita, Switzerland, 2000.

[BS03] E. Börger, R. Stärk. Abstract State Machines: A Method for High-Level
        System Design and Analysis, Springer-Verlag, 2003.

[GGV02] U. Glässer, Y. Gurevich and M. Veanes. An Abstract Communication
        Model, Micro-soft Research Technical Report MSR-TR-2002-55, May 2002.

[EK04] V. Ermolayev, N. Keberle. Towards a Framework for Agent-Enabled Semantic Web Service Composition, International Journal of Web Service Research,
        Volume X, No. X, 2004.

[UPnP] UPnP Device Architecture V1.0. Microsoft Universal Plug and Play Summit, Seattle 2000, Microsoft Corporation, Jan. 2000.

Metric Based Software Quality Assurance System

Dong Xu, Zongtian Liu, Bin Zhu, and Dahong Xing

School of Computer Engineering and Science, Shanghai University, Shanghai 200072, P.R. China. d.xuu@163.com

Abstract. A software quality assurance system named PMSQA[1] is introduced in this paper. A multi-dimension process template library is designed in the sense that through simple tailoring, it can be applied directly to various types of organization. A generic process modeling tool is contained in PMSQA, which can meantime support products model, functional model, resource model and role model. Some metric entities can be selected into the model and some useful data can be collected automatically when the process is executed. The whole system is implemented to support distributed development and collaboration working.

Keyword process modeling, software metrics, quality assurance.

1 Introduction

Among the available quality improvement models, the most influential ones include: the capability maturity model (CMM)[MK93], ISO/IEC 12207[ISO95], ISO 9001/9000-3[GB00], and the SPICE model[DA93]. Prominent among them are the CMM and ISO 9001. However, ISO 9001:2000 describe the elements of a quality assurance system in general terms, while CMM present key process areas and practice for software process improvement. Consequently, the challenge lies in designing and implementing a quality assurance system that meets the standard and fits the products, services, and culture of a specific organization. On the foundation of our previous research[SJ02],[QP01], [SW99],[LK01], a quality assurance system named PMSQA, which is based on ISO 9001 and CMM, is illustrated in this paper.

---

[1]This project is supported by the youth fund of Shanghai high-education bureau

2 Process-supported subsystem

The PMSQA contains three models which are process-supported subsystem, multi-dimension process template library and enterprise application subsystem. They are shown as figure 1. The process-supported subsystem constitutes the foundation of PMSQA. This process subsystem integrates version management, multi-user operation and cooperation, long transaction handling together so that it may address better the characteristics of software process.

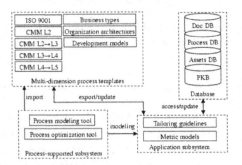

Fig. 1. system hierarchy model

The process modeling tool contains visual process modeling languages named ASTSPML. It can be used to define basic process elements visually, such as product, activity, relation, role and resource. There are a set of constraints mechanism on the operations of activities.The ASTSPML has various notations. There are some common process elements, such as activity, product, activity, relation, role, agent and resource. Some of the graphical notations of those elements are shown as figure 2:

activity	milestone	Association	Initial state	Terminal state	product	and	or
⬯	⬡	↗	●	◉	▭	⌓	⌓

Fig. 2. Some of the graphical notations of process elements

Process conception model includes function model, product model, cooperation model and resources model. The process concept model is shown as figure 3. Moreover, process database is included into the system.

3 Multi-dimension process templates

Multi-dimension application templates can import standard software process model or reusable process model that have been practiced in some organizations,

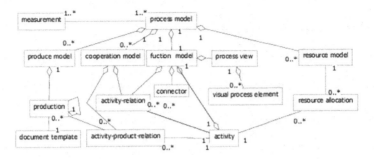

Fig. 3. process concept model

from which most software enterprise can export process model directly in terms of their own quality system, business type, development model and organization structure. The multi-dimension application templates have four perspectives. The profile of the multi-dimension application templates is shown in table 1.

Table 1. The profile of the multi-dimension application templates

Quality system	Business type	Architecture	Model
Importing ISO9001	Developing product	Function type	Waterfall model
ISO9001 to CMM	Developing Project	Line type	Increment model
CMM L2 to CMM L3	Contractor	Matrix type	Spiral model
CMM L3 to CMM L4	Providing service	Line-function	Iteration model

4 Application platform

No process is adequate to all organizations or projects. The key practices of each key process area must be tailored to meet requirement of organizations. The application platform may instantiate process model at two levels which are organization level and project level in light of the tailoring guidelines. Three types of metrics, product metric, intermediate product metric and process metric are included in this system. So, both the software process and products are quantitatively understood and controlled. Process instantiation is done by process tailoring guidelines. A sample of process model tailoring is show as figure 4.

Formally, a tailoring guideline can be defined as a 4-tuple: TG = (PF, TA, TI, TD). Here, TG denotes tailoring guideline; PF are process models or process elements which are CMM levels, agile process model, products, activities,

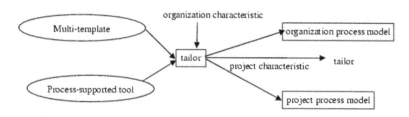

Fig. 4. process instantiation

such as analysis, design, testing, etc. TA is tailoring attributes, such as execution, measurement, archive, reviews, detailed levels, etc. TI is tailoring items that are making decision of selecting "yes", "No" or "detailed level" for each tailoring attribute. TD is tailoring instructions that explain in details why make so tailoring. A sample of tailoring guideline for coding is shown as table 2: In this paper, the Goal-Question-Metric (GQM) model is extended to be

Table 2. Tailoring guideline for coding

Elements	Attributes	Tailoring items	Tailoring instructions
Code review	Review	1.group review	1.pool skill
		2.single review	2.good skill
		3.not review	3.simple program
Code defect	Record	1.Record review	1.higher defect
		2.Record defects	2.lower defect
Unit testing	Execute	1.execute	1.complicated program
		2.not execute	2.good skill

GQM/SSC[8] metric framework by which the metric grains are fractionalized in the sense that each process step can be measured.

Metric tool classify the questions by the types of goals and decompose the metric. Metric model library includes product (containing intermediate product) metric, process metric and process area metric, i.e., measuring the execution degree of key practices in CMM.

5 The implementation of the system PMSQA

In order to use the system by internet, it is designed as B/S architecture. Browsing it easily in client through standard web explorer, the server port consists of web sever, application server and database server. All are implemented according to J2EE programming model, so the system is independent of platform. It shows as Fig 5.

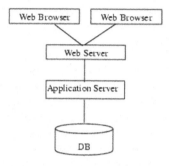

Fig. 5. process instantiation

6 Conclusion and Further Work

This paper introduces a software quantity assurance system-PMSQA that supports the whole software life cycle. We propose a practical and operational approach to set up process models corresponding to ISO 9001 and CMM to meet different requirements of a variety of organizations. The PMSQA will continue to evolve as experience is gained with improving software process and product quality.

References

[DA93]   Dorling, A.: SPICE: Software Process Improvement and Capability Determination. Software Quality Journal, Vol. (2) (1993).

[GB00]   GB/ T 19000(1,4)-2000, the quantity manages the system standard the [S],(2000,12)

[ISO95] ISO/IEC 12207, Software Life Cycle Process. 1st Ed., Aug (1995).

[LK01]   Li Xinke, Liu Zongtian A Measurement Tool for OO Software and Measure Experiments with itJournal of computers. (2000.11. pp1220-1225).

[MK93]   Mark C Paulk, B Curtis, M B Chrissis, C V Weber, Capability Maturity Model for software[R]. Version 1.1, CMU/SEI-93-TR-24, Software Engineering Institute, Camegie Mellon University, February, (1993).

[QP01]   QiuLing PAN, ZongTian Liu. Research on software process technology and environment, Journal of electronics, (2001.11, pp1575-1577).

[SJ02]   Shen bei-jun, cheng cheng, ju de-hua, criteria-based software process transaction model. Journal of software. (2002.1,pp24-32).

[SW99]   SU Weimin, ZHU Sanyuan, Journal of Software. (1999.8. pp843-849).

A Strategy for Data Replication in Data Grids*

Liutong Xu[1], Bai Wang[1], and Bo Ai[2]

[1] School of Computer Science and Technology,
   Beijing University of Posts and Telecommunications, Beijing 100876, China
   {xliutong, wangbai}@bupt.edu.cn
[2] Information Systems Division of China Unicom, Beijing 100032, China
   aibo@chinaunicom.com.cn

Abstract. Data replication strategy and replica selection are the most important topics in the data grid. This paper presents a new strategy for data replication in the data grid environment. The strategy combines data replication algorithm with job scheduling policy and is able to select best data replicas in a dynamic grid environment by monitoring the data replication processes, keeping the efficient ones and abandoning the inefficient ones. Grid community-based data transfer and multiple replications of fragmented data from multiple data sources greatly improve the replication efficiency. The strategy can also support various data replication and job scheduling algorithms.

Key words: data grid, grid community, replication, schedule, best replica

1 Introduction

Grid computing [FK03] is a new emerging distributed computing paradigm. Its goal is to unite geographically distributed computing resources for providing computing power as utilities and allowing resource sharing efficiently.

In an increasing number of scientific disciplines, large data collections are emerging as important community resources. Data grid [CFK01] is suitable for solving large-scale data-intensive scientific problems. Data grid is also the basis for the other computational grids or application grids [XA04]. In fact, many undergoing grid projects are data grid based. In most situations, the data that user application requested are not located at the local nodes in the data grid. Therefore, some replication mechanisms are needed to transfer the data from one node to another around the data grid. Efficient data replications are required to deal with the lower level data movement in the data grids and in many grid applications such as FTPGrid [XA03]

---

*This work is supported by the National Natural Science Foundation of China (Grant No. 90104024), and China National Science Fund for Distinguished Young Scholars (Grant No. 60125101).

This paper proposes a strategy for data replication in the data grid. We combine the data replication algorithm with job schedule policy in order to select best replicas in the dynamic running environment. The strategy supports multiple replications of fragmented data from multiple source nodes simultaneously in the data grid. It also supports various kinds of data replications and scheduling algorithms.

The remainder of the paper is organized as follows. Section 2 is about the related work; section 3 presents a new strategy for data replication and some discussions on related issues; and section 4 gives the conclusions.

2 Related Work

Data replication is one of the most important topics in the data grid. Many literatures have addressed the related issues.

The SDSC Storage Resource Broker (SRB) [BMR98] is intended to provides a uniform interface for connecting to heterogeneous data resources over a network and accessing replicated data sets. The GridFTP [ABB01] provides secure and reliable data transfer between grid hosts. It also provides many features such as third-party data transfer, parallel data transfer using multiple TCP streams, striped data transfer, partial file transfer etc. Various examples of replication and caching strategies are discussed and tested within a simulated Grid environment in [RF01], while their combination with scheduling algorithms is studied in [RF02]. Evaluations about different replication strategies, focusing on selecting the optimal replicas of data files and a prediction function to make informed decision about local data replication are also discussed in [BCC03].

In the literatures, many efforts are focused on selecting best replicas. But, what does it mean by *best*? Actually, the so-called best or optimal data replicas are based on some kinds of metadata or historical data access information. All these information are static. However, the grid system is a dynamic environment. A replica that is best for some grid node at certain period of time is not necessary to be best for another grid node at some different time, because of the changes in network, CPU capacity, workload, etc. Therefore, how to select *best* replicas in a dynamic grid environment is still a problem.

In this paper, we try to take a combination of data replication algorithm and job scheduling policy, and propose a new strategy for data replication in the data grid. The strategy will be able to select best replicas in the dynamic grid environment by monitoring and adjusting all replication job processes.

3 A New Strategy Combined with Job Scheduling

Grid system is usually built directly upon the Internet, it may cover wide area over the Internet. Generally, a grid system consists of lots of grid nodes and each grid node has one or any combination of following three functionalities: computational capacity, data storage, and directory service. In the data grid, we focus on the grid nodes that have data storage functionality. Data in the data grid are usually partitioned into small pieces, and each pieces of data have several copies resides in different grid nodes in order to utilize the storage resource and get ready for computational tasks. In each data node, there is a main module called replica manager as in [EUDG]. The replica manager is responsible for data replication according to certain strategies.

3.1 Grid Community

It is obvious that data replication from a neighboring grid node is always much more efficient than from some distant grid nodes. That means the locality information is very important. In order to simplify the data resource management, we introduce the concept of *grid community* which consists of neighboring grid nodes. The neighborhood is measured by some kinds of distances such as network hops, network bandwidth, and data transfer costs based on some economic models as in [BU02]. Of course, a grid node may belong to several communities at the same time.

Our discussions will be based on the following scenario: a data grid consists of many communities and the data resource management will be based on the grid community. For each grid community, there is a directory server to keep the metadata of data resources. Community-based information exchange implements a decentralized metadata synchronization in order to get the uniform view of all data resources in the grid.

3.2 A New Strategy for Replication

When an application requests some non-local data at some grid node, the replica manager will fragment the requested data into several small pieces, retrieve the metadata about all data pieces from community directory server, determine the data sources for every pieces and then launch a replication job for every pieces of data. After finishing replications of all data pieces, replica manager will assemble the fragmented data pieces back into the original form of the data. The key step here is to monitor the replication processes to identify the dynamic best replicas in the running environment.

Algorithm Data Replication Strategy
1. Split data into small data pieces if necessary. Simultaneous replications of small data piece will raise the replication efficiency.
2. For each piece of data:
    a) retrieve the community directory service and get its metadata.
    b) select one or more *static* best replicas according to the metadata. The criteria for *best* are mainly based on neighborhood, network bandwidth, availability, authority, access costs, etc.

c) build a data replication job for each replica.
3. Submit the job list to start data replication and monitor the replication processes to identify the best replicas as in Sect. 3.4.
4. After finishing all replication job, assemble all pieces into the original form of the requested data.

Remark 1. The job list in step3 consists of many small and maybe duplicated data piece replication jobs from multiple sources. The number of duplicated data replications and of simultaneous replications can be adjusted in the runtime to achieve a good balance between the consumption of network resources and the replication efficiency.

Remark 2. For the non-realtime applications, the replication jobs might be scheduled at off-peak hours to prevent traffic jams.

3.3 Data Splitting and Replica Selecting

When user application requests data–sometimes part of data – at some grid node, the replica manager will split the requested data into small pieces in order to fully utilize the network bandwidth and the data sources in the grid. Information about each piece of data such as data offset and data length will be used for locating data replicas through directory services. The replica manager will also evaluate several *static* best replicas for all pieces of data based on metadata such as locality, network bandwidth, availability, authority, access costs, etc.

3.4 Data Replication Job Monitoring

The replica manager will submit a replication job list that may contain duplicated data piece replication. How to schedule these replication jobs is the key that our strategy concerns. The replica manager will initiate several duplicated data replication job, and monitor their execution status. According to replication processes, the replica manager will be able to judge which replica is really the best one in the running environment, and then determine to keep some replication processes and to cancel the others.

Algorithm Job Scheduling and Monitoring
1. If one data piece replication job is completed, then all other replication jobs concerning the same data piece could be cancelled and deleted from the job list.
2. If some data piece replication job fails, then
    a) Cancel and delete it if there are one or more jobs concerning the same piece of data;
    b) Resubmit it if there are no job concerning the same piece of data.
3. If some data piece replication process is too slow, then

a) Cancel and delete it if one or more jobs concerning the same piece of data are in process or in the job list;
b) If it is the only job concerning the data piece, then
  i. Retain the data replication; or
  ii. Rebuild the data replication job for the piece of data using the algorithm in Sect. 3.2.

3.5 Data Assembling

After all replication jobs have been done, the replica manager of the grid node will assemble all these fragmented data pieces into the original form of the requested data according to the data offsets. Now, the new data replica is ready for use by the user applications.

3.6 Data Caching

The new data replica will be stored in the local systems for possible subsequent use. At the same time, the new replica should also be registered to the local community directory service as non-authority.

3.7 Advantages of the Strategy

The characteristics of our data replication strategy lead to many advantages:

- Grid community-based data replication will speed up the data transfer and reduce the traffic jams over the backbone of the network.
- Replications of multiple fragmented data will leverage the replication efficiency.
- Data replications from multiple sources will fully utilize network bandwidth and balance the workloads of data sources.
- Duplicated data replication will raise efficiency. If one replication fails, the others will continue. There is no need to reschedule the failed data replication.
- Duplicated data replication can also be used to obtain a high accuracy by comparing the multiple copies of the same data pieces.
- Job scheduling monitors duplicated data replication in order to select real best data replicas in the running environment.
- The strategy can support all kinds of data replication and job scheduling algorithms.

On the other hand, the disadvantage of the strategy is also obvious. It consumes more network resources because of duplicated data replications. However, it can also be carefully adjusted by reducing the number of duplicated data replicas to prevent overconsuming the network bandwidth.

Data replication is one of the key issues in data grid. The selection of data replica and the replication strategies have been studied widely. In this paper, we present a novel strategy for data replication with job scheduling. The major characteristics include: a) it can select best data replicas in a dynamic grid environment; b) it supports multiple replications of fragmented data from multiple data sources; and c) it also supports different data replication and job scheduling algorithms.

References

[ABB01] Allcock, W., Bester, J., Bresnahan, J., Chervenak, A., Liming, L. and Tuecke, S.: GridFTP: Protocol Extension to FTP for the Grid. Grid Forum Internet-Draft (2001)

[BMR98] Baru, C., Moore, R., Rajasekar, A. and Wan, M.: The SDSC Storage Resource Broker. In: Proc. CASCON '98 Conference (1998)

[BCC03] Bell, W.H., Cameron, D.G., Carvajal-Schiaffino, R., Millar, A.P., Stockinger, K. and Zini, F.: Evaluation of an Economy-Based File Replication Strategy for a Data Grid. In: Proc. of 3rd IEEE/ACM Int. Symp. on Cluster Computing and the Grid. IEEE-CS Press, Tokyo (2003)

[BU02] Buyya, R.: Economic-based Distributed Resource Management and Scheduling for Grid Computing. PhD dissertation, Monash University, Melbourne (2002)

[CFK01] Chervenak, A., Foster, I., Kesselman, C., Salisbury, C. and Tuecke, S.: The Data Grid: Towards an Architecture for the Distributed Management and Analysis of Large Scientific Datasets. Journal of Network and Computer Applications, 23, 187-200 (2001)

[EUDG] European DataGrid at CERN Accelerator Center, http://eu-datagrid.Web.cern.ch/eu-datagrid/.

[FK03] Foster, I. and Kesselman, C. (eds): The Grid 2: Blueprint for a New Computing Infrastructure. Morgan Kaufmann Publishers, San Francisco, CA (2003)

[RF01] Ranganathana, K. and Foster, I.: Identifying Dynamic Replication Strategies for a High Performance Data Grid. In: Proc. of the Int. Grid Computing Workshop. Denver, Colorado (2001)

[RF02] Ranganathana, K. and Foster, I.: Decoupling Computation and Data Scheduling in Distributed Data-Intensive Applications. In: Int. Symp. of High Performance Distributed Computing. Edinburgh, Scotland (2002)

[XA03] Xu, L. and Ai, B.: FTPGrid: A New Paradigm for Distributed FTP System. In: Proc. of 2nd Int. Workshop on Grid and Cooperative Computing. LNCS Vol.3033, Springer-Verlag, Berlin Heidelberg (2004)

[XA04] Xu, L. and Ai, B.: Notes on Grid and its Attributes. In: Proc. of the 2004 Int. Conf. on Parallel and Distributed Processing Techniques and Applications. CSREA Press, Las Vegas, Nevada (2004)

Analysis of an IA-64 Based High Performance VIA Implementation

Feng Yang[1] *, Zhihui Du, Ziyu Zhu, and Ruichun Tang

Department of Computer Science and Technology
Tsinghua University, 100084, Beijing, China
[1]yfeng99@mails.tsinghua.edu.cn

Abstract. Virtual Interface Architecture (VIA), which establishes a communication model with low latency and high bandwidth, is the industry standard for user level communication. In this paper, THVIA64, a hardware support VIA implementation based on IA-64 platform, is analyzed from three aspects, compatibility, stability and performance. Compatibility test shows how well it supports both for traditional network protocol, such as TCP/IP and VIA. Stability test measures how well it can transfer a large number of packages for both small and large data within a very long test time. The performance data are tested and compared on two different kinds of networks, Gigabyte Ethernet and THNET64. The test results show that THVIA64 is a compatible, stable and high performance implementation of VIA based on IA-64 platform for cluster computing.

Key words: ULN (User Level Network), KLN (Kernel Level Network), THVIA, IA-64, Cluster

1 INTRODUCTION

Cluster computing is becoming more and more important and popular. High performance network is a key component of cluster system, not only the network hardware, but also, even more important, the network support software. ULN (User Level Network) Communication has been employed into cluster system widely. ULN communication aims to provide low latency, high bandwidth and high reliable communication for cluster system. VIA (Virtual Interface Architecture) is the industry standard for ULN communication.

THVIA [ZLZ2001] is a hardware-based implementation of VIA. It can provide all necessary VIA functions needed by the standard VIA specification and be compatible with TCP/IP protocol considering the current and former applications. With the popularization of IA-64 in cluster systems, we designed

*This project is supported by Beijing Natural Science Foundation (4042018) and National High-Tech Research and Development Plan of China (No. 2004AA104330).

THVIA64 system based on IA-64 platform from the original THVIA system (THVIA32). THVIA64 is similar to the basic characteristics of THVIA32. It is a half user-level implementation and is compatible with TCP/IP. This paper gives and analyzes the test results over THVIA64 on ULN and KLN test.

2 TEST DESIGN

2.1 Test Program and Environment

In this paper, we use two programs to measure THVIA64's performance. The first is the vpingpong program in M-VIA [Nat99] packet. We make some modifications to make it run on THVIA64 system. We use it to measure the THVIA64's ULN(VIA) performance. The second is the test_tcp program made by our group. It is a test program on socket in C language. We use it to measure THVIA64 and Gigabyte Ethernet's KLN(TCP) performance.

The test platform consists of HP rx2600 (Dual Itanium2 900) nodes. Each rx2600 has two kinds of network, Gigabyte Ethernet and THVIA64. The OS is Linux with kernel version 2.4.18-e.12. Gigabyte Ethernet works at 64-bit, 66 MHz mode. THVIA64 should work at 32-bit, 66 Mhz mode. In order to improve stability, we modified THVIA64's hardware and make it work at 32-bit, 33 Mhz mode. All the test results below are based on the modified situation without special explanation.

2.2 Model

The main test model we use is the $Ping - Pong$ model. This model uses two parameters to measure the performance of a given network.
- L ($Latency$): the time spent on network to transmit a message
- o ($overhead$): the time spent on local to send or receive a message

To get the parameters, first, it takes a step to measure the $Round - Trip - Time$ (RTT). The RTT time is from sender overhead o_s, receiver overhead o_r, sending latency L_s and receiving latency L_r constitute of.

$$RTT = L_s + L_r + o_s + o_r \tag{1}$$

For the socket test program, the model takes a second step to measure the overhead. The sender sends the same message as in first step to itself. It means the sender takes the lo device as the receiver. In this way, L_s and L_r are very small and can be ignored. So $RTT' = o_s + o_r$, both of o_s and o_r is measured in the sender's side. In the $Ping - Pong$ model, the sender or receiver only process one message per one time, and the environment and network on each side is just the same. So $o_s = o_r$, $L_s = L_r$ and o_r in sender's side is equal to o_r in receiver's side. Then $RTT = L_s + L_r + o_s + o_r = 2 \times L_s + RTT'$

$$L = L_r = L_s = (RTT - RTT')/2 \tag{2}$$

For the VIA test program, we can get o_s, o_r, L_s, L_r directly in the first step from the test program. As we say above, THVIA64 is a half ULN implementation. The sender can make access to network hardware directly from user space, but the receiver must take one interrupt operation to receive message. So L_s is the accurate time for THVIA64's performance and L_r contains the time for making interrupt operations ($o_{interrupt}$).

$$L_r = L_s + o_{interrupt}, \quad L = L_s \tag{3}$$

Suppose M is the size of message in bytes. Then the network bandwidth is M/L.

3 TEST RESULTS ANALYSIS

3.1 Compatibility and Stability Analysis

The results from a series of evaluations on model applications of TCP/IP such as FTP, TELNET, SSH suggest that the compatibility of THVIA64 is good. We use the CHECK program from M-VIA Packet on THVIA64 with different data packet size from 1 byte to 64K bytes. All the results are correct. In the full-load test (THVIA64 works at 66 MHz), with the maximum packet size (64K), the system has being run for more than 10 hours. The total number of transferred packets is more than 10^9, and the quantity of transferred data in full-duplex is about 128 TB. There is no error in all situations. This shows that THVIA64 is stable enough.

3.2 Performance Analysis

Fig.1 is the pingpong bandwidth result of THVIA64 and Gigabyte Ethernet. From Fig.1(a) we can find that the THVIA64's KLN peak bandwidth (49.1

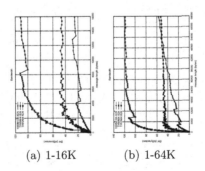

(a) 1-16K   (b) 1-64K

Fig. 1. Pingpong Bandwidth curves of THVIA64 and Gigabyte Ethernet

Mbytes/s) is only 44% of ULN peak bandwidth (112.3 Mbytes/s) over the same hardware. If THVIA64 works at 66 MHz, the difference should become larger. This test result indicates that a big part of the hardware network bandwidth can not be used if KLN communication is employed, so ULN communication is very necessary to achieve high performance communication in cluster system.

We adpot the N1/2 and HBWL concept here. N1/2 means the problem size for achieving half of maximal performance, HBWL (Half BandWidth Length) is the message length for achieving half of peak bandwidth. In Fig.1(a), THVIA64's KLN peak bandwidth is about 49 Mbytes/s, the HBWL is about 1000 bytes. THVIA64's ULN peak bandwidth is about 112 Mbytes/s, the HBWL is about 900 bytes. In Fig.1(b), THVIA64's KLN peak bandwidth is 53 Mbytes/s, the HBWL is about 1100 bytes. THVIA64's ULN peak bandwidth is 119 Mbytes/s, the HBWL is about 1100 bytes. Based on the research result of [Gus90], 80% messages are small message (less than 1K bytes), only 8% messages are bigger than 8K bytes, this means that both with ULN and KLN, THVIA64 can provide high bandwidth for most messages. However, in Fig.1(a), Gigabyte Ethernet's KLN peak bandwidth is 28 Mbytes/s, the HBWL is about 2000 bytes. In Fig.1(b), Gigabyte Ethernet's KLN peak bandwidth is 56 Mbytes/s, the HBWL is over 12000 bytes. This means that not only the bandwidth of KLN is very limited, but also the peak bandwidth is not easy to achieve for most messages.

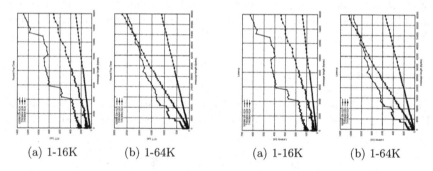

| (a) 1-16K | (b) 1-64K | (a) 1-16K | (b) 1-64K |

Fig. 2. Pingpong RTT curves of THVIA64 and Gigabyte Ethernet

Fig. 3. Pingpong Latency curves of THVIA64 and Gigabyte Ethernet

Fig.2 and Fig.3 include the pingpong RTT and Latency curves of THVIA64 and Gigabyte Ethernet. In Fig.2, THVIA64's ULN RTT is about 27 us for 8 bytes message, but for KLN, the corresponding RTT is 89 us. The performance of many applications is very sensitive to short message latency, so the RTT results also indicate that the ULN is superior to KLN. Otherwise, in Fig.2, THVIA64's KLN RTT for 8 bytes message is 89 us, the Gigabyte Ethernet's corresponding parameter is 238 us. It indicates the superiority of THVIA64 as a communication device. In general, the RTT should be linear with message length, just as the THVIA64's RTT curves show in Fig.2.

Table 1. Comparison of THVIA32 and THVIA64

Parameters	THVIA32	THVIA64 (original)	THVIA64 (modified)
BW (Mbytes/s)	103	>220	119
Latency (us)	8.95	<9	11
HBWL(Bytes)	1000	<1000	1100

In the analysis above we mainly analyses the ULN and KLN performance of THVIA64. With the former test results (Dual PIII 1G platform), Table 1 gives the comparison of THVIA32 (33 Mhz), original (66 Mhz) THVIA64 and modified (33 Mhz) THVIA64's bandwidth, latency and HBWL parameters from vpingpong test results. It shows that in same mode (32-bit, 33 Mhz), the maximum bandwidth of modified THVIA64 for message length 64K bytes is 119 Mbytes/s that is 15% better than THVIA32's. This is the predominance of IA-64 platform.

3.3 Smoothness Analysis

In this analysis we provide another aspect to assess THVIA64's performance, it is smoothness. In Fig.1, the trend of THVIA64 ULN bandwidth curve is increasing, but it decreases at 8K point. It's because THVIA64 splits messages at upper level in order to fit the hardware resource. The splitting point is 8K bytes. So the bandwidth curve decreases at there.

We can find that there are more sharp points in THVIA64 KLN curve than in THVIA64 ULN curve in Fig.1. This is because the KLN communication needs OS intervention. It would be affected by CPU or Network contention. The ULN communication eliminates OS intervention, so the smoothness of ULN curve is better. Smoothness reflects if the communication performance can be provided steadily. The result indicates that ULN is superior to KLN and THVIA64's performance is steady.

4 RELATED WORKS

There are many projects on ULN protocol, such as SHRIMP [BDF95], U-Net [TAV95] and VMMC [DBL97]. When the standard of ULN–VIA occurs, ULN research focuses on VIA specification and implementation, NERSC's M-VIA [Nat99], Berkeley's B-VIA [PAD98] and IBM [BMH2000] all have their VIA implementation. Most of the researches on VIA are focused on the platform with IA-32 architecture and Myrinet [BCF95]. This paper assesses the ULN and KLN performance of new VIA implementation on the platform with IA-64 architecture and VIA hardware support.

In this paper, we mainly use a Ping-Pong model to analyze THVIA64's KLN and ULN performance. Form the test results, the benefit of THVIA64/ULN is obvious in many aspects. So we conclude that high performance cluster system must adopt ULN to improve its performance and THVIA64 is fine as an ULN implementation and network communication device. Without ULN, cluster applications can not get even half of the hardware network performance. A cluster system connected by hardware support user level network, such as THVIA64, can provide high performance close to the traditional supercomputer or MPP (Massively Parallel Processors) systems. Cluster computing is promising and ULN will help cluster computing to be more powerful.

References

[BMH2000] Banikazemi, Mohammad; Moorthy, Vijay; Herger, Lorraine; Panda, Dhabaleswar K.; Abali, Bulent.: Efficient virtual interface architecture (VIA) support for the IBM SP switch-connected NT clusters. In Proc. of the Intern. Parallel Processing Symposium, IPPS, 33-42, 2000.

[BDF95] Blumrich, M.A.; Dubnicki, C.; Felten, E.W.; Li, K.: Virtual-memory-mapped network interfaces. IEEE Micro, 15(1), 21-28, 1995.

[BCF95] Boden, N.J.; Cohen, D.; Felderman, R.E.; Kulawik, A.E.; Seitz, C.L.; Seizovic, J.N.; Su, W.-K.: Myrinet A gigabit-per-second local area network. IEEE Micro, 15(1), 29-36, 1995.

[DBL97] Dubnicki, C., Bilas, A., Li, K., Philbin, J.: Design and Implementation of Virtual Memory-Mapped Communication on Myrinet. In Proc. Intern. Parallel Processing Symp., IPPS, 388-396, 1997.

[Gus90] Gusella, Riccardo.: A measurement study of diskless workstation traffic on an Ethernet. IEEE Transac. Communic., 38(9), 1557-1568, 1990.

[Nat99] National Energy Research Scientific Computing Center.: M-VIA: A high performance modular via for linux. 1999.

[PAD98] P. Buonadonna, A. Geweke, D. Culler: An implementation and analysis of the virtual interface architecture. In Proc. of Superc., 7-13, 1998.

[TAV95] T. von Eicken, A. Basu, V. Buch, and W. Vogels: U-net: A user-level network interface for parallel and distributed computing. Proc. of the 15th ACM Symp. on Operation Systems Principles, 40-53, 1995.

[ZLZ2001] Z. Du, L. Mai, Z. Zhu, H. Liu, R. Tang and S. Li.: Hardware Based THVIA User Level Communication System Supporting Linux Cluster Connected by Gigabit THNet. 2001 Intern. Symp. on Distributed Computing and Applications to Business, Engineering and Science, DCABES 2001, 134-138, 2001.

An Application and Performance Analysis Based on NetSolve

Linfeng Yang[1], Wu Zhang[1], Chaojiang Fu[1], Anping Song[1], Jie Li[2], and Xiaobing Zhang[1]

[1] School of Computer Engineering and Science, Shanghai University, Shanghai 200072, PR China LFYang98@163.com,{wzhang,fcj}@mail.shu.edu.cn
[2] Department of Computer Science, Guangxi Polytechnic, Chinajanelee@gxzjy.com

Abstract. NetSolve is a kind of grid middleware used for science computing. In this article, the architecture and operational principle of NetSolve system are firstly analyzed. Moreover a numerical experiment is given. NetSolve-parallel PCG for solving linear equation $Ax = b$ has been implemented in this experiment, the performance of it is also presented.

1 Introduction

As we know, two statements have been consistently true in the realm of computer science: (1) the need for computational power is always greater than what is available at any given point, and (2) to access our resources, we always want the simplest, yet most complete interface. With these two considerations in mind, researchers have directed considerable attention in recent years to the area of "grid computing". The NetSolve System, designed by the University of Tennessee and Oak Ridge National Laboratory, is a representative implementation of Grid. It is a software system founded on the concepts of remote procedure call (RPC) that allows users to access computational resources, such as hardware and software, distributed in both geography and ownership across the network[NW03].

This article focused on the implementation of NetSolve and its extensions in integrating PCG as its service for engineers to do Finite elements method analysis, and this implementation is based on the dell workstation cluster in Shanghai University.

2 The NetSolve System

There are three main components in the NetSolve system: agent, server, and client.

The agent maintains a database of NetSolve servers along with their capabilities and dynamic usage statistics for use in scheduling decisions. It attempts to find the server that will service the request, balance the load amongst its servers, and keep track of failed servers. The NetSolve server is a daemon process that awaits requests from client . It can run on single workstation, clusters of workstations, SMPs, or MPPs. One key component of the server is the ability to wrap software library routines into NetSolve services by using a Problems Definition File (PDF). The NetSolve client uses application programming interfaces (APIs) to send request to NetSolve system, along with the specific details required[CD97].

3 Preconditioned Conjugate Gradients

In this section we will discuss the solution of large sparse linear systems in the form of

$$Ax = b$$

where A is an $n \times n$, symmetric and positive definite matrix by the preconditioned conjugate gradient method (PCG)[AN96].

Table 1. Sequential PCG Algorithm

(1)% Initializations	(7)$d = dot(\tilde{r}, \tilde{r})$
(2)$a = c/dot(p, z)$	(8)if(sqrt(d)<tol) break
(3)$z = Ap$	(9)$\beta = d/c$
(4)$x = x + dot(a, p)$	(10)$p = \tilde{r} + dot(\beta, p)$
(5)$r = r - dot(a, p)$	(11)$c = d$
(6)$\tilde{r} = r \cdot M^{-1}$	(12)goto(1)

We can see, there are so many products of vector in Algorithm,which is show in Table 1. It is very important for us to implement NetSolve-parallelism.

4 Experiences

4.1 Parallel Platform

We had installed new release of NetSolve version-2.0 in our dell workstation cluster. This cluster has one four-processor node (Dell Power edge 6650) with 2.4GHz Intel Xeon chips (1MB cache), and four dual-processor nodes(Power edge 1600) with 2.0GHz Intel Xeon chips (512KB cache). The first one has 2048MB memory, the others 1024MB. These nodes are connected via a 100.0Mbps Ethernet. NS_server, NS_agent and NS_client have been installed on the first workstation, while only NS_server on the other four.

4.2 The Implementation of NetSolve-parallel PCG

As we knew, in NetSolve call, the routine service will be done in remote NetSolve server. The client will wait there until the routine returns. This calling mode is called Blocking Call. It may cost long time to wait. Another calling mode, Nonblocking Call, will not act like this. It returns control to program as soon as the request is sent out.

From here we can see that Nonblocking Call in NetSolve system allow users to achieve some NetSolve-parallelism. It can obviously improve the performance of user's code if the next few statements of the code have no relation to the result of NetSolve request.

In sequential algorithm,which be shown in Table 1, there are many products of vectors or matrix-vector products, most of the operations in getting product of vectors are independent. Also, the matrix-vector products can be decomposed into independent products of vectors. All these independent products of vectors are core operations in PCG. After the Nonblocking Call in NetSolve is discussed, we turn to PCG with NetSolve-parallelism, which can be designed as Figure 1.

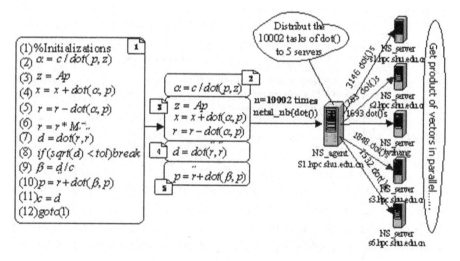

Fig. 1. parallel PCG with service of dot() in NetSolve

From Figure 1 we can see that the sequential algorithm of PCG is listed in frame 1. Picking all of the products from frame 1, and arranging them into other four frames based on their independency. In NetSolve system, we developed a product program, namely dot(vector v1, vector v2, int dim), and integrated it into NetSolve as a service.After this, the program of PCG can send the 10002 products of vectors to NetSolve in the form of nonblocking call (netsl_nb("dot()", v1, v2, dim)), And these jobs will been done in parallel.

4.3 Performance of NetSolve-Parallelism

In order to analyze the performance of NetSolve-parallelism, the problems solved by NetSolve should be designed as big as possible. So we integrate the whole PCG used to solve $Ax = b$ into NetSolve as a big service. In this experiment, A is a 2000 × 2000 sparse matrix, and 8% of its elements are nonzero. NetSolve uses the compressed Row Storage (CRS) to store sparse matrices [DGL89]. It will take 2000 × 2000 × (8%) × (8 + 2) + 2000 × 2 Byte, that is 3.204MB, to store sparse matrix A.

The problem $Ax = b$ was expanded to $AX = B$ in order to make the problems much bigger, Vector x and b have been extended into Matrx X and B with m cols.

We has done 9 nonblocking call for solving $AX = B$ (X and B all are matrices which col=1–9) in four different architectures:

(1) solving $AX = B$ in a single station directly;
(2) solving the equation in NetSolve with only one local NS_server;
(3) solving the equation in NetSolve with only a remote NS_server;
(4) solving the equation in NetSolve with all the five NS_servers in dell workstation cluster.

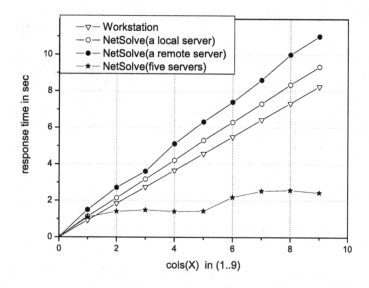

Fig. 2. 2000*2000 AX=B(cols(X)=cols(B)=(1..9)) equation solving

The response time in seconds was presented in Figure 2. In our experiment, the size of sparse matrix A is 3.204MB, and it will take about 0.3 sec to send

A from client to a remote NS_server in our cluster connecting via 100Mbps Ethernet card. A conclusion can be drawn: Good parallelism and speed ratio can be achieved if we solve many independency large problems with NetSolve Nonblocking call.

5 Conclusions and Future Work

In this paper, we discussed some of the issues involved in parallelize a PCG solver in an FEM program. PCG solvers are ideal for large problems and parallel computing is also ideal for large problems - and will, in the future, be the most cost effective way of analyzing structures with large computational models. An implementation of parallel PCG with nonblocking call in NetSolve had been presented, and the performance of it had been analyzed.

Future areas if research, that need to be addressed are improve the performance of the PCG algorithm on NetSolve [AGZ92]. And the most important job of ours is to integrate another bigger service, Finite Element Method, into NetSolve. At present, NetSolve system solves problem on the level of library function. The developing of a thinner granularity solution that accomplishes those functions with several servers will be our jobs if we want improve the performance of the PCG and FEM on NetSolve.

Acknowledgements. This paper was supported by "SEC E-Institute: Shanghai High Institutions Grid" project,the fourth key subject construction of Shanghai,and the Key Project of Shanghai Educational Committee (No.03AZ03).

References

[NW03]   NetSolve Web site http://icl.cs.utk.edu/netsolve/.
[CD97]   Casanova, H., Dongarra, J.: NetSolve: A network-enabled server for solving computational science problems. The International Journal of Supercomputer Applications and High Performance Computing, 11(3), 212-223, 1997.
[AN96]   Adams, L., Nazareth, J.: Linear and Nonlinear Conjugate Gradient-Related Methods,SIAM, 1996.
[DGL89] Duff, I., Grimes, R., Lewis, J.: Sparse matrix test problems, ACM Trans, Math. Soft., 15:1–14, 1989.
[AGZ92] Agarwal, R.C., Gustavson, F.G., Zubair, M.: A high performance algorithm using pre-processing for sparse matrix vector multiplication, Proceedings of Supercomputing '92, pp. 32-41.

FE Numerical Simulation of Temperature Distribution in CWR Process

Fuqiang Ying[1] and Jin Liu[2]

[1] School of Mechatronical Engineering and Automation, Shanghai University, Shanghai 200072, PRC motor@zjut.edu.cn
[2] liujin@public1.sta.net.cn

Abstract. A thermal-mechanical coupling finite element method (FEM) model is built up for the cross wedge rolling (CWR) process, based on the large elastic-plastic deformation FEM and the heat transfer theories. The influence of die parameters and rolling initial conditions on the temperature distribution and variation of the workpiece is described in the model. The overall temperature field of the rolled workpiece, the influence of the reduction rate of sectional area, the stretching angle, the forming angle, the initial rolling temperature and the temperature of both die and coral deformed zone of the workpiece are analyzed and discussed in detail. It is demonstrated that in the rolling process, the temperature of the workpiece in the deformed zone is affected mainly by the contact heat-conduction and plastic deformation heat. The simulating results coincide well with the experimental ones given by former researchers.

Keywords. Cross Wedge Rolling, Temperature Distribution, FEM

1 Introduction

The heat transfer in the cross wedge rolling process is a very complicated thermodynamic problem[Hu95]. During the deformation of the workpiece, its free surface exchanges heat with external environment in manner of convection and radiation, and meanwhile the contact surface conducts heat to the die. With further deformation, free surface reduces gradually while the contact surface increases constantly[GLP02]. Therefore, the heat dissipation condition of the workpiece changes. As a result the temperature of inner workpiece drops, various in different positions, due to the heat loss of the outer surface, and the temperature field changes eventually[LL99]. Consequently, most of the plasticity distortion energy of the workpiece is converted into the heat energy and the temperature of the workpiece rises, which also depends on the deformation distribution within workpiece. The interaction of the above-mentioned two factors makes it difficult to acquire the temperature field accurately with analytical method[SZH02], while the numerical method, especially finite element method, is feasible and effective.

2 Finite element analysis on the coupling of deformation and heat conduction

Assume the plasticity energy transformation in deformation to be the interior heat source, the workpiece to be isotropic in heat conduction. The energy balance equation is

$$kT_{ii} + \dot{q} - \rho c \dot{T} = 0 \tag{1}$$

To solve a temperature distribution problem is to solve the Eq. (1) under the boundary conditions and initial condition, which can be expressed equivalently as finding out the minimum value of the function I.

$$I = \frac{1}{2}\int_{v}[kT_{ii}T_{ii} - 2(\dot{q} - \rho c \dot{T})T]dV + \frac{1}{2}\int_{S_3} h(T - T_\infty)^2 dS - \int_{S_2} qT dS \tag{2}$$

Where kT_{ii} is heat conduction rate, k is the heat-conduction coefficient of the workpiece, $\rho c \dot{T}$ is internal energy rate, ρ is the density of the workpiece, c is the specific volume, $\dot{q}$ is the interior heat source rate, temperature T is a function of time t, h is the heat loss coefficient, T_∞ is the environment temperature.

According to the heat conduction variation theory, the temperature field obtained is the real solution of the question only if it satisfies initial condition formula, the boundary condition formula, and the variation of functional formula (2) equal to zero.

3 The emulation model

3.1 The geometrical parameters of the model

The emulation model of cross wedge rolling process is built up, with the main parameters: wedging angle $\alpha = 20° \sim 30°$, stretching angle $\beta = 5° \sim 7°$, reduction of area $\psi = 1 - (d/d_0)^2 = 30\% \sim 70\%$, the initial diameter of the blank $d_0 = 20 \sim 60mm$, and roll slenderness is 60mm, and the length of four stages including knifing zone, stretching zone, forming zone and sizing zone can be obtained by computation.

3.2 The definition of other relevant geometrical and physical parameters[SZH02]

The finite element model is composed of a pair of dies and a cylinder workpiece, the two dies are parallel and in relative movement with the velocity of

400 mm/s. The distance is the same at vertical and horizontal direction between dies and workpiece. Assume three kinds of diameter of rolled workpiece to be 20mm, 40mm and 60mm respectively and the length is 120mm. The number of elements in a same section is constant when mesh is plotted. The hexahedrons eight node is adopted, and the number of element is 4620, 2592 and 2100. And the workpiece of diameter 60mm is used to analyze the effect of other parameters on the cross wedge rolling process[DTL00].

3.3 The material model

The unit is the default of MSC.Superform[MSC02], the die is regarded as a rigid body (speed controlling), and the material of the workpiece is C_{45}. The model of the elastic-plastic material is as follows: Poisson ratio 0.3, the density 7. 85e-9, the conducting rate changing with the temperature and Youngs modulus, specific heat, and expansion coefficient also changing with the temperature.

3.4 The definition of contact

The workpiece is defined as deformable body, the upper and lower dies are the rigid body, and the frictional factor, the velocity of the die, and the initial and final positions are known.

3.5 Boundary conditions

Assume that the environment temperature is $20°C$, heat-dissipating coefficient of the surface is $20W/(m^2 \cdot ° C)$, heat-conduction coefficient by contact is $20KW/(m^2 \cdot ° C)$, and thermal conductivity is $0.17KW/(m^2 \cdot ° C)$.

4 Results and discussions

During metal forming, temperature varies as the workpiece deforming and the mechanical properties of the metal material vary obviously with the temperature. Because of the anisotropic of the workpiece, heat produced by the plastic deformation inside the workpiece is uneven either. If the temperature of the die is relatively low, the heat exchange on the interface of the work piece and the die will cause a dramatic decline on local temperature of the workpiece, at the same time the radiation and convection between the workpiece and surrounding environment will cause heat loss, therefore a great temperature gradient inside the work piece appears. The maldistribution of the temperature has a tremendous influence on the deformation of the material, and consequently influence the metal flow characteristic further.

4.1 The overall temperature field of the rolled workpiece

In deforming, because of the heat exchange between the workpiece and the external environment or the heat loss (such as convection, radiation and heat-conduction due to contact), and the uneven interior heat source caused by the uneven deformation of the workpiece, the inner temperature gradient may be sometimes very large. Assume the initial temperature is $950°C$, at the wedging zone, the temperature of deformed section is higher than that undeformed, and the temperature of both sides contacted with the die is lower than that below the die. It is because friction and plastic deformation produce heat, but the contact is the surface contact here and the temperature difference between the die and the workpiece is relatively large, then the temperature gradient and obvious heat exchange appears. In stretching, the temperature below the die is higher than that of both sides. At the end of the processing, the temperature gradient of the die and the workpiece is relatively small, the contact is line contact with large contact area, so the temperature in the middle of the workpiece is higher than both sides.

4.2 The influence of the die on the core temperature of deformed zone of the workpiece

In rolling, the temperature gradient of the workpiece and the die, environment temperature, workpiece itself exists. As the reduction of area increases, the core temperature of deformed zone of the workpiece rises obviously. This is because with deformation going on, the resistance increases, and the energy needed to resist the deformation increases, moreover heat caused by friction and plastic deformation increases as contact area increasing. So the core temperature rises as the temperature of deformed zone rises.

4.3 The influence of the initial rolling temperature on the core temperature of deformed zone of the workpiece

Under different initial rolling temperature, the core temperature of the deformed zone of the workpiece changes differently. As the initial rolling temperature varies from $900°C$ to $1100°C$, there is an obvious falling trend in the core temperature. When the initial rolling temperature is $900°C$, the core temperature of the cross section rises in general with the maximum amplitude up to $46.6°C(5.18\%)$, while descends in early process with the maximum amplitude up to $12°C(1.3\%)$. When the initial rolling temperature is $950°C$, the core temperature stays at about $950°C$, varying stably with a falling trend of $12°C(1.26\%)$ and rising trend of $4°C(0.4\%)$. When the initial rolling temperature is $1000°C$, the core temperature varies between $950°C$ and $1100°C$, descends in general, with a small average temperature and a greater amplitude of 1.8% than in initial temperature $950°C$. When the initial rolling temperature is $1100°C$, the core temperature descends in general with the maximum amplitude up to $32°C(2.9\%)$.

The following conclusions are obtained by analyzing the above:

(1)There is great plastic deformation and deformation energy at wedging zone and stretching zone, so the temperature rises due to the heat effect.

(2) Because of the higher temperature of the workpiece, heat exchange exists between the die and environment, there is no enough time to exchange heat with the outside at the beginning of simulation, but as the simulation going on, external temperature of the workpiece descends. So there is a temperature gradient inside and outside of the workpiece, the heat is transferred to the outside.

(3) When the initial rolling temperature is high (e.g.1100°C), the core temperature descends because the effect of heat transfer is larger than of heat produced by plastic deformation.

(4) When initial rolling temperature is low (900°C for instance), the surface temperature is larger than the core temperature because the deformation and friction heat is larger than heat transferred from the surface to the environment. And then heat transferred from the surface to the core, so the core temperature rises.

5 Conclusions

(1) A thermal-mechanical coupling FEM model of cross wedge rolling is put forward based on the large elastic-plastic deformation FEM and the heat transfer theories. The model can be used to analyze the influence of mould parameter and rolling initial conditions, etc. on the temperature distribution of the workpiece, reproduce accurately the rolling process, the temperature distribution and how the temperature varies.

(2) Simulate the temperature field with finite element method under different die parameter, initial rolling temperature, die temperature, and obtain the temperature distribution. It is shown that in rolling, temperature of the workpiece in the deformed zone is affected mainly by the contact heat-conduction and plastic deformation heat.

(3) The simulation results accords better with the past experimental results, which indicates that the simulation has high precision. Therefore the proposed method can be used to analyze the structure transform and property prediction of the rolled metals.

References

[Hu95]   monograph Hu Zhenghuan. Present state and perspectives of cross wedge rolling technology. Smithcraft ,5:25-27(1995)

[GLP02] G.Fang, L.P.Lei, P.Zen. Three-dimensional rigid-plastic finite element simulation for the two-roll cross wedge rolling process. Journal of Materials Processing Technology, 129:245-249(2002)

[LL99]   L.M.Galantucci, L.Tricarico. Thermo-mechanical simulation of a rolling process with an FEM approach, Journal of Materials Processing Technology ,92-93:494-501(1999)

[SZH02]   Su Xuedao, Zhang Kangsheng,Hu Zhenghuan,Yang Cuiping. Factors affecting parameters of force and energy in cross wedge rolling. Heavy Machine, 4:29-33(2002)

[DTL00]   Dong Y, Tagavi K A, Lovell M R, Deng Z. Analysis of stress in cross wedge rolling with application to failure. International Journal of Mechanical Science , 42:1233-1253(2000)

[MSC02]   MSC.Superform Version 2002,Users Guide, MSC.Software Corporation, South Coast Metro, USA

Contents Service Platforms:
Stage of Symmetrical and Active Services

Jie Yuan[1], Maogou Chen[1], Gengfeng Wu[1], James Zhang[2], Yu Liu[2], and Zhisong Chen[3]

[1] School of Computer Engineering and Science, Shanghai University, Shanghai, China jyuan@staff.shu.edu.cn
[2] Shanghai University-DingTech Software Co., Ltd, Shanghai, China
[3] Denso Corporation, Japan

Abstract The traditional information service model hinders the special service system from meeting the requirements in future. Therefore, a Symmetrical and Active Service Model (SASM) is proposed in the paper. As an example, Contents Service Platforms (CSP), which is based on the SASM and Grid technique, is presented to provide future car life with stage of symmetrical and active services. Two important abilities of the ClubSite that acts as the service information center in the CSP - "active discovery of potential demand" and "active promotion of service"- are described.
Keywords Information Grid, ITS, Service model, Active service

1 Introduction

The information grid is a kind of hinge which can integrates a great number of services into a special service system. However the roles of service providers and served objects are fixed, moreover the providers used to respond to requests passively, if the traditional service model is still used. As the field of grid application extends increasingly, it is necessary to construct a new service model, in order to improve the quality of service greatly.

We take the transportation modernization for example. Intelligence Transportation Systems (ITS) have become the important signs of entering information era for transportation. Vehicle is used as probe in ADVANCE [ITS03]. Honda Company in Japan puts the concept of "Probe Car" in practice firstly in October 2003 [Wad02]. Both research results are the breakthrough of the traditional service mode in ITS. Since October 2002 we have been undertaking a Car Life project, which includes three phases: the Systematization on Car Life Information (CLIS), research on the Contents Service Platforms (CSP), and the Abilities Supporting Active Service (ASAS). We have pointed out in the CLIS that the objects accepting service are also service providers. The concept of *Symmetrical and Active Service Model* (SASM) is based upon the results of the CLIS, proposed in the CSP research [YCW04], and is being improved in the ASAS.

2 Symmetrical and Active Service Model

In order to meet the three key requirements that future service systems face up: changing dynamically, two-way serving actively, and adjust the system automatically, the SASM was put forward [YCW04]. We briefed it as follows. If it is not pre-decided who will provide service or what service will be provided, the relation between the two nodes is defined as in dynamic relation. Two-way service relation is defined as if two nodes can provide services each other. Symmetrical service relation is defined as combination of both dynamic and two-way relations. If the service provider can actively find the served object and provide service to it, it is defined as active service. The model that can describe symmetrical service and active service is called the SASM.

The SASM has the following features: (1) Allowing the dynamic change of identification of service providers and service objects. (2) Allowing the dynamic insertion and modification of services. and (3) Facilitating service providers in their active search of service demanders, and vice versa.

Both passive service and active service are allowed in the SASM. In passive service, served objects search for the proper services; while in active service, service providers search for the possible served objects. In order to support the both effectively, the management of the service information must be considered in the SASM.

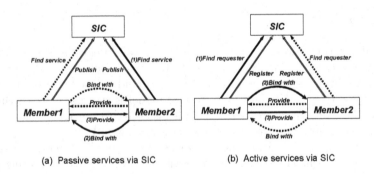

(a) Passive services via SIC   (b) Active services via SIC

Fig. 1. Services via SIC

It is a feasible way to increase the searching efficiency that a service information center (SIC) is set up in a service system. In the passive service, the SIC accepts and collects the service description information published by the service providers, and facilitates the served objects to search and get proper services. The numbered black lines in Fig.1(a) show the steps that *Member1* passively provides *Member2* with its service through the *SIC*. Similarly, *Member2* can also passively provide services to *Member1* (denoted by dotted lines). The numbered black lines in Fig.1(b) show the steps that *Member1* actively provides *Member2* with its service through the *SIC*, while *Member2* also can actively provides *Member1* with its services as well (denoted by dotted lines). From these two graphs, we can learn that the active service manner is different from the passive service manner both on the contents provided by the SIC and on the whole process. In a system that supports both passive services and active services, the SIC grasps the service description information and users' information simultaneously, so that it can respond to the users' search request of services or serviced objects.

3 Two Active Abilities in the SIC

Among the three features of the SASM, the first and second ones can be implemented directly using the Grid Service technologies, while the third one needs the support of new abilities in the SIC. The third feature means not only to actively seek served objects, but also to actively analyze out what is the potential demands of served objects. That a served object needs a service always originates from certain reasons, and these reasons are revealed in the object's basic attribution and instant status, more or less. The SIC can approach and analyze the served object's basic attribution and instant status to derive what kind of service is in need, and provides the information to service providers.

There are three steps for the SIC to discover the potential demands of served objects: (1)acquire potential status through capturing the action information of served objects. (2)reason the potential demands of served objects according to their potential status. (3)determine the served objects that need service with an utmost possibility.

The third feature of the SASM also means to actively extend the service scope to future served objects. Sales promotion is one of the most important measures to attract new served objects. It is necessary for a successful sales promotion that all the things such as promoted object, promotion message, and promotion method, are carefully prepared and very relevant. Owning plenty of information about services, service providers, served objects, promotion cases and service cases, the SIC can provide such information. Considering the integration and universality, and the special demands in the deduction process, the ontology [UG98] is used in conceptual design.

The process of ascertaining promotion object includes: acquiring the potential served objects and their requirements by using data mining technique, matching information based on semantic and vector model, and choosing promotion object according to the trusty degree and its trend. For generating the promotion message, Subjective Bayesian Probabilistic Inference is used. The process of promotion method reasoning is used mainly to determine concrete promotion method for special promotion object.

4 CSP: a Car-life-oriented Information Grid

As a kind of special service system, the ITS needs strong supports from high performance computing and distributed data management if it is expected to do the traffic dispatch with simulation technique and dynamical route selection with real-time traffic data. Therefore, a powerful ITS will be formed by combination of the SASM with Grid technique.

It is obvious that Grid technique is suitable for the implementation of the SASM. In an ITS which implements the SASM by using Grid technique, traffic dispatch system, parking inducement system, in-vehicle navigation system, and emergency assistance center etc may become a virtual organizations [FKT01]. There are great differences in computing capacity among these member nodes, but even for a vehicle equipped simply now no longer receives the information passively. The real-time information is collected actively by each vehicle through different kinds of sensors or by drivers directly, and is sent into shared database in the ITS. As owning a large amount of real-time traffic information and powerful computing capability, the ITS can provide more efficient services such as dynamic route selection and go-out scheme simulation. Such an ITS can be and should be called as *Transportation Information Grid*.

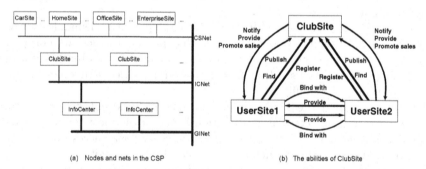

(a) Nodes and nets in the CSP   (b) The abilities of ClubSite

Fig. 2. Architecture of the CSP and abilities of the ClubSite

The CSP is based on the SASM and Grid technique to meet the requirements of future car life information services, and involves three kinds of nodes: the UserSite (US), the ClubSite (CS), and the InforCenter (IC), as shown in Fig.2(a) [YCW04]. In the CSP, the CS acts as a SIC, multiple USs are combined together through one CS to form a SASM. The CS has following functions (shown in Fig.2(b)):

- Register: Accepts the user registration from any US.

- Publish: Accepts the service or data units published by users.
- Find: Responds to requests for finding service, users, and data units.
- Notify: Actively notify the users after it discovers the potential demands.
- Provide: Provide service or data unit to user.
- Promote sales: Actively promote services.

The USs can also directly bind with and provide services to each other.
We developed a prototype of the CSP last year. Following cases can be demonstrated on the prototype. "Traffic Information Service" represents that the CSP supports dynamic service. "Tornado Warning" embodies how the CSP would provide effective service to insure the cars' safety in emergence. "Resources Information Adjusting" shows how IC does resources information adjusting among the CSs. We will further study the two active abilities. The other features of the SASM will be discussed also.

5 Acknowledgements

This paper is supported by "SEC E-Institute: Shanghai High Institutions Grid" project and the Fourth Key Subject Construction of Shanghai. The authors would like to extend their appreciation to Mr. Kenzo ITO and Mr. Akira Takahasi for their constant encouragement and insightful suggestions in the CSP project.

References

[ITS03]   U.S. Department of Transportation Intelligent Transportation Systems (ITS) Joint Program Office et al: Transportation Intelligent Transportation Systems (ITS) Projects Book (2003)

[Wad02]   Wada K.: Research, development and field testing of the probe car information system(ii). In: 9th World Congress on Intelligent Transport Systems (2002)

[YCW04]   Yuan Jie, Chen Maogou, Wu Gengfeng et.al: Contents service platforms: serving for future car life. In: 11th World Congress on ITS Nagoya Aichi2004. Japan, (2004)

[UG98]   Uschold M., Gruninder M: Ontologies: principles methods and application. Knowledge Engineering Review, 11, 93–155 (1998)

[FKT01]   Foster I., Kesselman C., Tuecke S.: The Anatomy of the Grid: enabling scalable virtual organizations. International J. Supercomputer Applications, 15, (2001)

An Efficient Planning System for Service Composition

Jianhong Zhang[1], Shensheng Zhang, and Jian Cao

CIT Lab, Computer Science Department, Shanghai Jiaotong University, Shanghai 200030, P.R. China zhang-jh@cs.sjtu.edu.cn

Abstract. This paper presents a planning-based framework for services composition. Workflow technology is used to abstract business process. Then a hybrid algorithm based on Hierarchical Task Network (HTN) planning and partial order planning (POP) is developed to do service planning through which each task of workflow becomes concrete. Furthermore a semantic type-matching algorithm is introduced to matching actions of plan to suitable operations of web services. Compared with other service composition methods, this system can deal with complex business process efficiently, possesses good executable ability and can satisfy the need of dynamic business process application.

Key words: Planning, Hierarchical task network, Workflow, Service composition

1 Introduction

The wide availability and standardization of Web services make it possible to compose basic Web services into composite services that provide more sophisticated functionality and create add-on values. However, business process, organization policies and applications change rapidly. It's necessary to construct an open, flexible and efficient system for service composition.

Service composition can be achieved by traditional workflow. There are also many service composition standards built such as BPEL4WS[BPEL], DAML-S[Ankolenkar02], WSFL, etc. But all the technology assumes that business process is static and predefined and needs vast programming work or human intervention. Now many researchers use AI technology to realize service composition. [cIlraith02] presents a way to compose e-Services based on planning under uncertainty and constraint satisfaction techniques. In [Kim], composition of e-Services is addressed by using the Situation Calculus-based programming language CONGOLOG and the client's needs are specified through suitable forms of constraints. [Evren] uses HTN planning system SHOP2 to realize service composition. But, these approaches assume to interact with services that are described in a standard and possibly formal manner and can't tackle heterogeneity of the service composition domain. This paper provides a domain independent service composition framework by combining workflow technology and planning technology in AI. Using the idea from hierarchical task network (HTN) planning, a partial-order planning (POP) based hybrid algorithm is developed. Furthermore, a type-structure matching algorithm is introduced to search appropriate service for each action of plan and finally an executable service dependent plan (SDP) is constructed.

2   Workflow and planning based service composition

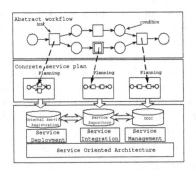

Fig. 1. Planning based Service Composition Framework

Through many years' efforts, workflow technology has improved in many aspects such as flexibility, robustness, etc. In this framework, workflow technology is used to represent the main steps of complex business process as abstract workflow(AW). Then each task of the AW is considered as a planning problem to search the state space of problem domain to find an action path, which changes the environment from initial states to goal states. When doing service composition, if there is no existing plan satisfying user's requirement, the planning algorithm is invoked to generate a plan composed of atomic action. Then type-matching algorithm is invoked to find appropriate service for each atomic action of the plan and an executable SDP is generated. The specification of services is stored in internal/internet UDDI. The definitions of atomic actions and plans are stored in the plan library. The service planning and invocation is based on the Service Oriented Architecture (SOA) including service deployment, service integration and service management. The description of service requirement, the planning algorithm and the type-matching algorithm are all based on the same domain term ontology.

Fig.2 represents the planning model. Action is the basic unit of the planning model. Atomic action is the basic reuse unit, which will be assigned with some

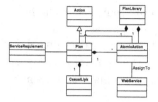

Fig. 2. Service planning model diagram

web service for execution. Plan action is composed of atomic actions to fulfill more complex service requirement.

Task of AW can be represent as (*CurrentInfor, Goals, Constraints*),where *CurrentInfor* denotes the basic information necessary for achieving the objective, *Goals* denotes states which the task wants to achieve, *Constraints* denotes the constraint set of service requirement. Following is an example of service requirement described in XML.

<ServiceRequirements Domain="Business_Process">
<SubObjectives>
<SubObjective Name="Plan_Travel">
<CurrentInfor>
<Item name="Departure_Place" type="LOCATION"
value="@Departure_Place"/>
...
<Goal>
<Item name="Flight" type="STRING" value=" "/>
...
<Constraints>
<Constraint Name="@Departure_Place" type="variable" Value="Beijing"/>
...

An action is specified in terms of the preconditions that must hold before it ca be executed, and the effects that ensue when it execute.

3 Hybrid planning algorithm for workflow task

Compared with other planning algorithm, HTN planning is very efficient. We take a hybrid approach in which action decompositions are used as plan refinements in partial-order planning, in addition to the standard operations of establishing an open condition and resolving conflicts by adding ordering constraints. This algorithm can decide whether a solution exists and can provide interface for user to complement necessary information, has high efficiency and avoid recursive problem.

When executing this planning algorithm, four rules should be obeyed:

Rule 1: A set of **actions** make up a plan. The "empty" plan contains just the *Start* and *Finish* actions.

Rule 2: Each **ordering constraint** is of the form $A \prec B$, which means that action A must be executed before action B.

Rule 3: A **causal link** between two actions A and B in the plan is written as $A \xrightarrow{P} B$, which asserts that p is an effect of the A and a precondition of B.It also asserts that P must remain true from the time of action A to the time of action B.

Rule 4: A set of **open preconditions**. A precondition is open if it is not achieved by some action in the plan.

Algorithm:

1. The initial plan contains *Start* and *Finish*, the ordering constraint *Start* $\prec$ *Finish*, no casual links, and all the preconditions in *Finish* are open preconditions.

2. The successor function arbitrarily picks one open precondition p on an action B, and generates a successor plan for ever possible consistent way of choosing an action A that achieve p. consistency is enforced as follows:

• The causal link $A \overrightarrow{p} B$ and the ordering constrain $A \prec B$ are added to the plan. Action A may be an existing action in the plan or a new one. If it is new, add it to the plan and also add Start A and A Finish.

• We resolve conflicts between the new causal link and all existing actions, and between the action A(if it is new) and all existing causal links. A conflict between $A \overrightarrow{p} B$ and C is resolved by making C occur at some time outside the protection interval, either by adding $B \overrightarrow{p} C$ or $C \overrightarrow{p} A$. We add successor states for either or both if they result in consistent plan.

3. The goal test checks the generated plan. Because only consistent plans are generated, the goal test just needs to check that there are no open preconditions.

4. Select some complicated action a' in Pl(the plan generated above)and for any Decomposition(a, d) method from the plan library such that a and a' unify with substitution Θ , we replace a' with $d' = SUBST(\Theta, d)$.

• First, the action a' is removed from Pl. Then, for each step s in the decomposition d, we need to choose an action to fill the role of s and add it to the plan.

• Hook up the ordering constraints for d in the original plan to the steps in d'.

• Hook up causal links. If $B \overrightarrow{p} a'$ was a causal link in the original plan, re-place it by a set of causal links from B to all the steps in d with precondition p that were supplied by the Start step in the decomposition d.

5. Run step4 repeatedly until there is no plan action in the plan.

6. Call type-matching algorithm to generate SDP.

4 Service matching algorithm

This paper developed a type-matching approach similar to[Carman03]. The definition of action can map to operation of service. For operation, its output message type, in some way describes the effects of action and its input message type can be seen as describing the preconditions of actions. This paper takes service matching as deciding whether one type structure is a more generic version of another type structure under a certain set of mapping. We use the semantic relations in WordNet[Fellbaum98] as the basis for our algorithm. For example, to match the action "BookHotel" with the service "ReserveHotel", there is:

$eaqual(BookHotel.Date,$
$ReserveHotel.Date)) \bigwedge generation(BookHotel.Location,$
$ReserveHotel.Address) \bigwedge generation(BookHotel.Room, ReserveHotel.$
$ResentedRoom) \Rightarrow generation(BookHotel, ReserveHotel).$ We can say they are matched.

This paper presents a workflow and planning based framework for services composition. A hybrid planning algorithm and semantic type-matching algorithm is developed to do service planning and matching efficiently. Compared with other service composition methods, this system can deal with complex business process efficiently and possesses good manageability.

References

[BPEL] Business Process Execution Language for Web Services, Version 1.1 ftp: //www.software.ibm.com/software/developer/library/ws-bpel11.pdf

[Ankolenkar02] Ankolenkar A., Burstein M., Hobbs J. R., Lassila O., Martin D. L., McDermott D.,McIlraith S. A., Narayanan S., Paolucci M., Payne T. R. and Sycara K.: DAML-S: Web Service Description for the Semantic Web. The First International Semantic Web Conference (ISWC), Sardinia (Italy), June, 2002.

[Arkin] Arkin   A.   :   Business   Process   Modeling   Language. http://www.bpmi.org/specifications.asp

[cIlraith02] S. McIlraith and T. Son. :Adapting golog for composition of semantic web services. In Proc. of KR02, 2002.

[Kim] Jihie Kim and Yolanda Gil:Towards Interactive Composition of Semantic Web Services.

[Fellbaum98] C. Fellbaum, editor. WordNet: An Electronic Lexical Database. The MIT Press, 1998.

[Evren] Evren Sirin, Bijan Parsia, Dan Wu, James Hendler, and Dana Nau. :HTN planning for web service composition using SHOP2. In Submitted to Journal of Web Semantics.

[Carman03] Carman, M.; Serafini, L.; and Traverso, P. :Web service composition as planning. In Workshop on Planning for Web Services, 13th International Conference on Automated Planning and Scheduling (ICAPS 2003).

A Sort-last Parallel Volume Rendering Based on Commodity PC Clusters

Jiawan Zhang[1], Zhou Jin[1], Jizhou Sun[1], Jiening Wang[2], and Yi Zhang[1], Qianqian Han[1]

[1] IBM Computer Technology Center, Department of Computer Science, School of Electronic and Information Engineering, Tianjin 300072, China
jwzhang@tju.edu.cn
[2] Civil Aviation University of China, Dongli District, Tianjin, China
jieningwang@cauc.edu.cn

Abstract. A sort-last parallel volume Splatting algorithm based on commodity PC clusters, which integrates hybrid data-space and image-space partitioning schemes, is proposed in this paper. Under the sort-last strategy, sorting is deferred until the end of the rendering pipeline. The approach also adopts a divide-and-conquer strategy and an object-order data decomposition scheme to distribute the volume among nodes in the PC clusters. The proposed method has excellent features including high scalability, predictable communication patterns and load balance in the rendering of large scale volume data sets on commercial PC clusters which have no high-end graphics adaptors equipped.

Key words: Sort-last, volume rendering, Splatting, cluster

1 Introduction

Direct volume rendering[LV1] is a useful technique to visualize the internal parts of three-dimensional data sets in applications such as medicine, computational fluid dynamics and finite element analysis. However, volume rendering is computational intensive. As the development of visualization techniques and the rapid progress in the applications, the scale of data sets becomes even larger. In general, medical diagnoses require both interactive and high-resolution rendering: rendering of 512^3 voxel volumes at least ten frames per second ($\geq$ 10fps). There exists an obvious contradiction between the real-time interaction requirements and the low rendering speed. When the scale of the data set becomes large or very large (such as the data set larger than 1024^3), it becomes impossible for a single PC or workstation to merely hold the data set into memory. One way to meet these requirements is to parallelize serial

volume rendering methods on parallel supercomputers and high-end workstations [AN1], which have been widely used for large-scale simulation, modeling
and visualization. In recent years, as the rapid growing of computational ability and increasing favorable price, PC clusters are now becoming widely used
in many scientific computation areas. So, we base our research work on PC
clusters.

A sort-last [MO1]parallel volume Splatting algorithm[WT1] based on commodity PC clusters is proposed in this paper. The main idea here is to take
full advantage of the PC clusters which have no high-end graphics adaptors
such as those with hardware texture mapping functions.

2 Sort-last Parallel Volume Rendering Sorting

Because the rapid growing of computational ability, PC-clusters are now becoming widely used in many scientific computation areas. In this Section, the
main ideas behind our sort-last parallel volume rendering method are discussed, and the rendering results are given out to prove the proposed method.

Data Partitioning and Distribution Strategy

Solving volume rendering of large scale data sets is the main target of our
research. In a parallel computing environment, distributing a large amount of
data to a large number of processors can take a tremendous amount of time.
Pure image-space partitioning scheme is discarded because one processor can
not hold the entire volume into memory. In our method, the cost of distributing data is reduced by loading data before rendering step, and therefore the
performance study could focus on the subsequent rendering steps, which are
performed repeatedly for users to explore the data.

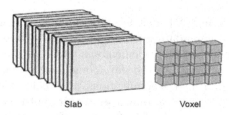

Slab   Voxel

Fig. 1. Data Partitioning the jth voxel block of the ith slab is assigned to $\lfloor j\text{-}i \rfloor^{th}$
nodes of the PC-clusters

As demonstrated in Fig. 1, we use data-space partitioning method to distribute the volume among nodes in the PC-clusters. If there are N nodes in the
PC-clusters, the volume is first divided into N parallel slabs. Then each slab

is subdivided into N voxel blocks. The size of the voxel block can be different; however, all the slabs must be subdivided in the same way to ensure that the corresponding voxel blocks along the slab perpendicular. According to the partitioning strategy, the volume is totally divided into N^2 voxel blocks.

Sort-last Strategy

The rendering of each voxel which is totally independent of other voxel blocks produces image-space ray segments by using Splatting rendering method. Using asynchronous communication, the transmission of ray segments are overlapped with rendering operation. In other words, these ray segments are delivered to processors responsible for compositing them immediately when the computing of jth voxel is finished. Our method (see Fig. 2) assigns each processor arbitrary subset of the volume data. Each render computes voxel values for its subset, no matter where they fall in the image screen. The advantage is that renderers operate independently until the compositing stage, the very end of the rendering pipeline. When ray segments arrive at a processor, they

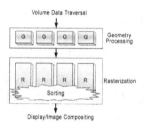

Fig. 2. Sort-last rendering

are sorted into the corresponding lists according to the depth order. We use the spatial partitioning tree to guide the rendering encourages premerging of ray segments, which helps keep the lists short and thus both the sorting cost and runtime memory consumption low.

Image Compositing Strategy

When all the nodes finish their work, the image compositing process begins. We use *Over* operator in the compositing process to obtain the corresponding part of screen under the binary-swap compositing strategy[MA1]. Our method could keep all processors busy during the whole course of the compositing. In Fig. 3, when all processors finish local resampling, each processor holds a partial image. Then each partial image is subdivided into two half-images by splitting along the X-axis. One processor keeps only the left half-image and sends its right half-image to its immediate-right sibling, the Processor which pair up with it. Both processors then composite the half image they keep

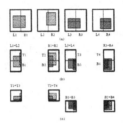

Fig. 3. Binary-swap compositing

with the half image they receive. After the first stage, each processor holds a partial image that is half the size of the original one. In the next stage, Processors alternate the image subdivision direction. This time it keeps the upper half-image and sends the lower half-image to its second-immediate-right sibling. After this stage, each processor holds only one-fourth of the original image. Then at last, the main node is responsible for the making up the final rendering image.

3 Experimental Results

Fig. 4 shows the rendered images in our experiments. These volumes are created by X-ray, CT and MR scan. We used IBM eServer 1350 cluster with 16 computing nodes in the experiments. The software platform is Linux (Red Hat 7.3),and MPI is equipped as parallel programming interfaces. Computing nodes connect one another by two independent intercommunication switch hubs (1000M, 16 ports) in TCP/IP Ethernet. Experiments show that high rendering speed and speedup can be obtained by adding nodes to the computation, which is shown in the Fig. 5.

4 Conclusion

Our sort-last parallel volume Splatting algorithm integrates hybrid data-space and image-space partitioning schemes, and adopts a divide-and-conquer strategy and an object-order data decomposition scheme. The method has excellent features including high scalability, predictable communication patterns and load balance in the rendering of large scale volume data sets on commercial PC clusters which have no high-end graphics adaptors equipped.

5 Acknowledgement

This paper is part work of Project - Research on Key Techniques and Supported Architectures for Real Time Volume Rendering, which is supported by the Natural Science Foundation of China(NSFC) under Grant No. 60373061.

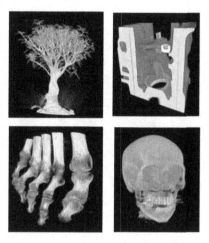

Fig. 4. Experiment Result

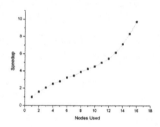

Fig. 5. The average speedup

References

[AN1]   Antonio Garcia, Han-wei Shen: An interleaved parallel volume renderer with PC-Clusters. In: Proceedings of 2002 Eurographic Workshop on Parallel Graphics and Visualization, 51-59 (2002)

[MA1]   Kwan-Liu Ma Painter, J.S. Hansen, C.D.: Parallel volume rendering using binary-swap compositing. IEEE Computer Graphics & Applications. (1994)

[LV1]   Levoy M. : Display of surfaces from volume data. IEEE Computer Graphics & Applications. **8**, 29–37 (1988)

[MO1]   Molnar, S. Cox, M. Ellsworth, D.: A sorting classification of parallel rendering. IEEE Computer Graphics & Applications. **14** (1994)

[WT1]   Westover L.: Footprint evaluation for volume rendering. Computer Graphics. **24**, 367–376 (1990)

Parallel Computation of Meshfree Methods for Extremely Large Deformation Analysis

Jifa Zhang and Yao Zheng

Center for Engineering and Scientific Computation, and College of Computer Science, Zhejiang University, Hangzhou, Zhejiang 310027, P. R. China
jifa_zhang@yahoo.com.cn; yao.zheng@zju.edu.cn

Abstract. Due to the heavier computation requirement than other competitive techniques and the essence of applications that are usually highly complex and computationally intensive, parallel computing is especially attractive for these meshfree methods. We only focus on the Reproducing Kernel Particle Method (RKPM), one of the meshfree methods for large strain elasto-plastic analysis of solid and structures, in considering with its ability to accurately model extremely large deformations without mesh distortion problems, and its ease of adaptive modeling by simply changing particle definitions for desired refinement regions. The parallel procedure primarily consists of a mesh partitioning pre-analysis phase, and a parallel analysis phase that includes explicit message passing among partitions on individual processors. With redefinition techniques applied to the shared zones of different geometrical parts, the graph-based procedure Metis, which is quite popular for mesh-based analysis, is used for partitioning in this meshfree analysis. Parallel simulations have been conducted on an SGI Onyx3900 supercomputer with MPI message passing statements. The effectiveness and performance with different partitions has then been compared, and a comparison of the meshfree method with finite element methods is also presented.

Key words: Meshfree Methods; Parallel Computation

1 Introduction

A variety of new meshfree modeling methods have recently emerged. These methods may be attractive for certain applications, since they possess various characteristics that overcome some of the shortcomings of more traditional methods. On the other hand, parallel computing is especially attractive for modern meshfree methods, since they usually produce heavier computational loads than other competitive techniques. In addition, the applications, where the new meshfree methods can be advantageous, are usually highly complex and thus computationally intensive.

The present work focuses on the Reproducing Kernel Particle Method [LJL95, LLB97, CPW96, CPC97] (RKPM) for elasto-plastic analysis of geotechnical problems. Several distinct advantages of RKPM include its ability to accurately model extremely large deformations without mesh distortion problems, and its ease to adaptive modelling by simply changing particle definitions for desired refinement regions. For effective parallel computing, it is critical to balance computational loads among processors while minimizing interprocessor communication. Therefore, separate pre-analysis software has been created to partition any general unstructured RKPM model.

In this paper, aspects of load distribution and message passing minimization are outlined, and data structures for processing on large scalable computing platforms are described. After a brief review of RKPM, the application of RKPM in geotechnical problems and corresponding comparison with FEM are addressed. Finally the difficulties in parallel implementation are also discussed.

2 Meshfree Discretization with RKPM

Spatially, RKPM formulations use a kernel approximation to the displacements $u(x)$ as

$$u^a(x) = \int_\Omega \bar{\Phi}_a(x; x - s) u(s) \, \mathrm{d}s \tag{1}$$

where $u^a(x)$ is called as the reproduced function of $u(x)$, and $\bar{\Phi}_a(x; x - s)$ the modified kernel function, that is expressed as

$$\bar{\Phi}_a(x; x - s) = C(x; x - s) \Phi_a(x - s) \tag{2}$$

$$\Phi_a(x - s) = a^{-1}\Phi((x - s)/a) \tag{3}$$

where a is the dilation parameter of kernel function $\Phi_a(x - s)$, and $C(x; x - s)$ is called as the correction function, which is constructed to avoid the difficulties resulting from finite domain effects, and to minimize the amplitude and phase error.

Similar to those in finite element methods, by using the expressions obtained above, we can discretize the weak forms of the small strain elasto-plastic problems

$$\int_V \delta\varepsilon_{ij}^{\mathrm{T}}\sigma_{ij}\mathrm{d}V - \int_V \delta u_i^{\mathrm{T}} f_i\mathrm{d}V - \int_{S_\sigma} \delta u_i^{\mathrm{T}} t_i\mathrm{d}S = 0 \tag{4}$$

to the following equations:

$$\Delta\Phi = \mathsf{K}_T\mathbf{G} - \int_\Omega \mathsf{P}^{\mathrm{T}}\Delta\mathbf{b}\mathrm{d}\Omega - \Delta\mathbf{f} \tag{5}$$

Based on this equation, the numerical procedure, such as Newton-Raphson Method and Tangential Stiffness Method (sometimes named as Generalized Newton-Raphson Method), can be directly applied.

3 Parallel Implementation and Numerical Examples

For the public availability, here we use graph-based partitioning code Metis [KK95] as the partitioner. The goal of a graph-based partitioner is to distribute the vertices so that each partition has equal amounts of vertex weight, while minimizing the amount of edge weight connecting partitions on different processors. The objective of the partitioning software is to provide partitions of nearly equal computational effort while also generally minimizing the size of the partition interfaces.

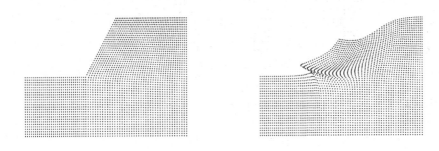

Fig. 1. Meshfree discretization (left) and computational result (right)

As an illustrative case, we study a plane strain problem of soil slope stability in geotechnical engineering. The geometrical model of the slope is shown in Fig. 1 with a slope of 5/3 and is discretized with 3029 particles/nodes (the left of Fig. 1). Under self-gravity, the slide will happen while the friction and cohesive strength of soil decreases.

Integration points in RKPM, different from that in FEM, are distributed to processors and nodes are shared by integration points on different processors (as shown in Fig. 2). The amount of computations will vary among integration points, since each may be related to a different Number of Particles/nodes (NP). RKPM integration points typically contribute to many more nodes than those of similar finite element models do. In the present RKPM implementation, vertex weighting is applied according to the number of nodes to which an integration point contributes. All graph edge weights are defined to be the same. The quality of a RKPM partitioning is thus defined by the balance of vertex weights and by the number of nodes shared by integration points on different processors. Thus, RKPM graphs can be very large with many edges. In the given example shown in Fig. 2, the graph has an average of 83 edges per vertex and the graph file requires space of more than 50 Mbytes. Using Metis, a typical partitioning could be performed for this model within about 1.5 CPU minutes on an SGI Onyx3900 processor.

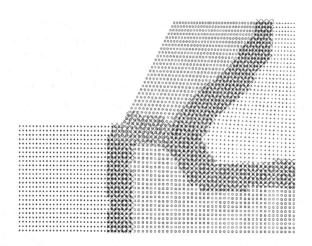

Fig. 2. 4 Partitions of the geometry configuration with shared nodes shown in shadow

Although the graph-based partitions, such as Metis, were quite good and the amounts of CPU time were reasonable for the 2D problem illustrated here, this approach may be unfeasible for parallel meshfree computing of large 3D models. In order to show this, we extrude the 2D geometrical model (Fig. 1 left) to a 3D model and discretized the model with 31036 nodes and 331278 nodes, respectively. For the case with 31036 nodes, the partitioning still works but cost more than 50 CPU minutes and the graph file is of about 600 Mbytes. The corresponding speedup factor and efficiency comparing to those of the 2D problem are shown in Fig. 3. However, for the case with 331278 nodes, the computing breaks off during the partitioning stage. Further investigation should be carried out towards this end.

The numerical computation shows that, while the deformation is large enough, with the same discretization and allowable errors, the FEM would result in divergence while RKPM resulting in convergence. As to the situation of extremely large deformation, such as the case of slope sliding we are discussing here, the FEM cannot tackle the problem due to the non-positiveness of the Jacobian, which is caused by the severe mesh distortion. Therefore, for extremely large deformation, the meshfree method RKPM is more appropriate. For the effective parallel computing with large amount of nodes, graph-based partitioners do not works well for RKPM while they are widely used in FEM analysis. In order to avoid the creation of large dense graphs by graph-based partitioners during the partitioning stage of parallel computation, geometric-based partitioners seem to be natural choices for RKPM. These approaches do

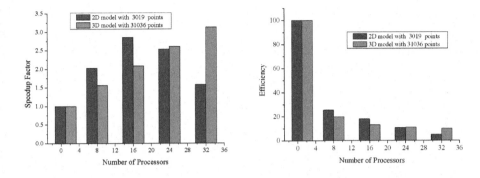

Fig. 3. The speedup (left) factor and efficiency (right) for the two models

not need a graph, as they only require coordinates of the integration points. The corresponding partitions are created by grouping points according to their spatial proximity.

References

[LJL95] Liu, W.K., Jun, S., Li, S., Adee, J., Belytschko, T.: Reproducing kernel particle methods for structural dynamics. Int. J. Numer. Methods Eng. **38** (1995) 1655–1679.

[LLB97] Liu, W.K., Li, S., Belytschko, T.: Moving least square reproducing kernel methods: (I) methodology and convergence. Comp. Methods Appl. Mech. Eng. **143** (1997) 113–154.

[CPW96] Chen, J.S., Pan, C., Wu, C.T., Liu, W.K.: Reproducing kernel particle methods for large deformation analysis of non-linear structures. Comp. Methods Appl. Mech. Eng. **139** (1996) 195–227.

[CPC97] Chen, J.S., Pan, C., Wu, C.T.: Large deformation analysis of rubber based on a reproducing kernel particle method. Comp. Mech. **19** (1997) 211–227.

[KK95] Karypis, G., Kumar, V.: A fast and high quality multilevel scheme for partitioning irregular graphs. Technical Report TR 95-035, Department of Computer Science, University of Minnesota (1995).

Solving PDEs using PETSc in NetSolve

Xiaobin Zhang, Wu Zhang, Guoyong Mao, and Linfeng Yang

School of Computer Engineering and Science, Shanghai University, Shanghai 200072, P.R. China zxb0412@hotmail.com

Abstract. A new way of combine the widely used PETSc math library and NetSolve (gird computing middleware) is explored in this paper. Using PDF file of NetSolve, applications using PETSc libraries can be called by NetSolve. However, the disadvantage of NetSolve Server causes the MPI's programs can not make good use of the Server's resource, which will lead to performance degradation of Server. A middle layer is added between the daemon module of the Server and the application module, which can be used to estimate the demand of application module. Based on the estimation, realtime configuration file can be created to enhance the efficiency of the Server.
Key words: PETSc, NetSolve, Grid Computing

1 Introduction

In last two decades, the rapid growth in the supply of basic computing resources, like PETSc[1](Portable, Extensible Toolkit for Scientific Computing), and advance in hardware, networking infrastructure and algorithms, let us successfully attack a wide range of computationally intensive problems using network scientific computing.

Grid computing middleware has supplied researchers with a better solution to implement modern computing technology. Netsolve[2] is specially designed to build a flexible, powerful, and easy computing environment. Based on Net-Solve, Moore has succeeded in implementing remote numerical computing education platform, and researches surrounding NetSolve are on the rise[3].

Additionally, the high-performance computing community has explored a variety of approaches for implementing parallel scientific applications[4].For large-scale PDE-based simulations, using existing math library - PETSc is a good choice.

It is of great significance to integrate PETSc into NetSolve, because existing good math library can be employed to speed up the developing of high efficiency programs, and the middleware provides a universal calling interface to directly use the computing resources. In this article, solving of PDEs problem is successfully implemented using PETSc, and this solution is integrated into an internal service of NetSolve.

The following paper can be divided into 5 parts: Section 2, grid computing middleware; Section 3, numerical computing library; Section 4, integrate PETSc into NetSolve; Section 5, testing platform and performance analysis; Section 6, conclusion.

2 Computing Middleware of Web Service

2.1 NetSolve and scientific computing environments (SCEs)

The NetSolve acts as glue layer that brings the application or user together with the hardware and/or software it needs to complete useful tasks.

The NetSolve can be divided three layers: Client(API), agent and server. The NetSolve client library is linked in with the user's application. Through Net-Solve's API, client-users gain access to aggregate resources without the users needing to know anything about computer networking or distributed computing.

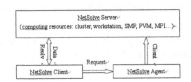

Fig. 1. The Architecture of NetSolve

Figure 1 helps to show what the programming code would look like before and after the NetSolve API has been integrated[2].

The NetSolve agent represents the gateway to the NetSolve system. The Net-Solve server is the computational backbone of the system. It is a deamon process that awaits client requests. The server can run on single worksta-tions, clusters of workstations, symmetric multi-processors or machines with massively parallel processors.

A key component of the NetSolve server problem is a source code generator which parses a NetSolve problem description file (PDF). This PDF contains information that allows the NetSolve system creates new modules and incor-porates new functionalities[2]. In essence, the PDF defines a wrapper that NetSolve uses to call the function being incorporated.

2.2 Expanding the Server Capabilities

NetSolve is highly scalable. PDF is the mechanism through which NetSolve enables services for the user. First, user must write PDF files if he want to add new functions to NetSolve. In addition, user should modify the server configuration file to enable or disable specific functionality to customize their server. At last, users recompile the server.

This feature of NetSolve makes it easy to incorporate PETSc applications.

2.3 NetSolve Pitfall

NetSolve supports programs of MPI type, which is popular in today's parallel programs. Therefore, it is possible for NetSolve to provide services of large scale computation. Several schemes about MPI applications are provided by NetSolve installation software packages. In PDF, section @PARALLEL is used to point out that MPI is needed in scheme, and users can call MPI instances easily in section @CODE. After the NetSolve server is reconstructed, with the help of MPI, this scheme can use multiple processors to solve problems.

However, only one MPI configuration file is allowed to use in one NetSolve server[2]. That is, many MPI schemes will use identical configuration, which can not be adjusted. For NetSolve server that can provide many services, this configuration will cause heavy performance degradation. For example, the job in Server is small, yet it is dispatched among multiple machines.

3 Problems, Algorithm and PETSc

3.1 Bratu's problem and Newton-Krylov-Schwarz

The Bratu problem is a classic PDE, the corresponding differential equation is[4]:

$$\nabla^2 u - \lambda e^u = 0, \tag{1}$$

In this equation, λ is a constant, u is the variable to be computed whose value equals 0 at the boundary. The solution to this problem exists in PETSc sample examples. Our test is based on this example.

We use 5-point stencil, a standardized approach[4], to discretize this problem, that is:

$$f(u) = 4u_{i,j} - u_{i-1,j} - u_{i+1,j} - u_{i,j-1} - u_{i,j+1} - h^2 \lambda e^{u_{i,j}}, \tag{2}$$

where f is the vector function of nonlinear residuals of the vector of discrete unknowns u, defined at interior and boundary grid point: $u_{i,j} \approx u(x_i, y_j); x_i \equiv ih, i = 0, 1, \cdots, n; y_j \equiv jh, j = 0, 1, \cdots, n; h \equiv 1/n$.

Newton iterative method can be used to solve Equation (2), but the precision may not be high due to an incomplete convergence employing the true Jacobian matrix or using an inexact or a 'lagged' Jacobian[4][5]. In Newton-Krylov method, Krylov method is used to solve the Newton correction equations. While in the Krylov method, matrices are partitioned into submatrices, the product of submatrices and vectors is computed and stored in each iteration. When the original matrix becomes Jocabian matrix after PDEs are discretized, the overhead of computation and communication for each matrix-vector product are similar.

3.2 PETSc Implementation

If Newton-Krylov method is used to solve PDEs using PETSc, solver and user's code will deal with data in turn in one iteration. In general, five modules should be written by user: master process (or daemon process) module, initialization module, nonlinear function computing module, Jacobian matrix computing module and post processing module.

Command line arguments can be used to select different way of solution and specifically set up parameters (i.e. times of iteration, convergence condition) when calling PETSc based applications.

4 Improve flexibility of PETSc applications

However, from section 2 we can see that the support of NetSolve to MPI is not very good. In order to overcome this disadvantage, we changed the sharing of one configuration file into the realtime creation of different configuration file in accord with different tasks.

A middle layer is put between NetSolve Server and solution modules of MPI based applications, the layer is to analyze the parameters send by client, estimates the amount of computation, and give out the demand of resources. Based on this estimation, realtime MPI configuration file is created. In our test, different numbers of processors are assigned only based on the size of matrices.

5 Testing environment and performance analysis

All performance data reported in this study are measured on ZQ2000 cluster computer of Shanghai University.

If there's no middle layer between Server and test application, when the number of subintervals is fixed, the execution time will decrease slack with the increasing of number of processors for 256×256 matrix. The research of Hayder et.al[4] on PETSc also proved this. However, this may not always be the case, the execution time became longer if the the number of processors exceeds certain point.

For equal number of processors, the execution time will increase almost linearly with the increasing of the number of subintervals.

Compared with the former two tests, if the ratio of matrix size to the number of processors is a constant, as shown in Fig. 2, the execution time will increase

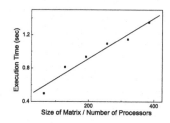

Fig. 2. Scaling of execution time

very slowly, which indicates that a smaller number of processors can be used for matrices with smaller size so as to improve the efficiency of Server.

In this paper, we proved that the realtime adjustment of MPI configuration file in accord with the amount of computation can significantly improve the overall performance of Server, and the resources can be saved.

Acknowledgements. This paper was supported by "SEC E-Institute: Shanghai High Institutions Grid" project, the fourth key subject construction of Shanghai, and the Key Project of Shanghai Educational Committee (No.03AZ03).

References

1. Satish Balay, Kris Buschelman, Victor Eijkhout, William Gropp et. al. PETSc Users Manual, 2004, http://www.mcs.anl.gov/petsc
2. Dorian Arnold, Sudesh Agrawal, Susan Blackford, Jack Dongarra et al., Users' Guide to NetSolve V2.0, 2003, http://icl.cs.utk.edu/netsolve/
3. Shirley Moore, A.J. Baker, Jack Dongarra, Christian Halloy, Active Netlib: An Active Mathematical Software Collection for Inquiry-based Computational Science and Engineering Education, 2002, http://www.cs.utk.edu / shirley /jodi /active-netlib.html
4. M. Ethesham Hayder, David E. Keyes, Piyush Mehrotra, A comparision of PETSc library and HPF implementations of an archetypal PDS computation, Advances in Engineering Software, Vol. 29(3-6), 415-423, 1997
5. J. Dennis, R. Schnabel, Numerical Methods for Unconstrained Optimization and Nonlinear Equations, Prentice Hall, 1983

Research on a Visual Collaborative Design System: A High Performance Solution Based on Web Service

Zhengde Zhao, Shardrom Johnson, Xiaobo Chen, Hongcan Ren, and Daniel Hsu

School of Computer Engineering and Science, Shanghai University, Shanghai 200072, China, zhdzhao@163.com

Abstract Collaborative Virtual Environment(CVE), which is the high performance combination of Computer Supported Cooperative Work(CSCW) and Virtual Reality(VR), makes it possible to support people working collaboratively in a visual way. Based on the conception of CVE, this paper presents an architecture on Internet for cooperative design supported by Web Services, and discusses the related key implementation technologies.

Keywords Collaborative Design, Visual Collaboration, Web Services

1 Introduction

Owing to lack of supporting communication and cooperative work among the designers, the traditional design project is always in a low efficiency. Especially, most modern projects need a lot of team members working collaboratively, the traditional method of CAD is obviously inadequate[PM01].

With the development of network technology and VR, we can present a visual reality interface for the CAD system with the collaborative design. Based on the current standard conditions of hardware and software, the immersed VR can not be implemented at low cost. But, with 3D graphics, some CAD applications based on spatial data can still benefit a lot from VR, since the user view can exactly consist with the problem area. Therefore, visual collaborative design can make this kind of CAD more efficient by both supporting people collaborative working and spatial visualization[AT00].

2 System Architecture and Features

Our prototype of visual collaborative design system applies the CVE concepts by combining the VR and CSCW technology.

Based on the openness of the system integration in the ISO standard, the VRML has been chosen for the presentation of 3D spatial data in the system.

In a CSCW system, there should be a solution for communication among all the distributed parts of the system. Many distributed technology are available, such as CORBA of OMG, RMI of SUN and DCOM of Microsoft. In our solution, we choose web service based on JAVA technology, not only for its flexibility to build loosely coupled system, also for its compatibility with VRML.

Figure 1 shows the rough architecture of the whole system.

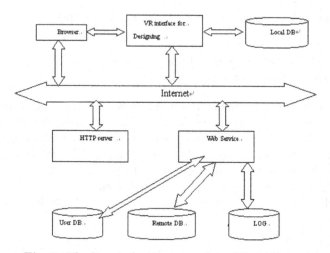

Fig. 1. The Rough Architecture of the Whole System

At first, the origin design models can be imported from some leading commercial software like 3DS MAX or AUTOCAD. Since our system apples VRML to present virtual world, the importing can be relatively easy for the openness of the VRML standard.

Secondly, each user of the system can interact with the system through a 3D graphic interface. The models under design can be modified as long as the user has been granted to do so. The key point to implement the collaborative work is to use web service technology. Many services are located in the server. Each action made by user which interfering the cooperation will call the proper web service. The service first does some proper cases about the server itself, and then broadcasting the action to every online user with the meta-data which can be used for building collaborative awareness.

There are three kinds of method for building collaborative systems[Pat01]. One is the centralized method, which makes all the cooperation messages be sent, received and managed by a center server. The second method is the pure distributed solution, which sends all cooperative messages peer to peer. Certainly, we can combine these two methods to present the third method to implement the collaborative communication mechanism. The first method can be relatively easy to manage messages, but the central bandwidth for messages could be the bottleneck of the whole system. Although the second method decreases the total amount of the messages, it could be not accepted for the complexity to implement. Obviously, the key point to apply the third method is to divide the cooperation messages into two groups. One of them should be sent to the central server for its mutuality with the server activity. Another messages that care about just one workstation can be sent directly to the proper host.

Figure 2 shows all the three methods to communication.

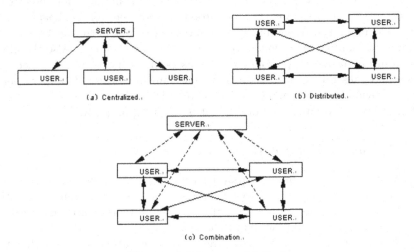

Fig. 2. Collaborative Methods

Here presents some specific features of this system solution:

The high performance solution combines VRML with JAVA technology including the openness of VRML and the flexibility of JAVA[YMJ99]. We apply VRML to implement the VR, and use JAVA to implement most of the application logic including the CSCW feature. Then the web service plays a vital role on across the firewall of the distributed system. CORBA, RMI and DCOM can work well inside an enterprise, but cannot go smoothly across the firewall.

3 The key Technology for Implementation

3.1 VRML and EAI

VRML is to implement interactive virtual reality application on web. It is supported by most popular web browsers. And with some 3rd part plugs-in, it can show some better render quality like more realistic shadow. External Authoring Interface(EAI) provides an interface to program for the VRML world with some programming languages. Java is chosen for our solution[YMJ99].

3.2 Communication Module

Three communication modes are used for a collaborative system: synchronous mode, asynchronous mode and communication out of the document. In the synchronous mode, each collaborative operation will be sent orderly and immediately. And the operations will not sent in time in the asynchronous mode. Otherwise, we can provide a communication method not directly in the document on designing, as a chat room outside the working space.

The communication module monitors the operations in the local system, if it finds a collaborative action, encodes the action and sends it. Also the module monitors the messages from the other users through network, then decodes them and calls the concurrency control mechanism to commit proper operations. Obviously, the communication module is vital for the CSCW system. Each collaborative operation depends it on working properly.

3.3 Concurrent Control Mechanism

In a multi-user system, many transactions are sent by different users. It needs a concurrent control mechanism to make the share objects consistent. Currently, three popular mechanisms are applied in CSCW systems: control mechanism, lock mechanism and transaction mechanism.

We use centralized control for synchronous operations. The one has the right to distribute token which means the priority to access the share object. The lock mechanism is for asynchronous operations. When an operation wants to access a shared object, first lock the object. Naturally, an object cannot be locked by locks which can cause conflicts. We have two kinds of lock mechanism in the system. They are share lock and exclude lock. The share lock is for reading a share object, and the exclude lock is for writing.

3.4 Collaborative Awareness

The purpose of the collaborative awareness is to provide an immersion cooperation work environment for each user. In our solution, two main implementations are provided. One is action awareness, which means that every need-to-be-known action should be aware by every online user. We can provide different color mark it. The second is customized viewpoint, which shows the status of each online user. It can provide a directly cooperative awareness for every user.

Figure 3 shows the effects.

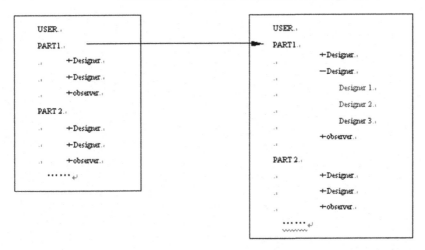

Fig. 3. Viewpoint Awareness

4 Conclusion

The distributed solution based on web service can be used to implement a design system with a spatial visual graphic interface for collaborative work in the Internet. This solution can be modified for being applied in lots of CAD area, such as mechanical design and interior design.

References

[PM01] Pettifer, S., Marsh, J.: A collaborative access model for shared virtual environments. Proceedings of IEEE WETICE 01. IEEE Computer Society, **6**,257–272 (2001)

[Pat01] Patrick, L.H.: Octopus: A study in collaborative virtual environment implementation. MA Thesis, Iowa State University , Ames (2001)

[AT00] Anrong Xie, Tianci Chen: Research on collaborative design based on CAD. Computer application and research, **10**,32–34 (2000)

[YMJ99] Yeqing Huang, Muqi Wulan, Jiati Deng: Research on distributed intelligent collaborative design system based on agent. Computer engineering and application, **8**,25–35 (1999)

Architectural Characteristics of Active Packets Workloads

Tieying Zhu[1,2] * and Jiubin Ju[1]

[1] School of Computer Science, Jilin University, China
[2] School of Computer Science, Northeast Normal University, China
zhuty@nenu.edu.cn

Abstract. Active networking needs the tradeoff of flexibility, safety and performance. Network processors targeted for safe active packets will provide customized packets processing at wire speed. In order to evaluate the architectural characteristics of active packets processing for network processors, this paper presents an architectural study of Safe and Nimble Active Packets (SNAP) processing. Two representative SNAP packets processing applications are taken as workloads and the characteristics of these two programs including instruction mix, instruction level parallelism (ILP), cache behavior are studied through simulation on a MIPS-like architecture. The measurements are compared to NetBench programs to highlight the different characteristics of active packets workloads.

Key words: Architecture simulation, Network processors, Active network, SNAP, Active packets processing

1 Introduction

Active networking [1] provides a powerful way to implement flexible packets processing. The flexible packets processing needs more powerful computational abilities than traditional packet forwarding, for the programmable routers must provide enough processing abilities at the interface to meet link speed.

Network processors are software programmable devices designed specifically to process packet data at wire speed [2]. While network processors have recently been developed commercially, not much effort has been put into systematically evaluating the processing requirements and design alternatives of these multiprocessors. In order to make network processors for active networking domains and evaluate the architectural and processing requirement, we study the Safe and Nimble Active Packets (SNAP) [3], which focuses on the balance of flexibility, safety and performance, and port SNAP user-space code

---

*Supported by Northeast Normal University Fund for Young Teachers.

[4] to the SimpleScalar simulation environment [5]. The architectural characteristics including the instruction mix, instruction level parallelism (ILP) and cache behavior are measured and compared with the average of NetBench programs[6], which are benchmarks for network processors.

The rest of the paper is organized as follows. Section 2 presents the related work. Section 3 gives an overview of SNAP and two representative active packets applications. Section 4 describes the parameters of the simulation environment and results of performance evaluation with SimpleScalar. Section 5 concludes the paper.

2 Relate work

In the research of active networking, many researchers make great efforts to improve the performance of active networks. Wolf et al. [7] put forward the design of high performance active router based on the port processors. Based on IBM PowerNP, the role of network processors in active networks and how to efficiently execute active code in data path are discussed by Kind et al. [8]. Kind et al.[9] also evaluate the performance of SNAP packets placed in the data path on NP cores and show that the performance can benefit from just-in-time compiler.The advanced five network applications on network processors including active networking applications using SNAP are described by Hass et al. [10]. PromethOS NP [11] provides a framework for dynamic service extension by plugins with integrated support of IBM PowerNP 4GS3.

The above work shows that SNAP is prospective and suited for network proces-sors capabilities. But there is less work to discuss the architectural and processing requirement of active networking applications.

3 SNAP packets

SNAP is a kind of byte-code language with specific safety characteristics. SNAP executes on a stack-based byte-code virtual machine. Besides payload and packet header, a SNAP packet consists of a series of byte-code instructions. For an SNAP interpreter-enabled node, these instructions are interpreted and executed. otherwise, the active packets are simply forwarded.

SNAP packets can implement the customized processing when they traverse the node. For example, in network management, SNAP-based management agents can roam around the managed objects, gather information, or even fix problems without reporting back [12]. In this paper, two representative SNAP packet processing programs, ping and node-resident services invocation, are taken as workloads for experiments. The former compares the packet's destination header to the current host's address. If they do not match, it creates a new packet with a copy of the current packet and forwards it to the destination. When the packet reaches it destination, a new packet is send back to the source. The latter allows a packet to invoke a service residing on the node named by a string. In this paper, a service getting the information in the direction "/proc/sys/net/ip-forward" is tested.

4 Characteristics of active packets workloads

4.1 Simulation environment

SimpleScalar Tool Set 3.0 is used to study the architectural characteristics of SNAP packets processing. It is a close derivative of MIPS architecture and wildly used as the simulation environment in the related study of network processors benchmarks. In this paper, the default processor architecture is 16K 4-way set associative L1 instruction cache and data cache with 2-cycle hit latency. Both have block sizes of 32 bytes and LRU as the replacement strategy. There is a 256K 4-way set associative unified L2 cache with 64-byte blocks and a 6-bycle hit latency. The processor has an instruction level parallelism of 4, and bimodal as the default branch predictor.

4.2 Instruction mix and ILP

The instruction mix is studied and the results are compared with that of Net-Bench programs in Figure 1. The SNAP packets program has a higher per-

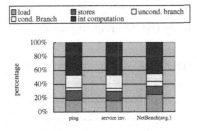

Fig. 1. Instruction mix

centage of conditional branch instructions compared to NetBench programs. The reason is that the interpretation of SNAP packets is not streamlined processing. There are many comparison operations and loops, for example in the un-marshalling and marshalling up a packet. The difference in conditional branch frequencies makes it necessary to explore the prediction mechanism. Other experiments shows that when complex prediction mechanisms such as bimodal or combined is taken, the branch predication hit rate significantly increases.

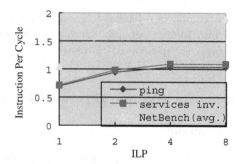

Fig. 2. Impact of ILP

The packet processing can be performed by highly parallel multiprocessor systems. Figure 2 shows the impact of ALU number on instruction level parallelism (ILP). Instruction per cycle is used to show ILP. It is observed that SNAP packets processing is less sensitive to instruction level parallelism, compared to NetBench programs. When ALU increase from 1 to 2 and 2 to 4, the instruction per cycle improve 36% and 11%, while that of NetBench programs improve 72% and 55%. The instruction level parallelism in SNAP packets processing is limited. So other techniques such as multi-threading are needed to speed up the processing.

4.3 Cache behavior

L1 instruction cache behavior is intensively studied by varying the parameter of cache size. Figure 3 shows the different impact of cache size of L1 instruction

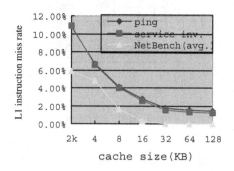

Fig. 3. Impact of L1 instruction cache

cache. The miss ratio of instruction cache decreases as cache size increases. Instruction cache is sensitive to cache size. SNAP packet benefits less from the cache size when the cache size increase from 64KB to 128KB. The reason is that the kernerl of SNAP is larger than NetBench programs. So larger cache size has less impact on SNAP packets processing.

Since the network environment is rapidly developed, new applications appear continuously. Active networking applications are examples. In this paper, the characteristics of typical active packets workloads are explored for network processors. The results show active networking has unique characteristics in instruction mix, ILP, cache behavior. It is observed that there are larger amount of branch operations and larger cache miss rate in active packets applications compared with programs of Netbench. To get more ILP, multithreading may be needed.

References

[1] David L. Tennenhouse, Jonathan M. Smith, W. David Sincoskie, David J. Wetherall and Gary J. Minden.: A Survey of Active Network Research. IEEE Communications, **35** 80–86 (1997)

[2] Andrew Heppel.: An Introduction to Network Processor. http://www.roke.co.uk/

[3] J. T. Moore, M. Hicks and S. Nettles.: Practical Programmable Packets. In: the Proceeding of the 20th annual Joint conference of the IEEE Computer and Communication Societies (INFOCOM '01). Anchorage, Alaska (2001)

[4] SNAP Web Site. http://www.cis.upenn.edu/ dsl/SNAP/(2000)

[5] D. Burger and T. Austin.: The SimpleScalar Tool Set 3.0. http://www.simplescalar.com (2003)

[6] G. Memik, W. H. Mangione-Smith and W. Hu. NetBench: a Bench-marking Suite for Network Processors. In: the Proceeding of International Conference on Computer-Aided Design. San Jose, California (2001)

[7] T. Wolf and J. Turner.: Design Issues for High Performance Active Routers. IEEE Journal on Selected Areas of Communication. **19** 404–409 (2001)

[8] A. Kind, R. Pletka and M. Waldvogel.: The Role of Network Processors in Active Net-works. In: Proceeding of IWAN 2003, Kyoto, Japan (2003)

[9] Andreas Kind, Roman Pletka and Burkhard Stiller.: The Potential of Just-in-Time Compilation in Active Networks based on Network Processors. In: the Proceeding of IEEE OpenArch 2002. New York City (2002)

[10] A. R. Haas, C. Jeffries, L. Kencl, A. Kind, B. Metzler, R. Pletka, M. Wald-vogel, L. Frelechoux and P. Droz.: Creating Advanced Functions on Network Processors: Experience and Perspectives. IEEE Network, **17** (2003)

[11] L. Ruf, R. Pletka, P. Erni, P. Droz, and B. Plattner.: Towards High-performance Active Networking. In: the Proceeding of IWAN'03. Kyoto, Japan (2003)

[12] W. Bruijn, H. Bos and J. T. Moore.: SNMP Plus a Lightweight API for SNAP Handling. In: the proceeding of the 9th IEEE/IFIP Network Operations and Management Symposium. Seoul, Korea (2004)

Design of Elliptic Curve Cryptography in GSI

Yanqin Zhu[1], Xia Lin[1], and Gang Wang[1]

Computer Science & Technology School, Soochow University, Soochow, China.
yqzhu@suda.edu.cn, linx166@163.com, trulywang@163.com

Abstract: In this paper, the main ideas of applications using elliptic curve cryptography in the grid security infrastructure are presented. The grid security infrastructure offers secure communication between varied elements within a grid, based on public key cryptography. The main attraction of ECC compared to RSA is that the best algorithm known for solving the elliptic curve discrete logarithm problem takes fully exponential time. This means that significantly smaller parameters can be used in ECC than in other systems such as RSA, but with equivalent levels of security. The schemes of ECC which include key generation, digital signature and key exchange are described in this paper. Several fast algorithms of implementation of ECC are also introduced.

KeyWords: Elliptic curve cryptography; Grid security infrastructure; Digital signature; Fast algorithms of implementation

1 Introduction

Grids have emerged as a common approach to constructing dynamic, inter-domain, distributed computing and data collaborations. The dynamic and multi-institutional nature of the environments introduces challenging security issues that demand new technical approaches. One of the best-known security approaches for grid computing can be found within the Globus Toolkit, a widely used set of components used for building grids. The Toolkit, developed by the Globus Project[FK98,FKT98], offers authentication, authorization and secure communications through its Grid Security Infrastructure (GSI).

Among the GSI's key purposes are to provide a single sign-on for multiple grid systems and applications, to offer security technology that can be implemented across varied organizations without requiring a central managing authority, and to offer secure communication between varied elements within a grid. A central concept in GSI authentication is the certificate. Every user and service on the Grid is identified via a certificate, which contains information vital to identifying and authenticating the user or service.

GSI certificates are encoded in the X.509 [CF99]certificate format, a standard data format for certificates established by the Internet Engineering Task Force (IETF). These certificates can be shared with other public key-based[IEE00] software, including commercial web browsers from Microsoft or Netscape. If two parties have certificates, and if both parties trust the CAs that signed each other's certificates, then the two parties can prove to each other that they are who they say they are. This is known as mutual authentication. The GSI uses the SSL/TLS for its mutual authentication protocol.Before mutual authentication can occur, the parties involved must first trust the CAs that signed each other's certificates. In practice, this means that they must have copies of the CAs' certificates–which contain the CAs' public keys—and that they must trust that these certificates really belong to the CAs.

In SSL/TLS protocol, RSA is recommended. As we have seen, the bit length for secure RSA use has increased over recent years, and has put a heavier processing load on applications using RSA. Elliptic curve cryptography(ECC) was introduced by Victor Miller[Mil86] and Neal Koblitz[Kob87] in 1985. ECC proposed as an alternative to established public-key systems such as RSA. The principal attraction of ECC compared to RSA is that it appears to offer equal security for a far smaller bit size, thereby reducing processing overhead.

In this paper, the main ideas of applications using elliptic curve cryptograph in the grid security infrastructure are presented. Then we will discuss several fast algorithms of implementation.

2 The main operations of ECC

There are several ways of defining equations for elliptic curves, which depend on whether the field is a prime finite field or a characteristic two finite field. In this paper, we will only discuss elliptic curve cryptography based on a prime finite field[Wil02,Cer00] which is defined as follows:

Let F_p be a prime finite field so that p is an odd prime number, and let $a, b \in F_p$,satisfy $4a^3 + 27b^2 \neq 0 \pmod p$.Then an elliptic curve $E(F_p)$ over F_p defined by the parameters $a, b \in F_p$ consists of the set of solutions or points $P = (x, y)$ for $x, y \in F_p$ to the equation:

$$y^2 = x^3 + ax + b \pmod p.$$

together with an extra point O called the point at infinity. For a given point $P = (x_p, y_p)$, x_p is called the coordinate of P,and y_p is called the coordinate of P.

2.1 Elliptic curve key pair generation

Elliptic curve domain parameters over F_p are a sextuple:

$$T = (p, a, b, G, n, h)$$

consisting of an integer p and two elements a, b(defined forward), a base point $P = (x_p, y_p)$ on $E(F_p)$, a prime n which is the order of G and an integer h which is the cofactor $h = \#E(F_p)/n$.

To generate a key pair, User U does the following:

1. Given some elliptic curve domain parameters.

2. Select a secret key d which is a random integer in the interval$[1, n-1]$.

3. Select an elliptic curve public key $Q = (x_p, y_p)$,which is the point $Q = dG$.

2.2 The operations of signing and verifying

The setup for generating and verifying signatures using the ECDSA[Ans99] is as follows. Suppose that signer U has domain parameters $T = (p, a, b, G, n, h)$ and a key pair (d, Q).

Signing operation:

To sign the message M, U does the following:

1. Select a key pair (k, R) with $R = (x_R, y_R)$ associated with the elliptic curve domain parameters T established during the setup procedure.

2. Convert the field element x_R to an integer $\overline{x_R}$ using a proper conversion routine.

3. Set $r = \overline{x_R}(\bmod\ n)$, if $r = 0$, return to step 1.

4. Use the hash function selected to compute the hash value:
$$e = hash(M).$$

5. Compute:
$$s = k^{-1}(e + rd)\ \bmod\ n.$$
If $s = 0$, return to step 1.

6. $U's$ signature for the message M is (r, s).

Verifying operation:

V should verify signed messages from U using the keys and parameters established during the setup procedure as following:

1. If r and s are not both integers in the interval $[1, n-1]$, the signature is invalid.

2. Use the hash function to compute the hash value:
$$e = hash(M).$$

3. Compute:
$$u_1 = s^{-1}e\bmod\ n\ \text{and}\ u_2 = s^{-1}r\ \bmod\ n.$$

4. Compute:
$$R = (x_R, y_R) = u_1 G + u_2 Q.$$

5. Convert the field element x_R to an integer $\overline{x_R}$ using the conversion routine.

6. Set $v = \overline{x_R}\ \bmod\ n$.

7. Accept the signature if and only if $v = r$.

2.3 Elliptic Curve Diffie-Hellman key exchange

The basic idea of key exchange using elliptic curves is to generate a shared secret value shared by user U and V. We assume that user U and V has domain parameters $T = (p, a, b, G, n, h)$. Key exchange between users U and V can be accomplished as follows[Cer00]:

1. User U selects an integer d_U less than n. This is $U's$ private key. U then generates a public key $Q_U = d_U \times G$.

2. User V similarly selects a private key d_V and computers a public key Q_V.

3. User U computes $K = d_U Q_V$.

4. User V computes $K = d_V Q_U$.

The two calculations in the last two steps produce the same result because

$$d_U \times Q_V = d_U \times (d_V \times G) = d_V \times (d_U \times G) = d_V \times Q_U.$$

3 The fast algorithms of scalar multiplication

The central operation of cryptographic schemes based on ECC is the elliptic scalar multiplication. It dominates the execution time of elliptic curve cryptographic schemes. Given an integer k and a point $P \in E(F_p)$, the elliptic scalar multiplication kP is the result of adding P to itself k times.

Now we consider methods for computing point kP. If $P = (x, y) \in E(F_p)$, then $-P = (x, -y)$. Thus point subtraction is as efficient as addition. This motivates using a signed digit representation $k = \sum_{i=0}^{l-1} k_i 2^i$, where $k_i \in \{0, \pm 1\}$. A particularly useful signed digit representation is the non-adjacent form(NAF) which has the property that no two consecutive coefficients k_i are non-zero. Every positive integer k has a unique NAF, denoted NAF(k). Moreover, NAF(k) has the fewest k non-zero coefficients of any signed digit representation of k, and can be efficiently computed using Algorithm 1[Sol00].

Algorithm 1. Computing the NAF of a positive integer

Input: A positive integer k

Output: NAF(k)

    1. $i \longleftarrow 0$.

    2. While $k \geq 1$ do

        If k is odd then

$$k_i \longleftarrow 2-(k \bmod 4), k \longleftarrow k - k_i;$$

        Else $k_i \longleftarrow 0$.

        $k_i \longleftarrow k/2, i \longleftarrow i + 1$.

    3. Return $((k_{i-1}, k_{i-2}, \cdots, k_1, k_0))$.

Algorithm 2 gives a binary NAF method for point multiplication.

Algorithm 2. Binary NAF method for point multiplication

Input: NAF(k)=$\sum_{i=0}^{l-1} k_i 2^i$

Output: kP

1. $Q \longleftarrow O$.
2. For i from $l - 1$ downto 0 do
$$Q \longleftarrow 2Q.$$
   If $k_i = 1$ then $Q \longleftarrow Q + P$.
   If $k_i = -1$ then $Q \longleftarrow Q - P$.
3. Return (Q).

The running time of Algorithm 2 can be decreased by using a window method[Sol00] which processes w digits of k at a time. One approach is to first compute width-w NAF of an integer k, then use it to compute point multiplication. A width-w NAF of an integer k is an expression $k = \sum_{i=0}^{l-1} k_i 2^i$, where each non-zero coefficient k_i is odd, $|k_i| < 2^{w-1}$, and at most one of any consecutive coefficients is nonzero. Every positive integer has a unique width-w NAF, denoted $NAF_w(k)$. $NAF_w(k)$ can be efficiently computed using Algorithm 1 modified as follows: in step 2 replace "$k_i \longleftarrow 2\text{-}(k \bmod 4)$" by "$k_i \longleftarrow k \bmod 2^w$", where $k \bmod 2^w$ denotes the integer u satisfying $u \equiv k(\bmod 2^w)$ and $-2^{w-1} \leq u < 2^{w-1}$. It is known that the length of $NAF_w(k)$ is at most one longer than the binary representation of k. Also, the average density of non-zero coefficients among all width-w NAFs of length l is approximately $1/(w + 1)$. The calculation of kP can be carried out by Algorithm 3.

Algorithm 3. Window NAF method for point multiplication
Input: Integers k and w, and a point $P = (x, y) \in E(F_p)$
Output: The point kP
   //Precomputation which is needed only once
   //Compute uP for u odd and $2 < u < 2^{w-1}$
   1. $P_0 \longleftarrow P, T \longleftarrow 2P$.
   2. For i from 1 to $2^{w-2}-1$ do
$$P_i \longleftarrow P_{i-1} + T.$$
   //Main computation
   3. Compute $NAF_w(k) = (u_{l-1}, u_{l-2}, \cdots, u_1, u_0)$.
   4. $Q \longleftarrow O$.
   5. For j from $l - 1$ downto 0 do
$$Q \longleftarrow 2Q.$$
   If $u_j \neq 0$ then
$$i \longleftarrow (|u_j| - 1)/2.$$
   If $u_j > 0$ then
$$Q \longleftarrow Q + P_i;$$
   Else $Q \longleftarrow Q - P_i$.
   6. Return(Q).

If some extra memory is available, the running time of Algorithm 3 can be decreased by using a window method which processes w digits of k at a time.

In this paper, we have presented the main ideas of applications using elliptic curve cryptograph in the grid security infrastructure. The grid security infrastructure offers secure communication between varied elements within a grid, based on public key cryptography. The principal attraction of ECC compared to other competitive public key cryptosystems is that it can offer equal security with a far smaller bit size. The benefits of having smaller key sizes include faster computations, and reductions in processing power, storage space and bandwidth.

We have focused on the schemes which include elliptic curve key pair generation, signing and verifying, key exchange using elliptic curves. As the central operation of cryptographic schemes based on ECC is the elliptic scalar multiplication, several fast algorithms in computing the point kP are also discussed in this paper.

Acknowledgements. This paper was supported by the Jiangsu Natural Science Foundation of China under Grant BK2004039.

References

[Ans99]   ANSI X9.62 : Public Key Cryptography for the Financial Services Industry: the Elliptic Curve Digital Signature Algorithm(ECDSA), (1999).

[Cer00]   Certicom Research : SEC1: Elliptic Curve Cryptography Version 1.0, (2000).

[CF99]   Chokhani,S., Ford,W. : Internet X.509 Public Key Infrastructure Certificate Policy and Certification Practices Framework , RFC 2527,(1999).

[FK98]   Foster,I.,Kesselman,C.: The Globus Project:A Status Report, Proc. IPPS/SPDP '98 Heterogeneous Computing Workshop, 4–18 (1998).

[FKT98]   Foster,I., Kesselman,C., Tsudik,G., Tuecke,S. : A security architecture for computational grids, Proceedings of the 5th ACM conference on computer and communications security, New York: ACM Press, 83–91 (1998).

[IEE00]   IEEE P1363 : Standard Specification for Public-key Cryptography, (2000).

[Kob87]   Koblitz,N. : Elliptic curve cryptosystems, Mathematics of Computation, vol.48, 203–209(1987).

[Mil86]   Miller,V.:Use of elliptic curves in cryptography, Advances in Cryptology-Crypto'85, LNCS 218, 417–426(1986).

[Sol00]   Solinas,J. : Efficient arithmetic on Koblitz curves, Designs, Codes and Cryptography,Vol.19, 195–249(2000).

[Wil02]   William Stallings. : Cryptography and Network Security: Principles and a Practice(Third Edition), Prentice Hall, (2002).

[WS03]   Welch,V., Siebenlist,F., Foster,I., Bresnahan,J., Czajkowski,K.,Gawor,J., Kesselman,C., Meder,S., Pearlman,L., Tuecke,S. : Security for Grid Services. Twelfth International Symposium on High Performance Distributed Computing (HPDC-12), IEEE Press (2003).

Policy-based Resource Provisioning in Optical Grid Service Network

Yonghua Zhu[1] and Rujian Lin[2]

[1] School of Computer Science and Engineering, Shanghai Unversity, Yanchang Rd.
149, Shanghai 200072, China yhzhu@mail.shu.edu.cn
[2] Communication and Inf. Eng. Tech. Lab, Shanghai Unversity, Yanchang Rd.
149, Shanghai 200072, China rujianlin@sina.com

Abstract. The use of the available fiber and DWDM infrastructure for the global Grid network is an attractive proposition ensuring global reach and huge amounts of cheap bandwidth. Fiber and DWDM networks have been great enablers of the World Wide Web fulfilling the capacity demand generated by Internet traffic and are expected to play an important role in creating an efficient infrastructure for supporting Grid applications. This article proposed a service level agreement applied to the optical domain which is worked as grid network service. We give an explanation of rationale behind an optical SLA and then put forward the parameters that could be included in this optical SLA. Different grid application and service type are distinguished when necessary. In this paper, we also propose a policy-based resource provisioning approach to services client in this optical SLA.

Keywords: optical grid network, service provisioning, policy-based, SLA, QoS

1 Introduction

Optical networks can be viewed as essential building blocks for a connectivity infrastructure for service architectures including the Open Grid Service Architecture (OGSA)[FKN02], or as "network resources" to be offered as services to the Grid like any other resources such as processing and storage devices.

QoS of an optical transport network will play an important role in the future of high-demand Grid computing. Optical connections in a Grid environment will be initiated on an as needed basis by the Grid applications, and that each connection request will have an associated set of optical transport QoS requirements. The following are potential QoS parameters for which a Grid application may request: i) optical layer restoration times, ii) priority and preemption of a connection, iii) physical layer signal degradation (application BER). The Grid applications connection request will contain the appropriate QoS parameters to meet the applications needs. Physical layer impairments are a key concern in high-datarate optical networks and will play a significant role in future Grid networks and SLAs. This section discusses some of the issues related to physical layer QoS [FFR04].

In the objective of convergence toward a unified network, a key feature is the capability to offer differentiated services in a single network, to accommodate the different requirements of the various clients.

A service level agreement (SLA) is a formal contract between a service provider and a subscriber that contains detailed technical specifications called service level specifications (SLSs). An SLS is a set of parameters and their values that together define the service offered to a traffic stream in a network [Tur01]. Until now, no standards for the contents of an SLS have been defined, but interesting proposals have been published as Internet drafts by the Internet Engineering Task Force (IETF) [Sal00].

Some work has been done in defining SLAs for traditional IP networks [God01, Sal00, Tur01], but these do not consider important issues involved in optical technologies, and do not meet the requirements and exigencies of the optical grid network service providers and service clients. This paper focuses on defining these SLAs specifically adapted to the relationship between optical network operators and their diverse clients. Moreover, we propose a policy-based provisioning approach to services described in this optical SLA.

2 Parameters for SLSs

In this article the relationships defined by an Optical SLA consider a service provider to be an optical carrier operator, and a service subscriber to be either an optical client or an IP or multiprotocol label switching (MPLS) client.

We have first a leased-line type of service where the bandwidth is not often changed and that consequently can tolerate a low level of connection automation. In the preprovisioned bandwidth case we supposed that bandwidth variations exist but have been scheduled in the SLA so that the carrier can easily preprovision the resources. Bandwidth on demand service is more constraining for the operator as it requires real-time provisioning of bandwidth without previous knowledge of demand variations. Finally, the optical virtual private network (O-VPN) [Ber031] is a multipoint-toCmultipoint service where the customer has at least visibility of the resources allocated to him and possibly the opportunity to partially directly manage them.

Besides the popular parameters such as service boundary, service schedule, and flow identifier, we propose some special parameters for optical SLS. They are connection setup time, service availability, routing constraints, service performance guarantees.

Connection setup time: The connection setup time specifies how long it will take for a service connection to be established once it has been negotiated and requested. Connection setup time might be expressed in seconds, minutes, hours, or even days, depending on the client demands and service characteristics.

Service Availability: The parameters for differentiation of service availability are proposed as followed:

1. Out-of-service criterion
2. Service recovery time
3. Recovery time with degraded performance
4. Service mean down time

The out-of-service criterion controls the triggering of the resilience mechanism. It can be a fault or degradation as some applications may tolerate degraded BER and others will not.

Routing Constraints include Routing stability, Route differentiation, Confidentiality and Classes of service and routing constraints. The process of routing connections within the network offers several possibilities for service differentiation.

For an IP/MPLS client the performance parameters will be those of a classical IP network, particularly impacted by the priorities given to the different clients in the routers performing the aggregation. These parameters are delay, jitter, throughput, and packet loss [God01].

3 Policy-Based Resource Provisioning

Network services assist a grid infrastructure in different ways. In the simplest setup, a grid application consults a network service as if it were an omniscient oracle, using a plain question/answer style of interaction. In more complex setups, network services interact with one another to realize one or more feedback loop cycles. Application requirements, policy considerations, and broker's directives are continuously injected into these feedback loops, via expressive languages and machine interfaces.

Fig. 1 shows an example of notional network services engaged in a fairly complex set of feedback loops, Applications demand, policy, and network's observed capacity are continuously mediated, resulting in provisioning actions upon the network as well as the system and middleware layers at the end-systems. The edge labeled 1 is meant to capture the following concept: There are mechanisms for the application to communicate with services factors like data rate profile, total data amount remaining, and other characteristics of the data stream to help the network fabric to optimize and predict load, which in turn may result in greater satisfaction to the end user. With regard to the edge labeled 2, a designated service must notify an application of those events that the application has negotiated and registered for. It must tell an application if it is admission-controlled out. It must timely notify an application of SLA violations. With regard to the edge labeled 3, when appropriate, credited services can dynamically (re)provision network aspects.

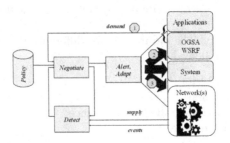

Fig. 1. Policy-based Grid Services

3.1 Resource Provisioning

Provisioning the optical network to meet the objectives defined by service contracts is a major concern for optical carrier operators. This process involves the setup of lightpaths subject to traffic requirements and current network state.

Up to now lightpath establishment in optical networks has been carried out manually via the management plane. Nonetheless, this approach suffers from several limitations, since the manual establishment of explicit LSPs with associated, QoS parameters would be slow, prone to error, and laborious to network administrators.

To cope with the limitations of the traditional architecture, a new one is being defined where a distributed control plane is appended [ITU01]. The main driver for the introduction of this control function is the need for automation in both traffic engineered optical path setup and fault handling.

It is then necessary to have a suite of control plane protocols that are flexible enough to support overlay, peer, and hybrid models and to provide routing, signaling, and efficient recovery techniques [Ban01].

3.2 Policy Usage

The optical SLA defined in this article provides guidelines on how to manage GMPLS-enabled optical grid networks. As such, the different policy categories that can be used for this purpose may be inferred from the SLS parameters. With regard to their impact on the functional plane, the SLS parameters defined in the optical SLA are classified into traffic-flow-related parameters and control-plane-related parameters.

Traffic-flow-related parameters include flow Id, traffic conformance, and excess treatment. This is normal since flow id identifies the traffic flow for which the service is to be provided. While traffic conformance indicates the profile based on which the traffic is classified as either in or out of profile, excess treatment precisely describes the treatment for out of profile traffic.

Control-plane-related parameters include routing constraints, service performance guarantees, and service availability and resilience. It is normal to find such parameters under this class, since these parameters characterize the lightpath that will be setup using the control plane.

Based on the above classification, the following policy categories are identified as crucial to ensure efficient management of GMPLS-enabled optical network:

1. Routing policies
2. LSP life cycle management policies
3. Flow management policies

The first category concerns routing policies. These policies offer the possibility to restrict the path taken by the lightpath and ensure the requested performance characteristics.

The performance of a lightpath is tightly related to the characteristics of the links assigned to it. Hence, route calculation is an important step during lightpath creation. In GMPLS networks, the path computation feature is fulfilled by a constraint-based routing (CBR) function that uses the following information as input:

1. SLS parameters characterizing the lightpath; for example, performance guarantee parameters
2. Attributes associated with resources; TE link attributes indicating resource availability in the optical network
3. Other topology information

Based on this information, a CBR process on each node computes explicit routes for lightpaths originating from that node. In this case, the explicit route is a specification of a path that satisfies the requirements expressed in the SLA, subject to constraints imposed by resource availability and other topology state information.

While the CBR function is capable of dealing with quantitative attributes, it is not for qualitative ones. In other words, how can we build a lightpath satisfying the confidentiality attribute? How can we avoid some links being associated with the lightpath in order to satisfy a certain route differentiation requirement? The answers to these two questions can be given in routing policies. In fact, resources are administratively assigned a certain color such that resources with the same color belong to the same class. This color concept is already defined as an attribute of TE links, so the idea is to provide each TE link a certain color based on routing policies.

LSP life cycle management policies deal with OXC configuration related to initiating, maintaining, and removing lightpaths. The third and last category of policies, flow management policies, deals with classification directives for mapping data flows onto lightpaths. It is important to filter flows that will use network resource based on the flow id directive defined in the optical SLA.

Once the traffic flow is identified, policing and shaping are applied on the traffic conformance and excess treatment directives.

After having analyzed these projects and studied the major trends, issues, and needs concerning optical networks and the technologies therein, the optical SLA described in this article has been proposed for the purpose of meeting both service clients and service providers needs, and to provide a strategy concerning service negotiations and agreements. In order to activate a service in the network, an optical SLA contracted between a client and a provider has to be provisioned. In this regard, several possible provisioning approaches in optical grid networks are discussed. A policy-based scheme has been retained as a suitable candidate to ensure service activation in a GMPLS-enabled optical network. Hence, different policy categories are introduced from the perspective of management architecture definition.

References

[Ban01]   A. Banerjee et al., Generalized Multiprotocol Label Switching: An Overview of Routing and Management Enhancements, IEEE Commun. Mag., Jan. 2001.

[Ber031] B. Berde, Contribution to a Management Plane for Optical Networks Operating a GMPLS Control Plane, Alcatel-CIT. Research and Innovation, NG-NSM

[Ber032] L. Berger, GMPLS Signaling RSVP-TE Extensions, RFC 3473, Jan. 2003.

[FFR04]   Foster, I., Fidler, M., Roy, A., V, Sander, Winkler, L., End-to-End Quality of Service for High-end Applications. Elsevier Computer Communications Journal, 2004.

[FKN02]   Foster, I., Kesselman, C., Nick, J., Tuecke, S., The Physiology of the Grid: An Open Grid Services Architecture for Distributed Systems Integration, Open Grid Service Infrastructure WG, Global Grid Forum, June 22, 2002

[God01]   D. Goderis et al., Service Level Specification Semantics and Parameters, Internet draft, draft-tequila-sis-01.txt, June 2001.

[ITU01]   ITU-T G.8080/Y1304, Architecture for the Automatically Switched Optical Network (ASON), Oct. 2001.

[Sal00]   S.Salsano et al., Definition and Usage of SLSs in the AQUILA Consortium, Internet draft, draft-salsano-aquila-sis-00.txt, Nov. 2000.

[Tur01]   F.De Turck et al., Desig and Implementation of a Generic Connection Management and Service Level Agreement Monitoring Platform Supporting the Virtual Private Network Service, IEEExplore 2001.

An Algorithm for Solving Computational Instability Problem in Simulation of Atmospheric Aerosol Growth and Evaporation

Yousuo J. Zou[1] and Chiren Wu[2]

[1] University of Guam, Mangilao, Guam 96923, USA. yjzou@guam.uog.edu
[2] Guangxi University of Technology, Liuzhou, Guangxi 545006, China
  chcldw@163.com

1 The Problem of Aerosol Growth and Evaporation

Aerosols are small particles suspending in the air that consist of air pollutants, fine dusts, small droplets, biological clusters, etc. Aerosol growth can be described as in Figure 1. An aerosol with radius, r, in the air with relative humility S, its growth or evaporation is decided by two processes: the vapor deposition on the aerosol's surface and heat that generated by condensation or evaporation diffuses from the aerosol surface [TIM03].

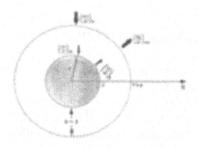

Fig. 1. Droplet/aerosol growth and diffusion-kinetic Process

If air is polluted, the aerosol/droplet will be containment and its surface energy changes. The surface property change will change the growth or evaporation rate that happens in the small layer (Figure 1) called collision free zone, Δ , which is about the thickness of free molecular length, $l = \Delta$. This modification has been made in the following growth equation,

$$\frac{dm}{dt} = \frac{4\pi r(S - 1 - \frac{a}{r} + \frac{b}{r^3})}{(\frac{L_C^2}{KR_vT^2} \cdot \frac{1}{f_\alpha} - \frac{1}{D\rho s} \cdot \frac{1}{f_\beta})} \tag{1}$$

where

$$f_\alpha = \frac{r}{r + l_\alpha} f_\beta = \frac{r}{r + l_\beta} \tag{2}$$

where $\frac{dm}{dt}$ is the growth rate of aerosol or droplet, D is diffusivity, K is heat conductivity, L_c is latent heat, m is mass of aerosol or droplet, r is aerosol or droplet radius, (S-1) is supersaturation of air, T is temperature, α is thermal accommodation coefficient, β is condensation coefficient, ρ_s is density of air mass.

The above equation is called diffusion-kinetic growth equation which was first developed by Fukuta and Walter [FW70]. Since aerosols are in the atmosphere the growth or evaporation of aerosols are controlled by other atmospheric process equations. These include equations of energy conservation, water vapor conservation, cloud water conservation, rain water conservation and vertical motion equations. The variables in the above equations are also constrained by the following atmospheric relationship: temperature and potential temperature relationship, atmospheric pressure and height relationship, condensation water change relationship. These equations are the aerosol growth processes inside the fog or cloud. The processes outside and below the cloud or fog can be described by equations of supersaturation-vapor pressure relationship, relative humility-pressure relationship, vapor pressure-temperature relationship, state equation of atmosphere, state equation of water vapor, and cloud water conservation relationship[ZF99].

2 The Computational Instability

In actual computation, the differential equations described above will be simplified and derived [KGP03] into practical format with 16 equations to solve 16 variables. The initial conditions are: atmospheric pressure, $P_0 = 809.21mb$; temperature, $T_0 = 10.93°C$; Relative Humility, $(RH)_0 = 95\%$ or $(S - 1)_0 = -5\%$. The initial aerosol solute is ammonium sulfate, which is divided into 100 bins from $10^{-19} - 10^{-11}$ gram [NBM03] that are converted into the initial size of aerosols (the radii of sphere of dry solute in micrometers, see Table 1).

The computational schemes are finite difference or sinc functional method [S93]. To save CPU time and memory, we need to develop an optimal computational algorithm in which computational instability must be avoided. The key question is to find an optimal time-step length for integrating the above differential equations. Through theoretical analysis, a critical time-step length for computational stability in aerosol and droplet growth integration has been developed [AB71]:

$$\Delta t < \frac{1}{4\pi\rho_L GH} \cdot \frac{1}{N\bar{r}} \tag{3}$$

where

$$H = \frac{P_b}{\varepsilon e_b} + \frac{\varepsilon L_c^2}{C_p R_d T_b^2} \tag{4}$$

$$G = \frac{1}{\rho_L[(\frac{L_c}{RvT} - 1)\frac{L_c}{KT} + \frac{1}{\rho_s D}]} \tag{5}$$

where Δt, in second, is the time-step length for integrating equations (1).

Using the above formula, Arnason and Brown [AB71] estimated that $\Delta t <= 0.01$ second.

However, if Δt is too small, it will use too much CPU time for computation, and generate too much data that occupy too much memory. If Δt is too large, computation will not be convergent, computational instability may happen. In order to determine the real optimal time-step, we tried in our simulation using $\Delta t = 1.0, 0.5, 0.1, 0.08, 0.06, ..., 0.01$ second.

When $\Delta t = 0.01$ second, the computational results show that the computational instability happens at aerosol radius equal to or less than 0.0037 micrometers (see Table 1).

Table 1. Computational Results on Integration of Aerosol Growth Equations at $(S-1) = -5\%$, time step $\Delta t = 0.01s$ (Aerosol size all in micrometer)

Size bin number	Dry radius	Equilibrium radius	Condensation radius	Size bin number	Dry radius	Equilibrium radius	Condensation radius
1	1.1049	2.6846	1.1340	83	0.0072	0.0106	0.0226
11	0.5979	1.4317	0.6268	85	0.0064	0.0089	0.0249
21	0.3236	0.7658	0.3511	87	0.0056	0.0075	0.0305
31	0.1751	0.4072	0.2000	89	0.0050	0.0063	0.0446
41	0.0948	0.2145	0.1158	91	0.0044	0.0053	0.0872
51	0.0513	0.1113	0.0680	93	0.0039	0.0044	0.1731
61	0.0278	0.0563	0.0408	95	0.0034	0.0037	-0.2715
71	0.0150	0.0273	0.0264	97	0.0030	0.0031	-3.3242
81	0.0081	0.0125	0.0218	99	0.0027	0.0028	0.0027

3 A New Algorithm and Its Computational Results

Computational results in Table 1 indicate that the theoretically predicted time step-length, Eq.(3), for integrating the aerosol growth/evaporation equation is not correct. When aerosol is very small, the computational instability still occurs in real computer simulation.

Then, how to solve the computational instability problems? Let's consider the aerosol equilibrium equation that is a cubic algebraic equation [ZF96]. Solving the cubic algebraic equation does not need to integrate differential equations, it is easy. Fig.2 compares the computational results of the aerosol algebraic equilibrium equation and the numerical integration of aerosol condensational growth. Fig2a is computed at relative humidity 95% (or supersaturation at $(S - 1) = -5\%$). Fig2b. is at relative humidity 100% or supersaturation $(S - 1) = 0\%$. The solid line is the aerosol sizes from condensational growth and the dashed line represents the aerosol sizes from equilibrium equation. The figure shows that:

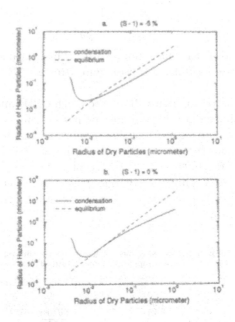

Fig. 2. Comparison of aerosol size solutions from aerosol growth equation and from aerosol size equilibrium equation

1). Compare with equilibrium aerosol size, condensational size has excessive growth. When aerosol size is too small, the integrated solution of condensational growth is not good while the equilibrium solution looks practical.

2). Compare to condensational growth, when aerosol size is too large, the equilibrium size is much larger that that of condensation growth size. It means the equilibrium size will never be reached and the solution of condensational growth is correct.

Therefore we may use an algorithm like the following: in order to deal with the computational instability in simulation of aerosol growth or evaporation, we compute both the condensational growth size and equilibrium aerosol size. When computational instability occurs at a smaller size of aerosol, stop the integration of growth equation but use the equilibrium size to replace the condensational growth size. In this way, we can use even higher integration time-step length, for example, $\Delta t = 0.1$ second, and the simulation goes well (see Fig. 2).

4 Conclusion

In simulation of evaporation and growth processes of atmospheric aerosols, a group of differential equations need to be solved numerically. One of the difficult problems in aerosol and small droplet modeling is the computational instability for integrating aerosol and droplet growth equations, particularly in the stages of aerosol nucleation, haze phase, and the growth at the maximum supersaturation. In this stage, the particle growth is so rapid that even if one uses a very small time step for the integration the computational instability may still occur. A smaller time-step could be employed to overcome the instability but this is limited by computer memory capacity and the CPU speed. When the CPU time is close to or longer than the characteristic time of particle's growth, the computational results become no significance. In order to develop a high-performance algorithm for the integration, an optimal value of time-step length should be found. This paper discussed the methods to determine such an optimal time-step length, and, also through our computational practice and numerical experiments, a high-performance computation scheme has been developed that is very efficiency in dealing with the computational instability of integrating differential equations for aerosol growth and evaporation.

References

[TIM03] Tanoue, K., Inoue, Y. and Masuda, H: Aerosol Scie. Tech., 37: 1-14 (2003)
[FW70]   Fukuta, N. and Walter, L. A: J. Atmos. Sci., 27: 1160-1172(1970)
[ZF99]   Zou, Y. and Fukuta, N.: Atmospheric Research, 52: 115 -141(1999)
[KGP03] Koo, B., Gaydos, T. M., and Pandis, S. P.: Aerosol Scie. Tech. 37:53-64(2003)
[NBM03] Nadeau, P., Berk, D., Munz, R. J.: Aerosol Sci. Tech., 37: 82-9(2003)
[S93]   Stenger, F.z: Numerical methods based on sinc and analytical functions. Springer-Verlag, New York(1993)
[AB71]   Arnason, G. and Brown, P. S. Jr.: J. Atmos. Scie., 28: 72-77(1971)
[ZF96]   Zou, Y., and Fukuta, N.: Nucleation and growth interaction in clouds and its effect on climate. In: Kulmala, M. and Wagner, P.E. (eds), Nucleation and Atmospheric(1996) Aerosols, Pergamon